Developments in Paleoenvironmental Research

Volume 21

Series Editor
John P. Smol
Department of Biology
Queen's University
Kingston, ON, Canada

Paleoenvironmental research continues to enjoy tremendous interest and progress in the scientific community. The overall aims and scope of the *Developments in Paleoenvironmental Research* book series is to capture this excitement and document these developments. Volumes related to any aspect of paleoenvironmental research, encompassing any time period, are within the scope of the series. For example, relevant topics include studies focused on terrestrial, peatland, lacustrine, riverine, estuarine, and marine systems, ice cores, cave deposits, palynology, isotopes, geochemistry, sedimentology, paleontology, etc. Methodological and taxonomic volumes relevant to paleoenvironmental research are also encouraged. The series will include edited volumes on a particular subject, geographic region, or time period, conference and workshop proceedings, as well as monographs.
Prospective authors and/or editors should consult the Series Editor John P. Smol for more details. Any comments or suggestions for future volumes are welcomed;

Eric Capo • Cécilia Barouillet • John P. Smol
Editors

Tracking Environmental Change Using Lake Sediments

Volume 6: Sedimentary DNA

Springer

Editors
Eric Capo
Department of Ecology and
Environmental Science
Umeå University
Umeå, Sweden

Cécilia Barouillet
INRAE, Université Savoie Mont
Blanc CARRTEL
Thonon-les-Bains, France

John P. Smol
Paleoecological Environmental Assessment
and Research Lab (PEARL)
Department of Biology
Queen's University
Kingston, ON, Canada

ISSN 1571-5299 ISSN 2215-1672 (electronic)
Developments in Paleoenvironmental Research
ISBN 978-3-031-43798-4 ISBN 978-3-031-43799-1 (eBook)
https://doi.org/10.1007/978-3-031-43799-1

This Springer imprint is published by the registered company Springer Nature Switzerland AG
The registered company address is: Gewerbestrasse 11, 6330 Cham, Switzerland

Paper in this product is recyclable.

The sedaDNA scientific society

Dedicated to the memory of Dr. Sarah E. Crump (1987–2022)

Preface

Just over two decades ago, the Developments in Paleoenvironmental Research (DPER) book series was initiated with a goal of providing an outlet for syntheses, monographs, and methodological books dedicated to the paleoenvironmental sciences. DPER is also the official book series of the International Paleolimnology Association. The first four volumes, under the title of *Tracking Environmental Change Using Lake Sediments*, were published simultaneously in 2001, and focussed on formalizing and synthesizing approaches used in paleolimnological research: Vol 1 on *Basin Analysis, Coring, and Chronological Techniques* (Last and Smol 2001); Vol 2 on *Physical and Geochemical Methods* (Last and Smol 2001); Vol 3 on *Terrestrial, Algal, and Siliceous Indicators* (Smol et al. 2001); and Vol 4 on *Zoological Indicators* (Smol et al. 2001). The fifth volume, *Data Handling and Numerical Techniques* (Birks et al. 2012), was published about a decade later. These books are still routinely cited today, as many of the field, lab, and numerical methods have not changed substantially in the intervening years. Nonetheless, there has been an explosion of new applications using these methodologies, as is evident by the large number of paleolimnology papers being published (and cited!). Although the first four *DPER* volumes represent over 1500 pages of text, DNA was never even mentioned.

Over the last decade, and especially over the last few years, sedimentary DNA (sedDNA) analyses have become prevalent in paleoenvironmental research, often forming the backbone of many high-profile publications. While sedDNA analyses pose many challenges (which are often highlighted in this volume), the potential for reconstructing a large suite of terrestrial and aquatic organisms has resulted in a revolution of new insights and applications addressing a remarkably large and diverse set of research questions. This new *DPER* volume attempts to synthesize some of these many recent advancements. To accomplish this, we have recruited 49 experts to author 13 chapters addressing a wide spectrum of sedDNA research. The book opens with a short introduction (Chap. 1), followed by an examination of the sources and fates of lake sedDNA (Chap. 2), and then an overview of the general workflow that is often employed for this research (Chap. 3). Chapters 4, 5, 6, 7, 8, 9, 10, and 11 summarize sedDNA approaches for various groups of organisms, from bacteria to humans. Chapter 12 focuses on biodiversity and network modeling approaches, and the final chapter provides a perspective for future developments within sedDNA research. A Glossary and an Index conclude the volume.

In addition to thanking the authors for their diligent work in preparing these chapters, we would also like to acknowledge the many colleagues who provided constructive comments on various chapter drafts.

These are exciting times for paleoenvironmental work, and especially for sedDNA studies. We hope this volume captures some of this excitement and helps propel this research towards addressing new questions.

Umeå, Sweden — Eric Capo
Thonon-les-Bains, France — Cécilia Barouillet
Kingston, ON, Canada — John P. Smol

Aims and Scope of Developments in Paleoenvironmental Research Series

Paleoenvironmental research continues to enjoy tremendous interest and progress in the scientific community. The overall aim and scope of the Developments in Paleoenvironmental Research book series is to capture this excitement and document these developments. Volumes related to any aspect of paleoenvironmental research, encompassing any time period, are within the scope of the series. For example, relevant topics include studies focused on terrestrial, peatland, lacustrine, riverine, estuarine, and marine systems, ice cores, cave deposits, palynology, isotopes, geochemistry, sedimentology, paleontology, etc. Methodological and taxonomic volumes relevant to paleoenvironmental research are also encouraged. The series will include edited volumes on a particular subject, geographic region, or time period, conference and workshop proceedings, as well as monographs. Prospective authors and/or editors should consult the series editor for more details. The series editor also welcomes any comments or suggestions for future volumes.

Editor and Board of Advisors of *Developments in Paleoenvironmental Research* Book Series

Professor Helen Bennion
Environmental Change Research Centre
Department of Geography
University College London
North-West Wing, Gower Street,
London WC1E 6BT, UK
e-mail: h.bennion@ucl.ac.uk

Professor Glen MacDonald
Department of Geography, Department of Ecology and Evolutionary Biology
Bunche Hall 1152
Los Angeles, CA 90095, USA
e-mail: macdonal@geog.ucla.edu

Professor Guangjie Chen
Yunnan Key Lab of Plateau Geographical Processes & Environmental Change
Faculty of Geography
Yunnan Normal University
Kunming, Yunnan 650500, China
e-mail: guangjiechen@ynnu.edu.cn

Safety Considerations and Caution

Paleolimnology has grown into a vast scientific pursuit with many branches and subdivisions. It should not be surprising, therefore, that the tools used by paleolimnologists are equally diverse. Virtually every one of the techniques described in this book requires some familiarity with standard laboratory or field safety procedures. The responsibility for safe and careful application of these methods is yours. Never underestimate the personal risk factor when undertaking either field or laboratory investigations. Researchers are strongly advised to obtain all safety information available for the techniques they will be using and to explicitly follow appropriate safety procedures. This is particularly important when using strong acids, alkalies, or oxidizing reagents in the laboratory or many of the analytical and sample collection/preparation instruments described in this volume. Most manufacturers of laboratory equipment and chemical supply companies provide this safety information, and many Internet and other library resources contain additional safety protocols. Researchers are also advised to discuss their procedures with colleagues who are familiar with these approaches, and so obtain further advice on safety and other considerations.

The editors and publisher do not necessarily endorse or recommend any specific product, procedure, or commercial service that may be cited in this publication.

Contents

Contributors

Inger G. Alsos The Arctic University Museum of Norway, UiT—The Arctic University of Norway, Tromsø, Norway

Linda Armbrecht Institute for Marine and Antarctic Studies (IMAS), Ecology and Biodiversity Centre, University of Tasmania, Hobart, Australia

Femke Augustijns Department of Earth and Environmental Sciences, Katholieke Universiteit Leuven, Leuven, Belgium

Cécilia Barouillet Pôle R&D ECLA, INRAE, CARRTEL, Thonon les Bains, France

INRAE, Université Savoie Mont Blanc, CARRTEL, Thonon-les-bains, France

Stefan Bertilsson Department of Aquatic Sciences and Assessment, Swedish University of Agricultural Sciences, Uppsala, Sweden

Kristine Bohmann Section for Molecular Ecology and Evolution, Globe Institute, University of Copenhagen, Copenhagen, Denmark

Eric Capo Department of Ecology and Environmental Science, Umeå University, Umeå, Sweden

Marco J. L. Coolen School of Earth and Planetary Sciences, Curtin University, Bentley, Australia

Sarah E. Crump Department of Geology & Geophysics, University of Utah, Salt Lake City, UT, USA

Isabelle Domaizon Pôle R&D ECLA, INRAE, CARRTEL, Thonon les Bains, France

INRAE, Université Savoie Mont Blanc, CARRTEL, Thonon-les-bains, France

Katharina Dulias Institute of Geosystems and Bioindication, Technische Universität Braunschweig, Braunschweig, Germany

Laura S. Epp Limnological Institute, Department of Biology, University of Konstanz, Konstanz, Germany

Anthony Foucher Laboratoire des Sciences du Climat et de l'Environnement, UMR8212 (CEA/CNRS/UVSQ), Université Paris-Saclay, Gif-sur-Yvette, France

Sandra Garcés-Pastor Marine Sciences Institute, ICM-CSIC, Barcelona, Spain

Department of Evolutionary Biology, Ecology and Environmental Sciences, University of Barcelona, Barcelona, Spain

Ludovic Gielly Université Grenoble Alpes, Université Savoie Mont Blanc, CNRS, LECA, Grenoble, France

Charline Giguet-Covex UMR 5204 EDYTEM, Université Savoie Mont Blanc, CNRS, Le Bourget du Lac, France

EDYTEM, CNRS, Université Savoie Mont Blanc, Le Bourget-du-Lac, France

Irene Gregory-Eaves Department of Biology, McGill University, Montreal, Quebec, Canada

Groupe de Recherche Interuniversitaire en Limnologie, Montreal, Quebec, Canada

Peter D. Heintzman Department of Geological Sciences, Centre for Palaeogenetics, Stockholm University, Stockholm, Sweden

Luke E. Holman Section for Molecular Ecology and Evolution, Globe Institute, University of Copenhagen, Copenhagen, Denmark

Stanislav Jelavić Université Grenoble Alpes, Université Savoie Mont Blanc, CNRS, IRD, Université Gustave Eiffel, ISTerre, Grenoble, France

Weihan Jia Polar Terrestrial Environmental Systems Research Group, Alfred Wegener Institute Helmholtz Centre for Polar and Marine Research, Potsdam, Germany

Institute of Environmental Science and Geography, University of Potsdam, Potsdam, Germany

Jens Kallmeyer GFZ German Research Centre for Geosciences, Section Geomicrobiology, Potsdam, Germany

Youri Lammers The Arctic University Museum of Norway, UiT – The Arctic University of Norway, Tromsø, Norway

Maria Leunda Department of Plant Biology and Ecology, University of the Basque Country (UPV/EHU), Leioa, Spain

Swiss Federal Institute for Forest, Snow and Landscape Research, Birmensdorf, Switzerland

Institute of Plant Sciences, Oeschger Centre for Climate Change Research, University of Bern, Bern, Switzerland

Susanne Liebner GFZ German Research Centre for Geosciences, Section Geomicrobiology, Potsdam, Germany

Sisi Liu Polar Terrestrial Environmental Systems Research Group, Alfred Wegener Institute Helmholtz Centre for Polar and Marine Research, Potsdam, Germany

Marie-Eve Monchamp Department of Biology, McGill University, Montreal, QC, Canada

Groupe de Recherche Interuniversitaire en Limnologie, Montreal, QC, Canada

Marina A. Morlock Department of Ecology and Environmental Science, Umeå University, Umeå, Sweden

Tyler J. Murchie McMaster Ancient DNA Centre, Department of Anthropology, McMaster University, Hamilton, ON, Canada

Hakai Institute, Heriot Bay, BC, Canada

Kevin Nota Department of Evolutionary Genetics, Max Planck Institute for Evolutionary Anthropology, Leipzig, Germany

Laura Parducci Department of Environmental Biology, Sapienza University of Rome, Rome, Italy

Department of Ecology and Genetics, Uppsala University, Uppsala, Sweden

Mikkel Winther Pedersen Centre for Ancient Environmental Genomics, Globe Institute, University of Copenhagen, Copenhagen, Denmark

Frances R. Pick Department of Biology, University of Ottawa, Ottawa, ON, Canada

Aloïs Revéret The Arctic University Museum of Norway, UiT—The Arctic University of Norway, Tromsø, Norway

Dilli P. Rijal The Arctic University Museum of Norway, UiT the Arctic University of Norway, Tromsø, Norway

Alexandra Rouillard Department of Geosciences, UiT – The Arctic University of Norway, Tromsø, Norway

Section for GeoGenetics, Globe Institute, University of Copenhagen, Copenhagen, Denmark

Rikai Sawafuji School of Advanced Sciences, The Graduate University for Advanced Studies (SOKENDAI), Kanagawa, Japan

Christoph Schwörer Institute of Plant Sciences, Oeschger Centre for Climate Change Research, University of Bern, Bern, Switzerland

Viviane Slon Department of Anatomy and Anthropology, Department of Human Molecular Genetics and Biochemistry, Faculty of Medicine, The Dan David Center for Human Evolution and Biohistory Research, Tel Aviv University, Tel Aviv, Israel

John P. Smol Paleoecological Environmental Assessment and Research Lab (PEARL) Department of Biology, Queen's University, Kingston, ON, Canada

Kathleen R. Stoof-Leichsenring Division Geosciences|Polar Terrestrial Environmental Systems, Alfred-Wegener-Institute, Potsdam, Germany

Zofia E. Taranu Environment and Climate Change Canada, Aquatic Contaminants Research Division, McGill University, Montreal, QC, Canada

Groupe de Recherche Interuniversitaire en Limnologie, Montreal, QC, Canada

Department of Biology, McGill University, Montreal, QC, Canada

Benjamin Vernot Department of Evolutionary Genetics, Max Planck Institute for Evolutionary Anthropology, Leipzig, Germany

Aurèle Vuillemin GFZ German Research Centre for Geosciences, Section Geomicrobiology, Potsdam, Germany

Yi Wang Limnological Institute, Department of Biology, University of Konstanz, Konstanz, Germany

Yucheng Wang Department of Zoology, University of Cambridge, Cambridge, UK

State Key Laboratory of Tibetan Plateau Earth System, Resources and Environment (TPESRE), Institute of Tibetan Plateau Research, Chinese Academy of Sciences, Beijing, China

Susanna A. Wood Cawthron Institute, Nelson, New Zealand

About the Editors

Eric Capo is an assistant professor at the Department of Ecology and Environmental Science at Umeå University (Umeå, Sweden). Dr Capo is the founder and the coordinator of the *sedaDNA Scientific Society*.

Cécilia Barouillet is a postdoctoral researcher at UMR CARRTEL INRAE (Thonon-les-Bains, France). Dr Barouillet is vice-president – communication of the *International Society of Limnology* (SIL) and on the board of the *sedaDNA Scientific Society*.

John P. Smol is a distinguished university professor at the Department of Biology at Queen's University (Kingston, Ontario, Canada), with a cross-appointment at the School of Environmental Studies. He co-directs the Paleoecological Environmental Assessment and Research Lab (PEARL). Prof Smol is editor of the journal *Environmental Reviews* and is the founding editor of the *Journal of Paleolimnology*.

Chapter 1
Using Lake Sedimentary DNA to Reconstruct Biodiversity Changes

Eric Capo, Cécilia Barouillet, and John P. Smol

Keywords Sedimentary DNA · Environmental DNA · Biodiversity · Paleolimnology · Lakes

Lake sediments consist of organic and inorganic matter originating from autochthonous (in-lake) and allochthonous (from the catchment and beyond) sources. Thus far, paleolimnological studies of biological communities have focussed primarily on morphological subfossils preserved in sediments, and as such most analyses are based on a relatively limited number of aquatic and terrestrial groups that leave well-preserved and readily identifiable indicators (Last and Smol 2001a, b; Smol et al. 2001a, b; Birks et al. 2012). These morphological subfossils often include among others the siliceous remains of diatoms and chrysophytes, calcified remains, organic walled or calcified dinoflagellates, cladoceran exoskeletons, *Chaoborus* mandibles, chironomid head capsules, fungal spores, Lepidoptera wing scales, as well as pollen and plant macrofossils. For some biological groups for which morphological remains are not well preserved or cannot be taxonomically identified to the species level, alternative proxies have been sought to provide a more detailed and comprehensive understanding of past biological diversity in a broader range of environments (Birks 1991; Birks and Birks 2016; Coolen et al. 2013). For example, the recovery and study of pigments, lipids, and other biogeochemical proxies have

E. Capo (✉)
Department of Ecology and Environmental Science, Umeå University, Umeå, Sweden
e-mail: eric.capo@umu.se

C. Barouillet
INRAE, Université Savoie Mont Blanc, CARRTEL, Thonon les Bains, France

J. P. Smol
Paleoecological Environmental Assessment and Research Lab (PEARL) Department of Biology, Queen's University, Kingston, ON, Canada
e-mail: smolj@queensu.ca

E. Capo et al. (eds.), *Tracking Environmental Change Using Lake Sediments*, Developments in Paleoenvironmental Research 21,
https://doi.org/10.1007/978-3-031-43799-1_1

been used to provide important insights into past ecosystem functions (Castañeda and Schouten 2011; Gregory-Eaves and Smol 2024). However, despite advances in the number of available proxies, the vast majority of organisms cannot be inferred from mainstream methods available to paleolimnologists. Fortunately, ongoing work on the application of DNA-based approaches to sedimentary archives, as described in this volume, provides many exciting opportunities for palaeosciences (Williams et al. 2023).

Sedimentary DNA (sedDNA) is the DNA that is deposited and preserved in sediment and soil archives, such as freshwater and marine sediments, paleo-soils, caves and permafrost, over relatively long time periods (i.e. from decades to hundreds of thousands of years). SedDNA degrades over time and accumulates post-mortem damages. The pool of ancient degraded DNA fragments (Orlando et al. 2021) found in sediments is called sedimentary ancient DNA (sedaDNA) (Crump 2021; Capo et al. 2021). The sedimentary DNA approach overcomes some limitations of the above-listed classical paleolimnological proxies by providing novel information on historical biodiversity changes from the genetic information naturally preserved in environmental archives. Lake sedDNA captures biological changes of both the aquatic and surrounding terrestrial ecosystems including vegetation, human-induced changes in land uses and food webs in pelagic and benthic communities of aquatic systems (Fig. 1.1). As such, past genetic signals have the potential to provide insights into landscape development and the internal ecological dynamics of any lake system. Importantly, the DNA signal from sedimentary archives provides a more integrated view of biological dynamics as it can be used to reconstruct changes in multiple trophic levels (Domaizon et al. 2017; Alsos et al. 2022) and to track the recent effects of anthropogenic pressures on aquatic food webs (Barouillet et al. 2023). Similarly, sedDNA can reveal long-term changes in terrestrial biodiversity (Parducci et al. 2012; Willerslev et al. 2014; Rijal et al. 2021; Wang et al. 2021; Kjær et al. 2022), the response of terrestrial organisms to anthropogenic perturbations (Giguet-Covex et al. 2014; Brown et al. 2022), and the impact of introduced alien species on ecosystems (Ficetola et al. 2018), amongst many other applications outlined in this volume.

SedDNA approaches are also of interest to archaeologists and paleo-anthrophologists, as these data may provide insights into human history (Pedersen et al. 2016; Slon et al. 2017) and the interactions of past societies with their environment, such as the development of agriculture and urbanization (Pansu et al. 2015; Garcés-Pastor et al. 2022). SedDNA can also be used by geologists and geomorphologists to trace sediment depositional processes (Collins et al. 2020) and historical events in lake catchments (Giosan et al. 2012; Szczuciński et al. 2016), as well as by evolutionary biologists to study population changes over time (Lammers et al. 2021; Meucci et al. 2021). Overall, the DNA of numerous biological groups of aquatic and terrestrial origin has been identified in sedimentary archives over the last few decades (Fig. 1.2). The rapid expansion of this research field has led to the creation of the "sedaDNA Scientific Society" in 2021, with over 300 members joining within the first year.

Fig. 1.1 Illustration of the sources of sedimentary DNA derived from watersheds (vegetation, cultivated plants, wildlife, livestock, human populations) and lakes (hydrophytes, benthic and pelagic food webs). Environmental pressures are depicted by the farming and the industries as well as the sun and the clouds representing global climate change. (This figure was produced by https://mikimostudio.com/)

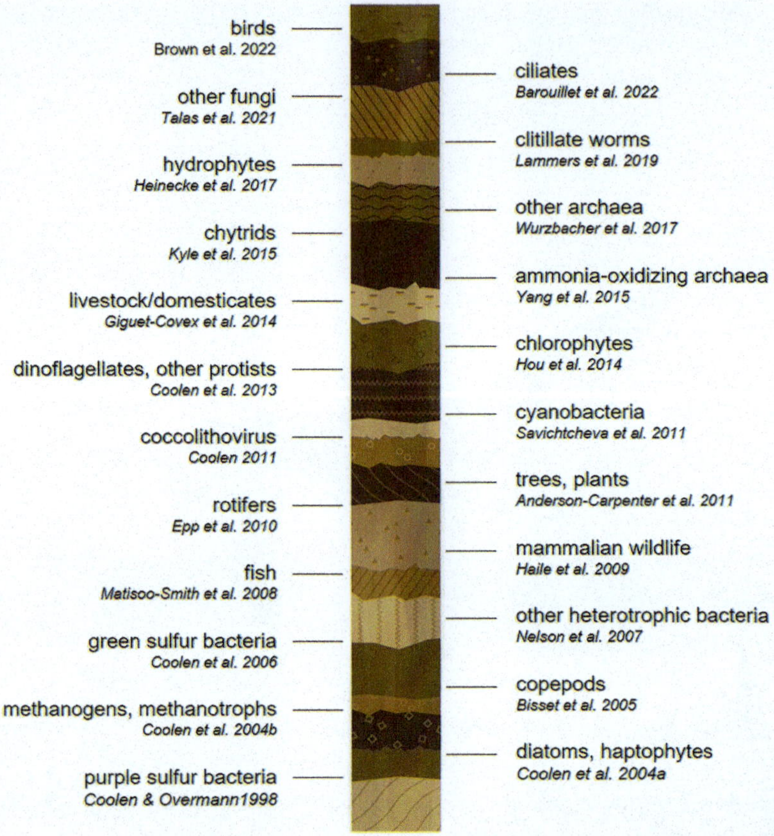

Fig. 1.2 Chronological recovery of past signals of specific biological groups using sedimentary DNA from lake sedimentary archives. Most groups are from organisms living outside the sediments, whereas some live and die in sediments resulting in a DNA signal relevant to both their past and current stages (see Chap. 4 for insights). This figure was produced by https://mikimostudio.com/)

This book is a continuation of the previous five volumes in the *Tracking Environmental Change Using Lake Sediments - Developments in Paleoenvironmental Research* (DPER) book series (Last and Smol 2001a, b; Smol et al. 2001a, b; Birks et al. 2012), which collectively attempt to summarize the major approaches and techniques used in paleolimnology to track long-term environmental changes. The first two books in this series dealt with physical and chemical techniques. Volume 1 (Last and Smol 2001a) included chapters on field work, such as core collection and sectioning, basin analysis techniques, and dating sedimentary profiles. Volume 2 (Last and Smol 2001b) summarized the many physical and chemical techniques that paleolimnologists use to describe and interpret sedimentary profiles. Volumes 3 and 4 focussed on the biological information that can be gleaned from sediments from archived morphological fossils, with some example applications. Specifically, volume 3 (Smol et al. 2001a) contained 15 chapters that dealt primarily with terrestrial, algal, and siliceous indicators, and included chapters on topics such as pollen, stomates, macrofossils, charcoal, fungi and non-pollen palynomorphs, as well as semi-terrestrial indicators, such as testate amoebae. This was followed by six chapters dealing with siliceous microfossils, inclusive of diatoms, chrysophytes, ebrideans, phytoliths, sponges, and siliceous protozoan plates. Volume 3 concluded with two chapters on how to make use of chemical techniques to track biological populations (i.e., biogenic silica, fossil pigments). Volume 4 (Smol et al. 2001b) focussed primarily on zoological indicators preserved in lake sediments, with chapters addressing the preserved remains of cladocerans, insects, mites, ostracods, molluscs, and fishes. Meanwhile, Volume 5 (Birks et al. 2012) provided a summary of the various numerical and data handling approaches commonly used in paleolimnology.

This sixth volume in the series attempts to summarize recent advances in methodology and applications of the rapidly developing field of sedDNA. We include chapters discussing the fate and sources of lake sedDNA (Chap. 2) and the general workflow used for sedDNA related-analyses (Chap. 3). This is followed by 8 chapters (Chaps. 4, 5, 6, 7, 8, 9, 10, and 11) that focus on reconstructing the past dynamics of various biological groups, including bacteria, Archaea, protists, animals and plants of both terrestrial and aquatic origins with sedDNA. The book concludes with a chapter about the use of network modeling approaches in sedDNA research (Chap. 12) and perspectives on the future development of sedDNA approaches (Chap. 13).

References

Alsos IG, Rijal DP, Ehrich D, Karger DN, Yoccoz NG, Heintzman PD, Brown AG, Lammers Y, Pellissier L, Alm T, Bråthen KA, Coissac E, Merkel MKF, Alberti A, Denoeud F, Bakke J, Phylonorway Consortium (2022) Postglacial species arrival and diversity buildup of northern ecosystems took millennia. Sci Adv 8(39):eabo7434. https://doi.org/10.1126/sciadv.abo7434

Barouillet C, Monchamp M-E, Bertilsson S, Brasell K, Domaizon I, Epp LS, Ibrahim A, Mejbel H, Nwosu EC, Pearman JK, Picard M, Thomson-Laing G, Tsugeki N, Von Eggers J, Gregory-Eaves I, Pick FR, Wood SA, Capo E (2023) Investigating the effects of anthropogenic stressors on lake biota using sedimentary DNA. Freshw Biol n/a(n/a). https://doi.org/10.1111/fwb.14027

Birks HH (1991) Holocene vegetational history and climatic change in West Spitsbergen - plant macrofossils from Skardtjørna, an Arctic lake. The Holocene 1(3):209–218. https://doi.org/10.1177/095968369100100303

Birks HJB, Birks HH (2016) How have studies of ancient DNA from sediments contributed to the reconstruction of quaternary floras? New Phytol 209(2):499–506. https://doi.org/10.1111/nph.13657

Birks HJB, Lotter AF, Juggins S, Smol JP (2012) Tracking environmental change using Lake sediments: data handling and numerical techniques. Springer. https://doi.org/10.1007/978-94-007-2745-8

Brown T, Rijal DP, Heintzman PD, Clarke CL, Blankholm H-P, Høeg HI, Lammers Y, Brathen KA, Edwards M, Alsos IG (2022) Paleoeconomy more than demography determined prehistoric human impact in Arctic Norway. PNAS Nexus 1(5):pgac209. https://doi.org/10.1093/pnasnexus/pgac209

Capo E, Giguet-Covex C, Rouillard A, Nota K, Heintzman PD, Vuillemin A, Ariztegui D, Arnaud F, Belle S, Bertilsson S, Bigler C, Bindler R, Brown AG, Clarke CL, Crump SE, Debroas D, Englund G, Ficetola GF, Garner RE, Gauthier J, Gregory-Eaves I, Heinecke L, Herzschuh U, Ibrahim A, Kisand V, Kjær KH, Lammers Y, Littlefair J, Messager E, Monchamp M-E, Olajos F, Orsi W, Pedersen MW, Rijal DP, Rydberg J, Spanbauer T, Stoof-Leichsenring KR, Taberlet P, Talas L, Thomas C, Walsh DA, Wang Y, Willerslev E, van Woerkom A, Zimmermann HH, Coolen MJL, Epp LS, Domaizon IG, Alsos I, Parducci L (2021) Lake sedimentary DNA research on past terrestrial and aquatic biodiversity: overview and recommendations. Quaternary 4(1):6. https://doi.org/10.3390/quat4010006

Castañeda IS, Schouten S (2011) A review of molecular organic proxies for examining modern and ancient lacustrine environments. Quat Sci Rev 30(21):2851–2891. https://doi.org/10.1016/j.quascirev.2011.07.009

Collins AL, Blackwell M, Boeckx P, Chivers C-A, Emelko M, Evrard O, Foster I, Gellis A, Gholami H, Granger S, Harris P, Horowitz AJ, Laceby JP, Martinez-Carreras N, Minella J, Mol L, Nosrati K, Pulley S, Silins U, da Silva YJ, Stone M, Tiecher T, Upadhayay HR, Zhang Y (2020) Sediment source fingerprinting: benchmarking recent outputs, remaining challenges and emerging themes. J Soils Sediments 20(12):4160–4193. https://doi.org/10.1007/s11368-020-02755-4

Coolen MJL, Orsi WD, Balkema C, Quince C, Harris K, Sylva SP, Filipova-Marinova M, Giosan L (2013) Evolution of the plankton paleome in the Black Sea from the Deglacial to Anthropocene. Proc Natl Acad Sci 110(21):8609–8614. https://doi.org/10.1073/pnas.1219283110

Crump SE (2021) Sedimentary ancient DNA as a tool in paleoecology. Nat Rev Earth Environ 2(4):229–229. https://doi.org/10.1038/s43017-021-00158-8

Domaizon I, Winegardner A, Capo E, Gauthier J, Gregory-Eaves I (2017) DNA-based methods in paleolimnology: new opportunities for investigating long-term dynamics of lacustrine biodiversity. J Paleolimnol 58(1):1–21. https://doi.org/10.1007/s10933-017-9958-y

Ficetola GF, Poulenard J, Sabatier P, Messager E, Gielly L, Leloup A, Etienne D, Bakke J, Malet E, Fanget B, Støren E, Reyss J-L, Taberlet P, Arnaud F (2018) DNA from lake sediments reveals long-term ecosystem changes after a biological invasion. Sci Adv 4(5):eaar4292. https://doi.org/10.1126/sciadv.aar4292

Garcés-Pastor S, Coissac E, Lavergne S, Schwörer C, Theurillat J-P, Heintzman PD, Wangensteen OS, Tinner W, Rey F, Heer M, Rutzer A, Walsh K, Lammers Y, Brown AG, Goslar T, Rijal DP, Karger DN, Pellissier L, Heiri O, Alsos IG (2022) High resolution ancient sedimentary DNA shows that alpine plant diversity is associated with human land use and climate change. Nat Commun 13(1):6559. https://doi.org/10.1038/s41467-022-34010-4

Giguet-Covex C, Pansu J, Arnaud F, Rey P-J, Griggo C, Gielly L, Domaizon I, Coissac E, David F, Choler P, Poulenard J, Taberlet P (2014) Long livestock farming history and human landscape shaping revealed by lake sediment DNA. Nat Commun 5(1):3211. https://doi.org/10.1038/ncomms4211

Giosan L, Coolen MJL, Kaplan JO, Constantinescu S, Filip F, Filipova-Marinova M, Kettner AJ, Thom N (2012) Early anthropogenic transformation of the Danube-Black Sea system. Sci Rep 2(1):582. https://doi.org/10.1038/srep00582

Gregory-Eaves I, Smol JP (2024) Paleolimnology: approaches and applications. Chapter 30. In: Jones I, Smol JP (eds) Wetzel's limnology: Lake and river ecosystems, 4th, Elsevier edn, Oxford

Kjær KH, Bjørk AA, Kjeldsen KK, Hansen ES, Andresen CS, Siggaard-Andersen M-L, Khan SA, Søndergaard AS, Colgan W, Schomacker A, Woodroffe S, Funder S, Rouillard A, Jensen JF, Larsen NK (2022) Glacier response to the little ice age during the Neoglacial cooling in Greenland. Earth Sci Rev 227:103984. https://doi.org/10.1016/j.earscirev.2022.103984

Lammers Y, Heintzman PD, Alsos IG (2021) Environmental palaeogenomic reconstruction of an ice age algal population. Commun Biol 4(1):1–11. https://doi.org/10.1038/s42003-021-01710-4

Last WM, Smol JP (2001a) Tracking environmental change using Lake sediments: basin analysis, coring, and chronological techniques. Springer, Dordrecht. https://doi.org/10.1007/0-306-47669-X

Last WM, Smol JP (2001b) Tracking environmental change using Lake sediments: physical and geochemical methods. Kluwer Academic Publishers. https://doi.org/10.1007/0-306-47670-3

Meucci S, Schulte L, Zimmermann HH, Stoof-Leichsenring KR, Epp L, Bronken Eidesen P, Herzschuh U (2021) Holocene chloroplast genetic variation of shrubs (Alnus alnobetula, Betula nana, Salix sp.) at the siberian tundra-taiga ecotone inferred from modern chloroplast genome assembly and sedimentary ancient DNA analyses. Ecol Evol 11(5):2173–2193. https://doi.org/10.1002/ece3.7183

Orlando L, Allaby R, Skoglund P, Der Sarkissian C, Stockhammer PW, Ávila-Arcos MC, Fu Q, Krause J, Willerslev E, Stone AC, Warinner C (2021) Ancient DNA analysis. Nat Rev Methods Primers 1(1):1–26. https://doi.org/10.1038/s43586-020-00011-0

Pansu J, Giguet-Covex C, Ficetola GF, Gielly L, Boyer F, Zinger L, Arnaud F, Poulenard J, Taberlet P, Choler P (2015) Reconstructing long-term human impacts on plant communities: an ecological approach based on lake sediment DNA. Mol Ecol 24(7):1485–1498. https://doi.org/10.1111/mec.13136

Parducci L, Jørgensen T, Tollefsrud MM, Elverland E, Alm T, Fontana SL, Bennett KD, Haile J, Matetovici I, Suyama Y, Edwards ME, Andersen K, Rasmussen M, Boessenkool S, Coissac E, Brochmann C, Taberlet P, Houmark-Nielsen M, Larsen NK, Orlando L, Gilbert MTP, Kjær KH, Alsos IG, Willerslev E (2012) Glacial survival of boreal trees in northern Scandinavia. Science 335(6072):1083–1086. https://doi.org/10.1126/science.1216043

Pedersen MW, Ruter A, Schweger C, Friebe H, Staff RA, Kjeldsen KK, Mendoza MLZ, Beaudoin AB, Zutter C, Larsen NK, Potter BA, Nielsen R, Rainville RA, Orlando L, Meltzer DJ, Kjær KH, Willerslev E (2016) Postglacial viability and colonization in North America's ice-free corridor. Nature 537(7618):45–49. https://doi.org/10.1038/nature19085

Rijal DP, Heintzman PD, Lammers Y, Yoccoz NG, Lorberau KE, Pitelkova I, Goslar T, Murguzur FJA, Salonen JS, Helmens KF, Bakke J, Edwards ME, Alm T, Bråthen KA, Brown AG, Alsos IG (2021) Sedimentary ancient DNA shows terrestrial plant richness continuously increased over the Holocene in northern Fennoscandia. Science. Advances 7(31):eabf9557. https://doi.org/10.1126/sciadv.abf9557

Slon V, Hopfe C, Weiß CL, Mafessoni F, de la Rasilla M, Lalueza-Fox C, Rosas A, Soressi M, Knul MV, Miller R, Stewart JR, Derevianko AP, Jacobs Z, Li B, Roberts RG, Shunkov MV, de Lumley H, Perrenoud C, Gušić I, Kućan Ž, Rudan P, Aximu-Petri A, Essel E, Nagel S, Nickel B, Schmidt A, Prüfer K, Kelso J, Burbano HA, Pääbo S, Meyer M (2017) Neandertal and Denisovan DNA from Pleistocene sediments. Science 356(6338):605–608. https://doi.org/10.1126/science.aam9695

Smol JP, Birks HJB, Last WM (eds) (2001a) Tracking environmental change using Lake sediments: zoological indicators. Springer, Dordrecht. https://doi.org/10.1007/0-306-47671-1

Smol JP, Birks HJB, Last WM, Bradley RS, Alverson K (eds) (2001b) Tracking environmental change using Lake sediments: terrestrial, algal, and siliceous indicators. Springer, Dordrecht. https://doi.org/10.1007/0-306-47668-1

Szczuciński W, Pawłowska J, Lejzerowicz F, Nishimura Y, Kokociński M, Majewski W, Nakamura Y, Pawlowski J (2016) Ancient sedimentary DNA reveals past tsunami deposits. Mar Geol 381:29–33. https://doi.org/10.1016/j.margeo.2016.08.006

Wang Y, Pedersen MW, Alsos IG, De Sanctis B, Racimo F, Prohaska A, Coissac E, Owens HL, Merkel MKF, Fernandez-Guerra A, Rouillard A, Lammers Y, Alberti A, Denoeud F, Money D, Ruter AH, McColl H, Larsen NK, Cherezova AA, Edwards ME, Fedorov GB, Haile J, Orlando L, Vinner L, Korneliussen TS, Beilman DW, Bjørk AA, Cao J, Dockter C, Esdale J, Gusarova G, Kjeldsen KK, Mangerud J, Rasic JT, Skadhauge B, Svendsen JI, Tikhonov A, Wincker P, Xing Y, Zhang Y, Froese DG, Rahbek C, Bravo DN, Holden PB, Edwards NR, Durbin R, Meltzer DJ, Kjær KH, Möller P, Willerslev E (2021) Late quaternary dynamics of Arctic biota from ancient environmental genomics. Nature 600(7887):86–92. https://doi.org/10.1038/s41586-021-04016-x

Willerslev E, Davison J, Moora M, Zobel M, Coissac E, Edwards ME, Lorenzen ED, Vestergård M, Gussarova G, Haile J, Craine J, Gielly L, Boessenkool S, Epp LS, Pearman PB, Cheddadi R, Murray D, Bråthen KA, Yoccoz N, Binney H, Cruaud C, Wincker P, Goslar T, Alsos IG, Bellemain E, Brysting AK, Elven R, Sønstebø JH, Murton J, Sher A, Rasmussen M, Rønn R, Mourier T, Cooper A, Austin J, Möller P, Froese D, Zazula G, Pompanon F, Rioux D, Niderkorn V, Tikhonov A, Savvinov G, Roberts RG, MacPhee RDE, Gilbert MTP, Kjær KH, Orlando L, Brochmann C, Taberlet P (2014) Fifty thousand years of Arctic vegetation and megafaunal diet. Nature 506(7486):47–51. https://doi.org/10.1038/nature12921

Williams JW, Spanbauer TL, Heintzman PD, Blois J, Capo E, Goring SJ, Monchamp M-E, Parducci L, Eggers JMV, Alsos IG, Bowler C, Coolen MJL, Cullen N, Crump S, Epp LS, Fernandez-Guerra A, Grimm E, Herzschuh U, Mereghetti A, Meyer RS, Nota K, Pedersen MW, Pérez V, Shapiro B, Stoof-Leichsenring KR, Wood J (2023) Strengthening global-change science by integrating aeDNA with paleoecoinformatics. Trends Ecol Evol 0(0). https://doi.org/10.1016/j.tree.2023.04.016

Chapter 2
The Sources and Fates of Lake Sedimentary DNA

Charline Giguet-Covex, Stanislav Jelavić, Anthony Foucher, Marina A. Morlock, Susanna A. Wood, Femke Augustijns, Isabelle Domaizon, Ludovic Gielly, and Eric Capo

Keywords Sedimentary DNA · DNA taphonomy · DNA sources · DNA transfer · DNA degradation · DNA preservation · Lakes

Introduction

For over two decades, ancient DNA (aDNA) from various organisms has been successfully recovered from lake sediments from all over the world, ranging from decades to hundreds of thousands of years old (Fig. 2.1). Analysis of lake

C. Giguet-Covex (✉)
UMR 5204 EDYTEM, Université Savoie Mont Blanc, CNRS, Le Bourget du Lac, France
e-mail: charline.giguet-covex@univ-smb.fr

S. Jelavić
Université Grenoble Alpes, Université Savoie Mont Blanc, CNRS, IRD, Université Gustave Eiffel, ISTerre, Grenoble, France
e-mail: stanislav.jelavic@univ-grenoble-alpes.fr

A. Foucher
Laboratoire des Sciences du Climat et de l'Environnement, UMR8212 (CEA/CNRS/UVSQ), Université Paris-Saclay, Gif-sur-Yvette, France
e-mail: anthony.foucher@lsce.ipsl.fr

M. A. Morlock · E. Capo
Department of Ecology and Environmental Science, Umeå University, Umeå, Sweden
e-mail: marina.morlock@umu.se; eric.capo@umu.se

S. A. Wood
Cawthron Institute, Nelson, New Zealand
e-mail: susie.wood@cawthron.org.nz

E. Capo et al. (eds.), *Tracking Environmental Change Using Lake Sediments*, Developments in Paleoenvironmental Research 21,
https://doi.org/10.1007/978-3-031-43799-1_2

sedimentary DNA (sedDNA) has provided new information about past aquatic and terrestrial biodiversity and the trajectories of socio-ecosystems in various biomes (Crump 2021). These data have enabled scientists to answer ecological/environmental questions about organisms that do not leave visible and identifiable remains in sedimentary archives and that are therefore overlooked in traditional paleolimnology studies, e.g. fish and terrestrial mammals (e.g., Matisoo-Smith et al. 2008; Graham et al. 2016; Domaizon et al. 2017). SedDNA analyses also provided complementary information on taxonomic groups that are traditionally studied via morphological analysis, e.g. diatom cysts and pollen grains (Stoof-Leichsenring et al. 2012; Parducci et al. 2015).

SedDNA based-approaches have a great potential to address key and novel questions in ecology, such as on ecosystem functioning and community structure (Keck et al. 2020), phylogenetic and functional diversity and species richness (Huang et al. 2021; Alsos et al. 2022) and biogeographic patterns, for instance in response to glacial retreat or human movements (i.e., colonisation, migrations; Pedersen et al. 2016; Ficetola et al. 2018). These approaches are also of high interest for archaeologists to identify past human occupations and activities (Giguet-Covex et al. 2014; Brown et al. 2021) and their impacts on landscapes and ecosystems (Pansu et al. 2015a; Messager et al. 2022). Determining to what extent sedDNA records represent the past composition of living organisms is crucial to propose robust models and provide reliable answers to questions of interest (Birks and Birks 2016; Alsos et al. 2018; Capo et al. 2022). The reliability of DNA records is not only linked to methodological aspects (see Chap. 3 of this volume) but also to taphonomic processes leading to the record of organisms in the sediments (Giguet-Covex et al. 2019; Zinger et al. 2019; Capo et al. 2021). Taphonomy – "taphos'' for burial and "nomos" for law – as applied to lake sediments include the exploration of processes of production at the source (on land or in the water [and sediment] column), of transportation and deposition, as well as of preservation at each step of this chain of processes (Fig. 2.2).

The understanding of the sources and fates of lake sedDNA encompasses a broad range of research fields, from bio-geochemistry, sedimentology, biology and archaeology to forensic analyses, and is based on theoretical as well as empirical experimental and field studies. More specifically, knowledge can be gained through the

F. Augustijns
Department of Earth and Environmental Sciences, Katholieke Universiteit Leuven, Leuven, Belgium
e-mail: femke.augustijns@kuleuven.be

I. Domaizon
Pôle R&D ECLA, INRAE, CARRTEL, Thonon les Bains, France
e-mail: isabelle.domaizon@inrae.fr

L. Gielly
Université Grenoble Alpes, Université Savoie Mont Blanc, CNRS, LECA, Grenoble, France
e-mail: ludovic.gielly@univ-grenoble-alpes.fr

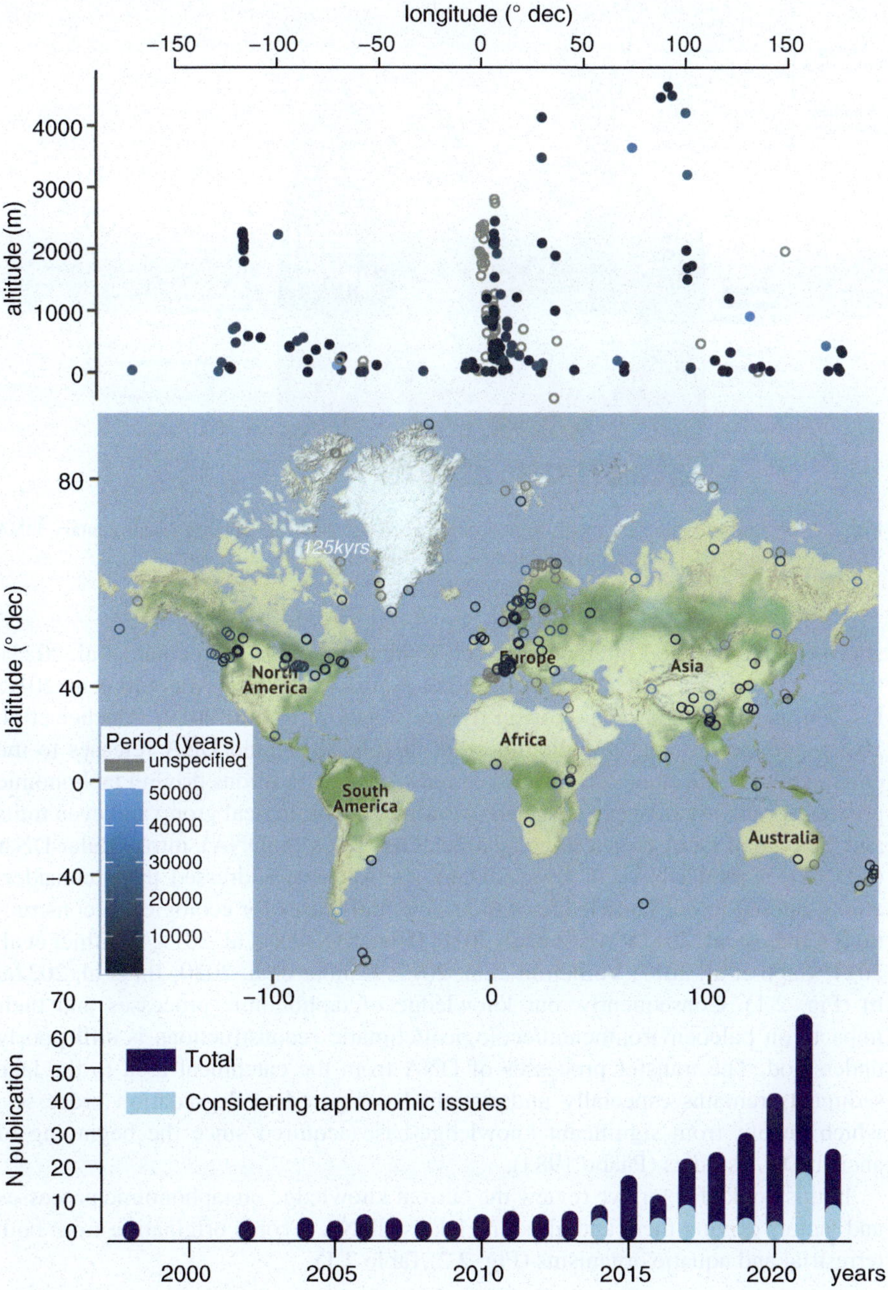

Fig. 2.1 Location and altitude of lake sedimentary DNA studies and their temporality (period in years, grey dots represent studies where the temporality is not reported in the database used, i.e. Von Eggers et al. 2022). Trends of the number of publications, including those considering potential taphonomic issues in the DNA records, are based on the following key words (used in the appropriate context) in the publications: taphonomy/ic, source, origin, production, transfer/transport, preservation, degradation, diagenesis and reliable/bility. (Modified from Von Eggers et al. 2022)

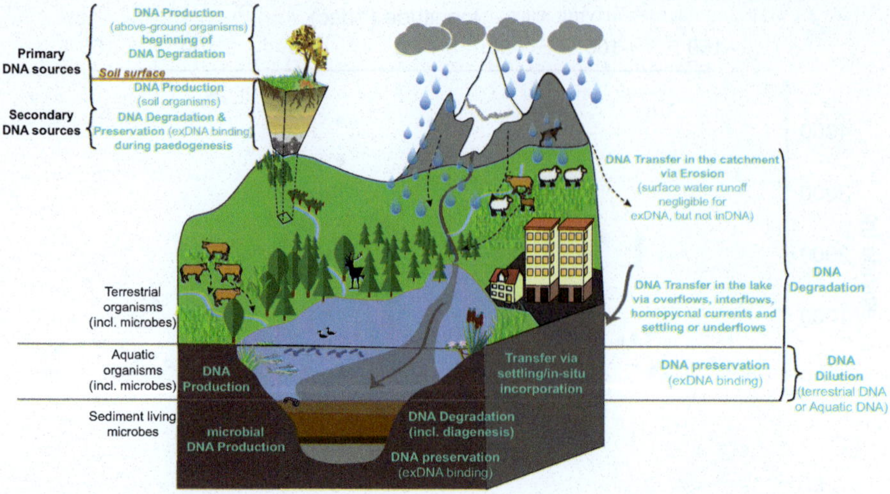

Fig. 2.2 Synthesis of the chain of taphonomic processes (in teal) affecting sedimentary DNA signals from terrestrial, water and sedimentary environments

study of DNA-mineral interactions (e.g., Kanbar et al. 2020; Freeman et al. 2023), theoretical and experimental DNA life-time (Smith et al. 2001; Allentoft et al. 2012; Lindahl and Nyberg 1972), erosion dynamics (Evrard et al. 2019; Foucher et al. 2020), geomicrobiology (Vuillemin et al. 2016a, 2017) and DNA releases to the environment (Poté et al. 2009; Rourke et al. 2022). Conditions driving taphonomic processes can vary between sites and over time, the biological group and even traits and the type of DNA targeted (i.e., extracellular DNA (exDNA), intracellular DNA (inDNA) or total DNA). A few sedDNA studies have addressed these considerations, adding to our knowledge of their potential biases for ecological reconstructions (Boere et al. 2011a; Alsos et al. 2018; Giguet-Covex et al. 2019; Gauthier et al. 2021; Capo et al. 2017; Vuillemin et al. 2017; Kanbar et al. 2020; Jia et al. 2022a, b) (Fig. 2.1). Consequently, our knowledge of taphonomic processes and their impacts on paleoenvironmental/ecological/climatic reconstructions is still poorly understood. The transfer processes of DNA from the catchment area to the lake sediments remains especially under-studied compared to degradation processes, which benefit from significant knowledge-base acquired since the beginning of ancient DNA studies (Pääbo 1984).

In this book chapter, we review the current knowledge on taphonomic processes and factors driving them and affecting lake sedDNA records originating from both terrestrial and aquatic organisms (Fig. 2.2; Table 2.1).

Table 2.1 Forcing factors driving each taphonomic process affecting terrestrial, aquatic and sediment-living organisms. Expected consequences associated with each of these processes in terms of spatial variability, interspecies variability and exDNA vs inDNA variability and potential issues for sedDNA reconstructions are also described

DNA origin	Taphonomic processes	Forcing factors	Spatial variability	Inter-species variability	exDNA/inDNA variability	Potential issues for sedaDNA reconstructions
Terrestrial organisms	DNA production (*primary source*)	Biomass and copy number of DNA in plasts, mitochondries…, type of biological tissue	Yes, linked to habitats	Yes	Yes	Over/underestimation (or even absence) of taxa
	DNA degradation (incl. during pedogenesis, and diagenesis)	Physico-chemical conditions, UV & biological activity, litter turnover rates, time	Yes	Yes	Yes	Loss or poor detection of taxa
	exDNA preservation via DNA binding in soils (secondary source)	Physico-chemical conditions, mineralogy, organic components and DNA conformation and size	Yes	Yes ?		
	DNA transfer from the catchment to the lake mostly via erosion (surface water runoff negligible for exDNA, probably not for inDNA)	Precipitation, temperature, bedrock, soil thickness, vegetation cover and human activities	Yes, depend on the connectivity between the DNA sources and the lake which depend on both the development of the hydrographic web and of soil erodibility	Yes	Yes	Over/underestimation (or even absence) of taxa, integration of "relic DNA", changes of DNA sources over time not only spatially in the catchment area but also in soil horizons affected by erosion
	DNA transfer to the sediment via overflow, interflow, homopycnal current and settling or underflow	Lake water density vs density of sedimentary inputs and lake thermal stratification	Can be, especially in large catchment and lake, with several tributaries		Yes probably partly	
	DNA dilution	Depend on allochtonous inputs relative to autochtonous production	Yes	Yes	No	Loss of the rarest taxa
Aquatic organisms	DNA production	Biomass and copy number of DNA in plasts, mitochondries…, type of biological tissue	Yes, linked to habitats	Yes	Yes	Over/underestimation (or even absence) of taxa
	DNA transfer	Settling/in-situ incorporation	Yes, linked to currents	Yes	Yes	
	DNA dilution	Depend on autochtonous inputs relative to allochtonous production	Yes	Yes	No	Loss of the rarest taxa
	exDNA preservation via DNA binding to particles	Physico-chemical conditions, mineralogy, organic components and DNA conformation and size, availability of binding sites	Yes	Yes ?		Over/underestimation (or even absence) of taxa
	DNA degradation (incl. during diagenesis)	Physico-chemical conditions, UV, speed of sedimentation of dead cells/carcasses in the water column and grazing by predators	Yes	Yes	Yes	Loss or poor detection of taxa (may be season dependent)
Sediment-living microbes	DNA production	Biomass and copy number of DNA in plasts, mitochondries…, type of biological tissue	Yes, linked to habitats	Yes	Yes	Over/underestimation (or even absence) of taxa, challenges the understanding of microbial records and paleoecological reconstructions
	exDNA preservation via DNA binding to particles	Physico-chemical conditions, mineralogy, organic components and DNA conformation and size, availability of binding sites	Yes	Yes ?		
	DNA degradation (from early diagenesis to deep burial)	Physico-chemical conditions, biological activity & time	Yes	Yes	Yes	Loss or poor detection of taxa

Effects of Taphonomic Processes on DNA Fractions Archived in Lake Sediments

The total sedDNA pool is composed of intracellular DNA (inDNA) and extracellular DNA (exDNA) fractions (Torti et al. 2015; Ellegaard et al. 2020; Capo et al. 2021). InDNA is defined as being inside the cells, whether they are dead or alive. ExDNA is released into the environment after cell lysis, also irrespective of whether the cells come from dead organisms or organisms still alive. Although exDNA can remain free in the environment, it usually get quickly adsorbed on minerals, organo-mineral complexes and organic compounds such as humic substances, which significantly increase DNA lifetime (Lorenz et al. 1981; Cai et al. 2006a; Saeki et al.

2011). Free DNA thus should not represent a significant fraction in the sedDNA pool. Knowledge about taphonomic processes affecting each of these DNA fractions is still limited, although some studies have shown variability between exDNA and inDNA fractions in terms of quantity and taxonomic composition in both lacustrine and marine sediments (Corinaldesi et al. 2005; Torti et al. 2018; Vuillemin et al. 2017; Case studies A3, 4, 5 in Capo et al. 2021; Gauthier et al. 2022).

Most sedDNA studies use protocols to extract the total DNA pool, but some studies have focused on recovering specifically the exDNA and inDNA fractions (see syntheses in Armbrecht et al. 2019; Capo et al. 2021). Regarding the exDNA extraction protocols, only the soluble fraction is extracted by washing with alkaline phosphate buffers, while the insoluble fraction, corresponding to organically/inorganically complexed DNA, remains inaccessible with this method (Torti et al. 2015). In the sedDNA literature, the most used protocol to extract exDNA is the one developed by Taberlet et al. (2012). More recently, a modified version of this protocol has been developed to increase the rate of recovery (Giguet-Covex et al. 2020; Capo et al. 2021). For total DNA extractions, more protocols are available and used in publications but the most popular are those from the PowerMax Soil® and PowerSoil® (Qiagen) kits (Capo et al. 2021). Here, we report on how both DNA quantity and the taxonomic composition obtained from the inDNA (or total DNA) and exDNA fractions may be differentially affected by taphonomic processes due to different sources, transfer processes or preservation processes/conditions.

Concentrations of inDNA and exDNA Fractions in Lake Sediments

Studies have shown that the geochemical composition of sediments, especially with different proportions and sources of the organic matter, have an influence on the proportions of inDNA and exDNA fractions in the total DNA pool. For instance, in recent (post-1950) sediments from two alpine and shallow lakes and two deep hardwater lakes, a linear positive correlation ($r^2 = 0.98$, $p = 0.0072$) has been observed between the total/exDNA ratio and the organic matter content (Fig. 2.3a, case study A3 from Capo et al. 2021). In this case study, sediments with less than 10% of organic matter, showed similar concentrations of total DNA extracted with the NucleoSpin® Soil kit (Macherey-Nagel, Düren, Germany) and of exDNA extracted following the Taberlet et al. (2012) protocol. In contrast, the sediments with the highest organic matter content (60%) were 7.5 times more concentrated in total DNA than in exDNA. These results suggest that organic-rich sediments are enriched in inDNA and are thus more suitable environments for preserving dead cells and for supporting dormant and live cells, but not necessarily for preserving the soluble fraction of exDNA. In that study, the researchers also observed that the most organic-rich sediments are not the richest in DNA, regardless of the fraction targeted (Fig. 2.3a; Capo et al. 2021). This lack of correlation has also been evidenced

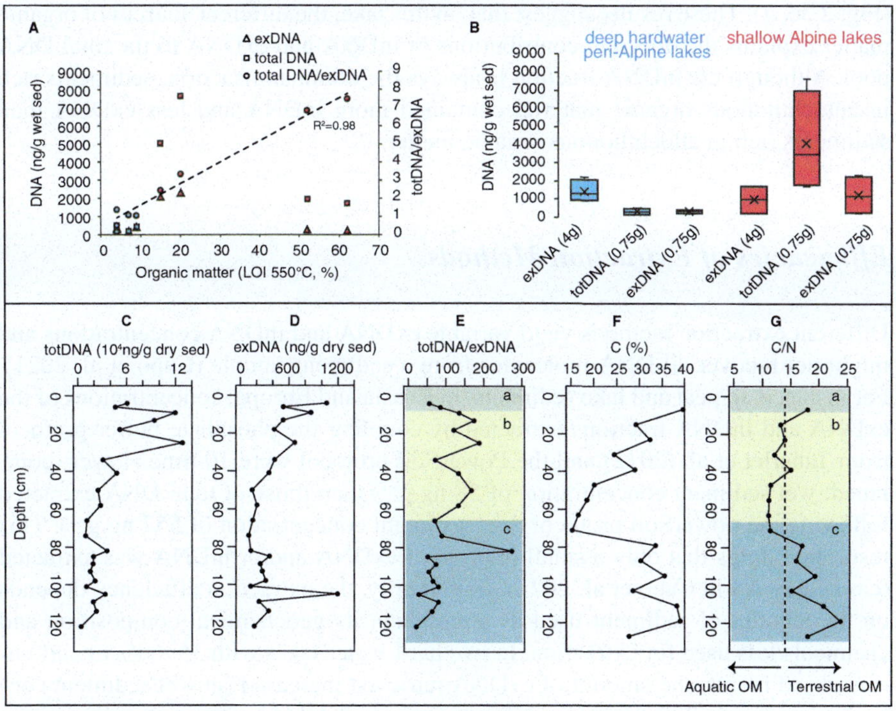

Fig. 2.3 Relationships between the different DNA pools (total DNA (totDNA) vs. extracellular DNA (exDNA) concentrations in ng.g^{-1} of wet or dry sediment (sed)) and organic matter (OM) quantity and/or quality in temperate lakes (deep hard-water lakes Léman and Le Bourget in blue and Alpine shallow lakes Serre de l'Homme and Lauzanier in pink; Capo et al. 2021) and in a tropical organic-rich and shallow lake (Gelba in Ethiopia, original data). A and B present the relationship between DNA concentrations and organic matter content (loss on ignition at 550 °C) from the temperate lake sediments, and C to G represent the DNA concentrations and ratio, the organic matter concentration (C% from elemental analyses) and the C/N atomic ratio (from elemental analyses) from tropical lake sediments. Phases a, b and c correspond to phases discussed in the text

from tropical peat and lake sediments containing 31 to 80% of organic matter and analysed following a similar extraction protocol as in the previous study (Fig. 2.3c, d, f; Giguet-Covex et al. 2023). However, in this study (Giguet-Covex et al. 2023), a decreasing trend with depth of both DNA fractions can be observed suggesting a temporal degradation over the last 2500 years. As for organic-rich recent sediments studied in Capo et al. (2021), total DNA concentrations were higher than exDNA concentrations (21–278 times higher), suggesting that inDNA accounts for most of the total DNA pool. However, in this case the total DNA/exDNA ratio was not correlated with the organic matter content. Three distinct deposition phases were defined based on the total DNA/exDNA ratio (phases a, b, c; Fig. 2.3d). Phase b, characterised by a high ratio, mostly contains organic matter of aquatic origin (low C/N atomic ratio; Bertrand et al. 2010; Thevenon et al. 2012), while phases a and c, characterised by a low ratio, are enriched in terrestrial organic matter (high C/N)

(Fig. 2.3e, g). These results suggest that, in this lake, the different sources of organic matter explains the different contributions of inDNA and exDNA to the total DNA pool. Although the inDNA fraction is always the dominant fraction, sediments rich in autochthonous organic material contained more inDNA and less exDNA than sediments rich in allochthonous organic matter.

Efficiencies of Extraction Methods

Different extraction methods yield variable exDNA and inDNA concentrations and might not recover all DNA molecules from a sediment sample (Capo et al. 2021). For instance, in peat and lake sediments in Russia and Europe, concentrations of the exDNA and inDNA fractions extracted by coupling the phosphate buffer protocol from Taberlet et al. (2012) and the PowerSoil protocol were 10 times lower (combined: wet sediment concentration of 58 ng.g^{-1}) than those of total DNA extracted following the powerSoil protocol (wet sediment concentration of 557 ng.g^{-1}). This result highlights that only a small fraction of exDNA and/or inDNA was extracted (case study A4 in Capo et al. 2021). Importantly, the extraction efficiency depends on the amount of sediment used as a template, its geochemical composition and chemicals/kits used for extraction. In two hard-water lakes with 35–60% of carbonates in sediments, the amount of exDNA retrieved increased (in wet sediment concentration, from ~200–1000 and 2000 ng.g^{-1} in each lake) by increasing the amount of sediment used for the extraction (from 0.75 to 4 g of wet sediments), while no increase was observed for the organic-rich lakes characterised by 2.3–3.5% of carbonates in the sediments (Fig. 2.3b, case study A3 in Capo et al. 2021). Carbonates, such as calcite, are probably effective at storing exDNA because of the strong adsorption of DNA at the edges of calcite crystals (Freeman et al. 2023). This DNA is thus harder to extract (case study A6 from Capo et al. (2021). Defining the taphonomic processes affecting the different DNA fractions remains challenging, in part due to these methodological shortcomings. As such, future studies are required to improve analytical and interpretative frameworks.

Molecular Inventories from inDNA and exDNA Fractions

Based on qPCR analyses, Nota and Parducci (case study A4 in Capo et al. 2021) assessed differences between the PCR amplification efficiency of inDNA and exDNA from various taxonomic groups (i.e., arthropods, bacteria, diatoms, eukaryotes, plants and vertebrates). For bacteria, no difference between inDNA and exDNA amplification efficiency was observed. In contrast, total eukaryotes were better amplified from the exDNA fraction, while the amplification of arthropods, diatoms, plants and vertebrates was more efficient with the inDNA fraction. Among these groups, only diatoms showed similar melting temperature profiles (suggesting similar diversity) in both DNA fractions. The differences observed in the melting temperatures of PCR amplification assays between the inDNA and exDNA fractions for arthropods, plants and vertebrates suggest the amplifications of DNA from different taxa. These results may reflect differences in the taphonomic processes that affect each DNA fraction from these groups, although it is difficult to assess this presumption in the absence of sequencing analyses and information about sediment types or mineralogical compositions. The different presumed taphonomic mechanisms are summarised below.

Some aquatic organisms, such as diatoms and chrysophytes, leave siliceous remains (i.e. diatom valves and chrysophyte scales and cysts) and dormant cells that are preserved for long periods of time in sediments while other organisms do not. This likely affects the taxonomic composition of DNA inventories from inDNA compared to exDNA fractions, where protected cells contribute more to the inDNA compared to the exDNA fraction. This might explain the better amplification (suggesting higher quantity) of diatoms in the inDNA fraction than in the exDNA fraction (case study A4 in Capo et al. 2021). Similarly, some polymers of terrestrial organisms, such as lignin, are better preserved than others (Boere et al. 2011a; Yoccoz et al. 2012; Foucher et al. 2020). Woody species might thus be more represented in inDNA compared to exDNA than herbaceous species, which might explain, at least partly, the different melting temperatures found from inDNA vs exDNA in case study A4 from Nota and Parducci in Capo et al. (2021). It is also possible that the transfer of inDNA is different from the transfer of exDNA from terrestrial environments, leading to different species identified in each DNA fraction. This hypothesis is based on the known interactions between exDNA and minerals, organic compounds, and organo-mineral particles, suggesting erosion as a likely key transfer process of exDNA (see next sections on DNA binding and terrestrial transfer DNA), whereas aeolian transport, direct deposition into the lake, and overland-flow might be more important for the inDNA transfer (Fig. 2.2).

Two other studies compared archaeal and bacterial communities recovered from inDNA and exDNA fractions in different sedimentary contexts: the ferruginous sediments of Lake Towuti, Indonesia (Vuillemin et al. 2017) and marine sediments from Aarhus Bay, Denmark (Torti et al. 2018). The authors compared the overlap in Operational Taxonomic Units (OTUs) obtained from the two DNA fractions. OTUs only found in the exDNA fraction were considered exogenous to the sediment layer analysed, while OTUs only detected in the inDNA were considered endogenous. In Lake Towuti, OTUs specific to the exDNA represented 40–50% of all sequences of the exDNA pool (Vuillemin et al. 2017). They included taxa from soils, such as a majority of Actinobacteria and some Verrucomicrobia, Solibacteres and Alphaproteobacteria, including *Pseudonocardia* and *Pedomicrobium* (Delgado-Baquerizo et al. 2018), but also primary (e.g., Cyanobacteria) and secondary producers (e.g., other Alpha- and Betaproteobacteria species) in the water column. In Aarhus Bay, a lower proportion of OTUs shared between the two fractions and a higher number of OTUs unique to the exDNA fraction was observed in the upper sediments (<34 cm-depth) compared to deeper sediments. These results were interpreted as the consequence of the introduction of exogenous exDNA in the sediment layer of interest, which are not coming from soils or the water column as in Lake Towuti, but coming from bioturbation in the surface sediments (Torti et al. 2018). In significantly bioturbated sediments, reconstructions of the paleo-community composition and diversity from total DNA or exDNA will be thus distorted (Torti et al. 2018).

In addition to the importance of source type (i.e., more or less resistant cells), origin of the organisms (terrestrial, aquatic or sedimentary for microbes) and transfer mechanisms, it is crucial to consider the differential degradation effects as a function of time. With time, it is likely that inDNA increasingly contributes to the exDNA fraction. This hypothesis is in line with the increase in depth of microbial DNA from taxa related to known spore- and cyst-formers, including Actinobacteria, Clostridia and Planctomycetes, observed in Lake Towuti (Vuillemin et al. 2017).

The synthesis of studies comparing the composition of each DNA fraction for different groups of organisms highlights the complexity of the taphonomic processes behind the fossilisation of each fraction. Given the low number of such studies, our knowledge is still limited, and the identification of general patterns is challenging. More systematic comparisons in different contexts and including other sedimentological and geochemical data, are necessary to assess the complementarity of the signals from these DNA pools. This knowledge can also increase the range of information offered by sedDNA analyses and its robustness by improving the selection of the DNA fraction to be targeted (inDNA, exDNA or both DNA fractions) according to the scientific question.

DNA Degradation and Preservation in the Environment

After cellular death, DNA repair mechanisms do not operate, leading to the degradation of inDNA by chemical reactions (oxidation, hydrolysis, alkylation and the Maillard reaction), processes that continue outside cells, affecting the exDNA fraction as well (Willerslev and Cooper 2005). The exDNA fraction is also actively damaged by enzymes (DNases) produced by microbes or can be directly consumed by microbes as a source of nutrients. However, exDNA can bind to particles such as minerals, carbonaceous materials (e.g., soot and charcoal), and complex organic compounds such as humic substances or organo-mineral complexes. The adsorption of DNA to a surface is known to reduce its degradation (Lorenz et al. 1991; Romanowski et al. 1991; Khanna and Stotzky 1992; Crecchio and Stotzky 1998; Cai et al. 2006a; Pietramellara et al. 2009). The mechanisms which protect the adsorbed DNA from degradation are not entirely clear. Protection against enzymatic activity might be related to the modification of DNA conformation in proximity of solid surfaces (a transition between A- and B-forms and/or modification in coiling, *i.e.* quaternary DNA structure), which inhibits enzymatic recognition, and/or the adsorption of enzymes on the mineral surfaces, effectively inactivating and preventing them from attacking the DNA (Levy-Booth et al. 2007 and references therein). The modification of conformation upon adsorption could also protect DNA against oxidative damage (Scappini et al. 2004). Protection is possible only if DNA is adsorbed to particles faster than any biological, physical, or chemical degradation process and if no desorption occurs during the lifetime of a DNA-mineral/organic complex.

The processes of DNA degradation can theoretically take place in the DNA production area (source), during the DNA transfer from the source to the sediment (in the catchment and water column for terrestrial DNA and in the water column for aquatic DNA), at the water-sediment interface and upper sediments (where microbial activity is relatively high and partially led to what we call "early diagenesis") and during deep burial in sediments (Fig. 2.2). There is no consensus about whether DNA degradation rates are faster in the water column, at the water-sediment interface, or in upper or deeper sediment layers, as this seems to depend on the environmental conditions. Indeed, as presented in the following sections, physico-chemical conditions are known to affect the rates of chemical and microbial DNA degradation and the DNA binding to particles. The mineral and organic composition and the DNA conformation are also important factors influencing the DNA binding.

Effects of Physico-Chemical Conditions on DNA Degradation

Modelling and statistical approaches have revealed a relationship between the environmental temperature and the length of DNA fragments (Smith et al. 2001; Geggier et al. 2011; Hofreiter et al. 2015). The rate of depurination, *i.e.* the breaking of the bond between a purine [adenine or guanine] and the deoxyribose due to hydrolysis, increases as temperature increases (Lindahl and Nyberg 1972). Based on the depurination rate and a thermal history of sediment, the DNA preservation can be predicted (Smith et al. 2001). For instance, at neutral pH, physiological salt concentration and at 15 ° C, DNA is expected to degrade after 100,000 years, but in colder environments this theoretical limit is extended to ~400,000 years (Hofreiter et al. 2001; Willerslev et al. 2004). High temperatures have also an indirect effect on DNA degradation by favouring the activity of microorganisms that recycle the exDNA buried in sediments (Dell'Anno and Danovaro 2005). The respective contribution of the direct and indirect effects of temperature on DNA degradation is still unclear and further work is required. In lake sediments, the lowest DNA decay rates were associated with cold temperatures (i.e. ~4 ° C, the typical lake bottom water temperature of deep temperate lakes) and anoxic conditions because anoxia reduces microbial activity, thus microbial degradation (Mejbel et al. 2022). This notion of better DNA preservation in cold conditions is so deeply embedded in the scientific community that most sedDNA studies have been conducted in high latitude/altitude areas or in temperate environments, with little work reported from tropical regions (Fig. 2.1). Nonetheless, recent lake sedDNA studies showed that DNA preservation in tropical areas is possible as well. Indeed, in sediments from low-altitude lakes, DNA as old as 5000 years ago was successfully extracted (Vuillemin et al. 2016a; Bremond et al. 2017) and perhaps as old as 1 million years (Ekram et al. 2021).

Other sedDNA studies in African high-altitude lakes (from 1200 to 4127 m a.s.l.) also reported successful sedDNA recovery from sediments up to several millennia old (Epp et al. 2010, 2011; Stoof-Leichsenring et al. 2012; Boessenkool et al. 2014; Dommain et al. 2020; Tabares et al. 2020).

In addition to the chemical and biological effects of temperature and oxygen on DNA degradation, laboratory experiments revealed that high salt concentration (high ionic strength) and high pH limit hydrolysis, which favour DNA preservation (Lindahl and Nyberg 1972, 1974; Strickler et al. 2015). Furthermore, environments protected from ultraviolet radiation such as sediment pores favour DNA preservation because UV radiation damages DNA (Rastogi et al. 2010; Strickler et al. 2015). A few field-based studies have explored potential effects of these environmental factors on the preservation of lake sedDNA. For instance, poor DNA preservation in acidic environments has been proposed to explain the low recovery of terrestrial plant DNA from sediments of two small alpine lakes (Giguet-Covex et al. 2019). The analysis of 219 lake surface sediment samples from China and Siberia suggested that neutral to slightly alkaline water (i.e. pH = ~7–9) and intermediate water conductivity (100–500 μS.cm^{-1}) facilitated plant DNA preservation in the sediments (Jia et al. 2022a).

DNA Binding on Particles: Implications for Lake sedDNA Studies, Mechanisms and Factors

The adsorption of DNA at mineral surfaces and organic particles, and its effect on DNA preservation, is an important aspect to consider for the analysis of lake sedDNA. Adsorbed DNA is more difficult to degrade, suggesting that it can survive in sediments across time and space (Sand and Jelavić 2018), likely even in warm and humid environments depending on the mineral substrate and the nature of binding. For instance clay minerals, recognized as "DNA hotspots' 'in lake sediments (Kanbar et al. 2020), in combination with low ambient temperatures, are likely responsible for the persistence of ~2 million years old DNA in ancient shallow marine near-shore sediments in Northern Greenland (Kjær et al. 2022). This study extended by more than 1.5 Ma the previous estimates of the theoretical limit of DNA preservation based on temperature as the only determining factor. In permafrost and cave sediments, minerals are also suspected as important contributors to the persistence of hundreds of thousands year old DNA (Slon et al. 2017; Zavala et al. 2021; Wang et al. 2021). The success of lake sedDNA studies reported from tropical areas may also relate to the presence of appropriate substrate favouring the DNA binding and thus preservation.

For terrestrial organisms, the adsorption of DNA on particles is important to consider because this process is expected to determine the mode of exDNA transfer to the lake bottom. More specifically, the adsorption of DNA on particles implies that allochthonous exDNA is mostly likely to be transported to lake sediment during erosion events and not (or to a much lesser extent) during surface water runoff events (Fig. 2.2; Giguet-Covex et al. 2019; Morlock et al. 2021). This involves extracellular sedDNA reflecting biological communities from areas where erosion occurs. This has to be considered when interpreting past communities in terms of paleoenvironmental/ecological changes.

Binding Mechanisms Between DNA and Minerals

The properties of both suspended minerals and dissolved DNA vary as a function of the chemical conditions (eg. pH and ionic strength), which affect the adsorption and binding between them. Here, we first discuss basic principles governing the development of surface charging on minerals and DNA, arguably the most important parameter controlling their binding. Later, we discuss how the charging affects adsorption and introduce other important parameters such as DNA conformation and composition of ambiental solutions.

We focus the discussion on the charging properties of double stranded DNA because the degradation rate of single stranded DNA is orders of magnitude higher than that of double stranded DNA (Lindahl 1993), thereby suggesting that the majority of single stranded exDNA quickly disappears from the archive. Theoretical and experimental studies demonstrated that double stranded DNA adsorbs to common minerals via the phosphate backbone (Okazaki et al. 2001; Pietramellara et al. 2001; Saeki et al. 2010). The phosphate backbone of DNA being negatively charged above a pH of ~2, the binding to positively charged minerals can happen directly through electrostatic attraction, thereby resulting in the formation of strong bonds. However, the isoelectric point of DNA, denoting the pH at which the whole molecule is neutral in an electric field, is somewhat higher, between a pH of 2–4.4 (Sherbet et al. 1983; Yetgin and Balkose 2015). Because the backbone surface of double stranded DNA is dominated by phosphates, the much higher observed values of isoelectric point compared to the pK_a of phosphate are likely the result of either protein impurities or exposure of nucleobases during the measurement. Such impurities or exposure of nucleobases would be due to depurination of DNA at pH < 5. Thus, we hold that the true value of DNA's isoelectric point is at the lower end of reported range, closer to the pK_a of phosphate.

The electrical charge of a mineral is a result of the finite sizes of crystals and various crystal defects that reach the surface. With a few exceptions (e.g., a special surface of clay minerals called the basal plane), the charge of mineral surfaces is a function of the pH solution and ion concentration. Depending on the pH, mineral charge can be either positive, negative or neutral. The pH at which the surface of minerals is neutral is referred to as the Point of Zero Charge (PZC). At pH < PZC, the mineral surface is positively charged and at pH > PZC it is negatively charged. Depending on the ion composition and concentration in the solution, the surface charge can, however, be neutralised or even reversed. For example, in karstic lakes where the concentration of Ca^{2+} is high because of dissolution of calcite ($CaCO_3$), when the pH is neutral, the usually negatively charged quartz can adsorb high enough amounts of Ca^{2+} for its surface to effectively become neutral or even positively charged. Because the surface charge of minerals is largely controlled by environmental factors such as water hardness and pH, the binding of DNA at mineral surfaces will also be influenced by the composition of environmental solutions.

Binding of DNA to a mineral can also take place on a negatively charged surface. In this case DNA and a mineral must first overcome the repulsive forces by adsorption of cations to the mineral surface, which then screens the negative charge and allows the phosphate backbone to bind. Such an effect is often called "cation bridging" where a cation makes a "bridge" between a negatively charged mineral surface and a negatively charged organic surface. Cation bridging implies that divalent cations common in freshwater such as Ca^{2+} and Mg^{2+} are more effective in charge screening than monovalent such as Na^{+} or K^{+} since one ion can bind directly to both the mineral surface and the phosphate backbone at the same time.

Factors Controlling the Binding on Minerals

The adsorption of DNA on minerals is a function of: (a) molecular weight and conformation (circular, linear) of DNA (Franchi et al. 1999); (b) composition and surface properties of minerals such as defects and charge (Fig. 2.4) (Jelavić et al. 2022; Freeman et al. 2023); (c) composition and pH of the solution (Paget et al. 1992) (Fig. 2.4); and (d) presence of other organic compounds in the system (Cai et al. 2006a) (Table 2.1).

The adsorption capacity of low molecular weight DNA is greater than for high molecular weight DNA because of the lower steric hindrances between shorter strands (Ogram et al. 1988; Franchi et al. 1999). This suggests that modern environmental DNA molecules adsorb to a lesser extent to minerals than older, shorter, and degraded DNA. The adsorption capacity of circular (coiled) DNA is higher than that of linear DNA (Poly et al. 2000; Pietramellara et al. 2001), likely because of the higher charge density of coiled DNA and the contribution of non-electrostatic forces in the interaction between coiled DNA and minerals.

Clay minerals absorb an order of magnitude more DNA than other common soil and sediment silicates (Lorenz et al. 1981; Kjær et al. 2022). One reason is that the small (colloidal) size of clay minerals results in significantly higher specific surface area compared to minerals with larger particle size. Among clay minerals, some smectites (so-called swelling or expansive clay minerals) have higher adsorption capacity than kaolinite, illite and chlorites (non-swelling clay minerals) (Demanèche et al. 2001; Pietramellara et al. 2001; Cai et al. 2006b; Kjær et al. 2022), likely because they can adsorb DNA inside of their crystal structure, a process known as intercalation (Greaves and Wilson 1969). However, only smectites with relatively low surface charge are able to intercalate DNA (Jelavić et al. 2022). Iron and aluminium oxides, such as goethite and gibbsite, are less widespread and less common than silicates (Ito and Wagai 2017) but have large adsorption capacities for DNA (Saeki et al. 2010; Cao et al. 2011). Interestingly, these minerals are commonly found in tropical soils (Schwertmann and Taylor 1989; Duchaufour et al. 2020),

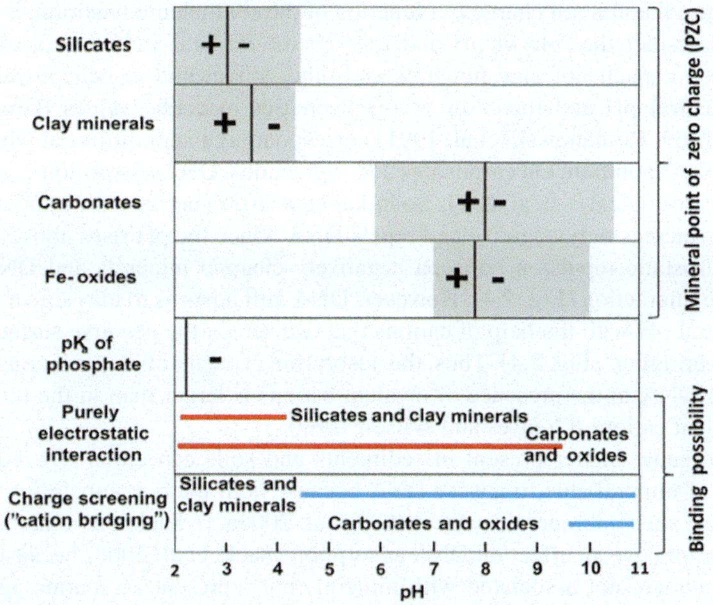

Fig. 2.4 Schindler diagram for DNA binding: net charge of mineral surfaces and their binding modalities with DNA as a function of pH. An electrostatic attraction (red horizontal line) between the negatively charged phosphate backbone of DNA and positively charged minerals is possible at low pH (pH < ~4.5) for silicates (such as quartz, feldspars, and zeolites) and clay minerals (such as montmorillonite, chlorites, kaolins, illites, smectites, and vermiculites) and at a large pH range (pH < ~10) for carbonates (such as calcites, magnesites, and dolomites) and Fe-oxides (such as hematites, goethites, magnetites, and ferrihydrites). The screening of repulsion (blue horizontal line), *i.e.* the neutralisation or charge reversal of negative mineral surface charges and negatively charged phosphate backbone of DNA by cations ("cation bridge"), is possible at a pH above ~4.5 for silicates and clay minerals and at a pH above ~9.5 for carbonates and Fe-oxides. The median value (vertical black line) and the range for point-of-zero charge of minerals (grey shaded areas) are taken from Kosmulski (2011) and Oelkers et al. (2009)

where sedDNA studies are still poorly reported (Fig. 2.1). Unlike clay minerals, goethite and gibbsite are often positively charged at neutral pH, so they bind directly to negatively charged DNA (Fig. 2.4), likely resulting in much stronger binding compared to the "cation bridge" interaction. Unlike surfaces that vary in charge as a function of solution composition, structural defects or crystal steps represent areas of permanent charge that also affect DNA binding (Freeman et al. 2023). Crystal defects and crystal steps are features on crystal surfaces that usually exhibit larger charge density compared to the surrounding, pristine surface. Freeman et al. (2023) observed preferential binding of DNA to the crystal steps of calcite in NaCl, $MgCl_2$ and $NiCl_2$ solutions but no significant interaction between DNA with the surrounding surface; this observation emphasises the importance the magnitude of surface charge on DNA adsorption, in addition to its sign (positive or negative as a function of pH).

Because the mineral charge is a function of the solution composition, it is important to consider the role of pH and ions on the mineral-DNA interaction. DNA adsorption capacity of clay minerals and other silicates is in general minimal at circumneutral pH and increases as pH decreases to acidic values (Greaves and Wilson 1969; Romanowski et al. 1991) corresponding to conditions at which DNA is unstable in solution. On carbonates and iron oxides, DNA adsorption decreases as pH increases (Cleaves et al. 2011; Sodnikar et al. 2021) because the mineral surface becomes increasingly negative and repels DNA. Once the pH rises above the PZC, the electrostatic repulsion between negatively charged minerals and DNA dominates the interaction (Fig. 2.4). However, DNA still adsorbs to clay minerals at circumneutral pH with the help of cations that can screen the negative surface charge ("cation bridging", Fig. 2.4). Thus, the adsorption capacity of clay minerals and silicates for DNA in the presence of divalent cations is larger than in the presence of monovalent cations (Greaves and Wilson 1969).

The organic matter present in sediments and soils can either: (i) reduces the amount of mineral able to adsorb DNA because both likely compete for the same adsorption sites on mineral surfaces (Cai et al. 2006a, b; Saeki and Sakai 2009); or (ii) does not have an effect on DNA adsorption (Saeki et al. 2008) because organic compounds are not associated with minerals but represent a separate aggregated phase within the soil or sediment. Such an explanation, however, is difficult to reconcile with the observation that the DNA adsorption capacity of humic substances is high (Saeki et al. 2011). These discrepancies might be explained by the extreme molecular diversity of organic compounds in soils and sediments (Christl et al. 2000; Sutton and Sposito 2005) and the prospect that different organic compounds will interact in different ways with minerals and DNA.

DNA Binding on Organic Compounds and Carbonaceous Materials

Practical issues hinder a comprehensive insight into the interaction between DNA and organic compounds commonly found in sediments, such as humic substances. Humic and fulvic acids absorb light at similar wavelengths as DNA (200–300 nm), preventing a reliable spectrophotometric determination of DNA concentration in solution with humic substances. Determination of DNA concentration based on fluorescence resolves this issue (*e.g.* Singer et al. 1997) but the specificity of various fluorophores is unknown in a range of pH values, concentrations of humic substances and ionic strengths. In addition, the solubility of humic substances abruptly changes as a function of pH (Kipton et al. 1992), much more than the solubility of minerals, the majority of which is considered insoluble at the timescale of a single adsorption experiment, which usually lasts a few days.

A few carefully designed studies have shown that the adsorption capacity of humic acids is high (Saeki et al. 2011), comparable to clay minerals that cannot intercalate DNA. The interaction is largely electrostatic in nature (Nguyen and Elimelech 2007), although some hydrophobic forces might also play a role (Saeki et al. 2011). The adsorption is likely irreversible in solution buffered at a pH of 5.8 (Nguyen and Elimelech 2007), suggesting good preservation potential for DNA over geological timescales, but it is unclear how a variation in pH would affect the nature and reversibility of binding.

Pyrolysis of organic compounds in wildfires produces carbonaceous materials such as soot and charcoal. These materials are generally low in abundance but widespread in soils and sediments (Schmidt and Noack 2000). The capacity of carbonaceous materials to adsorb DNA is higher at a pH ranging from 3 to 8 and when compared to clay minerals that can intercalate DNA (Fang et al. 2021; Jelavić et al. 2022). Both electrostatic and hydrophobic forces control the binding, which is likely irreversible (Fang et al. 2021; Jelavić et al. 2022), and thus conducive for good DNA preservation. The significant contribution of hydrophobic forces in DNA binding to carbonaceous materials calls into question the suitability of common DNA extraction techniques that are tailored for sequestration of electrostatically bound DNA, revealing carbonaceous materials as a likely overlooked reservoir of sedDNA.

Early Diagenesis in the Sedimentary Column: Effects on DNA Concentration and Taxonomic Diversity

Early diagenesis refers to physical, chemical and biological modifications of the sediments that occur soon after their depositions, i.e. at the sediment-water interface and in the upper sediments. The thickness of this zone varies as a function of environmental conditions, and is especially influenced by oxygen concentration along the sediment depth, which can vary from a few millimetres in productive hypoxic systems to several centimetres in oligotrophic oxygenated systems (Kristensen 2000; Haglund et al. 2003; Vuillemin et al. 2013). Burial rates and frequency of water mixing are other variables controlling the sediment thickness affected by the early diagenesis.

Microbial activity directly influences early diagenesis (Vuillemin et al. 2016b), as such degradation rates of organic matter and DNA are significant in sediments affected by the early diagenesis. This is reflected by a sudden decrease in the concentration of extracted DNA in the upper sediments (Vuillemin et al. 2017; Dommain et al. 2020; Fig. 2.3c, d). Early diagenesis is also expected to modify the DNA composition, i.e. the taxonomic diversity recovered by DNA analyses. The number of sequences from some taxa may decrease and even be lost, while microbes taxa growing in the sediments are expected to increase in the DNA record. For instance, Vuillemin et al. (2017) analysed the changes in the relative abundance of shared OTUs by both inDNA and exDNA fractions. They observed that in the shared inDNA fraction there are three groups of taxa with different evolution patterns in the upper ~30 cm of sediment. The first group containing Deltaproteobacteria, Clostridia, Elusimicrobia, Bathyarchaeota, and Hadesarchaea increased in abundance with depth. In contrast, the second group composed of Chlorobi, Anaerolineae and Omnitrophica decreased in abundance, whereas the third group with Dehalococcoidia, Nitrospirae, Aminicenantes, Aenigmarchaeota, and Thermoplasmata first increased before quickly decreasing in abundance. This succession of microbial populations was interpreted as reflecting variations in biogeochemical conditions during early diagenesis (Vuillemin et al. 2016b). However, the authors acknowledge that this interpretation assumes that the shared inDNA only reflects the growth, or decline of populations living in the sediments and does not account for the accumulation of inactive intact cells in the sediments. Another study tested the impact of early diagenesis on the protist DNA signal by analysing laminated sediments from two cores sampled in a boreal lake with a time lag of 6 years (Capo et al. 2017). In the deeper sediment layers (age ~ 15–40 years), the molecular signal was very similar in the two cores (except for fungi) and was considered stable. However, for the uppermost strata (age < 15 years; upper 7 cm), moderate modifications in the composition of the protist community were detected probably due to the biogeochemical conditions.

Taxa-Dependent Preservation of Sedimentary DNA

The potential for DNA preservation in lake sediments depends on features that vary between different species, some having protective resting stages (e.g., cysts, spores, akinetes; Ellegaard and Ribeiro 2018) or resistant cell walls (e.g., diatom frustule, lignin, pollen), thus being more resistant to degradation processes (Setlow 2007; Boere et al. 2011a, b; Capo et al. 2015; Mejbel et al. 2022). In a 1.4 million years (Ma) marine record, the number of chloroplast DNA sequences was found to decrease with depth, with a strong decay within 100–200,000 years of deposition (Kirkpatrick et al. 2016). However, a few diatom phylotypes persisted for up to 1.4 Ma, matching fossil diatom records. This result suggested that the frustules favoured the preservation of diatom DNA. In addition, in fjord sediments from eastern Antarctica, dinoflagellate DNA was found to degrade faster than diatoms (Boere et al. 2011a). In 80–125 ka Mediterranean sapropels, with mostly marine organic matter, more terrestrial plant totDNA was recovered compared to planktonic taxa (Boere et al. 2011b). In this case, most planktonic taxa were dinoflagellates that may leave low amounts of remains (i.e., protective resting stages), while terrestrial plant DNA might have originated from pollen or other cells protected from degradation, which could explain the lower preservation of DNA from aquatic species relative to terrestrial plant DNA and the apparent discrepancy with the main origin of the organic matter. However, based on the review of taphonomic processes in this chapter, two other processes might have contributed to the preservation of the terrestrial plant DNA relative to planktonic DNA. The binding of plant DNA to detrital mineral or organic particles likely promoted its preservation. These DNA-bearing particles carried by the Nile River to the marine environment probably encountered aquatic exDNA but could not adsorb it (and protect it) because of the lack of additional adsorption sites and/or due to unfavourable chemical conditions for the sorption. In addition, the potential contribution of the aquatic inDNA to the exDNA fraction as a function of time and sedimentary depth was prevented because of high autochthonous organic matter content which quickly metabolised newly released exDNA. These alternative explanations highlight our lack of knowledge about the taphonomic processes that can bring bias into the reconstruction of community composition. More specifically, a better understanding of the influence of geochemical and mineralogical sediment properties on the recovery of exDNA from different groups over time is still needed (Sand et al. 2023). More systematic multi-proxy approaches and analyses of the different DNA fractions are probably key to answering the following questions: (1) where to look (i.e., targeted lake and coring site based on mineralogic composition and geochemical properties of the sediments); (2) what to look for (i.e., targeted organisms, DNA fraction(s)); and (3) over which time scale?

Terrestrial DNA in Lake Sediments

Plants and mammals are the main terrestrial organisms targeted by DNA analyses from lake sediments (Anderson-Carpenter et al. 2011; Parducci et al. 2013; Giguet-Covex et al. 2014; Pedersen et al. 2016). However, in a study focused on mammals, an unexpected detection of earthworms was also reported (Lammers et al. 2019). Furthermore, soil-derived microbial communities (Vuillemin et al. 2017) and human and animal faecal-specific bacteria (Madeja et al. 2009, 2010; Madeja 2015) have been investigated in certain lakes. In terms of biomass in catchments, mammals are generally less abundant than plants or invertebrates like earthworms (Lammers et al. 2019), except in specific situations where there are high concentrations of mammals such as on migration routes, waterholes or due to human management (livestock farming and associated practices such as gathering in pens). This low biomass may at least partly explain several reported failures in the detection of animals (Giguet-Covex et al. 2019; Lammers et al. 2019). In some lakes, the poor detections of terrestrial plant DNA (Alsos et al. 2018; Giguet-Covex et al. 2019), which were not attributed to degradation issues, highlight the need to improve our understanding of taphonomic processes related to the sources and transport pathways of DNA and sediments, in order to reliably interpret DNA-based ecosystem reconstructions. Below, we synthesise the knowledge and uncertainties on these processes and their impacts on the sedDNA records.

DNA Production and Sources

"Primary Source" of DNA

At the level of organisms, for both plants and mammals, the sources of DNA released to the environment remain largely unknown. For plants, potential sources are above-ground (shoot systems like leaves, seeds) and below-ground (root system) organs (Parducci et al. 2017). For plants, the relatively poor overlap between pollen and DNA records from lake sediments suggest that pollen does not significantly contribute to the sedDNA pool, and sedDNA from plants likely originated from other plant tissues (Parducci et al. 2017; Messager et al. 2022). For mammals, faeces, hair, skin flakes, urine (Willerslev et al. 2003; Lydolph et al. 2005; Haile et al. 2007) and decaying tissue from dead animals (Ficetola et al. 2018) are all assumed to contribute to the DNA release in the environment, while a study from archaeological soils in Greenland suggest that bones do not, or minimally contribute to the sedDNA pool (Hebsgaard et al. 2009). However, other studies argue for the opposite. For instance, inputs of fresh bones from crannogs into a lake were suggested as a source contributing to the sedDNA signal (Brown et al. 2021). In Arctic environments, where cold temperatures significantly increase bone persistence and where burial can take a long time, Miller and Simpson (2022) argue that bones might be a continuous source of environmental DNA. This continuous release of DNA may explain the time lags in estimating the extinction of mammoths from bones compared to sedimentary DNA analyses (Miller and Simpson 2022).

Soil DNA, a "Secondary and Primary Source" of DNA

DNA from plant and animal tissues is incorporated into soils through pedogenesis processes (Poté et al. 2005). This "soil DNA" represents a secondary source of lake sedDNA. In addition, soils are a primary source of DNA for soil-derived microbial communities (Vuillemin et al. 2017) and other soil-living organisms such as earthworms (Pansu et al. 2015a). Soil environmental DNA (eDNA) studies thus provide useful knowledge about these primary and secondary sources of DNA for lake sediments. Importantly, they may provide answers to questions pertaining to: (1) the spatial representativeness of the DNA signal at the scale of the earth surface; (2) the vertical distribution of DNA in soil profiles; (3) the potential for keeping quantitative information, i.e. information linked to biomass; and (4) DNA persistence in soils.

Spatial Representativeness of the DNA Signals

Plants DNA metabarcoding analyses conducted on soils from high altitude/latitude areas, temperate agricultural lowlands and tropical areas suggested that most soil DNA is of local origin. For instance, in boreal ecosystems from Norway, which were characterised by a mosaic of dwarf shrub-dominated heath and forb- and grass-rich meadows, the plant community composition identified by DNA corresponded with the above-ground diversity estimated from floristic surveys (Yoccoz et al. 2012). In temperate agricultural lowland sites in France, the last crops to be cultivated were detected in 96–100% of the studied soils and were among the most abundant species, together with a large variety of weeds (Foucher et al. 2020). A primarily local origin of plant DNA in soils was also identified in the Arctic ecosystem in Svalbard (Edwards et al. 2018). However, in that study, as in Norway, distant minor contributions, through the transport by wind, water or animals, were also identified (Yoccoz et al. 2012; Edwards et al. 2018).

DNA Distribution in a Soil Profile

Few studies on microbial and mammal DNA investigated this question. In forest soil profiles, a higher diversity of microbial communities and a higher density of prokaryotes have been found in the upper organic soil horizons compared to the deeper organo-mineral and mineral horizons (Fig. 2.5; Agnelli et al. 2004; Kandeler et al. 2009). Differences were evident between the totDNA and exDNA fractions in terms of microbial composition (Agnelli et al. 2004). Mammal DNA is heterogeneously distributed along the soil profile with generally higher content at the surface. This distribution and, more precisely, the depth at which DNA is still detectable, depends on a combination of factors: animal behaviour, physiological differences, population density and soil texture, which can favour or limit leaching processes and bioturbation (Haile et al. 2007; Hebsgaard et al. 2009; Andersen et al. 2012). For example, sandy soils are more prone to leaching while clay acts as a barrier to leaching. Areas used for a latrine showed a deeper DNA leaching compared to areas only used, even regularly, as trails (Andersen et al. 2012).

Quantitative information from soil DNA analyses: In high latitude ecosystems, relationships between the relative abundance or biomass of above-ground plants and the relative abundance of the corresponding DNA sequences have been observed (Yoccoz et al. 2012; Pansu et al. 2015b). This result shows that soil DNA analysis

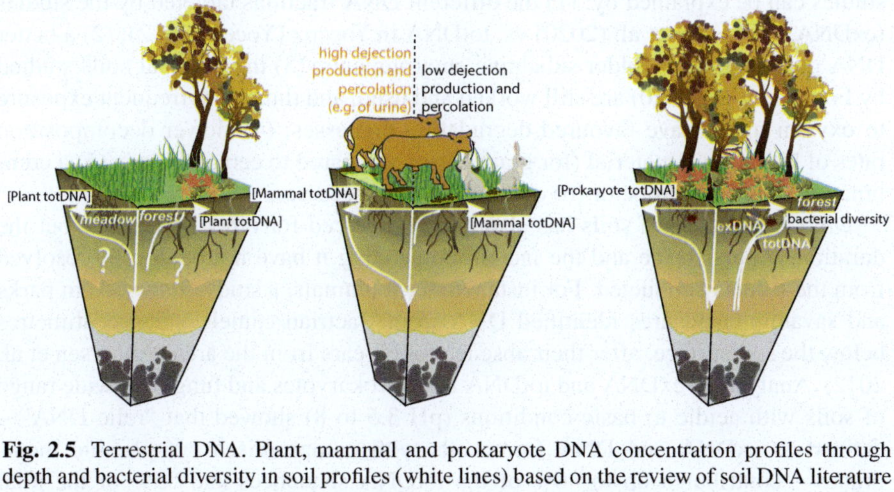

Fig. 2.5 Terrestrial DNA. Plant, mammal and prokaryote DNA concentration profiles through depth and bacterial diversity in soil profiles (white lines) based on the review of soil DNA literature and knowledge on soil carbon. For plant DNA, soil surface concentrations proposed are based on Yoccoz et al. (2012). The deeper concentrations are unknown (?), but inspired by knowledge on soil carbon. Mammal DNA concentration profiles are inspired by data from Andersen et al. (2012) and profiles for microbial DNA are made from data presented in Agnelli et al. (2004) and Kandeler et al. (2009). The soil interface design is similar to Chorover et al. (2007)

yields quantitative information. However, Yoccoz et al. (2012) demonstrated that the relationships are different according to the plant functional groups. An almost 1:1 relationship was found between the biomass and DNA sequence proportions for graminoids, whereas for woody plants, the DNA proportion was underestimated compared to the biomass. In contrast to woody plants, the DNA proportion from forbs was overestimated compared to the biomass. These differences might be due to varying litter turnover rates, different root/shoot biomass ratios or different genome sizes, copy numbers and chloroplast numbers between species. For mammals, a study on soil surface layers from safari parks, zoological gardens and farms also reported that the DNA concentrations were positively correlated with animal biomass (Andersen et al. 2012).

DNA persistence in soils: Yoccoz et al. (2012) and Foucher et al. (2020) investigated the issue of persistence of plant DNA in soils from formerly cultivated terraces in subalpine meadows and from agricultural soils in temperate lowland sites, respectively. In subalpine meadows, DNA from crops (cereals and potatoes) cultivated five decades earlier was still detected. In these soils, the DNA sequence frequency declined with time, with a very poor detection after ~50 years of crop abandonment (Yoccoz et al. 2012). In the temperate agricultural soils, the DNA persistence of similar crops (e.g., cereals, rapes, peas, potatoes, sugarbeets, sunflowers, hemps, raddish) appeared to be lower, i.e. only 8 years (Foucher et al. 2020). A declining trend with time was also evidenced as plants cultivated in the last 3 years were found in higher proportion than plants cultivated 8 years ago. However, in currently cultivated plots and grasslands where grapevines were grown about 60 years before the study, DNA from grapevines was still detected (Foucher et al. 2020). The differences in the soil DNA signal persistence found in the different case studies can be explained by: (1) the different DNA fractions targeted by the studies (exDNA in Foucher et al. (2020) vs. totDNA in Yoccoz (Yoccoz 2012); (2) a better DNA preservation in colder subalpine environments; (3) the fact that soils studied by Foucher et al. (2020) are still worked and used, and thus more frequent exposure to oxygen might have favoured degradation processes; (4) slower decomposition rates of high-lignin material (for grapevines) compared to cereal tissues; (5) a combination of these factors; or (6) other unknown processes.

DNA persistence in soils has also been evidenced for other organisms, but the duration of persistence and the factors influencing it have not been fully resolved from the studies conducted. For instance, for mammals, a study from soils in parks and savanna enclosures identified DNA from Bactrian camels a few centimetres below the soil surface, after their absence for 6 years from the area (Andersen et al. 2012). Analysis of exDNA and totDNA from prokaryotes and fungi in a wide range of soils with acidic to basic conditions (pH 3.5 to 8) showed that "relic DNA" – defined as exDNA and DNA from cells with compromised cytoplasmic membranes – contributed up to 40% of the totDNA sequences and 55% of the total taxonomic richness (Carini et al. 2016). Soils with low pH, electrical conductivity and cation exchange capacity (low Ca^{2+}, K^{+}) showed higher contributions of this relic DNA to the DNA pool. This finding agrees with the current understanding about the conditions that favour binding of exDNA on minerals and its potential long-term preservation. However, it is not known how long this relic DNA can persist in soils. This question is of primary importance to assess potential biases in the reconstructions of past living soil microbial communities from lake sedDNA analyses. In other words – is there an important contribution of non-contemporaneous microbial DNA to a sediment layer? Agnelli et al. (2007) investigated this question from radiocarbon analyses of exDNA, but without success. Indeed, based on the analyses of stable isotopes (^{13}C and ^{15}N) on both soil organic matter (SOM) and exDNA, radiocarbon results were assumed to be affected by inefficient purification procedures of the DNA extracts (Agnelli et al. 2007). Two purification procedures were tested. In one case, the stable isotopic signatures of exDNA were the same as for SOM, suggesting that the purification procedure was not efficient for the removal of SOM. In the second case, the low $^{\Delta 14}C$ values suggested the presence of fossil fuel-derived substances, which were used for the purification procedure and not effectively removed. Consequently, it was not possible to estimate the age of soil exDNA and thus estimate the DNA persistence.

What are the implications of the different knowledge acquired on soils for lake sediment studies? The results on DNA persistence in soils, especially from plant DNA studies, suggest that the transfer of old DNA molecules in lake sediments at the scale of a few decades should be limited compared to the transfer of DNA molecules corresponding to the modern biomass. Consequently, the temporal sedDNA reconstructions of land use and land cover should not be significantly skewed. This conclusion was also drawn from the comparison of a recent lake sedDNA record and historical records of forestry plantation (Sjögren et al. 2017). Nevertheless, higher persistence of some species, especially tree and shrub species (Foucher et al. 2020) and environmental conditions favourable for the preservation of cells and DNA (such as Arctic environments, Miller and Simpson 2022), might lead to temporal inconsistency in paleoenvironmental reconstructions, at least on the scale of a few years to decades and possibly much more in specific environments as suggested by Miller and Simpson (2022). Time lags of a few years to decades are not expected to significantly affect paleoenvironmental interpretations for long-term trend studies, because it also corresponds to the range of radiocarbon dating uncertainties.

The variations in DNA concentration and richness in soil profiles involve that the DNA signature found in the lake sediment ought to be different depending on the soil horizon(s) that is/are eroded. In addition, the targeted DNA fraction (totDNA or exDNA) may provide different signatures, especially regarding microbial communities (Agnelli et al. 2004).

Studies on plants and mammals showed that biomass-DNA quantity relationships are existing in soils. Do similar relationships also exist in lake sediments? Alsos et al. (2018) showed that 61% of dominant taxa, 47% of common taxa, 25% of scattered taxa and 15% of rare taxa are detected in modern sediments from high latitude (boreal and alpine) lakes. These results indicate that quantitative estimates of plant abundance can be obtained using sedDNA approaches like metabarcoding. However, this study also identified the role of the distance to the coring site on the rate of DNA detection (see next section). Indirectly, this result argues for the importance to consider and understand the DNA transfer processes as another driver of the sedDNA records.

Transfer of Terrestrial DNA and Spatial Representativeness of the Records

DNA transfer mechanisms play an important role on the spatial representativeness of the sediment DNA record. One common question when we explore community composition from DNA analyses is "where did the taxa come from?" A possible bias in past reconstructions is an artificial over-estimation of taxonomic richness due to an increase in the area contributing to the DNA inputs, which may be due to the activation of DNA transfer mechanisms in new areas. As such, identifying transfer mechanisms and their potential evolution is crucial to ensure a correct interpretation of the sedDNA data.

In Alsos et al. (2018)'s study investigating 11 small (0.04 to 27 ha) boreal and alpine lakes located in northern Norway, the overall detection of taxa was 1.5 times higher for <2 m vegetation surveys than extended vegetation surveys (within 50 m from the lakeshore), suggesting that distance to the lake influences the plant detection. In this study, a correlation between the catchment relief (i.e. the altitudinal difference between the top of the catchment and the lake) and the total number of plant taxa detected was also observed, which can be explained by a higher plant diversity in catchment covering larger altitudinal gradients and/or a better efficiency in DNA transfer due to higher runoff and/or erosion processes in catchments with more relief.

Two other studies targeting both plant and mammal DNA examined alpine lakes with contrasting physical settings (e.g., topography, geology, hydrography, vegetation cover and land use), resulting in sediments either dominated by in-lake production or by detrital and mostly mineral inputs (Giguet-Covex et al. 2019; Morlock et al. 2021). The results suggest that catchments characterised by easily erodible bedrock and steep slopes, are typically exposed to pronounced erosion dynamics and present well-developed hydrographic web, which facilitate the transfer of terrestrial DNA to lake sediments. In addition, the study by Morlock et al. (2021), which analysed continuously deposited background sediments and high-precipitation event layers of similar age, showed that DNA-inferred species composition was significantly different between the two contrasting erosion regimes and that taxonomic richness was higher in event layers. Because it is unlikely that species communities changed systematically across the catchment on yearly to decadal timescales, the switch in taxonomic composition was likely caused by a different sediment provenance between the two contrasting erosion regimes. In other words, different parts of the catchment contributed DNA to the lake during continuous annual sedimentation with low erosion inputs and during heavy precipitation events leading to high erosion events (Fig. 2.6). Erosion as a mechanism of terrestrial DNA transfer has also been proposed in a study on swamp sediments in Uganda (Dommain et al. 2020). In this record, based on a different analytical approach (shotgun analyses compared to metabarcoding for the other studies),

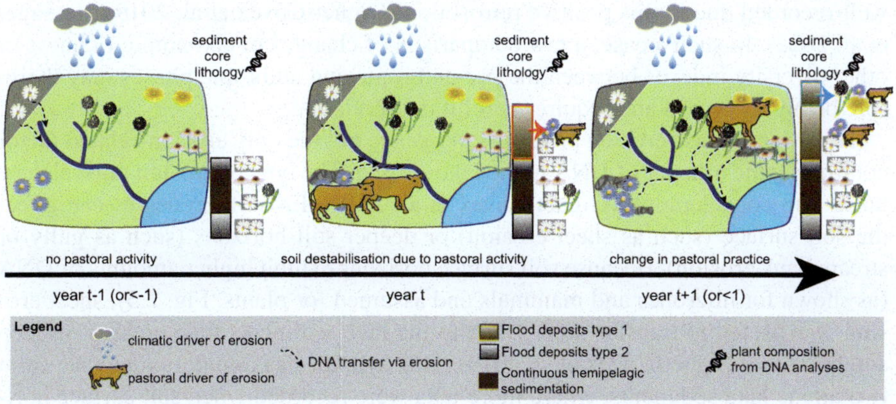

Fig. 2.6 Scenario illustrating that the same landscape can be represented in the sediment archive by different taxonomic composition, as shown in the two upper successive flood deposits. This is because different areas can contribute to erosion over time (e.g., changes in grazing areas). Similarly, in large catchments, we can expect variability in meteorological conditions which can affect the DNA composition. Flood deposits type 1 and 2 track floods transporting sediment from different sources (more developed soils vs leptosols, respectively) and thus containing different DNA compositions. The continuous hemipelagic sedimentation consists of in-lake production and fine detrital particles that settle, forming the paleoenvironmental record

higher proportions of plant DNA and higher richness were detected in time periods of higher detrital inputs as supported by higher sedimentation rate and input of clastic elements (Al and Ti).

Findings from Evrard et al. (2019) and Giguet-Covex et al. (2022) from river and floodplain sediments in mountainous areas, and the opinion paper by Frankl et al. (2022), also suggest that environmental DNA analysis may help to identify areas of intense soil erosion, as different source regions host specific plant communities, which can function to create DNA 'fingerprints' and back-track erosion hotspots in a landscape. *Vice versa*, if the geological and/or pedological contexts are contrasting, mineralogy and/or organic geochemical analyses can be used to independently trace the origin of the sediments (that is the source area(s) of erosion) and assess the catchment representativity of the DNA signal in paleo-reconstructions (Giguet-Covex et al. 2019). The larger the studied catchment area and the more differentiated the habitats, the more biases can be expected in the reconstructed landscape because erosion is unlikely to affect all habitats or not with the same magnitude. However, in such a context we can expect to obtain information on the source(s) contributing to erosion (and potentially of the drivers, e.g. Rapuc et al. 2021) and thus, to identify the erosion hotspots through sedDNA analysis. Erosion processes can strongly change over time according to the vegetation cover, soil evolution, climate, and with the development of human activities (e.g. land use, and urbanisation). Organisms associated with human activities are thus expected to be well-recorded due to this positive retroaction (Giguet-Covex et al. 2019; Messager et al. 2022). In such a case, a raw comparison of changes in taxonomic richness or other diversity indexes between the pre-anthropic and anthropic phases may therefore be highly biased and requires careful interpretation.

However, intense erosion processes *per se* are not the only erosion-related factor controlling the transfer of DNA to the lake system. The type of erosion must also be taken into consideration (Giguet-Covex et al. 2019). Erosion processes can affect the soil surface (such as sheet erosion) or deeper soil horizons (such as gully or stream bank erosion). Because soil surface horizons exhibit higher amounts of DNA (as shown for microbes and mammals and assumed for plants, Fig. 2.5), sheet erosion is expected to transfer more DNA to the lake sediments than erosion of deep soil horizons (Giguet-Covex et al. 2019). This hypothesis is supported by data from two alpine lake sediments, where more taxa were recorded when soil surface horizons mostly contributed to the sediments (Pansu et al. 2015b; Bajard et al. 2017; Giguet-Covex et al. 2019). Conversely, less taxa were found when deep soil horizons contributed to the sediments (Fig. 2.7).

Animal DNA is not uniformly distributed across the animal habitat, due to their mobility in the catchment and the species behaviour (Andersen et al. 2012). Animals kept in enclosures or folds are more likely to be detected in the sedDNA record when compared to a more diffuse source when kept in low abundances or scattered distributions Giguet-Covex et al. (2019). Carnivores seem more challenging to detect through eDNA analysis due to their lifestyle and comparatively low biomass (Lyet et al. 2021). Many terrestrial animals access water bodies actively for drinking and swimming and may also urinate, defecate or die in the water, depositing their DNA in more or less high quantity (Harper et al. 2019; Ushio et al. 2017). For instance, an eDNA study on wild pigs in small waterers showed that the number of individuals shedding into the water system can affect the DNA persistence (Williams et al. 2018), which consequently may be reflected in the sedDNA record. However, an animal dying in the water is expected to shed a large quantity of DNA into the water and sediments, which may not reflect the actual biomass present in the catchment. Yet, even though plants and animals have very different mobilities and distributions, the detection probability for terrestrial mammals in river water samples also correlated with catchment area and precipitation intensity (Lyet et al. 2021), again highlighting the importance of runoff and erosion for eDNA detection. The hydrological connection and/or erosion from the animal's location or, more precisely, from locations where high animal DNA quantities are released (e.g., latrines, trails, enclosures, etc.) is thus critical for their detection in the sedDNA record (Graham et al. 2016; Ficetola et al. 2018; Giguet-Covex et al. 2019; Morlock et al. 2021). Importantly, animals, especially domestic herds/flocks, often promote the erosion process and consequently can increase the probability of DNA transfer and detection in lake sediments.

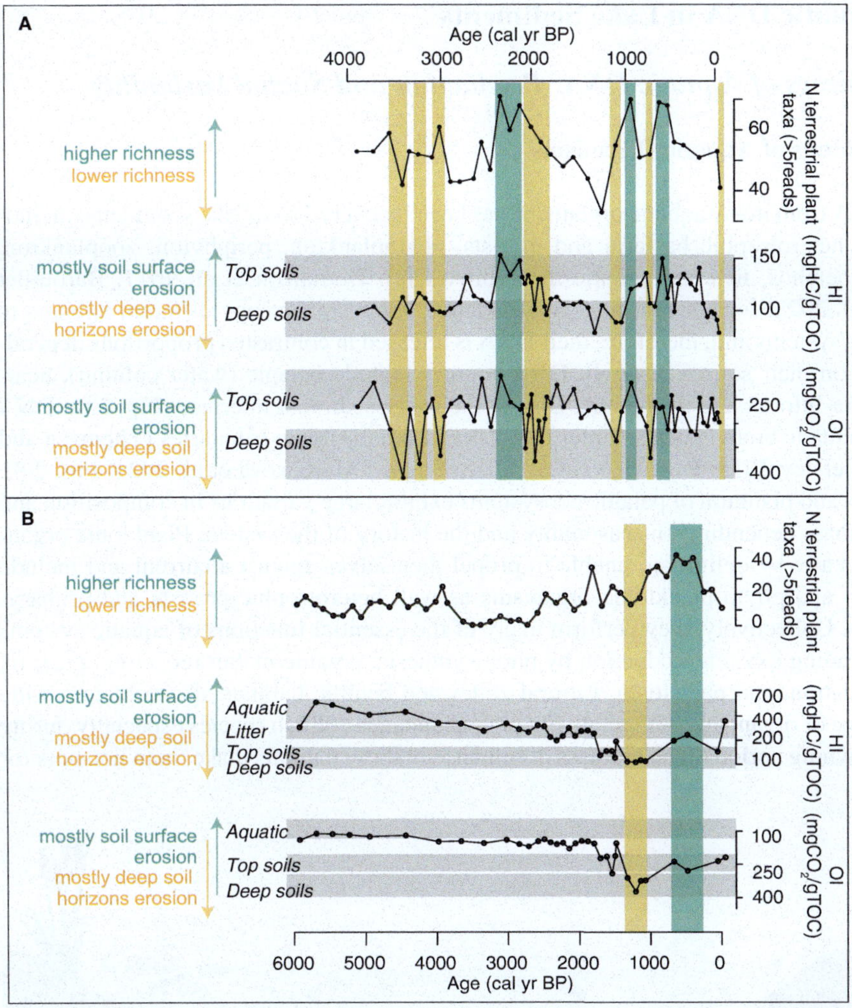

Fig. 2.7 Holocene sediment records from two lakes ((**a**) Lake Anterne and (**b**) Lake La Thuile) in the Northern French Alps highlight the influence of different erosion processes, surface soil erosion and deeper soil erosion, on the taxonomic richness of terrestrial plant DNA record. Erosion sources are identified by pyrolysis Rock Eval analysis, which provides two indexes. The combination of values in Hydrogen Index (HI) and Oxygen Index (OI) allows distinguishing different sources of organic matter as highlighted on the fig. HI is the best indicator for identifying the aquatic organic matter (highest values) and organic matter coming from top soils (medium values), while OI is the best indicator for distinguishing the organic matter coming from deep soil horizons (highest values). When HI is around 150 and OI around 250, erosion mostly affects the soil surface, which corresponds to the highest plant DNA richness recorded (green areas). On the opposite, when the erosion mostly affects the deep soil horizons (highest OI values and low HI), lower taxonomic richness is observed (orange areas). This suggests that the number of plant taxa detected in lake sediments may be modulated by the thickness of soil affected by erosion. (Data from Pansu et al. (2015b); Bajard et al. (2017) and Giguet-Covex et al. (2019))

Aquatic DNA in Lake Sediments

Sources of Aquatic DNA: Production and Spatial Variability

Habitats of Aquatic Organisms

DNA from many aquatic organisms has been recovered from lake sediments, including heterotrophic bacteria and protists, phytoplankton, periphyton, zooplankton, zoobenthos, fish and aquatic macrophytes (cf. Domaizon et al. 2017; Barouillet et al. 2023, for a synthesis). These aquatic organisms occupy different habitats in lake systems and, therefore, their DNA is released in contrasted proportions depending on their source. Overall, DNA sources include pelagic (water column), nearshore-littoral, and the benthic zone at the water-sediment interface (Fig. 2.8). DNA of aquatic biota may also enter the system from upstream tributaries (Thomsen and Willerslev 2015; Pawlowski et al. 2018; Sagova-Mareckova et al. 2021, Fig. 2.8). Fish and plankton of pelagic ecosystems exhibit large variations in composition and biomass depending on seasonality and the history of the system. Planktonic organisms are, by definition, unable to propel themselves against a current and include zoo- and phyto-planktonic organisms as well heterotrophic protists and prokaryotes. Collectively, they perform many of the essential functions of aquatic systems including oxygen production by photosynthesis, organic matter recycling, grazing, predation, and parasitism. Littoral zones and benthic habitats also contain a wide range of organisms with relatively high abundances of fish species especially during spawning periods. The biomass dominance in these habitats make them hotspots for

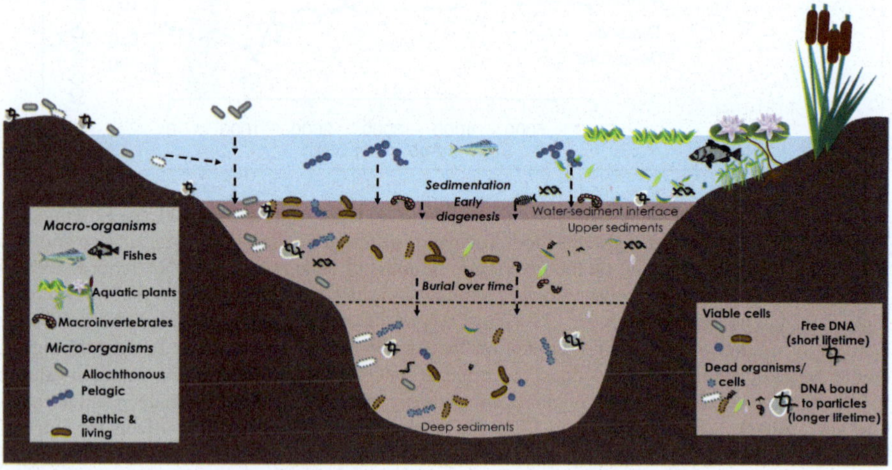

Fig. 2.8 Illustration of the origin (allochthonous sources, and water and sediment living organisms) and fate (after sedimentation) of microbial and aquatic DNA. (Modified from Capo et al. 2022). Both viable and dead cells contain DNA (inDNA). ExDNA may be found as free DNA molecules and in higher amounts as DNA bound to particles

DNA production in lake systems. The littoral zone is also the habitat for aquatic macrophytes, which are well detected in plant sedDNA records (Sjögren et al. 2017; Alsos et al. 2018; Ibrahim et al. 2022). In addition, many protists and prokaryotic taxa and macroinvertebrates (e.g., oligochaetes, nematodes, mussels, and crayfish) can be found in benthic environments (Vivien et al. 2019; Schenk et al. 2020).

Variability in DNA Detection Linked to Biomass and Spatio-Temporal Distribution of Organisms and Abiotic Factors

The success in the detection of environmental DNA from aquatic organisms depends on many factors related to the characteristics of the studied systems (e.g, water residence time, abiotic conditions) as well as the type of organisms studied with different types of distribution and abundance. For instance, the recovery of fish DNA from sedimentary archives is challenging and only a few studies to date have successfully achieved this (Pedersen et al. 2016; Nelson-Chorney et al. 2019; Kuwae et al. 2020; Thomson-Laing et al. 2022; Brown et al. 2022; Giguet-covex et al. 2022). This is likely related to their heterogeneous in-lake distributions and low biomass (compared to other aquatic organisms), but may also be due to unknown methodological issues (Huston et al. 2023; Chap. 9 of this volume). Considerable research efforts have focused on understanding the relationship between fish biomass and DNA concentrations extracted from water samples (for a synthesis, see Barnes et al. 2014; Rourke et al. 2022) to enable historical reconstruction of fish stocks using molecular methods. This approach has proved to be challenging because multiple, sometimes interacting, biotic and abiotic factors preclude the successful detection of fish DNA in sediments. For instance, the taxon examined, its body size, distribution, reproduction, and migration patterns may all impact the amount and location of DNA shed by the fish. Abiotic factors, like hydrological processes, stratification and temperature, also affect the dispersal and persistence of eDNA. Additionally, the biomass of aquatic populations varies between seasons, with the increased temperature and organic matter release during spring having consequences on their growth.

As an exercise we compare the biomass of different aquatic groups from lake water columns. In lakes with oligotrophic to mesotrophic conditions, bacterial biomass has been estimated between 17 and 530 $\mu g{\cdot}L^{-1}$ while phytoplankton biomass ranged between 2 and 4600 $\mu g.L^{-1}$ (Simon et al. 1992). Fish production data obtained for 100 populations from 38 lakes and reservoirs ranged from 0.02 to 771 $kg{\cdot}ha^{-1}$ (Downing and Plante 1993). In systems with limited external sources of food for fish populations, their biomass should not, in theory, be higher than the biomass of the plankton that serve as an important source of food for many species. There is, however, no study that compares the fish biomass with other biological groups in lakes, thus it is not possible to draw firm conclusions about which groups contribute most to lake biomass. The generally patchy distributions of fish populations in lake systems result in spatial variations in their DNA concentrations in the water column and the sediment (Itakura et al. 2019).

In contrast, planktonic organisms are generally relatively evenly spread in the water column and have a more homogeneous distribution. However, mixing, wind and lake basin morphology can impact the spatial distribution of planktonic organisms. Unlike planktonic organisms, where the DNA is more prone to be mixed by water movement, the DNA of benthic organisms is expected to be localised especially for large organisms. Pawlowski et al. (2022) reported that DNA extracted from sediment samples is inefficient at describing the diversity of macroinvertebrates because their DNA typically represents only a small fraction of the total DNA extracted from sediment, and only a limited subset of the macroinvertebrates diversity is revealed by sedDNA metabarcoding analyses in comparison to results obtained by conventional methods (morpho-taxonomic identification). Nevertheless, recent improvements in molecular genetic methods offer reasonable chances in the coming years to monitor meiobenthos from different environmental matrices (Gielings et al. 2021).

Microbial DNA found in sedimentary archives is a mix of DNA originating from external sources (terrestrial and tributaries), the water column and the sediment (Fig. 2.2). Tracing the (planktonic, benthic or others) origins of microbial DNA is difficult because of the metabolic versatility of certain microorganisms and their capacities to thrive, or at least remain viable, in a wide range of changing environmental conditions. Molecular investigations of water samples revealed that certain prokaryotic and eukaryotic microorganisms recorded in freshwater systems originated from surrounding catchment soils (Crump et al. 2012; Ruiz-González et al. 2015). A total of 58% and 43% of the bacteria and archaeal taxa from soil waters could be detected downstream in lake waters (Crump et al. 2012). In contrast, in this freshwater system, only 18% of the soil microbial eukaryotic taxa were recorded in the lake waters. Similarly, prokaryotes commonly found in soils, i.e., belonging to Actinobacteria, Verrucomicrobia, Alphaproteobacteria and Solibacteres, were detected in lake sediments (Vuillemin et al. 2017). Certain groups of microbial eukaryotes were more abundant in soils (i.e., cercozoans, amoebozoans and apicomplexans), while other in freshwater biomes (i.e., ciliates, chrysophytes, dinophytes, synurophytes, chytrids, and diatoms) (Grossmann et al. 2016). DNA from both pelagic and benthic ciliates can also be retrieved in sedimentary archives (e.g., Stoeck et al. 2018; Barouillet et al. 2022).

Transfer of Aquatic DNA to Sediments

The persistence and transfer of DNA molecules in the water column is impacted by a range of abiotic and biotic factors (Rourke et al. 2022). For fish populations, their DNA is released in the water during their life by shedding and excretion, and at their death, with fish carcasses breaking down and sinking to the lake bottom. Turner et al. (2015) showed that higher concentrations of fish DNA can be found in surface sediments compared to the water column in experimental ponds and natural rivers. However, higher detection of fish eDNA in water compared to sediment has been previously observed (Valdez-Moreno et al. 2019). Although reasons for this discrepancy remains unclear, fish life histories, biogeochemical factors, the nature and origin of the sediment(i.e., allochthonous inputs vs autochthonous production) and erosion events are likely important influencing factors (Alsos et al. 2018; Giguet-Covex et al. 2019). A commonly overlooked parameter is the composition of sediments and endogenous particles (mineral or organic) that can provide a preliminary insight into the preservation potential of adsorbed exDNA. For example, in Lake Bacalar (Mexico), Valdez-Moreno et al. (2019) detected higher eDNA concentrations only during spring sampling when intense algal photosynthesis removes CO_2 from the lake water and triggers calcite precipitation. Calcite might quickly adsorb exDNA fragments and incorporate it in the lake sediments.

For planktonic organisms, the transfer of their DNA from the water column to the sediments happens either when they die and sink while still within their own carcasses (especially for larger taxa) or as exDNA. Free exDNA is small enough to be controlled by the water current, thus not necessarily prone to sedimentation. In contrast, given that the solution conditions favour adsorption, adsorbed exDNA will settle with particles to the lake bottom at a faster rate, promoting its sedimentation. Thus, differences are expected in the efficiency of the DNA transfer from the water to the sediments, depending on species morphologies and geochemical conditions in the water column, bioturbation at the mud-water interface and the pore water within sediments. Grazing or cell lysis of specific taxa during sedimentation into the water column could be a cause of differential recovery in the sedDNA signal. Capo et al. (2015) compared microbial eukaryotic inventories obtained from the water column at 2 m and 130 m depth and from upper sediments of Lake Bourget (France). This comparison showed that the poor efficiency in recovering DNA from cryptophytes and haptophytes from sediments was attributed to the early diagenesis of these taxa DNA within sediments. Supporting these results, comparison of DNA inventories from the water column with sediment traps showed a higher detection of cryptophytes in exDNA compared to inDNA and poor recovery of cryptophytes and haptophytes in sediment trap samples (Gauthier et al. 2021). The poor detection of these taxa in the sediments might be because they are soft-bodied algae, thus more susceptible to cell lysis. Differential recovery of DNA from cyanobacteria was also reported in Nwosu et al. (2021) when they compared DNA inventories from the water column, sediment traps and sedimentary archives. Their findings show a poor representation of filamentous *Planktothrix* in sediment samples, despite their relatively high abundance at 10 m water depth contrasting with higher sedimentary signal from aggregate-forming *Aphanizomenon*. Monchamp et al. (2016) compared the richness of cyanobacterial taxa identified in water (microscopy) and sediments (metabarcoding) samples corresponding to the same time period and showed that the congruence between richness estimates varied between cyanobacterial orders. The richness of Chroococcales and Synechococcales was well represented in the sediments, but not well-captured by microscopy counts performed on water samples. Synechococcales and picocyanobacteria, in general, are well represented in the sedimentary archives of deep hard-water lakes, probably due to their dominance in this type of lakes and/or due to the calcite precipitation process that occurs during spring and summer and allows a rapid sedimentation and burying of organic matter associated with the authigenic calcite flux (Domaizon et al. 2013). Overall, aside from sediment sources and difference in sedimentation rates of allochthonous vs autochthonous matters, as mentioned above, the speed of sedimentation of dead cells/carcasses in the water column, grazing by predators, and DNA degradation related to their morphological features are factors that likely explain the differential recovery of the sedDNA signal from specific organisms.

Transfer mechanisms of DNA from aquatic macrophytes are poorly documented. A common observation is that aquatic plants are often better detected compared to terrestrial plants, either based on totDNA or exDNA analyses (Sjögren et al. 2017; Alsos et al. 2018; Giguet-Covex et al. 2019; Ibrahim et al. 2022). This might be related with the proximity, thus more direct connection, with the sediments. A study based on a core taken from the deepest part of a relatively large peri-alpine lake at least 500 m from the shore showed different amounts of DNA sequences and PCR replicates according to the categories of aquatic macrophytes (Giguet-covex et al. 2022). Free-floating plants were the most represented, followed by submerged, floating leaf and lastly helophyte/immersed plants. This result was interpreted as reflecting the different dispersion capacities of each category of macrophyte because of their rooting system as well as their mobility with currents and the size of their habitat impacting the biomass production.

Summary

Two decades of research on sedDNA have revealed that lake sediments from all over the world, from the tropics to the poles, preserve genetic information about past aquatic and terrestrial organisms. DNA storage processes are different if we consider the extracellular DNA or intracellular DNA fractions. Important biotic factors include biomass, habitats, physiological features, and mobility of the organisms, as well as abiotic factors such as erosion dynamic, water currents, mineralogical properties, geochemical conditions and the age of sediments. How these parameters influence DNA records in terms of their representativeness is still not fully understood and the question of their relative importance remains unanswered.

Several uncertainties remain about the processes that control the fossilisation of DNA from terrestrial and aquatic organisms. Identifying the sources, improving our understanding of the transfer processes, and the fate of the DNA signal in sedimentary archives is crucial for evaluating the reliability of sedDNA to reconstruct past dynamics of aquatic and terrestrial organisms and ecosystems in general. Studies specifically dedicated to the understanding of all taphonomic processes will be key over the next decades to assist in the robust analysis of sedDNA retrieved from lakes. For this goal, more field studies based on modern sediment core samples or sediment traps, as well as experimental and modelling approaches, should be conducted. The lake sedDNA record currently goes back to hundreds of thousands of years, but we believe this can be pushed even further back in time with targeted sampling based on the mineralogic properties of sediments and the geochemical conditions of their formation. A more thorough understanding of DNA taphonomy in sedimentary systems will help build a solid foundation for this tool to spatially reconstruct past trajectories of terrestrial and aquatic ecosystems and their biodiversity, and will help in the identification of drivers of ecosystem change from both natural and anthropogenic stressors.

Acknowledgements We are very thankful to Marie-Eve Monchamp, Tyler Murchie, Weihan Jia, Antony G Brown, Inger Greave Alsos and the editors for their comments on our manuscript. SAW was supported by the New Zealand Ministry of Business, Innovation and Employment research programme – Our lakes' health; past, present, future (C05X1707). MAM was supported by the Swiss National Science Foundation (SNSF, grant no. 188256). SJ was funded by the French Government through MOPGA Postdoctoral Programme (reference number 3-5402234721) and a postdoctoral fellowship from Labex OSUG@2020 (investissements d'avenir, ANR10-LABX56).

References

Agnelli A, Ascher J, Corti G, Ceccherini MT, Nannipieri P, Pietramellara G (2004) Distribution of microbial communities in a forest soil profile investigated by microbial biomass, soil respiration and DGGE of total and extracellular DNA. Soil Biol Biochem 36(5):859–868. https://doi.org/10.1016/j.soilbio.2004.02.004

Agnelli A, Ascher J, Corti G, Ceccherini MT, Pietramellara G, Nannipieri P (2007) Purification and isotopic signatures (δ13C, δ15N, Δ14C) of soil extracellular DNA. Biol Fertil Soils 44(2):353–361. https://doi.org/10.1007/s00374-007-0213-y

Allentoft ME, Collins M, Harker D, Haile J, Oskam CL, Hale ML, Campos PF, Samaniego JA, Gilbert MTP, Willerslev E, Zhang G, Scofield RP, Holdaway RN, Bunce M (2012) The half-life of DNA in bone: measuring decay kinetics in 158 dated fossils. Proc R Soc B Biol Sci 279(1748):4724–4733. https://doi.org/10.1098/rspb.2012.1745

Alsos IG, Lammers Y, Yoccoz NG, Jørgensen T, Sjögren P, Gielly L, Edwards ME (2018) Plant DNA metabarcoding of lake sediments: how does it represent the contemporary vegetation. PLoS One 13(4):e0195403. https://doi.org/10.1371/journal.pone.0195403

Alsos IG, Rijal DP, Ehrich D, Karger DN, Yoccoz NG, Heintzman PD, Brown AG, Lammers Y, Pellissier L, Alm T, Bråthen KA, Coissac E, Merkel MKF, Alberti A, Denoeud F, Bakke J, Phylonorway Consortium (2022) Postglacial species arrival and diversity buildup of northern ecosystems took millennia. Science. Advances 8(39):eabo7434. https://doi.org/10.1126/sciadv.abo7434

Andersen K, Bird KL, Rasmussen M, Haile J, Breuning-Madsen H, Kjaer KH, Orlando L, Gilbert MTP, Willerslev E (2012) Meta-barcoding of "dirt" DNA from soil reflects vertebrate biodiversity. Mol Ecol 21(8):1966–1979. https://doi.org/10.1111/j.1365-294X.2011.05261.x

Anderson-Carpenter LL, McLachlan JS, Jackson ST, Kuch M, Lumibao CY, Poinar HN (2011) Ancient DNA from lake sediments: bridging the gap between paleoecology and genetics. BMC Evol Biol 11(1):30. https://doi.org/10.1186/1471-2148-11-30

Armbrecht LH, Coolen MJL, Lejzerowicz F, George SC, Negandhi K, Suzuki Y, Young J, Foster NR, Armand LK, Cooper A, Ostrowski M, Focardi A, Stat M, Moreau JW, Weyrich LS (2019) Ancient DNA from marine sediments: precautions and considerations for seafloor coring, sample handling and data generation. Earth Sci Rev 196:102887. https://doi.org/10.1016/j.earscirev.2019.102887

Bajard M, Poulenard J, Sabatier P, Etienne D, Ficetola F, Chen W, Gielly L, Taberlet P, Develle A-L, Rey P-J, Moulin B, de Beaulieu J-L, Arnaud F (2017) Long-term changes in alpine pedogenetic processes: effect of millennial agro-pastoralism activities (French-Italian Alps). Geoderma 306:217–236. https://doi.org/10.1016/j.geoderma.2017.07.005

Barnes MA, Turner CR, Jerde CL, Renshaw MA, Chadderton WL, Lodge DM (2014) Environmental conditions influence eDNA persistence in aquatic systems. Environ Sci Technol 48(3):1819–1827. https://doi.org/10.1021/es404734p

Barouillet C, Vasselon V, Keck F, Millet L, Etienne D, Galop D, Rius D, Domaizon I (2022) Paleoreconstructions of ciliate communities reveal long-term ecological changes in temperate lakes. Sci Rep 12(1):7899. https://doi.org/10.1038/s41598-022-12041-7

Barouillet C, Monchamp M-E, Bertilsson S, Brasell K, Domaizon I, Epp LS, Ibrahim A, Mejbel H, Nwosu EC, Pearman JK, Picard M, Thomson-Laing G, Tsugeki N, Von Eggers J, Gregory-Eaves I, Pick FR, Wood SA, Capo E (2023) Investigating the effects of anthropogenic stressors on lake biota using sedimentary DNA. Freshw Biol. https://doi.org/10.1111/fwb.14027

Bertrand S, Sterken M, Vargas-Ramirez L, De Batist M, Vyverman W, Lepoint G, Fagel N (2010) Bulk organic geochemistry of sediments from Puyehue Lake and its watershed (Chile, 40°S): implications for paleoenvironmental reconstructions. Palaeogeogr Palaeoclimatol Palaeoecol 294(1):56–71. https://doi.org/10.1016/j.palaeo.2009.03.012

Birks HJB, Birks HH (2016) How have studies of ancient DNA from sediments contributed to the reconstruction of quaternary floras? New Phytol 209(2):499–506. https://doi.org/10.1111/nph.13657

Boere AC, Sinninghe Damsté JS, Rijpstra WIC, Volkman JK, Coolen MJL (2011a) Source-specific variability in post-depositional DNA preservation with potential implications for DNA based paleoecological records. Org Geochem 42(10):1216–1225. https://doi.org/10.1016/j.orggeochem.2011.08.005

Boere AC, Rijpstra WIC, De Lange GJ, Sinninghe Damsté JS, Coolen MJL (2011b) Preservation potential of ancient plankton DNA in Pleistocene marine sediments. Geobiology 9(5):377–393. https://doi.org/10.1111/j.1472-4669.2011.00290.x

Boessenkool S, Mcglynn G, Epp LS, Taylor D, Pimentel M, Gizaw A, Nemomissa S, Brochmann C, Popp M (2014) Use of ancient sedimentary DNA as a novel conservation tool for high-altitude tropical biodiversity. Conserv Biol 28(2):446–455. https://doi.org/10.1111/cobi.12195

Bremond L, Favier C, Ficetola GF, Tossou MG, Akouégninou A, Gielly L, Giguet-Covex C, Oslisly R, Salzmann U (2017) Five thousand years of tropical lake sediment DNA records from Benin. Quat Sci Rev 170:203–211. https://doi.org/10.1016/j.quascirev.2017.06.025

Brown AG, Van Hardenbroek M, Fonville T, Davies K, Mackay H, Murray E, Head K, Barratt P, McCormick F, Ficetola GF, Gielly L, Henderson ACG, Crone A, Cavers G, Langdon PG, Whitehouse NJ, Pirrie D, Alsos IG (2021) Ancient DNA, lipid biomarkers and palaeoecological evidence reveals construction and life on early medieval lake settlements. Sci Rep 11(1):11807. https://doi.org/10.1038/s41598-021-91057-x

Brown T, Rijal DP, Heintzman PD, Clarke CL, Blankholm H-P, Høeg HI, Lammers Y, Bråthen KA, Edwards M, Alsos IG (2022) Paleoeconomy more than demography determined prehistoric human impact in Arctic Norway. PNAS Nexus 1(5):pgac209. https://doi.org/10.1093/pnasnexus/pgac209

Cai P, Huang Q, Zhang X, Chen H (2006a) Adsorption of DNA on clay minerals and various colloidal particles from an Alfisol. Soil Biol Biochem 38(3):471–476. https://doi.org/10.1016/j.soilbio.2005.05.019

Cai P, Huang Q-Y, Zhang X-W (2006b) Interactions of DNA with clay minerals and soil colloidal particles and protection against degradation by DNase. Environ Sci Technol 40(9):2971–2976. https://doi.org/10.1021/es0522985

Cao Y, Wei X, Cai P, Huang Q, Rong X, Liang W (2011) Preferential adsorption of extracellular polymeric substances from bacteria on clay minerals and iron oxide. Colloids Surf B Biointerfaces 83(1):122–127. https://doi.org/10.1016/j.colsurfb.2010.11.018

Capo E, Debroas D, Arnaud F, Domaizon I (2015) Is planktonic diversity well recorded in sedimentary DNA? Toward the reconstruction of past protistan diversity. Microb Ecol 70(4):865–875. https://doi.org/10.1007/s00248-015-0627-2

Capo E, Domaizon I, Maier D, Debroas D, Bigler C (2017) To what extent is the DNA of microbial eukaryotes modified during burying into lake sediments? A repeat-coring approach on annually laminated sediments. J Paleolimnol 58(4):479–495. https://doi.org/10.1007/s10933-017-0005-9

Capo E, Giguet-Covex C, Rouillard A, Nota K, Heintzman PD, Vuillemin A, Ariztegui D, Arnaud F, Belle S, Bertilsson S, Bigler C, Bindler R, Brown AG, Clarke CL, Crump SE, Debroas D, Englund G, Ficetola GF, Garner RE, Gauthier J, Gregory-Eaves I, Heinecke L, Herzschuh U, Ibrahim A, Kisand V, Kjær KH, Lammers Y, Littlefair J, Messager E, Monchamp M-E, Olajos F, Orsi W, Pedersen MW, Rijal DP, Rydberg J, Spanbauer T, Stoof-Leichsenring KR, Taberlet P, Talas L, Thomas C, Walsh DA, Wang Y, Willerslev E, van Woerkom A, Zimmermann HH, Coolen MJL, Epp LS, Domaizon IG, Alsos I, Parducci L (2021) Lake sedimentary DNA research on past terrestrial and aquatic biodiversity: overview and recommendations. Quaternary 4(1):6. https://doi.org/10.3390/quat4010006

Capo E, Monchamp M-E, Coolen MJL, Domaizon I, Armbrecht L, Bertilsson S (2022) Environmental paleomicrobiology: using DNA preserved in aquatic sediments to its full potential. Environ Microbiol 24(5):2201–2209. https://doi.org/10.1111/1462-2920.15913

Carini P, Marsden PJ, Leff JW, Morgan EE, Strickland MS, Fierer N (2016) Relic DNA is abundant in soil and obscures estimates of soil microbial diversity. Nat Microbiol 2(3):1–6. https://doi.org/10.1038/nmicrobiol.2016.242

Chorover J, Kretzschmar R, Garcia-Pichel F, Sparks DL (2007) Soil biogeochemical processes within the critical zone. Elements 3(5):321–326. https://doi.org/10.2113/gselements.3.5.321

Christl I, Knicker H, Kögel-Knabner I, Kretzschmar R (2000) Chemical heterogeneity of humic substances: characterization of size fractions obtained by hollow-fibre ultrafiltration. Eur J Soil Sci 51:617–625. https://doi.org/10.1111/j.1365-2389.2000.00352.x

Cleaves HJ, Crapster-Pregont E, Jonsson CM, Jonsson CL, Sverjensky DA, Hazen RA (2011) The adsorption of short single-stranded DNA oligomers to mineral surfaces. Chemosphere 83(11):1560–1567. https://doi.org/10.1016/j.chemosphere.2011.01.023

Corinaldesi C, Danovaro R, Dell'Anno A (2005) Simultaneous recovery of extracellular and intracellular DNA suitable for molecular studies from marine sediments. Appl Environ Microbiol 71(1):46–50. https://doi.org/10.1128/AEM.71.1.46-50.2005

Crecchio C, Stotzky G (1998) Binding of DNA on humic acids: effect on transformation of Bacillus subtilis and resistance to DNase. Soil Biol Biochem 30(8):1061–1067. https://doi.org/10.1016/S0038-0717(97)00248-4

Crump SE (2021) Sedimentary ancient DNA as a tool in paleoecology. Nat Rev Earth Environ 2(4):229–229. https://doi.org/10.1038/s43017-021-00158-8

Crump BC, Amaral-Zettler LA, Kling GW (2012) Microbial diversity in arctic freshwaters is structured by inoculation of microbes from soils. ISME J 6(9):1629–1639. https://doi.org/10.1038/ismej.2012.9

Delgado-Baquerizo M, Oliverio AM, Brewer TE, Benavent-González A, Eldridge DJ, Bardgett RD, Maestre FT, Singh BK, Fierer N (2018) A global atlas of the dominant bacteria found in soil. Science 359(6373):320–325. https://doi.org/10.1126/science.aap9516

Dell'Anno A, Danovaro R (2005) Extracellular DNA plays a key role in Deep-Sea ecosystem functioning. Science 309(5744):2179–2179. https://doi.org/10.1126/science.1117475

Demanèche S, Jocteur-Monrozier L, Quiquampoix H, Simonet P (2001) Evaluation of biological and physical protection against nuclease degradation of clay-bound plasmid DNA. Appl Environ Microbiol 67(1):293–299. https://doi.org/10.1128/AEM.67.1.293-299.2001

Domaizon I, Savichtcheva O, Debroas D, Arnaud F, Villar C, Pignol C, Alric B, Perga ME (2013) DNA from lake sediments reveals the long-term dynamics and diversity of *Synechococcus* assemblages. Biogeosciences 10(6):3817–3838. https://doi.org/10.5194/bg-10-3817-2013

Domaizon I, Winegardner A, Capo E, Gauthier J, Gregory-Eaves I (2017) DNA-based methods in paleolimnology: new opportunities for investigating long-term dynamics of lacustrine biodiversity. J Paleolimnol 58(1):1–21. https://doi.org/10.1007/s10933-017-9958-y

Dommain R, Andama M, McDonough MM, Prado NA, Goldhammer T, Potts R, Maldonado JE, Nkurunungi JB, Campana MG (2020) The challenges of reconstructing tropical biodiversity with sedimentary ancient DNA: a 2200-year-long metagenomic record from Bwindi impenetrable Forest, Uganda. Front Ecol Evol 8

Downing JA, Plante C (1993) Production of fish populations in lakes. Can J Fish Aquat Sci 50(1):110–120. https://doi.org/10.1139/f93-013

Duchaufour P, Faivre P, Poulenard J, Gury M (2020) Introduction à la science des sols. 7ème édition. Dunod

Edwards ME, Alsos IG, Yoccoz N, Coissac E, Goslar T, Gielly L, Haile J, Langdon CT, Tribsch A, Binney HA, von Stedingk H, Taberlet P (2018) Metabarcoding of modern soil DNA gives a highly local vegetation signal in Svalbard tundra. The Holocene 28(12):2006–2016. https://doi.org/10.1177/0959683618798095

Ekram A-E, Hamilton R, Campbell M, Plett C, Kose S, Russell J, Stevenson J, Coolen M (2021) A 1 Ma record of climate-induced vegetation changes using sed aDNA and pollen in a biodiversity hotspot: Lake Towuti, Sulawesi, Indonesia. :EGU21–4003. https://doi.org/10.5194/egusphere-egu21-4003

Ellegaard M, Ribeiro S (2018) The long-term persistence of phytoplankton resting stages in aquatic "seed banks". Biol Rev Camb Philos Soc 93(1):166–183. https://doi.org/10.1111/brv.12338

Ellegaard M, Clokie MRJ, Czypionka T, Frisch D, Godhe A, Kremp A, Letarov A, McGenity TJ, Ribeiro S, John Anderson N (2020) Dead or alive: sediment DNA archives as tools for tracking aquatic evolution and adaptation. Commun Biol 3(1):1–11. https://doi.org/10.1038/s42003-020-0899-z

Epp LS, Stoof KR, Trauth MH, Tiedemann R (2010) Historical genetics on a sediment core from a Kenyan lake: intraspecific genotype turnover in a tropical rotifer is related to past environmental changes. J Paleolimnol 43(4):939–954. https://doi.org/10.1007/s10933-009-9379-7

Epp LS, Stoof-Leichsenring KR, Trauth MH, Tiedemann R (2011) Molecular profiling of diatom assemblages in tropical lake sediments using taxon-specific PCR and denaturing high-performance liquid chromatography (PCR-DHPLC). Mol Ecol Resour 11(5):842–853. https://doi.org/10.1111/j.1755-0998.2011.03022.x

Evrard O, Laceby JP, Ficetola GF, Gielly L, Huon S, Lefèvre I, Onda Y, Poulenard J (2019) Environmental DNA provides information on sediment sources: a study in catchments affected by Fukushima radioactive fallout. Sci Total Environ 665:873–881. https://doi.org/10.1016/j.scitotenv.2019.02.191

Fang J, Jin L, Meng Q, Wang D, Lin D (2021) Interactions of extracellular DNA with aromatized biochar and protection against degradation by DNase I. J Environ Sci 101:205–216. https://doi.org/10.1016/j.jes.2020.08.017

Ficetola GF, Poulenard J, Sabatier P, Messager E, Gielly L, Leloup A, Etienne D, Bakke J, Malet E, Fanget B, Støren E, Reyss J-L, Taberlet P, Arnaud F (2018) DNA from lake sediments reveals long-term ecosystem changes after a biological invasion. Science. Advances 4(5):eaar4292. https://doi.org/10.1126/sciadv.aar4292

Foucher A, Evrard O, Ficetola GF, Gielly L, Poulain J, Giguet-Covex C, Laceby JP, Salvador-Blanes S, Cerdan O, Poulenard J (2020) Persistence of environmental DNA in cultivated soils: implication of this memory effect for reconstructing the dynamics of land use and cover changes. Sci Rep 10(1):10502. https://doi.org/10.1038/s41598-020-67452-1

Franchi M, Bramanti E, Morassi Bonzi L, Luigi Orioli P, Vettori C, Gallori E (1999) Clay-nucleic acid complexes: characteristics and implications for the preservation of genetic material in primeval habitats. Orig Life Evol Biosph 29(3):297–315. https://doi.org/10.1023/A:1006557832574

Frankl A, Evrard O, Cammeraat E, Tytgat B, Verleyen E, Stokes A (2022) Tracing hotspots of soil erosion in high mountain environments: how forensic science based on plant eDNA can lead the way. An opinion Plant Soil 476(1):729–742. https://doi.org/10.1007/s11104-021-05261-9

Freeman CL, Dieudonné L, Agbaje OBA, Zure M, Sanz JQ, Collins M, Sand KK (2023) Survival of environmental DNA in sediments: Mineralogic control on DNA taphonomy. 2020.01.28.922997

Gauthier J, Walsh D, Selbie DT, Bourgeois A, Griffiths K, Domaizon I, Gregory-Eaves I (2021) Evaluating the congruence between DNA-based and morphological taxonomic approaches in water and sediment trap samples: analyses of a 36-month time series from a temperate monomictic lake. Limnol Oceanogr 66(8):3020–3039. https://doi.org/10.1002/lno.11856

Gauthier J, Walsh D, Selbie DT, Domaizon I, Gregory-Eaves I (2022) Sedimentary DNA of a human-impacted lake in Western Canada (Cultus Lake) reveals changes in micro-eukaryotic diversity over the past 200 years. Environmental DNA 4(5):1106–1119. https://doi.org/10.1002/edn3.310

Geggier S, Kotlyar A, Vologodskii A (2011) Temperature dependence of DNA persistence length. Nucleic Acids Res 39(4):1419–1426. https://doi.org/10.1093/nar/gkq932

Giguet-Covex C, Taberlet P, Ficetola FG (2020) Extracellular DNA extraction 6. https://doi.org/dx.doi.org/10.17504/protocols.io.bdwsi7ee

Gielings R, Fais M, Fontaneto D, Creer S, Costa FO, Renema W, Macher J-N (2021) DNA Metabarcoding methods for the study of marine benthic Meiofauna: a review. Front Mar Sci 8

Giguet-Covex C, Pansu J, Arnaud F, Rey P-J, Griggo C, Gielly L, Domaizon I, Coissac E, David F, Choler P, Poulenard J, Taberlet P (2014) Long livestock farming history and human landscape shaping revealed by lake sediment DNA. Nat Commun 5(1):3211. https://doi.org/10.1038/ncomms4211

Giguet-Covex C, Ficetola GF, Walsh K, Poulenard J, Bajard M, Fouinat L, Sabatier P, Gielly L, Messager E, Develle AL, David F, Taberlet P, Brisset E, Guiter F, Sinet R, Arnaud F (2019) New insights on lake sediment DNA from the catchment: importance of taphonomic and analytical issues on the record quality. Sci Rep 9:14676. https://doi.org/10.1038/s41598-019-50339-1

Giguet-covex C, Messager E, Arthaud F, Gielly L, Jenny J-P (2022) Aquatic plant dynamic and fishes over the last 1800 years in the Lake Aiguebelette (Northern French Alps). ScienceOpen Posters. https://doi.org/10.14293/S2199-1006.1.SOR-.PPHTSYA.v1

Giguet-Covex C, Bajard M, Chen W, Walsh KJ, Rey P-J, Messager E, Etienne D, Sabatier P, Ficetola FG, Gielly L, Blanchet C, Guffond C, Chiquet P, Arnaud F, Poulenard J (2023) Long-term trajectories of mountain agro-ecosystems in the North-Western Alps. Reg Environ Change 23, 58. https://doi.org/10.1007/s10113-023-02030-5

Graham RW, Belmecheri S, Choy K, Culleton BJ, Davies LJ, Froese D, Heintzman PD, Hritz C, Kapp JD, Newsom LA, Rawcliffe R, Saulnier-Talbot É, Shapiro B, Wang Y, Williams JW, Wooller MJ (2016) Timing and causes of mid-Holocene mammoth extinction on St. Paul Island, Alaska. Proc Natl Acad Sci 113(33):9310–9314. https://doi.org/10.1073/pnas.1604903113

Greaves MP, Wilson MJ (1969) The adsorption of nucleic acids by montmorillonite. Soil Biol Biochem 1(4):317–323. https://doi.org/10.1016/0038-0717(69)90014-5

Grossmann L, Jensen M, Heider D, Jost S, Glücksman E, Hartikainen H, Mahamdallie SS, Gardner M, Hoffmann D, Bass D, Boenigk J (2016) Protistan community analysis: key findings of a large-scale molecular sampling. ISME J 10(9):2269–2279. https://doi.org/10.1038/ismej.2016.10

Haglund A-L, Lantz P, Törnblom E, Tranvik L (2003) Depth distribution of active bacteria and bacterial activity in lake sediment. FEMS Microbiol Ecol 46(1):31–38. https://doi.org/10.1016/S0168-6496(03)00190-9

Haile J, Holdaway R, Oliver K, Bunce M, Gilbert MTP, Nielsen R, Munch K, Ho SYW, Shapiro B, Willerslev E (2007) Ancient DNA chronology within sediment deposits: are paleobiological reconstructions possible and is DNA leaching a factor? Mol Biol Evol 24(4):982–989. https://doi.org/10.1093/molbev/msm016

Harper LR, Buxton AS, Rees HC, Bruce K, Brys R, Halfmaerten D, Read DS, Watson HV, Sayer CD, Jones EP, Priestley V, Mächler E, Múrria C, Garcés-Pastor S, Medupin C, Burgess K, Benson G, Boonham N, Griffiths RA, Handley LL, Hänfling B (2019) Prospects and challenges of environmental DNA (eDNA) monitoring in freshwater ponds. Hydrobiologia 826:25–41

Hebsgaard MB, Gilbert MTP, Arneborg J, Heyn P, Allentoft ME, Bunce M, Munch K, Schweger C, Willerslev E (2009) 'The farm beneath the Sand' – an archaeological case study on ancient 'dirt' DNA. Antiquity 83(320):430–444. https://doi.org/10.1017/S0003598X00098537
Hofreiter M, Jaenicke V, Serre D, von Haeseler A, Pääbo S (2001) DNA sequences from multiple amplifications reveal artifacts induced by cytosine deamination in ancient DNA. Nucleic Acids Res 29(23):4793–4799
Hofreiter M, Paijmans JLA, Goodchild H, Speller CF, Barlow A, Fortes GG, Thomas JA, Ludwig A, Collins MJ (2015) The future of ancient DNA: technical advances and conceptual shifts. BioEssays 37(3):284–293. https://doi.org/10.1002/bies.201400160
Huang S, Stoof-Leichsenring KR, Liu S, Courtin J, Andreev AA, Pestryakova A, Herzschuh U (2021) Plant sedimentary ancient DNA from Far East Russia covering the last 28,000 years reveals different assembly rules in cold and warm climates. Front Ecol Evol 9
Huston GP, Lopez MLD, Cheng Y, King L, Duxbury LC, Picard M, Thomson-Laing G, Myler E, Helbing CC, Kinnison MT, Saros JE, Gregory-Eaves I, Monchamp M-E, Wood SA, Armbrecht L, Ficetola GF, Kurte L, Von Eggers J, Brahney J, Parent G, Sakata MK, Doi H, Capo E (2023) Detection of fish sedimentary DNA in aquatic systems: A review of methodological challenges and future opportunities. Environmental DNA n/a(n/a). https://doi.org/10.1002/edn3.467
Ibrahim A, Höckendorff S, Schleheck D, Epp L, van Kleunen M, Meyer A (2022) Vegetation changes over the last centuries in the lower lake constance region reconstructed from sediment-core environmental DNA. Environmental DNA 4(4):830–845. https://doi.org/10.1002/edn3.292
Itakura H, Wakiya R, Yamamoto S, Kaifu K, Sato T, Minamoto T (2019) Environmental DNA analysis reveals the spatial distribution, abundance, and biomass of Japanese eels at the river-basin scale. Aquat Conserv Mar Freshwat Ecosyst 29(3):361–373. https://doi.org/10.1002/aqc.3058
Ito A, Wagai R (2017) Global distribution of clay-size minerals on land surface for biogeochemical and climatological studies. Sci Data 4(1):170103. https://doi.org/10.1038/sdata.2017.103
Jelavić S, Thygesen LG, Magnin V, Findling N, Müller S, Meklesh V, Sand KK (2022) Soot and charcoal as reservoirs of extracellular DNA
Jia W, Liu X, Stoof-Leichsenring KR, Liu S, Li K, Herzschuh U (2022a) Preservation of sedimentary plant DNA is related to lake water chemistry. Environmental DNA 4(2):425–439. https://doi.org/10.1002/edn3.259
Jia W, Anslan S, Chen F, Cao X, Dong H, Dulias K, Gu Z, Heinecke L, Jiang H, Kruse S, Kang W, Li K, Liu S, Liu X, Liu Y, Ni J, Schwalb A, Stoof-Leichsenring KR, Shen W, Tian F, Wang J, Wang Y, Wang Y, Xu H, Yang X, Zhang D, Herzschuh U (2022b) Sedimentary ancient DNA reveals past ecosystem and biodiversity changes on the Tibetan Plateau: overview and prospects. Quat Sci Rev 293:107703. https://doi.org/10.1016/j.quascirev.2022.107703
Kanbar HJ, Olajos F, Englund G, Holmboe M (2020) Geochemical identification of potential DNA-hotspots and DNA-infrared fingerprints in lake sediments. Appl Geochem 122:104728. https://doi.org/10.1016/j.apgeochem.2020.104728
Kandeler E, Brune T, Enowashu E, Dörr N, Guggenberger G, Lamersdorf N, Philippot L (2009) Response of total and nitrate-dissimilating bacteria to reduced N deposition in a spruce forest soil profile. FEMS Microbiol Ecol 67(3):444–454. https://doi.org/10.1111/j.1574-6941.2008.00632.x
Keck F, Millet L, Debroas D, Etienne D, Galop D, Rius D, Domaizon I (2020) Assessing the response of micro-eukaryotic diversity to the great acceleration using lake sedimentary DNA. Nat Commun 11(1):3831. https://doi.org/10.1038/s41467-020-17682-8
Khanna M, Stotzky G (1992) Transformation of Bacillus subtilis by DNA bound on montmorillonite and effect of DNase on the transforming ability of bound DNA. Appl Environ Microbiol 58(6):1930–1939. https://doi.org/10.1128/aem.58.6.1930-1939.1992
Kipton H, Powell J, Town RM (1992) Solubility and fractionation of humic acid; effect of pH and ionic medium. Anal Chim Acta 267(1):47–54. https://doi.org/10.1016/0003-2670(92)85005-Q

Kirkpatrick J, Walsh E, D'Hondt S (2016) Fossil DNA persistence and decay in marine sediment over hundred-thousand-year to million-year time scales. Graduate School of Oceanography Faculty Publications https://doi.org/10.1130/G37933.1

Kjær KH, Winther Pedersen M, De Sanctis B, De Cahsan B, Korneliussen TS, Michelsen CS, Sand KK, Jelavić S, Ruter AH, Schmidt AMA, Kjeldsen KK, Tesakov AS, Snowball I, Gosse JC, Alsos IG, Wang Y, Dockter C, Rasmussen M, Jørgensen ME, Skadhauge B, Prohaska A, Kristensen JÅ, Bjerager M, Allentoft ME, Coissac E, Rouillard A, Simakova A, Fernandez-Guerra A, Bowler C, Macias-Fauria M, Vinner L, Welch JJ, Hidy AJ, Sikora M, Collins MJ, Durbin R, Larsen NK, Willerslev E (2022) A 2-million-year-old ecosystem in Greenland uncovered by environmental DNA. Nature 612(7939):283–291. https://doi.org/10.1038/s41586-022-05453-y

Kosmulski M (2011) The pH-dependent surface charging and points of zero charge: V update. J Colloid Interface Sci 353(1):1–15. https://doi.org/10.1016/j.jcis.2010.08.023

Kristensen E (2000) Organic matter diagenesis at the oxic/anoxic interface in coastal marine sediments, with emphasis on the role of burrowing animals. Hydrobiologia 426(1):1–24. https://doi.org/10.1023/A:1003980226194

Kuwae M, Tamai H, Doi H, Sakata MK, Minamoto T, Suzuki Y (2020) Sedimentary DNA tracks decadal-centennial changes in fish abundance. Commun Biol 3(1):1–12. https://doi.org/10.1038/s42003-020-01282-9

Lammers Y, Clarke CL, Erséus C, Brown AG, Edwards ME, Gielly L, Haflidason H, Mangerud J, Rota E, Svendsen JI, Alsos IG (2019) Clitellate worms (Annelida) in lateglacial and Holocene sedimentary DNA records from the polar Urals and northern Norway. Boreas 48(2):317–329. https://doi.org/10.1111/bor.12363

Levy-Booth DJ, Campbell RG, Gulden RH, Hart MM, Powell JR, Klironomos JN, Pauls KP, Swanton CJ, Trevors JT, Dunfield KE (2007) Cycling of extracellular DNA in the soil environment. Soil Biol Biochem

Lindahl T (1993) Instability and decay of the primary structure of DNA. Nature 362(6422):709–715. https://doi.org/10.1038/362709a0

Lindahl T, Nyberg B (1972) Rate of depurination of native deoxyribonucleic acid. Biochemistry 11(19):3610–3618. https://doi.org/10.1021/bi00769a018

Lindahl T, Nyberg B (1974) Heat-induced deamination of cytosine residues in deoxyribonucleic acid. Biochem 13:3405–3410. https://doi.org/10.1021/bi00713a035

Lorenz MG, Aardema BW, Krumbein WE (1981) Interaction of marine sediment with DNA and DNA availability to nucleases. Mar Biol 64(2):225–230. https://doi.org/10.1007/BF00397113

Lorenz MG, Gerjets D, Wackernagel W (1991) Release of transforming plasmid and chromosomal DNA from two cultured soil bacteria. Arch Microbiol 156(4):319–326. https://doi.org/10.1007/BF00263005

Lydolph MC, Jacobsen J, Arctander P, Gilbert MTP, Gilichinsky DA, Hansen AJ, Willerslev E, Lange L (2005) Beringian paleoecology inferred from permafrost-preserved fungal DNA. Appl Environ Microbiol 71(2):1012–1017. https://doi.org/10.1128/AEM.71.2.1012-1017.2005

Lyet A, Pellissier L, Valentini A, Dejean T, Hehmeyer A, Naidoo R (2021) eDNA sampled from stream networks correlates with camera trap detection rates of terrestrial mammals. Sci Rep 11(1):11362. https://doi.org/10.1038/s41598-021-90598-5

Madeja J (2015) A new tool to trace past human presence from lake sediments: the human-specific molecular marker Bacteroides strain HF 183. J Quat Sci 30(4):349–354. https://doi.org/10.1002/jqs.2783

Madeja J, Wacnik A, Zyga A, Stankiewicz E, Wypasek E, Guminski W, Harmata K (2009) Bacterial ancient DNA as an indicator of human presence in the past: its correlation with palynological and archaeological data. J Quat Sci 24(4):317–321. https://doi.org/10.1002/jqs.1237

Madeja J, Wacnik A, Wypasek E, Chandran A, Stankiewicz E (2010) Integrated palynological and molecular analyses of late Holocene deposits from Lake Miłkowskie (NE Poland): verification of local human impact on environment. Quat Int 220(1):147–152. https://doi.org/10.1016/j.quaint.2009.09.008

Matisoo-Smith E, Roberts K, Welikala N, Tannock G, Chester P, Feek D, Flenley J (2008) Recovery of DNA and pollen from New Zealand lake sediments. Quat Int 184(1):139–149. https://doi.org/10.1016/j.quaint.2007.09.013

Mejbel HS, Dodsworth W, Pick FR (2022) Effects of temperature and oxygen on cyanobacterial DNA preservation in sediments: a comparison study of major taxa. Environmental DNA 4(4):717–731. https://doi.org/10.1002/edn3.289

Messager E, Giguet-Covex C, Doyen E, Etienne D, Gielly L, Sabatier P, Banjan M, Develle A-L, Didier J, Poulenard J, Julien A, Arnaud F (2022) Two millennia of complexity and variability in a perialpine socioecological system (Savoie, France): the contribution of palynology and sedaDNA analysis. Front Ecol Evol 10

Miller JH, Simpson C (2022) When did mammoths go extinct? Nature 612(7938):E1–E3. https://doi.org/10.1038/s41586-022-05416-3

Monchamp M-E, Walser J-C, Pomati F, Spaak P (2016) Sedimentary DNA reveals cyanobacterial community diversity over 200 years in two perialpine lakes. Appl Environ Microbiol 82(21):6472–6482. https://doi.org/10.1128/AEM.02174-16

Morlock MA, Rodriguez-Martinez S, Yu-Tuan Huang D, Glaus N, Vogel H, Anselmetti FS, Klaminder J (2021) Holocene human-environment interactions and their link to erosion in the Eastern Alps, inferred from sedaDNA. In: EGU general assembly conference abstracts. https://doi.org/10.5194/EGUSPHERE-EGU21-14979

Nelson-Chorney HT, Davis CS, Poesch MS, Vinebrooke RD, Carli CM, Taylor MK (2019) Environmental DNA in lake sediment reveals biogeography of native genetic diversity. Front Ecol Environ 17(6):313–318. https://doi.org/10.1002/fee.2073

Nguyen TH, Elimelech M (2007) Adsorption of plasmid DNA to a natural organic matter-coated silica surface: kinetics, conformation, and reversibility. Langmuir 23(6):3273–3279. https://doi.org/10.1021/la0622525

Nwosu EC, Roeser P, Yang S, Ganzert L, Dellwig O, Pinkerneil S, Brauer A, Dittmann E, Wagner D, Liebner S (2021) From water into sediment – tracing freshwater cyanobacteria via DNA analyses. Microorganisms 9(8):1778. https://doi.org/10.3390/microorganisms9081778

Oelkers EH, Golubev SV, Chairat C, Pokrovsky OS, Schott J (2009) The surface chemistry of multi-oxide silicates. Geochim Cosmochim Acta 73(16):4617–4634

Ogram A, Sayler GS, Gustin D, Lewis RJ (1988) DNA adsorption to soils and sediments. Environ Sci Technol 22(8):982–984. https://doi.org/10.1021/es00173a020

Okazaki M, Yoshida Y, Yamaguchi S, Kaneno M, Elliott JC (2001) Affinity binding phenomena of DNA onto apatite crystals. Biomaterials 22(18):2459–2464. https://doi.org/10.1016/S0142-9612(00)00433-6

Pääbo S (1984) Über den Nachweis von DNA in altägyptischen Mumien. Das Altertum 30:213–218

Paget E, Monrozier LJ, Simonet P (1992) Adsorption of DNA on clay minerals: protection against DNaseI and influence on gene transfer. FEMS Microbiol Lett 97(1–2):31–39. https://doi.org/10.1111/j.1574-6968.1992.tb05435.x

Pansu J, De Danieli S, Puissant J, Gonzalez J-M, Gielly L, Cordonnier T, Zinger L, Brun J-J, Choler P, Taberlet P, Cécillon L (2015a) Landscape-scale distribution patterns of earthworms inferred from soil DNA. Soil Biol Biochem 83:100–105. https://doi.org/10.1016/j.soilbio.2015.01.004

Pansu J, Giguet-Covex C, Ficetola GF, Gielly L, Boyer F, Zinger L, Arnaud F, Poulenard J, Taberlet P, Choler P (2015b) Reconstructing long-term human impacts on plant communities: an ecological approach based on lake sediment DNA. Mol Ecol 24(7):1485–1498. https://doi.org/10.1111/mec.13136

Parducci L, Matetovici I, Fontana SL, Bennett KD, Suyama Y, Haile J, Kjaer KH, Larsen NK, Drouzas AD, Willerslev E (2013) Molecular- and pollen-based vegetation analysis in lake sediments from Central Scandinavia. Mol Ecol 22(13):3511–3524. https://doi.org/10.1111/mec.12298

Parducci L, Väliranta M, Salonen JS, Ronkainen T, Matetovici I, Fontana SL, Eskola T, Sarala P, Suyama Y (2015) Proxy comparison in ancient peat sediments: pollen, macrofossil and plant DNA. Philos Trans R Soc B, Biol Sci 370(1660):20130382. https://doi.org/10.1098/rstb.2013.0382

Parducci L, Bennett KD, Ficetola GF, Alsos IG, Suyama Y, Wood JR, Pedersen MW (2017) Ancient plant DNA in lake sediments. New Phytol 214(3):924–942. https://doi.org/10.1111/nph.14470

Pawlowski J, Kelly-Quinn M, Altermatt F, Apothéloz-Perret-Gentil L, Beja P, Boggero A, Borja A, Bouchez A, Cordier T, Domaizon I, Feio MJ, Filipe AF, Fornaroli R, Graf W, Herder J, van der Hoorn B, Iwan Jones J, Sagova-Mareckova M, Moritz C, Barquín J, Piggott JJ, Pinna M, Rimet F, Rinkevich B, Sousa-Santos C, Specchia V, Trobajo R, Vasselon V, Vitecek S, Zimmerman J, Weigand A, Leese F, Kahlert M (2018) The future of biotic indices in the ecogenomic era: integrating (e)DNA metabarcoding in biological assessment of aquatic ecosystems. Sci Total Environ 637–638:1295–1310. https://doi.org/10.1016/j.scitotenv.2018.05.002

Pawlowski J, Bruce K, Panksep K, Aguirre FI, Amalfitano S, Apothéloz-Perret-Gentil L, Baussant T, Bouchez A, Carugati L, Cermakova K, Cordier T, Corinaldesi C, Costa FO, Danovaro R, Dell'Anno A, Duarte S, Eisendle U, Ferrari BJD, Frontalini F, Frühe L, Haegerbaeumer A, Kisand V, Krolicka A, Lanzén A, Leese F, Lejzerowicz F, Lyautey E, Maček I, Sagova-Marečková M, Pearman JK, Pochon X, Stoeck T, Vivien R, Weigand A, Fazi S (2022) Environmental DNA metabarcoding for benthic monitoring: a review of sediment sampling and DNA extraction methods. Sci Total Environ 818:151783. https://doi.org/10.1016/j.scitotenv.2021.151783

Pedersen MW, Ruter A, Schweger C, Friebe H, Staff RA, Kjeldsen KK, Mendoza MLZ, Beaudoin AB, Zutter C, Larsen NK, Potter BA, Nielsen R, Rainville RA, Orlando L, Meltzer DJ, Kjær KH, Willerslev E (2016) Postglacial viability and colonization in North America's ice-free corridor. Nature 537(7618):45–49. https://doi.org/10.1038/nature19085

Pietramellara G, Franchi M, Gallori E, Nannipieri P (2001) Effect of molecular characteristics of DNA on its adsorption and binding on homoionic montmorillonite and kaolinite. Biol Fertil Soils 33(5):402–409. https://doi.org/10.1007/s003740100341

Pietramellara G, Ascher J, Borgogni F, Ceccherini MT, Guerri G, Nannipieri P (2009) Extracellular DNA in soil and sediment: fate and ecological relevance. Biol Fertil Soils 45(3):219–235. https://doi.org/10.1007/s00374-008-0345-8

Poly F, Chenu C, Simonet P, Rouiller J, Jocteur Monrozier L (2000) Differences between linear chromosomal and supercoiled plasmid DNA in their mechanisms and extent of adsorption on clay minerals. Langmuir 16(3):1233–1238. https://doi.org/10.1021/la990506z

Poté J, Rossé P, Rosselli W, Van VT, Wildi W (2005) Kinetics of mass and DNA decomposition in tomato leaves. Chemosphere 61(5):677–684. https://doi.org/10.1016/j.chemosphere.2005.03.030

Poté J, Ackermann R, Wildi W (2009) Plant leaf mass loss and DNA release in freshwater sediments. Ecotoxicol Environ Saf 72(5):1378–1383. https://doi.org/10.1016/j.ecoenv.2009.04.010

Rapuc W, Bouchez J, Sabatier P, Genuite K, Poulenard J, Gaillardet J, Arnaud F (2021) Quantitative evaluation of human and climate forcing on erosion in the alpine critical zone over the last 2000 years. Quat Sci Rev 268:107127. https://doi.org/10.1016/j.quascirev.2021.107127

Rastogi RP, Richa KA, Tyagi MB, Sinha RP (2010) Molecular mechanisms of ultraviolet radiation-induced DNA damage and repair. J Nucleic Acids 2010:592980. https://doi.org/10.4061/2010/592980

Romanowski G, Lorenz MG, Wackernagel W (1991) Adsorption of plasmid DNA to mineral surfaces and protection against DNase I. Appl Environ Microbiol 57(4):1057–1061. https://doi.org/10.1128/aem.57.4.1057-1061.1991

Rourke ML, Fowler AM, Hughes JM, Broadhurst MK, DiBattista JD, Fielder S, Wilkes Walburn J, Furlan EM (2022) Environmental DNA (eDNA) as a tool for assessing fish biomass: a review of approaches and future considerations for resource surveys. Environmental DNA 4(1):9–33. https://doi.org/10.1002/edn3.185

Ruiz-González C, Niño-García JP, del Giorgio PA (2015) Terrestrial origin of bacterial communities in complex boreal freshwater networks. Ecol Lett 18(11):1198–1206. https://doi.org/10.1111/ele.12499

Saeki K, Sakai M (2009) The influence of soil organic matter on DNA adsorptions on andosols. Microbes Environ 24:175. https://doi.org/10.1264/jsme2.ME09117

Saeki K, Morisaki M, Sakai M (2008) The contribution of soil constituents to adsorption of extracellular DNA by soils. Microb Environ 23(4):353–355. https://doi.org/10.1264/jsme2.ME08531
Saeki K, Sakai M, Wada S-I (2010) DNA adsorption on synthetic and natural allophanes. Appl Clay Sci 50(4):493–497. https://doi.org/10.1016/j.clay.2010.09.015
Saeki K, Ihyo Y, Sakai M, Kunito T (2011) Strong adsorption of DNA molecules on humic acids. Environ Chem Lett 9(4):505–509. https://doi.org/10.1007/s10311-011-0310-x
Sagova-Mareckova M, Boenigk J, Bouchez A, Cermakova K, Chonova T, Cordier T, Eisendle U, Elersek T, Fazi S, Fleituch T, Frühe L, Gajdosova M, Graupner N, Haegerbaeumer A, Kelly A-M, Kopecky J, Leese F, Nõges P, Orlic S, Panksep K, Pawlowski J, Petrusek A, Piggott JJ, Rusch JC, Salis R, Schenk J, Simek K, Stovicek A, Strand DA, Vasquez MI, Vrålstad T, Zlatkovic S, Zupancic M, Stoeck T (2021) Expanding ecological assessment by integrating microorganisms into routine freshwater biomonitoring. Water Res 191:116767. https://doi.org/10.1016/j.watres.2020.116767
Sand KK, Jelavić S (2018) Mineral facilitated horizontal gene transfer: a new principle for evolution of life? Front Microbiol 9:2217. https://doi.org/10.3389/fmicb.2018.02217
Sand KK, Jelavić S, Kjær KH, Prohaska A (2023) Importance of eDNA taphonomy and provenance for robust ecological inference: insights from interfacial geochemistry (preprint). Biochem https://doi.org/10.1101/2023.01.24.525431
Scappini F, Casadei F, Zamboni R, Franchi M, Gallori E, Monti S (2004) Protective effect of clay minerals on adsorbed nucleic acid against UV radiation: possible role in the origin of life. Int J Astrobiol 3(1):17–19. https://doi.org/10.1017/S147355040400179X
Schenk J, Kleinbölting N, Traunspurger W (2020) Comparison of morphological, DNA barcoding, and metabarcoding characterizations of freshwater nematode communities. Ecol Evol 10(6):2885–2899. https://doi.org/10.1002/ece3.6104
Schmidt MWI, Noack AG (2000) Black carbon in soils and sediments: analysis, distribution, implications, and current challenges. Glob Biogeochem Cycles 14(3):777–793. https://doi.org/10.1029/1999GB001208
Schwertmann U, Taylor R (1989) Iron Oxides. In: Minerals in soil environments. Wiley, pp 379–438
Setlow P (2007) I will survive: DNA protection in bacterial spores. Trends Microbiol 15(4):172–180. https://doi.org/10.1016/j.tim.2007.02.004
Sherbet GV, Lakshmi MS, Cajone F (1983) Isoelectric characteristics and the secondary structure of some nucleic acids. Biophys Struct Mechanism 10, 121–128. https://doi.org/10.1007/BF00537554
Simon M, Cho BC, Azam F (1992) Significance of bacterial biomass in lakes and the ocean: comparison to phytoplankton biomass and biogeochemical implications. Mar Ecol Prog Ser 86(2):103–110
Singer VL, Jones LJ, Yue ST, Haugland RP (1997) Characterization of PicoGreen reagent and development of a fluorescence-based solution assay for double-stranded DNA quantitation. Analytical biochemistry, 249(2):228–238
Sjögren P, Edwards ME, Gielly L, Langdon CT, Croudace IW, Merkel MKF, Fonville T, Alsos IG (2017) Lake sedimentary DNA accurately records 20th century introductions of exotic conifers in Scotland. New Phytol 213(2):929–941. https://doi.org/10.1111/nph.14199
Slon V, Hopfe C, Weiß CL, Mafessoni F, de la Rasilla M, Lalueza-Fox C, Rosas A, Soressi M, Knul MV, Miller R, Stewart JR, Derevianko AP, Jacobs Z, Li B, Roberts RG, Shunkov MV, de Lumley H, Perrenoud C, Gušić I, Kućan Ž, Rudan P, Aximu-Petri A, Essel E, Nagel S, Nickel B, Schmidt A, Prüfer K, Kelso J, Burbano HA, Pääbo S, Meyer M (2017) Neandertal and Denisovan DNA from Pleistocene sediments. Science 356(6338):605–608. https://doi.org/10.1126/science.aam9695
Smith CI, Chamberlain AT, Riley MS, Cooper A, Stringer CB, Collins MJ (2001) Not just old but old and cold? Nature 410(6830):771–772. https://doi.org/10.1038/35071177
Sodnikar K, Parker KM, Stump SR, ThomasArrigo LK, Sander M (2021) Adsorption of double-stranded ribonucleic acids (dsRNA) to iron (oxyhydr-)oxide surfaces: comparative analysis of

model dsRNA molecules and deoxyribonucleic acids (DNA). Environ Sci: Processes Impacts 23(4):605–620. https://doi.org/10.1039/D1EM00010A

Stoeck T, Kochems R, Forster D, Lejzerowicz F, Pawlowski J (2018) Metabarcoding of benthic ciliate communities shows high potential for environmental monitoring in salmon aquaculture. Ecol Indic 85:153–164. https://doi.org/10.1016/j.ecolind.2017.10.041

Stoof-Leichsenring KR, Epp LS, Trauth MH, Tiedemann R (2012) Hidden diversity in diatoms of Kenyan Lake Naivasha: a genetic approach detects temporal variation. Mol Ecol 21(8):1918–1930. https://doi.org/10.1111/j.1365-294X.2011.05412.x

Strickler KM, Fremier AK, Goldberg CS (2015) Quantifying effects of UV-B, temperature, and pH on eDNA degradation in aquatic microcosms. Biol Conserv 183:85–92. https://doi.org/10.1016/j.biocon.2014.11.038

Sutton R, Sposito G (2005) Molecular structure in soil humic substances: the new view. Environ Sci Technol 39(23):9009–9015. https://doi.org/10.1021/es050778q

Tabares X, Zimmermann H, Dietze E, Ratzmann G, Belz L, Vieth-Hillebrand A, Dupont L, Wilkes H, Mapani B, Herzschuh U (2020) Vegetation state changes in the course of shrub encroachment in an African savanna since about 1850 CE and their potential drivers. Ecol Evol 10(2):962–979. https://doi.org/10.1002/ece3.5955

Taberlet P, Prud'homme SM, Campione E, Roy J, Miquel C, Shehzad W, Gielly L, Rioux D, Choler P, Clément J-C, Melodelima C, Pompanon F, Coissac E (2012) Soil sampling and isolation of extracellular DNA from large amount of starting material suitable for metabarcoding studies. Mol Ecol 21(8):1816–1820. https://doi.org/10.1111/j.1365-294X.2011.05317.x

Thevenon F, Adatte T, Spangenberg JE, Anselmetti FS (2012) Elemental (C/N ratios) and isotopic (δ15Norg, δ13Corg) compositions of sedimentary organic matter from a high-altitude mountain lake (Meidsee, 2661 m a.s.l., Switzerland): implications for Lateglacial and Holocene alpine landscape evolution. The Holocene:1135–1142. https://doi.org/10.1177/0959683612441841

Thomsen PF, Willerslev E (2015) Environmental DNA – an emerging tool in conservation for monitoring past and present biodiversity. Biol Conserv 183:4–18. https://doi.org/10.1016/j.biocon.2014.11.019

Thomson-Laing G, Howarth JD, Vandergoes MJ, Wood SA (2022) Optimised protocol for the extraction of fish DNA from freshwater sediments. Freshw Biol 67(9):1584–1603. https://doi.org/10.1111/fwb.13962

Torti A, Lever MA, Jørgensen BB (2015) Origin, dynamics, and implications of extracellular DNA pools in marine sediments. Mar Genomics 24:185–196. https://doi.org/10.1016/j.margen.2015.08.007

Torti A, Jørgensen BB, Lever MA (2018) Preservation of microbial DNA in marine sediments: insights from extracellular DNA pools. Environ Microbiol 20(12):4526–4542. https://doi.org/10.1111/1462-2920.14401

Turner CR, Uy KL, Everhart RC (2015) Fish environmental DNA is more concentrated in aquatic sediments than surface water. Biol Conserv 183:93–102. https://doi.org/10.1016/j.biocon.2014.11.017

Ushio M, Fukuda H, Inoue T, Makoto K, Kishida O, Sato K, Murata K, Nikaido M, Sado T, Sato Y, Takeshita M, Iwasaki W, Yamanaka H, Kondoh M, Miya M (2017) Environmental DNA enables detection of terrestrial mammals from forest pond water. Mol Ecol Resour 17(6):e63–e75. https://doi.org/10.1111/1755-0998.12690

Valdez-Moreno M, Ivanova NV, Elías-Gutiérrez M, Pedersen SL, Bessonov K, Hebert PDN (2019) Using eDNA to biomonitor the fish community in a tropical oligotrophic lake. PLoS One 14(4):e0215505. https://doi.org/10.1371/journal.pone.0215505

Vivien RL, Apotheloz-Perret-Gentil L, Pawlowski JW, Werner I, Ferrari BJD (2019) Testing different (e)DNA metabarcoding approaches to assess aquatic oligochaete diversity and the biological quality of sediments. Ecol Indic 106. https://doi.org/10.1016/j.ecolind.2019.105453

Von Eggers J, Monchamp M-E, Capo E, Giguet-Covex C, Spanbauer T, Heintzman PD (2022) Inventory of ancient environmental DNA from sedimentary archives: locations, methods, and target taxa. Zenodo. https://doi.org/10.5281/ZENODO.6847522

Vuillemin A, Ariztegui D, De Coninck AS, Lücke A, Mayr C, Schubert CJ, The PASADO Scientific Team (2013) Origin and significance of diagenetic concretions in sediments of Laguna Potrok Aike, southern Argentina. J Paleolimnol 50(3):275–291. https://doi.org/10.1007/s10933-013-9723-9

Vuillemin A, Friese A, Alawi M, Henny C, Nomosatryo S, Wagner D, Crowe SA, Kallmeyer J (2016a) Geomicrobiological features of ferruginous sediments from Lake Towuti. Indonesia Front Microbiol 7:1007. https://doi.org/10.3389/fmicb.2016.01007

Vuillemin A, Ariztegui D, Leavitt PR, Bunting L, The PASADO Science Team (2016b) Recording of climate and diagenesis through sedimentary DNA and fossil pigments at Laguna Potrok Aike, Argentina. Biogeosciences 13(8):2475–2492. https://doi.org/10.5194/bg-13-2475-2016

Vuillemin A, Horn F, Alawi M, Henny C, Wagner D, Crowe SA, Kallmeyer J (2017) Preservation and significance of extracellular DNA in ferruginous sediments from Lake Towuti, Indonesia. Front Microbiol 8

Wang Y, Pedersen MW, Alsos IG, De Sanctis B, Racimo F, Prohaska A, Coissac E, Owens HL, Merkel MKF, Fernandez-Guerra A, Rouillard A, Lammers Y, Alberti A, Denoeud F, Money D, Ruter AH, McColl H, Larsen NK, Cherezova AA, Edwards ME, Fedorov GB, Haile J, Orlando L, Vinner L, Korneliussen TS, Beilman DW, Bjørk AA, Cao J, Dockter C, Esdale J, Gusarova G, Kjeldsen KK, Mangerud J, Rasic JT, Skadhauge B, Svendsen JI, Tikhonov A, Wincker P, Xing Y, Zhang Y, Froese DG, Rahbek C, Bravo DN, Holden PB, Edwards NR, Durbin R, Meltzer DJ, Kjær KH, Möller P, Willerslev E (2021) Late quaternary dynamics of Arctic biota from ancient environmental genomics. Nature 600(7887):86–92. https://doi.org/10.1038/s41586-021-04016-x

Willerslev E, Cooper A (2005) Review paper. Ancient DNA. Proc R Soc B Biol Sci 272(1558):3–16. https://doi.org/10.1098/rspb.2004.2813

Willerslev E, Hansen AJ, Binladen J, Brand TB, Gilbert MTP, Shapiro B, Bunce M, Wiuf C, Gilichinsky DA, Cooper A (2003) Diverse plant and animal genetic records from Holocene and Pleistocene sediments. Science 300(5620):791–795. https://doi.org/10.1126/science.1084114

Willerslev E, Hansen AJ, Poinar HN (2004) Isolation of nucleic acids and cultures from fossil ice and permafrost. Trends Ecol Evol 19(3):141–147. https://doi.org/10.1016/j.tree.2003.11.010

Williams KE, Huyvaert KP, Vercauteren KC, Davis AJ, Piaggio AJ (2018) Detection and persistence of environmental DNA from an invasive, terrestrial mammal. Ecol Evol 8(1):688–695. https://doi.org/10.1002/ece3.3698

Yetgin S, Balkose D (2015) Calf thymus DNA characterization and its adsorption on different silica surfaces. RSC Adv 5(71):57950–57959. https://doi.org/10.1039/C5RA01810B

Yoccoz NG (2012) The future of environmental DNA in ecology. Mol Ecol 21(8):2031–2038. https://doi.org/10.1111/j.1365-294X.2012.05505.x

Yoccoz NG, Bråthen KA, Gielly L, Haile J, Edwards ME, Goslar T, Von STEDINGKH, Brysting AK, Coissac E, Pompanon F, Sønstebø JH, Miquel C, Valentini A, De BELLOF, Chave J, Thuiller W, Wincker P, Cruaud C, Gavory F, Rasmussen M, Gilbert MTP, Orlando L, Brochmann C, Willerslev E, Taberlet P (2012) DNA from soil mirrors plant taxonomic and growth form diversity. Mol Ecol 21(15):3647–3655. https://doi.org/10.1111/j.1365-294X.2012.05545.x

Zavala EI, Jacobs Z, Vernot B, Shunkov MV, Kozlikin MB, Derevianko AP, Essel E, de Fillipo C, Nagel S, Richter J, Romagné F, Schmidt A, Li B, O'Gorman K, Slon V, Kelso J, Pääbo S, Roberts RG, Meyer M (2021) Pleistocene sediment DNA reveals hominin and faunal turnovers at Denisova Cave. Nature 595(7867):399–403. https://doi.org/10.1038/s41586-021-03675-0

Zinger L, Bonin A, Alsos IG, Bálint M, Bik H, Boyer F, Chariton AA, Creer S, Coissac E, Deagle BE, De Barba M, Dickie IA, Dumbrell AJ, Ficetola GF, Fierer N, Fumagalli L, Gilbert MTP, Jarman S, Jumpponen A, Kauserud H, Orlando L, Pansu J, Pawlowski J, Tedersoo L, Thomsen PF, Willerslev E, Taberlet P (2019) DNA metabarcoding – need for robust experimental designs to draw sound ecological conclusions. Mol Ecol 28(8):1857–1862. https://doi.org/10.1111/mec.15060

Chapter 3
The Sedimentary Ancient DNA Workflow

Peter D. Heintzman, Kevin Nota, Alexandra Rouillard, Youri Lammers, Tyler J. Murchie, Linda Armbrecht, Sandra Garcés-Pastor, and Benjamin Vernot

Keywords Sedimentary ancient DNA · Paleogenomics · Ancient metagenomics · Metabarcoding · Contamination

Introduction

Sedimentary ancient DNA (*sed*aDNA) is continuing to revolutionise our understanding of past biological and geological processes by retrieving and analysing the ancient DNA preserved in lake, cave, open terrestrial, midden, permafrozen, and marine environments (Crump 2021). The study of *sed*aDNA began in the late 1990s (Coolen and Overmann 1998) with the first reports of extinct animal *sed*aDNA in 2003 (Hofreiter et al. 2003; Willerslev et al. 2003). Since then, it has been shown that *sed*aDNA can be recovered at high resolution from recent (10^1–10^2 year-old) (e.g., Capo et al. 2017) through to deep-time (10^5–10^6 year-old) sediments from a vast diversity of environments (Crump et al. 2021; Zavala et al. 2021; Armbrecht

P. D. Heintzman (✉)
Department of Geological Sciences, Stockholm University, Stockholm, Sweden

Centre for Palaeogenetics, Stockholm, Sweden
e-mail: peter.d.heintzman@geo.su.se

K. Nota · B. Vernot
Department of Evolutionary Genetics, Max Planck Institute for Evolutionary Anthropology, Leipzig, Germany
e-mail: kevin_nota@eva.mpg.de; benjamin_vernot@eva.mpg.de

A. Rouillard
Department of Geosciences, UiT – The Arctic University of Norway, Tromsø, Norway

Section for GeoGenetics, Globe Institute, University of Copenhagen, Copenhagen, Denmark
e-mail: alexandra.rouillard@uit.no

E. Capo et al. (eds.), *Tracking Environmental Change Using Lake Sediments*, Developments in Paleoenvironmental Research 21,
https://doi.org/10.1007/978-3-031-43799-1_3

et al. 2022; Kjær et al. 2022). Unlike traditional palaeoenvironmental and palaeoecological proxies, *sed*aDNA is unique in that it is derived from any type of organism that was present in the local area and that may contain population-level information. This latter characteristic means that, unlike any other comparable proxy, *sed*aDNA can be used for evolutionary analyses (Gelabert et al. 2021; Lammers et al. 2021; Pedersen et al. 2021; Vernot et al. 2021).

The advent of next-generation sequencing (NGS) provided massively enlarged, yet economically feasible, dataset sizes and increased analytical sensitivity that has allowed the field to flourish by generating robust datasets in which contamination can be detected and controlled, and hypotheses can be tested. Coupled with ongoing methodological innovations in both molecular data generation and bioinformatics analysis techniques, NGS has driven the exponential growth in *sed*aDNA research over the past two decades.

In this chapter, we present an overview of the state-of-the-art for the *sed*aDNA workflow. We do not present detailed methodologies and descriptions, as these have been published elsewhere (e.g., Armbrecht et al. 2019; Capo et al. 2021 and references therein). For each step in the workflow, from ethical considerations during experimental design to environmental and evolutionary inferences, we instead outline the general rationale for conducting the step, a brief overview of the approach and methods involved, pros and cons, key pitfalls, and how the current state-of-the-art is likely to develop in the near future. Importantly, we highlight that molecular and bioinformatic methods (i.e., steps presented from section 'DNA extraction' onwards) are still developing due to the relative infancy of the field.

Y. Lammers
The Arctic University Museum of Norway, UiT – The Arctic University of Norway, Tromsø, Norway
e-mail: youri.lammers@uit.no

T. J. Murchie
McMaster Ancient DNA Centre, Department of Anthropology, McMaster University, Hamilton, ON, Canada

Hakai Institute, Heriot Bay, BC, Canada
e-mail: murchiet@mcmaster.ca

L. Armbrecht
Institute for Marine and Antarctic Studies (IMAS), Ecology and Biodiversity Centre, University of Tasmania, Hobart, Australia
e-mail: linda.armbrecht@utas.edu.au

S. Garcés-Pastor
Marine Sciences Institute, ICM-CSIC, Barcelona, Spain
e-mail: sandragarces@ub.edu

Ethical Considerations

Sources of environmental sediments potentially amenable to *sed*aDNA analysis are ubiquitous and global, thereby providing ample material for a multitude of research questions. Importantly, however, there are ethical, political, and/or legal considerations that must be accounted for during the study design and data collection stages of a project (Bardill et al. 2018). This includes abiding by international agreements on the use and sharing of genetic resources such as the Nagoya protocol (Buck and Hamilton 2011).

Permits from national/local authorities and/or local landowners are often required to perform the fieldwork necessary to collect sediments. If sediments and/or the extracted DNA are to be shipped internationally, as is often the case in current *sed*aDNA research, then export and/or import permits will be required (e.g., from the United States Department of Agriculture or Australian Government). In many regions of the world, especially with legacies of colonialism, there is a strong need to engage with key stakeholders: the local indigenous communities (Handsley-Davis et al. 2021; Kowal et al. 2023). This is particularly important for *sed*aDNA research, given that questions often revolve around climate, environmental, and land use changes on scales that are of relevance to indigenous peoples.

One of the ethically problematic aspects of *sed*aDNA research that has been most discussed in the literature is the potential for recovering DNA from humans (*Homo sapiens*), of which many metagenomic and animal-targeting assays are both able and prone to detect (see section 'Data generation strategies'). Although most, if not all, human DNA recovered from some sedimentary archives, such as lake, marine, and permafrost, likely represents modern-day contamination, there is potential for authentic human *sed*aDNA to be recovered. In other sedimentary contexts, such as caves and archaeological sites, authenticated human *sed*aDNA sequences that are tens of thousands of years old have been reported (e.g., Braadbaart et al. 2020; Zavala et al. 2021). If there will be the possibility of recovering *sed*aDNA of humans, or other culturally important taxa, then, where applicable, it is crucial to consult with indigenous communities during the consultation stage of experimental design (Alpaslan-Roodenberg et al. 2021; Handsley-Davis et al. 2021; Kowal et al. 2023). A potential solution to dealing with the undesired recovery of human *sed*aDNA is to use a bioinformatic filtering step to remove human, or potentially human, sequences without further analysis. However, we emphasise that the ethics of *sed*aDNA, and particularly human DNA recovered from sediment, are an active area of discussion that is only in the early stages of conversation.

Fieldwork

The general logistics of fieldwork for a *sed*aDNA study are similar to those used for the collection of sediments for other, more traditional proxies. This includes excavations in natural and archaeological settings, horizontal coring from accessible

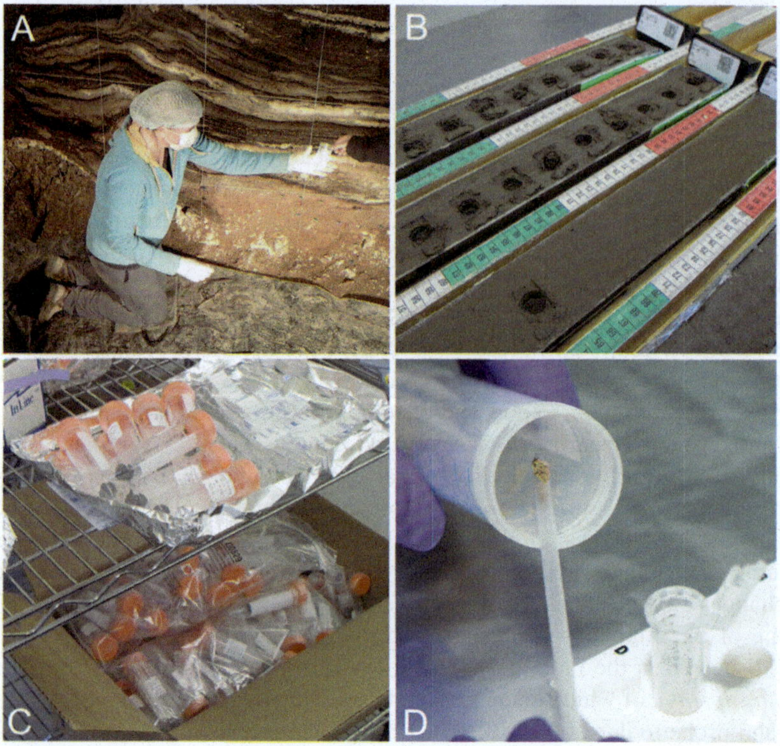

Fig. 3.1 Examples of (**a**) researchers sampling an accessible cave sediment profile in the field; (**b**) a freshly subsampled lake core, where the rectangular areas represent discarded sediment surface for access to underlying pristine sediment, and the circular holes correspond to the material that was sampled for DNA analysis. Alternative core sectioning and subsampling strategies can be used for higher resolution analyses; (**c**) subsamples that were taken in a non-cleanroom space; and (**d**) a subsample to be used as input for DNA extraction in a cleanroom space. Images are copyright of Dr. Richard G. Roberts (**a**), Dr. Peter D. Heintzman (**b**, **c**), and the Max Planck Institute for Evolutionary Anthropology (**d**)

profiles such as permafrost, soil, and cave settings (Fig. 3.1a), or vertical coring from inaccessible profiles such as lake, marine and ice systems. The key difference from traditional fieldwork is that steps must be taken to prevent or reduce contamination by human or environmental DNA during sample collection from the sediment profile. Due to the high sensitivity of the *sed*aDNA technique, inferences are particularly prone to sources of contamination, both by modern-day environmental DNA and from sediment cross-contamination that can occur during sample collection and handling.

Contamination from modern-day environmental DNA, including from the persons performing the fieldwork, is of paramount concern when taking sediments from excavation and accessible profile contexts. Standard ancient DNA precautions can be adopted to reduce this type of contamination, including the use of sterile utensils and storage bags/tubes, and personal protective equipment such as

Table 3.1 Taxon-specific methodological considerations for the *seda*DNA workflow, with signposts to other chapters of this volume

Group	Fresh cores recommended	Contamination potential	Suitable for Amplicon	Suitable for Metagenomics[d]	Further reading
Prokaryotes	Yes	High	No[b]	Yes	Chapters 4, 5
Microbial eukaryotes	Yes	High	No[b]	Yes	Chapters 6, 7
Fungi	Yes	High	Yes[c]	Yes	Chapter 6
Wildlife[a]	No	Low	Yes[c]	Yes	Chapters 8–11
Domesticated/ farmed taxa[a]	No	Intermediate	Yes[c]	Yes	Chapters 10, 11
Human	Yes	High	No	Yes	Chapter 11

aWildlife and domesticated/farmed taxa include both animal and plant taxa
bAmplicon methods may be appropriate for taxa that are unable to have colonised the sediment after collection or be present in the modern sampling or laboratory environment
cData are prone to contamination and so need to be interpreted carefully. See also Chap. 11
dThe feasibility of metagenomics may be limited by the availability of reference genome data for certain taxonomic groups

facemask, gloves, and a full-body hazmat suit (Fulton and Shapiro 2019). Modern-day environmental DNA contamination can be monitored by the use of field controls, whereby sterile tubes, either empty or containing a storage buffer/water, are opened during sediment collection. An alternative approach, especially for inaccessible profiles, is to pre-sterilise coring equipment and/or coat the inside of a coring tube with a known exotic DNA, or synthetic tracer, and monitor for its presence during downstream analyses (Armbrecht et al. 2019; Graham et al. 2016; Pedersen et al. 2016; Rijal et al. 2021). For highly sensitive applications, such as *seda*DNA analysis of environmental microorganisms (Table 3.1), a pipe-based coring system, such as a Livingstone, Nesje (Nesje 1992), UWITEC (https://www.uwitec.at/en/), or Kasten, is better suited to preventing modern-day microbes from colonising the sediment. All sediment cores or samples should be immediately sealed from the external environment after collection to prevent desiccation and ongoing contamination from modern-day environmental DNA, and to maintain a cold, dark and, if possible, anoxic environment to reduce post-sampling DNA degradation and secondary growth.

Cross-contamination, whereby sediment from a different part of the profile is incorporated into a sample during collection, can be limited by systematically sampling accessible profiles from bottom to top. This reduces the chance of loose sediments displaced during sampling from causing cross-contamination. In inaccessible profiles, cross-contamination can occur if cores are taken in a series of drives, as is standard with the Livingstone coring system, and so results from the tops of core drives should be interpreted cautiously. If possible, taking a single long-core, such as with a Nesje-style or UWITEC system, avoids this type of cross-contamination from inaccessible profiles (e.g., Armbrecht et al. 2019).

In excavations and accessible profiles, it is common practice to avoid both types of contamination in the field by removing the outer layer of exposed sediment, consisting of several mm to cm of the surface depending on context, using sterile implements (see also section 'Subsampling'). This allows for pristine sediment samples to be taken from inside the profile (ideally still frozen in the case of permafrost), which can be achieved using either a new set of sterile implements or by inserting a sampling tube directly into the sediment profile (e.g. Ardelean et al. 2020; Vernot et al. 2021).

Sediment Storage

The storage conditions for collected sediments depends on a variety of factors, including logistic availability and the intended downstream analyses. For intact sediment cores taken from lake or marine non-frozen contexts, refrigerated storage is preferred because freezing can adversely affect stratigraphic integrity as well as being impractical given the total sediment mass. Cores for *sed*aDNA analysis are therefore usually stored in cold rooms at 4 °C, with studies reporting successful *sed*aDNA retrieval from cores stored for >20 years at this temperature (e.g., Seeber et al. 2022). For some applications, freezing is necessary to maintain sediment integrity. Such occasions include, for example, the preservation of permafrost sediment plugs and when sediments are subsampled for *sed*aDNA analysis immediately after coring (e.g., on board research vessels, Armbrecht et al. 2019). However, in such circumstances, complementary archive cores are often taken to allow for future follow-up study (Selway et al. 2022). If cores are frozen, it is important to keep freeze-thaw cycles to a minimum as these can negatively impact the integrity of *sed*aDNA molecules by allowing for microbial-induced degradation (e.g., case study 2 in Capo et al. 2021) or through physical shearing during ice crystal formation (Shao et al. 2012; Ross et al. 1990). However, the degree to which freeze-thaw meaningfully degrades *sed*aDNA remains somewhat unresolved as others have found negligible impacts from repeated freeze/thaw cycles (Safarikova et al. 2021).

Evidence for the detrimental effects of X-rays on *sed*aDNA within sediments, such as those conducted during transport at airports and customs or on the intact cores for multi-proxy analyses of the sediment (ITRAX or μCT scanning), is currently lacking. To our knowledge, retrieval of *sed*aDNA has been successful following X-rays during transport, but should likely be minimised where possible. The impacts of X-rays on sub-fossil bone ancient DNA during μCT scanning have been found to be particularly damaging in wet and/or frozen materials (Immel et al. 2016). We therefore recommend limiting resolution and total exposure to X-rays prior to *sed*aDNA subsampling (Immel et al. 2016).

Sediment cores and samples are often stored in specialist facilities, such as the LacCore facility in Minnesota (Schnurrenberger et al. 2001), the Permafrost ArChives Science (PACS) Laboratory at the University of Alberta, core repositories of the International Ocean Discovery Program (IODP, https://www.iodp.org/resources/core-repositories), or various institutional archives. These archival facilities not only allow for the long-term preservation of these sediments, but provide material to allow for follow-up and reproducible research following the Findability, Accessibility, Interoperability, and Reusability (FAIR) principles (Wilkinson et al. 2016).

Sedimentary ancient DNA has been successfully recovered from sediments collected years, or even decades, prior to analysis (e.g., Massilani et al. 2022; Rijal et al. 2021; Wang et al. 2021), thereby demonstrating the value of utilising existing collections for *sed*aDNA research. However, caution should be taken when using such sediments for the study of microbes and fungi, due to the potential for secondary growth of modern-day contaminants on, and potentially within, the sediments (Selway et al. 2022). In some instances, these modern organisms can overwhelm the *sed*aDNA signal below meaningful detection levels (see also section 'Identification and validation: Validation'). Where possible, sediment cores should remain sealed with plastic caps and duct-tape until immediately prior to *sed*aDNA subsampling. This is to prevent microbial and fungal growth by maintaining either an air-tight or anoxic environment within the core.

Subsampling

The goal of subsampling is to remove a small portion of sediment core or sample, either to be used directly in a DNA extraction (~50 mg to 1 g) or to provide a larger stock of subsample suitable for multiple extractions (~10 to 50 g) (Fig. 3.1b,c).

Subsampling is the first step in the workflow to be conducted ideally in a specialised clean lab, and should always follow standard anti-contamination precautions (Epp et al. 2019; Fulton and Shapiro 2019). Recommended measures include bleach and ultraviolet (UV) treatment of work surfaces and equipment, use of sterile utensils and personal protective equipment (see also section 'Sediment storage'), working within a HEPA-filtered positive-pressured atmosphere, and the one-way movement of personnel and equipment from the clean-room environment to laboratories that use high-copy and/or amplification-based molecular biology procedures (see section 'Data generation strategies: Metagenomics'). If access to a clean lab is not possible for subsampling, then the work environment should be made as clean as is practical and located in an area without active ventilation and where no previous molecular biology procedures have been conducted. Subsampling negative controls should be used to monitor for modern-day environmental DNA contamination in a similar manner as described in the section 'Fieldwork', particularly when subsampling in non-cleanroom environments.

A detailed step-by-step guide for subsampling frozen and non-frozen sediment cores can be found in Epp et al. (2019). Prior to and immediately after entry into the cleanroom facility, the external surfaces of sediment sample bags/tubes and core tubes should be decontaminated with a bleach solution. We emphasise that, whilst UV may be used to decontaminate the surface of a subsample, chemical decontamination, such as the use of bleach, should not be used as *sed*aDNA within the subsample will not be protected from chemically-induced degradation. An alternative approach to remove surface contamination from blocks of sediment, such as unthawed permafrost or sediment cores, is to sequentially remove one or two exterior layers so that a previously undisturbed sector of the sediment is subsampled similarly to that described in section 'Fieldwork' (Fig. 3.1b; Graham et al. 2016; Pedersen et al. 2016; Parducci et al. 2017). The implements used to collect subsamples will depend on the sediment characteristics, such as hardness and consistency, but, in all cases, sediment in direct contact with tube edges should be avoided. If a tracer was applied during coring, then subsamples should be taken from the exterior of the core to confirm the presence of the tracer. Surface cores, such as those obtained by gravity coring devices, often contain unconsolidated and/or watery sediment that may prove challenging to subsample cleanly, but protocols are available (e.g., Pawlowski et al. 2022). Once taken, it is standard practice to store subsamples frozen at −20 °C in an ancient DNA laboratory in preparation for DNA extraction.

DNA Extraction

The isolation and purification of *sed*aDNA from a sediment subsample is known as DNA extraction. This should always be performed in cleanroom conditions and include negative controls consisting of all steps and reagents used in the extraction protocol but with no sediment input (Fig. 3.2). Sediment input masses vary between protocols, but can be as low as 50 mg and as high as 15 g (e.g., Ficetola et al. 2018; Massilani et al. 2022, Fig. 3.1d).

DNA extraction involves three major steps: disintegration, inhibitor removal, and purification (e.g., Epp et al. 2019). The disintegration step frees DNA molecules that are either mineral-bound (extracellular) or still associated with organic material in tissues or small clumps of cells (intracellular), using physical, chemical, and/or enzymatic reagents to break up the sediment and its organic components. Inhibitor removal steps use chemical or physical filtering to remove complex organic molecules, such as cellular debris, humic acids and polysaccharides, that can inhibit the downstream enzymatic molecular biology procedures outlined in section 'Data generation strategies'. Inhibitor removal can be non-trivial, as humic acids, for example, are biochemically similar to DNA for extraction purposes. This means that aggressive inhibitor removal can also remove *sed*aDNA molecules, whereas too lenient inhibitor removal results in the co-extraction of inhibitors (e.g., Murchie et al. 2021). Consequently, inhibitor removal may be omitted from the extraction of DNA from some sediment types, such as those with a low organic content where inhibitor content is expected to be low. Purification of the *sed*aDNA involves the removal of remaining cellular debris and reagents from previous steps. This can be achieved through binding the *sed*aDNA to silica, which is then washed and the purified DNA released (e.g., Epp et al. 2019), or through the use of organic phase separation (e.g., Wang et al. 2021). The resulting *sed*aDNA extract should be stored

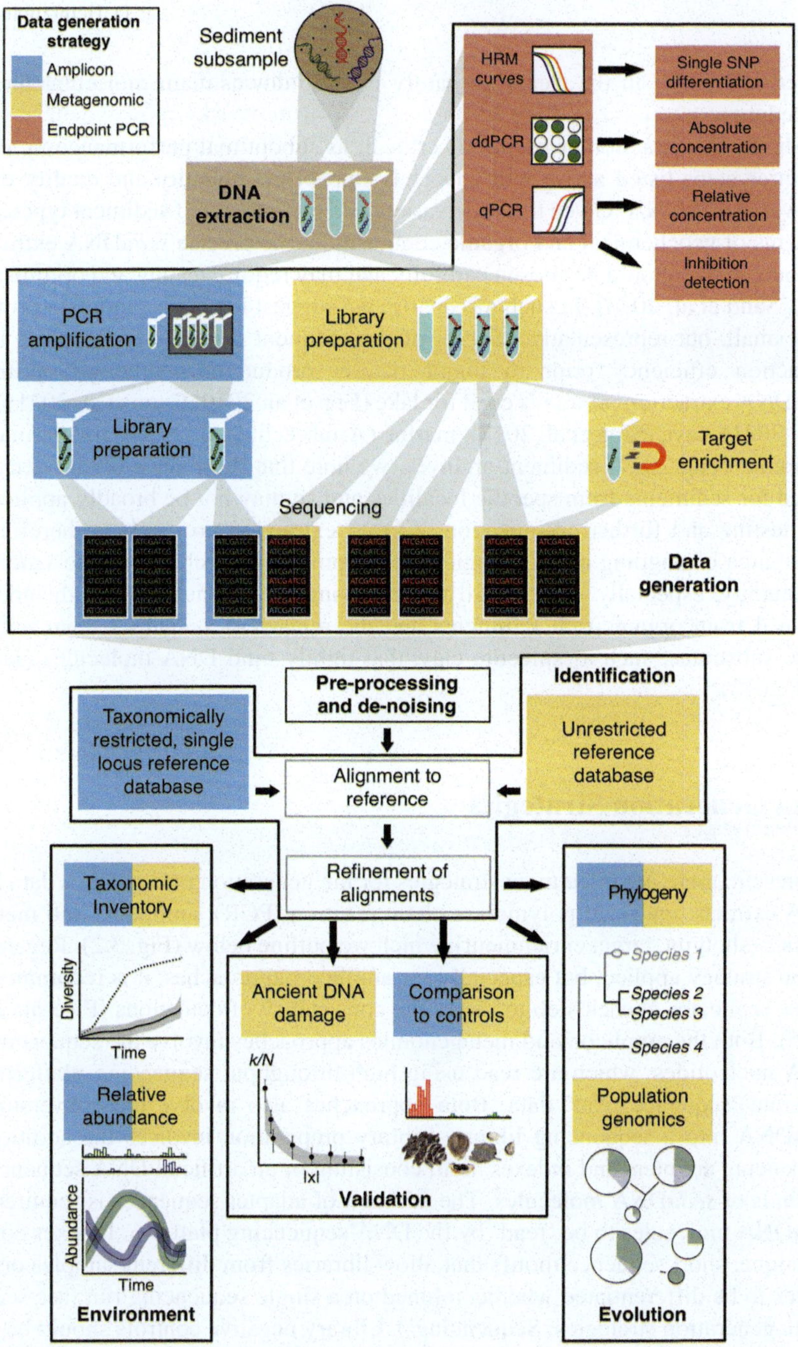

Fig. 3.2 A schematic of the complete *sed*aDNA workflow after subsampling. The main steps are listed in bold. (The figure is modified from Williams et al. (2023), with illustrations of validation, environmental, and evolutionary analyses inspired by Kjær et al. (2022), Rijal et al. (2021), and Wang et al. (2021). Note that library preparation involves an indexing PCR amplification step. Acronyms are defined in the text. Source credit for macrofossil images: Dr. Alexandra Rouillard)

frozen at −20 °C in the clean laboratory prior to downstream molecular biology procedures.

The DNA extraction procedure is critical, as suboptimal performance in any of the three steps listed above will adversely impact the quantity and quality of the recovered *sed*aDNA. Given the wide variety and complexity of sediment types, both in terms of geochemical and organic composition, the precise *sed*aDNA extraction protocol to use should be chosen carefully and may require testing and optimisation (e.g., Sand et al. 2023). In such a scenario, we suggest testing a candidate protocol on a small, but representative, subset of the sediment subsamples to assess DNA extraction efficiency prior to major dataset production. Although optimised *sed*aDNA extraction protocols exist for lake (Epp et al. 2019; Capo et al. 2021; Rijal et al. 2021), cave (Slon et al. 2017), marine (Armbrecht et al. 2020), and permafrost (Murchie et al. 2021) sediment archives, we note that these have often been optimised for sediments from specific localities and so may not be broadly applicable. The testing and further optimisation of DNA extraction protocols is therefore an active area of ongoing methodological development research within the *sed*aDNA community, especially with regard to improving the efficiency of both inhibitor removal from organic-rich sediments and the release of *sed*aDNA from minerogenic substrates, such as smectite clay, that tightly bind DNA molecules (Jelavic et al. 2023).

Data Generation Strategies

At present, there are three main strategies for the generation of *sed*aDNA data from DNA extracts: endpoint polymerase chain reaction (PCR), amplicon, and metagenomics (shotgun, target enrichment), which we outline below (Fig. 3.2). Regardless of the strategy applied, but especially for amplicon approaches, it is recommended to use replicates at each step to verify the consistency of detections (Ficetola et al. 2015). Both the amplicon and metagenomics approaches involve the sequencing of DNA nucleotides, which are read using high-throughput sequencing platforms to generate 'sequence read' data. Both approaches also involve the conversion of *sed*aDNA into a sequencing library. Library preparation involves the addition of sequencing adapters and indexes, both consisting of an artificial DNA sequence, to the ends of *sed*aDNA molecules. The addition of adapter sequences is required for *sed*aDNA molecules to be 'read' by the DNA sequencing platform. Indexes consist of unique, short sequence motifs that allow libraries from different samples or replicates to be differentiated when combined on a single sequencing run (see section 'Data generation strategies: Sequencing'). Library negative controls should be generated in parallel with libraries from DNA extracts.

Endpoint PCR

High-throughput sequencing may not be required for some research questions, such as quantifying template abundance, quantifying inhibition, detecting the presence of a single taxon, or discriminating between closely related genotypes in a *sed*aDNA extract. Instead, taxon- or gene function-specific detection assays, or assays that can discriminate between a small number of targets, might be more appropriate. Such methods are not frequently used for detecting higher organisms (Nota et al. 2022), but are relatively common for quantifying microorganisms (e.g., Pal et al. 2015; Pilon et al. 2019; Picard et al. 2022; Nwosu et al. 2023).

Endpoint PCR methods rely on designing taxon- or functional gene-specific primers (short pre-designed sequences used to initiate PCR) with high specificity during amplification. Using quantitative PCR (qPCR) or droplet digital PCR (ddPCR), it is possible to estimate the original number of template molecules present. The ddPCR approach has the advantage over qPCR in that standard curves are not required for quantification and might improve template amplification in inhibited samples (Capo et al. 2021). When the aim is to detect a single taxon or to distinguish a small number of taxa, qPCR has been used together with (high resolution) melting curve analysis (HRM). Melting curve analysis takes advantage of denaturing DNA from double to single-stranded molecules, which is a function of molecule length and GC content. For performing melting curve analysis, enough variation between amplicons is required to create distinct melting peaks. It has been shown that even single nucleotide polymorphisms (SNPs) can be genotyped using melting curve analysis and specific high-resolution instruments (Liew et al. 2004). For *sed*aDNA applications, melting curves have been used to assess the specificity of qPCR amplicons (Savichtcheva et al. 2011; Domaizon et al. 2013; Nwosu et al. 2023) and for genotyping (Nota et al. 2022).

For inhibition quantification, endpoint PCR can be used by spiking a *sed*aDNA extract into a PCR designed to amplify an unrelated template, such as a synthetic control or lambda phage (Warinner et al. 2014; Murchie et al. 2021). Assays with *sed*aDNA extract inhibition will exhibit an increased cycle threshold (Ct) value, due to impeded exponential amplification during qPCR. By comparing *sed*aDNA extracts in dilution series to positive controls (i.e., with no *sed*aDNA extract spike), relative inhibition levels can be determined between *sed*aDNA extracts.

Caution is required using endpoint PCR approaches for template quantification and SNP detection due to the potential amplification of non-target DNA from unrelated organisms. The specificity of endpoint PCR is generally tested on DNA from present-day environments whereby results can be validated by independent observations. This is not possible for *sed*aDNA. It is therefore strongly advised to verify that the signal is coming from the expected target by sequencing a subset of the positive PCR amplifications. The validation of qPCR results may be facilitated with melting curves, although different sequences can give rise to the same melting curve profile.

Amplicon

Amplicon methods involve the high-throughput sequencing of PCR amplicons that are either generated using a generic primer set, which amplify templates from a broad spectrum of taxa or functional genes (metabarcoding PCR), or PCRs that simultaneously use multiple primer sets (multiplex PCR). Primer sequences are often extended with unique, short motifs (tags) that allow for amplicons to be assignable to their original PCR in experiments where multiple PCR products are pooled together prior to library preparation (Binladen et al. 2007).

Metabarcoding PCR is used to simultaneously amplify *sed*aDNA templates from broad groups such as vascular plants or vertebrates, and thereby assess the diversity and composition of taxa represented within the *sed*aDNA extract (Fig. 3.2; Capo et al. 2021). This robust, relatively cheap, and widely used technique is based on the barcoding principle whereby a pair of generic primers bind to separate parts of the DNA template, which are conserved across the group of interest, and flank a variable region of the template, the 'barcode' region, that allows for discrimination of species or genera (Nichols et al. 2019; Taberlet et al. 2018). As metabarcoding requires intact *sed*aDNA fragments that include the primer binding sites, there is a trade-off between using a barcode that is short enough to be compatible with degraded ancient DNA (typically <150 bp) yet long enough to be taxonomically informative (e.g. Taberlet et al. 2007). Metabarcoding is a highly sensitive technique, which makes it ideal for detecting template molecules that occur at ultra-low concentrations in a *sed*aDNA extract. However, it is also prone to amplification biases (Ziesemer et al. 2015; Nichols et al. 2018; Capo et al. 2021) and is sensitive to contamination. Human contamination can be particularly problematic for animal metabarcoding assays, and so blocking primers, which compete with metabarcoding primers during binding to human template molecules, have been developed to limit the amplification of contaminating human DNA (e.g., Boessenkool et al. 2012). Contamination sensitivity coupled with the fact that PCR amplicons cannot be authenticated using deamination-based damage profiles (see also section 'Identification and validation: Validation') means that metabarcoding assays should not target bacteria or human *sed*aDNA and caution should be applied to studies of plants, fungi, and animals (Table 3.1, see also Chap. 11 of this volume).

Multiplex PCR uses multiple primer sets within a single PCR reaction to target multiple templates simultaneously. This greatly increases the amount of information obtained and eliminates the need to run multiple reactions, which deplete *sed*aDNA extracts and increase costs (Taberlet et al. 2018). Multiplex PCR has been applied to ancient DNA to improve the taxonomic resolution of key taxa that are otherwise difficult to resolve with a single primer set (Côté et al. 2016). Based on applications to modern environmental DNA, multiplex PCR could be applied to *sed*aDNA to study cryptic species (Brosseau et al. 2019), simultaneously target multiple broad taxonomic groups of interest (Weber et al. 2023), or recover population-wide genotyping data (Andres et al. 2021). However, the increased complexity of multiplex PCR can cause problems and therefore requires careful consideration during study design (Ficetola and Taberlet 2023). First, all primers need to have compatible annealing temperatures and amplify templates of comparable size, in order to ensure equal amplification. Secondly, the primers included have to be selected so as to avoid potential dimer formation during the PCR (Côté et al. 2016; Taberlet et al. 2018). These issues can be mitigated by dividing the primer sets into several multiplex PCRs with their own specific annealing temperatures, lengths, or lack of hybridising partner primers, though at increased cost (Côté et al. 2016; Brosseau et al. 2019). Another concern is that of different template concentrations when multiplexing metabarcode primers that target multiple taxonomic groups, which can result in uneven amplicon concentrations after PCR. Adjustments can be made to the primer concentrations to offset these issues, but optimisation can require considerable time (Taberlet et al. 2018). Lastly, the complexity multiplex PCR data may be challenging to bioinformatically analyse (Ficetola and Taberlet 2023).

Metagenomics

Metagenomics-based data generation approaches involve the conversion of total *sed*aDNA in an extract into a sequencing library followed by either direct sequencing (shotgun strategy) or by a targeted enrichment step to enrich the library for *sed*aDNA molecules of interest prior to sequencing. In both cases, entire *sed*aDNA molecules are sequenced, allowing both for molecules too short for amplicon approaches to be analysed and for the assessment of DNA damage profiles (see section 'Identification and validation').

Shotgun metagenomics is the most conceptually simple form of sequence data generation, whereby any of the *sed*aDNA molecules present in a DNA extract can potentially be sequenced. As *sed*aDNA molecules can exist in either single- or double-stranded form (i.e., a half or complete double helix), shotgun library preparation procedures can either target the double-stranded fraction (e.g., Meyer and Kircher 2010) or both the double- and single-stranded DNA molecular components (e.g., Gansauge et al. 2017). Traditionally, double-stranded DNA library preparation has been preferred due to simplicity and lower cost (e.g., Graham et al. 2016; Pedersen et al. 2016; Wang et al. 2021), but methods that also target the single-stranded DNA component are now economically competitive (Kapp et al. 2021) and are being used in an increasing number of studies (e.g., Schulte et al. 2021; Courtin et al. 2022). Once the *sed*aDNA extract has been converted into a library, it is directly sequenced without any form of enrichment. The resulting data therefore represent, relative to other methods mentioned in this section, an unbiased snapshot of the *sed*aDNA extract. Given that total *sed*aDNA is generally dominated by bacteria and fungi (or is unidentifiable), shotgun metagenomics data are often overwhelmed by sequences from these taxa (e.g., Pedersen et al. 2016; Lammers et al. 2021) (see also section 'Identification and validation: Identification'). If the aim is to study plants and animals, which are typically present in very low proportions of shotgun metagenomics data, then additional analyses beyond the detection of taxa and assessment of DNA damage may not be feasible or economically justifiable. For other questions, such as those requiring population-level data, then a targeted enrichment approach may be more appropriate.

Targeted enrichment, also referred to as targeted or hybridization capture, provides a middle ground for *sed*aDNA sequence data generation by targeting only a subset of the library molecules that align with the research question, akin to metabarcoding, but by capturing entire *sed*aDNA molecules, akin to shotgun metagenomics. This approach therefore allows for a greater proportion of *sed*aDNA of interest to be analysed and to be assessed via DNA damage analyses. Targeted enrichment uses pre-designed DNA or RNA 'bait' molecules, also known as probes, to capture complementary molecules in a shotgun metagenomics library. Uncaptured 'off-target' molecules are then removed, thereby enriching the library for *sed*aDNA molecules of interest (e.g., Soares 2019; Armbrecht et al. 2021; Murchie et al. 2021; Vernot et al. 2021). A weakness of the target enrichment approach is that it requires *a priori* knowledge of the reference sequences required to design baits, although, if uncontaminated specimens are available, probes can also be generated directly from modern DNA (e.g., Maricic et al. 2010). However, experimental parameters (i.e., hybridisation temperature) can be modified to reduce the specificity of the capture, which can allow for the enrichment of more distantly related desired sequences but also increase the likelihood of capturing undesirable off-target molecules. Targeted enrichment has been used in *sed*aDNA studies to capture barcode loci (Murchie et al. 2021), functional genic loci (Armbrecht et al. 2021), mitochondrial and chloroplast genomes (Slon et al. 2017; Schulte et al. 2021; Kjær et al. 2022), and genome-wide data (Vernot et al. 2021) from either single species or whole plant and animal communities. The application of target enrichment to generate genome-wide data is an area of anticipated growth that will allow for the investigation of evolutionary and population-level questions using *sed*aDNA (see section 'Identification and validation: Validation').

Sequencing

The final step of sequence data generation is the high-throughput sequencing of amplicon and metagenomics libraries, in which the library molecules are 'read' to produce sequencing reads that are used for downstream bioinformatics analyses. At present, *sed*aDNA sequencing is performed almost exclusively on short-read *Illumina* next-generation sequencers due to low cost and high-throughput, producing $>10^7$–10^9 reads in a single sequencing run. However, we note that emerging short-read sequencers, such as the *PacBio Onso* platform that has a lower error rate than *Illumina*, may become dominant in the future (Eisenstein 2023). Illumina sequencing can be performed either in single-end mode, whereby library molecules are read from one end only, or in paired-end mode, in which molecules are read from both ends. Single-end sequencing may be more economical whereas pair-end sequencing has a lower error rate due to parts of the library molecule being read twice.

Due to the high throughput of DNA sequencers, it is often desirable to multiplex libraries, whereby multiple samples are sequenced on the same run. DNA reads from individual libraries are then identified based on their library index (see introduction to this section) and demultiplexed during pre-processing (see section 'Data pre-processing and denoising'). Index-hopping, whereby a read is assigned the wrong index, will result in a form of cross-contamination between libraries, whereas tag-jumping is the within-library equivalent for amplicon experiments (Rodriguez-Martinez et al. 2023). These artefacts can be avoided by using a unique dual-indexing strategy, in which each library molecule has two indexes (Kircher et al. 2012), the use of unique dual-tags in amplicon experiments (Binladen et al. 2007), unique molecular identifiers in library preparation (MacConaill et al. 2018), and/or by not including genomic DNA libraries on the same sequencing runs as *sed*aDNA libraries (Graham et al. 2016). If unique dual-tagging/indexing is used, then the incidences of library index-hopping and amplicon tag-jumps can be checked using the prevalence of unexpected tag and/or index combinations (Zinger et al. 2019; Rodriguez-Martinez et al. 2023).

Lastly, it is crucial that negative controls from each step in the data generation pipeline (e.g., sampling, DNA extraction, PCR amplification or metagenomic library preparation) are sequenced to monitor for contamination, and, where possible, trace its origins in the workflow.

Data Pre-processing and Denoising

The first bioinformatics step in processing *sed*aDNA sequence data is to filter out sequences and reads that are either experimental artefacts or have insufficient information content to be confidently identified and authenticated. Pre-processing consists of demultiplexing, or 'separating', of libraries together with the removal of library adapter/primer sequences and, if applicable, merging of overlapping paired-end reads. While library demultiplexing is often performed by the sequencing platform's on-board software, read merging and adapter/primer trimming are achieved using tools such as fastp (Chen et al. 2018), AdapterRemoval2 (Schubert et al. 2016), or SeqPrep (https://github.com/jstjohn/SeqPrep). Denoising then removes reads that are of an incorrect length, have low information content, and/or are duplicated. Pre-processing steps are generally identical for both amplicon and metagenomics experiments, whereas denoising procedures diverge.

In an amplicon experiment, denoising involves the removal of reads that are likely to represent artefacts or off-target amplification. This includes reads that lack expected primer and tag sequences and/or are longer or shorter than the expected amplicon length (including primers and tags). Duplicated amplicon reads are then collapsed to a single representative read, often whilst also accounting for errors introduced by PCR or sequencing. After denoising, amplicons may be kept as amplicon sequence variants (ASVs) or collapsed by a pairwise-identity threshold (e.g., 95%) into molecular operational taxonomic units (MOTUs) prior to identification. The short lengths of *sed*aDNA amplicons generally result in reduced taxonomic precision with the MOTU approach, and so ASVs are more commonly used for identification. Pipelines for denoising of amplicon data include ObiTools (Boyer et al. 2016), Anacapa (Curd et al. 2019), and LULU (Frøslev et al. 2017).

Denoising of metagenomics data is more complex, as there is no *a priori* expectation regarding DNA fragment length and sequence complexity. Denoising procedures therefore focus on the removal of reads with low information content, such as those shorter than ~25–35 bp (Graham et al. 2016; de Filippo et al. 2018; Wang et al. 2021) and/or exhibiting low sequence complexity such as tandem repeats (e.g., Graham et al. 2016; Pedersen et al. 2016). Read duplicates can either be removed during denoising or after alignment (see section 'Identification and validation'). As metagenomics analysis of *sed*aDNA is a recent development, the denoising of these data is currently not standardised between studies.

While there are well-established tools for pre-processing and the denoising of amplicon data, the denoising of metagenomics data is likely to be further developed and optimised to allow for standardisation between data sets collected by different research groups. In addition, iterative testing of settings may be required to determine optimum filtering thresholds that are likely to be data set specific.

Identification and Validation

Identification

After data denoising, retained sequences are taxonomically assigned by comparison to a reference database, followed by refinement of these identifications based on the application of filtering thresholds and *a priori* knowledge (Fig. 3.2).

Taxonomic assignment can be achieved either by aligning whole reads (mapping) or comparing unique equal-size motifs (k-mers) within reads to a reference database. K-mer based assignments are rapid and scalable but can result in high proportions of false positives due to reduced specificity from the minimum k-mer-size limit. Mapping-based approaches have the opposite characteristics. For metagenomics data, widely used software for k-mer based analyses includes KRAKEN2 (Wood et al. 2019), whereas BWA (Li and Durbin 2009), Bowtie2 (Langmead and Salzberg 2012), and BLAST (Altschul et al. 1990) are used for mapping with Bowtie2 as the basis for the Holi pipeline (Pedersen et al. 2016). The relatively small size of amplicon datasets means that mapping approaches with software such as ObiTools (Boyer et al. 2016) and Anacapa (Curd et al. 2019) are commonly used.

Reference databases can consist of barcodes (e.g. BOLD (Ratnasingham and Hebert 2007), SILVA (https://www.arb-silva.de/), mitochondrial or chloroplast genomes (e.g. NCBI RefSeq (O'Leary et al. 2016), PhyloNorway for Arctic plants (Wang et al. 2021), single nuclear genomes, multiple genomes (e.g. NCBI RefSeq, the Genome Taxonomy Database (GTDB; https://gtdb.ecogenomic.org/) for Bacteria and Archaea (Parks et al. 2022)), or global nucleotide databases (e.g. EMBL, NCBI). High quality, complete, and well curated reference databases are essential for accurate taxonomic assignment. It is therefore important to have a representative reference database containing not only taxa expected from the region, but also common contaminants and taxa closely related to those that are expected (e.g., Graham et al. 2016; Kjær et al. 2022). Database incompleteness can lead to both false negatives, due to the absence of a reference close enough to the *sed*aDNA sequence in question, or false positives, due to the presence of a close relative to the *sed*aDNA sequence that is not the correct taxon. Although global nucleotide databases are often the most complete in terms of taxonomic coverage, they are minimally curated with errors occurring in taxonomic labelling and/or sequence quality within the database. Furthermore, global databases are generally over-represented by model and human-related organisms, such as animal husbandry, crop, and disease taxa, and so caution should be applied when identifying *sed*aDNA, especially if from natural environments, to avoid false positives. On the other hand, barcode databases for amplicon data are generally more straightforward, complete, and well-curated since the genomic region is constrained. Reference databases should therefore be selected carefully with consideration of the research question and data generation strategy used. We highlight that, at present, genomic reference databases are sparse for many groups of non-model organisms, such as invertebrates, fungi, and microbes.

The taxonomic resolution of a *sed*aDNA sequence identification is dependent on its length and whether it originates from a genomic region with diagnostic motifs for assignment at high taxonomic resolution such as the species or population level. For this reason, amplicon and target enrichment experiments are specifically designed to enrich taxonomically-informative genomic regions. On the other hand, taxonomic classification is usually less specific with shotgun metagenomics data, as these data originate from any part of the genome including from large genomic tracts that are shared between organisms. This reduced specificity may be exacerbated if reference genomes are lacking for the target taxa. Regardless of data generation strategy, there will be sequences, especially those that are short (~30–60 bp), that can only be identified to the genus level or higher. These identifications are achieved using lowest common ancestor (LCA) algorithms, which are automatically applied in the amplicon mapping and k-mer softwares listed above, but specialised software exists for other pipelines [e.g., MEGAN6 (Huson et al. 2007), ngsLCA (Wang et al. 2022), PIA (Cribdon et al. 2020)]. LCA algorithms assign the sequences to a taxonomic level that all share the same sequence at a given genetic region. It is therefore important that the used alignment software, such as Bowtie2, outputs all possible alignments and not only the selected best hit.

Setting a threshold can increase the confidence of detections by reducing false positives caused by errors and artefacts. These 'above the detection limit' thresholds, or cut-offs, can be based on read counts assigned to a certain taxon, the presence of taxa in the negative controls, or whether the detection is replicated. The latter is particularly applicable to amplicon experiments, where sporadic false-positive detections can be commonplace. However, threshold values can be rather arbitrary and experiment-specific, and so there is a need for better standardization and more objective approaches to setting cut-off values for *sed*aDNA-based detections.

The curation of identifications after bioinformatics-based assignments may be helpful in cases where there is an obvious or well characterised error in the assignment. For example, when using a global database, an assignment may be to an exotic taxon as the expected locally-occurring and closely-related taxon is absent from the database. While this step may be necessary at present, increased database completeness coupled with incorporating biogeographic data into assignment algorithms should reduce the need to curate data in future.

Validation

A critical stage of bioinformatic processing and interpretation is data set validation using *sed*aDNA authentication evidence (Fig. 3.2), comparison to independent proxies, and scrutiny of potential taphonomic impacts. For characterising little understood, difficult, or complex study systems, such as abyssal plain and/or deep-time sediments, then simulated or synthetic datasets can be used to test assignment specificity.

Given its propensity for results to include contamination, evidence for the existence of authentic *sed*aDNA molecules should always be presented as part of a study (Smith et al. 2015; Weiß et al. 2015; Kistler et al. 2015). For metagenomics datasets, authentication evidence includes assessment of DNA fragment lengths, which should average < 100 bp, and the existence of ancient DNA damage as exhibited by the deamination-induced excess of cytosine-to-thymine misincorporations at the ends of molecules (Dabney et al. 2013). Tools now exist to explore metagenomic *sed*aDNA damage profiles across multiple taxa simultaneously (Michelsen et al. 2022; Everett and Cribdon 2023). A basal or divergent phylogenetic placement, as compared to a well-sampled modern-day reference panel, may also provide supporting evidence for *sed*aDNA being ancient, but these analyses require genome-wide data and relatively large quantities of reads for a given taxon (Pedersen et al. 2021; Kjær et al. 2022). In both amplicon and metagenomics experiments, negative controls are used to detect contaminant taxa and therefore test the veracity of a sediment sample-derived detection. We note that some taxa, such as domesticated plants and animals, are often sporadically detected in negative controls due to contamination, but may be reproducibly detected in samples, and so results should therefore be interpreted cautiously (see also Chap. 11). We highlight that, unless the negative controls were also impacted, cross-contamination cannot necessarily be detected using these authentication approaches.

An independent source of validation is to compare occurrences of taxa in *sed*aDNA and other proxy data sets from the same sediment profile. For example, macrofossils, pollen and coprophilous spore records can be used to cross-validate plant and mammalian *sed*aDNA datasets, respectively (Graham et al. 2016; Clarke et al. 2020; ter Schure et al. 2021; van Vugt et al. 2022; Kjær et al. 2022; Garcés-Pastor et al. 2022). However, it is important to account for proxy-specific taphonomic factors when comparing multiple proxies in this manner. For instance, plant pollen provides a regional signal dominated by wind-dispersed taxa, whereas *sed*aDNA provides a local signal of both wind and pollinator-dispersed plant taxa. The two proxies therefore give different yet complementary patterns (Parducci et al. 2015; Clarke et al. 2020).

Sedimentary ancient DNA data sets are also prone to sources of error, including taphonomy-induced taxon dropout and leaching in some environments that can impact environmental and evolutionary inferences (see section 'Environmental and evolutionary analysis'; Sand et al. 2023). If DNA from a subset of taxa overwhelms the *sed*aDNA signal, also known as 'swamping', due to, for example, a taxonomically heterogenous distribution of *sed*aDNA within the sediment sample, then this could lead to drop out of the underrepresented taxa resulting in false negatives (Alsos et al. 2020). The vertical leaching of *sed*aDNA between sediment layers, through the movement of porewater, is a serious concern, as this could lead to erroneous interpretations of the timing of events inferred from the *sed*aDNA record (Haile et al. 2007). Leaching can be monitored by comparison to independent proxies and/or assessing the rapidity of turnovers, as a clear rapid turnover would not be apparent from a severely leached sediment profile. Although leaching may be an issue for sediments that allow the movement of porewater, the growing body of literature suggests leaching does not occur in waterlogged sediments (e.g., Graham et al. 2016; Sjögren et al. 2017).

Environmental and Evolutionary Analysis

Environmental Analysis

The environmental analysis of *sed*aDNA data follows the same principles as for other more traditional palaeoecological and archaeological proxies, such as pollen and spores, whereby detections, interpreted as either presence/absence or relative abundance data, are mapped by taxonomy and time. This allows for analysis of co-correlating taxa (communities), periods of community change (turnovers), within- and between-species biogeography (Schulte et al. 2022), and times of first or last occurrence (e.g. Graham et al. 2016; Ficetola et al. 2018; Alsos et al. 2022), as well as the calculation of biodiversity indices (e.g., Chen and Ficetola 2020) and the identification of independent drivers of changes such as climate or human impact (e.g. Clarke et al. 2019; ter Schure et al. 2021; Garcés-Pastor et al. 2022; Barouillet et al. 2023). However, it is important to account for false negatives in *sed*aDNA-derived environmental datasets (Chen and Ficetola 2020) and for heterogeneity when combining datasets from multiple sites (Rijal et al. 2021). The key difference over traditional proxies is the taxonomic precision of *sed*aDNA identifications, which can be to the genus, species, and even population level (e.g., Garcés-Pastor et al. 2022) coupled with the potential for scalable and simultaneous analysis of multiple ecosystem components (e.g., Wang et al. 2021; Courtin et al. 2022).

The use of *sed*aDNA abundance data, based on relative read counts assigned to taxa, to infer either relative biomass or relative abundance of individuals, both key measures for many environmental analyses, is controversial and poorly understood in palaeoecological and archaeological systems. This is because multiple biological, taphonomic, and experimental factors can theoretically cause severe deviation between *sed*aDNA abundance and the true past relative biomass or individual abundance of taxa (see also Chap. 2 of this volume; Capo et al. 2021). Controlled experiments (e.g., Alsos et al. 2018) coupled with a greater diversity of palaeoecological and comparative ecological datasets will be needed to understand whether, and if so in what contexts, *sed*aDNA abundance is indicative of past relative biomass or relative individual abundance.

A new and exciting application of *sed*aDNA for environmental inference is using trait analysis to examine the development of environmental, structural, and functional changes across communities through time. This is possible due to the species discriminating power of *sed*aDNA, but has so far only been applied to plants (Alsos et al. 2022; Revéret et al. 2023) and microorganisms (Keck et al. 2020). We anticipate that the use of ecological traits will be expanded across multiple taxonomic groups thereby allowing for integrated multi-trophic trait analysis to unveil palaeoecosystem reconstructions in unprecedented detail.

Evolutionary Analysis

Amongst sediment-derived proxies, *sed*aDNA is unique in its ability to simultaneously track evolutionary change in multiple taxa and components of ecosystems through time (e.g., Dussex et al. 2021; Gelabert et al. 2021; Capo et al. 2022). This includes reconstructing mitochondrial and chloroplast genomes for phylogenetics, detecting population turnovers, characterising population histories, tracking functional responses to past environmental stressors (Poulain et al. 2015), and reconstructing genome-wide data for dating the age of *sed*aDNA (Fig. 3.2).

As *sed*aDNA is derived from entire genomes of multiple taxa, the analysis of reconstructed environmental palaeogenomes or genome-wide datasets can allow for serially-sampled evolutionary analyses of multiple taxa simultaneously. Such analyses are possible if data cover multiple alleles, which include datasets generated by shotgun metagenomics and certain target enrichment or multiplex PCR assays. The major challenge of using these data for evolutionary analyses is that *sed*aDNA is a complex mixture derived from multiple individuals and taxa. This precludes the possibility of some evolutionary analyses, such as those requiring estimates of individual heterozygosity (e.g., inbreeding). However, population structure, turnover, and admixture analyses can be used (e.g., Gelabert et al. 2021; Pedersen et al. 2021;

Vernot et al. 2021), as the multi-individual composition of *seda*DNA (the "environmental genome") is effectively a population sample in a similar manner to a single genome derived from an individual organism (Pedersen et al. 2021). For the reconstruction of mitochondrial and chloroplast genomes (Seersholm et al. 2016; Schulte et al. 2021), which often consist of multiple haplotypes within a population and can be sequenced to high coverage, it is possible to tease apart haplotypes within a single *seda*DNA extract based on relative allele frequencies. Haplotypes can then be placed into phylogenies (Lammers et al. 2021; Murchie et al. 2022). By comparing the haplotypic and population structure composition of *seda*DNA extracts across a sediment profile i.e., through time, it is possible to detect population turnovers (e.g., Vernot et al. 2021).

Given *seda*DNA can be preserved over evolutionarily relevant timescales, molecular tip-dating approaches may be applied to reconstructed haplotypes to provide an independent age estimate of the sediment layer (e.g., Shapiro et al. 2011). Tip dating leverages the fact that ancient haplotypes are missing evolutionary changes that would have evolved during the time since the *seda*DNA was deposited. By using a reference panel of well-dated comparative sequences and assuming a molecular clock, it is possible to use this missing evolution to estimate an age for the haplotype and therefore the sediment layer. This has been applied to Early Pleistocene sediments from northern Greenland, which showed remarkable agreement between the tip-dated ages of both birch (*Betula*) and mastodon (*Mammut*) haplotypes (Kjær et al. 2022). Molecular clock dating has also been applied to reconstructed mammalian mitochondrial genomes from Late Pleistocene Eurasian cave sediments, including those derived from human (*Homo sapiens*), hominin (*Homo*), wolf (*Canis*), bison (*Bison*), and red deer (*Cervus*) (Gelabert et al. 2021; Vernot et al. 2021; Essel et al. 2023). However, tip dating methods are reliant on the accurate recovery of a single haplotype for each age estimate.

For any evolutionary analyses, it is crucial that *seda*DNA derived from other taxa is not mis-characterized as originating from the taxon under investigation. To avoid this, loci that are easily differentiated between related taxa can be targeted (Vernot et al. 2021) or competitive mapping approaches can be applied, whereby reads are simultaneously mapped against multiple related reference genomes to determine the most appropriate identity (e.g., Pedersen et al. 2021; Kjær et al. 2022). The ability to conduct population-level evolutionary analyses is dependent on available reference datasets for the taxa under investigation.

The application of evolutionary analyses to *seda*DNA data is a recent development, with the first mitochondrial and genome-wide reconstructions only published in 2016 and 2021, respectively (Seersholm et al. 2016; Vernot et al. 2021). However, the recovery of sufficient reads for the taxa of interest remains a bottleneck for genomic investigations in metagenomic studies, although target enrichment methods show promising results for specific taxa (e.g., Vernot et al. 2021). Consequently, we anticipate that there will be significant methodological innovation to fully leverage *seda*DNA for evolutionary analyses in the near future, in particular for the integrated analysis of multiple taxa (Dussex et al. 2021).

Data and Script Availability

The volume of *sed*aDNA datasets is growing exponentially both in terms of counts of publications (Capo et al. 2021; Von Eggers et al. 2022) and dataset size (e.g., >700 samples in Zavala et al. 2021). Regardless of data generation strategy, these datasets have huge potential for re-analysis, for example due to newly available reference databases exponentially expanding from increased global genome sequencing efforts (Lewin et al. 2018), improved or extended analytical methods (Williams et al. 2023), and/or basic reproducibility. It is therefore crucial that researchers continue to make their raw sequence data publicly and freely available, as is also common in the general ancient metagenomics field (Fellows Yates et al. 2021). Going forward as bioinformatic methods continue to develop, it is increasingly important that scripts, tools, and pipelines are also made publicly available, which will enable continued rapid innovation and discovery in the field (e.g., Cribdon et al. 2020; Lammers et al. 2021; Michelsen et al. 2022). International efforts are also underway to integrate *sed*aDNA data into existing public repositories of palaeoecological and archaeological proxies (e.g., Neotoma Paleoecology Database; Williams et al. 2023).

Future Directions

Development and optimisation of the molecular biology and bioinformatics steps of the *sed*aDNA workflow is an ongoing endeavour by the research community. Based on current trends, there is the distinct possibility that target enrichment will supersede the metabarcoding approach, and that shotgun metagenomics may eventually supersede target enrichment. This is due to the ongoing decline in sequencing costs, the increasingly routine automation of laboratory steps (Gansauge et al. 2020), coupled with the accelerating availability of reference genomes (Lewin et al. 2018), although new efficient computational pipelines for *sed*aDNA sequence identification will be required to leverage these increasingly large data sets. In addition to higher throughput in the laboratory, robotisation may contribute to the standardisation of sediment collection and sampling steps and further minimise contamination risks in the field. We speculate that these advances will lead to new applications for *sed*aDNA data, such as the detection of single alleles to infer population history at a site through time (Dussex et al. 2021; Vernot et al. 2021) or using *sed*aDNA damage profiles as a routine relative dating method within a sedimentary sequence.

Summary

Sedimentary ancient DNA is a relatively new and highly promising approach for deriving palaeoenvironmental and evolutionary information from sediments. However, the workflow used to collect, generate, and analyze *sed*aDNA data is complex and experiencing ongoing rapid innovations. In this chapter, we have provided an overview of the *sed*aDNA workflow currently employed in palaeoenvironmental research, by outlining the rationale, general approach, pros and cons, pitfalls, and future perspectives for each workflow step.

Acknowledgements We thank Eric Capo, Frances Pick, Gentile Francesco Ficetola, Yucheng Wang, and Mikkel Winther Pedersen for comments that greatly improved earlier drafts of this chapter. P.D.H. was supported by the Knut and Alice Wallenberg Foundation (KAW 2021.0048 and KAW 2022.0033). A.R. was supported by the Research Council of Norway (KLIMAFORSK, 294929). L.A. was supported by an Australian Research Council (ARC) Discovery Early Career Researcher (DECRA) Fellowship (DE210100929). S.G.P. was supported by the Beatriu de Pinós Programme (BP-2021-00131).

References

Alpaslan-Roodenberg S, Anthony D, Babiker H, Bánffy E, Booth T, Capone P, Deshpande-Mukherjee A, Eisenmann S, Fehren-Schmitz L, Frachetti M, Fujita R, Frieman CJ, Fu Q, Gibbon V, Haak W, Hajdinjak M, Hofmann KP, Holguin B, Inomata T, Kanzawa-Kiriyama H, Keegan W, Kelso J, Krause J, Kumaresan G, Kusimba C, Kusimba S, Lalueza-Fox C, Llamas B, MacEachern S, Mallick S, Matsumura H, Morales-Arce AY, Matuzeviciute GM, Mushrif-Tripathy V, Nakatsuka N, Nores R, Ogola C, Okumura M, Patterson N, Pinhasi R, Prasad SPR, Prendergast ME, Punzo JL, Reich D, Sawafuji R, Sawchuk E, Schiffels S, Sedig J, Shnaider S, Sirak K, Skoglund P, Slon V, Snow M, Soressi M, Spriggs M, Stockhammer PW, Szécsényi-Nagy A, Thangaraj K, Tiesler V, Tobler R, Wang C-C, Warinner C, Yasawardene S, Zahir M (2021) Ethics of DNA research on human remains: five globally applicable guidelines. Nature 599(7883):41–46. https://doi.org/10.1038/s41586-021-04008-x

Alsos IG, Lammers Y, Yoccoz NG, Jørgensen T, Sjögren P, Gielly L, Edwards ME (2018) Plant DNA metabarcoding of lake sediments: how does it represent the contemporary vegetation. PLoS One 13(4):e0195403. https://doi.org/10.1371/journal.pone.0195403

Alsos IG, Sjögren P, Brown AG, Gielly L, Merkel MKF, Paus A, Lammers Y, Edwards ME, Alm T, Leng M, Goslar T, Langdon CT, Bakke J, van der Bilt WGM (2020) Last Glacial Maximum environmental conditions at Andøya, northern Norway; evidence for a northern ice-edge ecological "hotspot.". Quat Sci Rev 239:106364. https://doi.org/10.1016/j.quascirev.2020.106364

Alsos IG, Rijal DP, Ehrich D, Karger DN, Yoccoz NG, Heintzman PD, Brown AG, Lammers Y, Pellissier L, Alm T, Bråthen KA, Coissac E, Merkel MKF, Alberti A, Denoeud F, Bakke J, Phylonorway Consortium (2022) Postglacial species arrival and diversity buildup of northern ecosystems took millennia. Sci Adv 8(39):eabo7434. https://doi.org/10.1126/sciadv.abo7434

Altschul SF, Gish W, Miller W, Myers EW, Lipman DJ (1990) Basic local alignment search tool. J Mol Biol 215(3):403–410. https://doi.org/10.1016/S0022-2836(05)80360-2

Andres KJ, Sethi SA, Lodge DM, Andrés J (2021) Nuclear eDNA estimates population allele frequencies and abundance in experimental mesocosms and field samples. Mol Ecol 30(3):685–697. https://doi.org/10.1111/mec.15765

Ardelean CF, Becerra-Valdivia L, Pedersen MW, Schwenninger JL, Oviatt CG, Macías-Quintero JI, Arroyo-Cabrales J, Sikora M, Ocampo-Díaz YZE, Rubio-Cisneros II, Watling JG, de Medeiros VB, De Oliveira PE, Barba-Pingarón L, Ortiz-Butrón A, Blancas-Vázquez J, Rivera-González I, Solís-Rosales C, Rodríguez-Ceja M, Gandy DA, Navarro-Gutierrez Z, De La Rosa-Díaz JJ, Huerta-Arellano V, Marroquín-Fernández MB, Martínez-Riojas LM, López-Jiménez A, Higham T, Willerslev E (2020) Evidence of human occupation in Mexico around the last glacial maximum. Nature 584(7819):87–92. https://doi.org/10.1038/s41586-020-2509-0

Armbrecht LH, Coolen MJL, Lejzerowicz F, George SC, Negandhi K, Suzuki Y, Young J, Foster NR, Armand LK, Cooper A, Ostrowski M, Focardi A, Stat M, Moreau JW, Weyrich LS (2019) Ancient DNA from marine sediments: precautions and considerations for seafloor coring, sample handling and data generation. Earth Sci Rev 196:102887. https://doi.org/10.1016/j.earscirev.2019.102887

Armbrecht L, Herrando-Pérez S, Eisenhofer R, Hallegraeff GM, Bolch CJS, Cooper A (2020) An optimized method for the extraction of ancient eukaryote DNA from marine sediments. Mol Ecol Resour 20(4):906–919. https://doi.org/10.1111/1755-0998.13162

Armbrecht L, Hallegraeff G, Bolch CJS, Woodward C, Cooper A (2021) Hybridisation capture allows DNA damage analysis of ancient marine eukaryotes. Sci Rep 11:3220. https://doi.org/10.1038/s41598-021-82578-6

Armbrecht L, Weber ME, Raymo ME, Peck VL, Williams T, Warnock J, Kato Y, Hernández-Almeida I, Hoem F, Reilly B, Hemming S, Bailey I, Martos YM, Gutjahr M, Percuoco V, Allen C, Brachfeld S, Cardillo FG, Du Z, Fauth G, Fogwill C, Garcia M, Glüder A, Guitard M, Hwang J-H, Iizuka M, Kenlee B, O'Connell S, Pérez LF, Ronge TA, Seki O, Tauxe L, Tripathi S, Zheng X (2022) Ancient marine sediment DNA reveals diatom transition in Antarctica. Nat Commun 13(1):5787. https://doi.org/10.1038/s41467-022-33494-4

Bardill J, Bader AC, Garrison NA, Bolnick DA, Raff JA, Walker A, Malhi RS, The Summer Internship for Indigenous Peoples in Genomics (Sing) Consortium (2018) Advancing the ethics of paleogenomics. Science 360(6387):384–385. https://doi.org/10.1126/science.aaq1131

Barouillet C, Monchamp M-E, Bertilsson S, Brasell K, Domaizon I, Epp LS, Ibrahim A, Mejbel H, Nwosu EC, Pearman JK, Picard M, Thomson-Laing G, Tsugeki N, Von Eggers J, Gregory-Eaves I, Pick FR, Wood SA, Capo E (2023) Investigating the effects of anthropogenic stressors on lake biota using sedimentary DNA. Freshw Biol. https://doi.org/10.1111/fwb.14027

Binladen J, Gilbert MTP, Bollback JP, Panitz F, Bendixen C, Nielsen R, Willerslev E (2007) The use of coded PCR primers enables high-throughput sequencing of multiple homolog amplification products by 454 parallel sequencing. PLoS One 2(2):e197. https://doi.org/10.1371/journal.pone.0000197

Boessenkool S, Epp LS, Haile J, Bellemain E, Edwards M, Coissac E, Willerslev E, Brochmann C (2012) Blocking human contaminant DNA during PCR allows amplification of rare mammal species from sedimentary ancient DNA. Mol Ecol 21(8):1806–1815. https://doi.org/10.1111/j.1365-294X.2011.05306.x

Boyer F, Mercier C, Bonin A, Le Bras Y, Taberlet P, Coissac E (2016) Obitools: a unix-inspired software package for DNA metabarcoding. Mol Ecol Resour 16(1):176–182. https://doi.org/10.1111/1755-0998.12428

Braadbaart F, Reidsma FH, Roebroeks W, Chiotti L, Slon V, Meyer M, Théry-Parisot I, van Hoesel A, Nierop KGJ, Kaal J, van Os B, Marquer L (2020) Heating histories and taphonomy of ancient fireplaces: a multi-proxy case study from the upper Palaeolithic sequence of Abri Pataud (les Eyzies-de-Tayac, France). J Archaeol Sci Rep 33:102468. https://doi.org/10.13039/501100004189

Brosseau L, Udom C, Sukkanon C, Chareonviriyaphap T, Bangs MJ, Saeung A, Manguin S (2019) A multiplex PCR assay for the identification of five species of the Anopheles barbirostris complex in Thailand. Parasit Vectors 12(1):223. https://doi.org/10.1186/s13071-019-3494-8

Buck M, Hamilton C (2011) The Nagoya protocol on access to genetic resources and the Fair and equitable sharing of benefits arising from their utilization to the convention on biological diversity. Rev Eur Commu Int Env Law 20(1):47–61. https://doi.org/10.1111/j.1467-9388.2011.00703.x

Capo E, Debroas D, Arnaud F, Perga M-E, Chardon C, Domaizon I (2017) Tracking a century of changes in microbial eukaryotic diversity in lakes driven by nutrient enrichment and climate warming. Environ Microbiol 19(7):2873–2892. https://doi.org/10.1111/1462-2920.13815

Capo E, Giguet-Covex C, Rouillard A, Nota K, Heintzman PD, Vuillemin A, Ariztegui D, Arnaud F, Belle S, Bertilsson S, Bigler C, Bindler R, Brown AG, Clarke CL, Crump SE, Debroas D, Englund G, Ficetola GF, Garner RE, Gauthier J, Gregory-Eaves I, Heinecke L, Herzschuh U, Ibrahim A, Kisand V, Kjær KH, Lammers Y, Littlefair J, Messager E, Monchamp M-E, Olajos

F, Orsi W, Pedersen MW, Rijal DP, Rydberg J, Spanbauer T, Stoof-Leichsenring KR, Taberlet P, Talas L, Thomas C, Walsh DA, Wang Y, Willerslev E, van Woerkom A, Zimmermann HH, Coolen MJL, Epp LS, Domaizon IG, Alsos I, Parducci L (2021) Lake sedimentary DNA research on past terrestrial and aquatic biodiversity: overview and recommendations. Quaternary 4(1):6. https://doi.org/10.3390/quat4010006

Capo E, Monchamp M-E, Coolen MJL, Domaizon I, Armbrecht L, Bertilsson S (2022) Environmental paleomicrobiology: using DNA preserved in aquatic sediments to its full potential. Environ Microbiol 24(5):2201–2209. https://doi.org/10.1111/1462-2920.15913

Chen S, Zhou Y, Chen Y, Gu J (2018) fastp: an ultra-fast all-in-one FASTQ preprocessor. Bioinformatics 34(17):i884–i890. https://doi.org/10.1093/bioinformatics/bty560

Chen W, Ficetola GF (2020) Numerical methods for sedimentary-ancient-DNA-based study on past biodiversity and ecosystem functioning. Environmental DNA 2(2):115–129. https://doi.org/10.1002/edn3.79

Clarke CL, Edwards ME, Gielly L, Ehrich D, Hughes PDM, Morozova LM, Haflidason H, Mangerud J, Svendsen JI, Alsos IG (2019) Persistence of arctic-alpine flora during 24,000 years of environmental change in the Polar Urals. Sci Rep 9(1):19613. https://doi.org/10.1038/s41598-019-55989-9

Clarke CL, Alsos IG, Edwards ME, Paus A, Gielly L, Haflidason H, Mangerud J, Regnéll C, Hughes PDM, Svendsen JI, Bjune AE (2020) A 24,000-year ancient DNA and pollen record from the Polar Urals reveals temporal dynamics of arctic and boreal plant communities. Quat Sci Rev 247:106564. https://doi.org/10.1016/j.quascirev.2020.106564

Coolen MJL, Overmann J (1998) Analysis of subfossil molecular remains of purple sulfur bacteria in a Lake Sediment. Appl Environ Microbiol 64(11):4513–4521. https://doi.org/10.1128/AEM.64.11.4513-4521.1998

Côté NML, Daligault J, Pruvost M, Bennett EA, Gorgé O, Guimaraes S, Capelli N, Bailly ML, Geigl E-M, Grange T (2016) A new high-throughput approach to genotype ancient human gastrointestinal parasites. PLoS One 11(1):e0146230. https://doi.org/10.1371/journal.pone.0146230

Courtin J, Perfumo A, Andreev AA, Opel T, Stoof-Leichsenring KR, Edwards ME, Murton JB, Herzschuh U (2022) Pleistocene glacial and interglacial ecosystems inferred from ancient DNA analyses of permafrost sediments from Batagay megaslump. East Siberia Env DNA 4(6):1265–1283. https://doi.org/10.1002/edn3.336

Cribdon B, Ware R, Smith O, Gaffney V, Allaby RG (2020) PIA: more accurate taxonomic assignment of metagenomic data demonstrated on sedaDNA from the North Sea. Front Ecol Evol 8

Crump SE (2021) Sedimentary ancient DNA as a tool in paleoecology. Nat Rev Earth Environ 2(4):229–229. https://doi.org/10.1038/s43017-021-00158-8

Crump SE, Fréchette B, Power M, Cutler S, de Wet G, Raynolds MK, Raberg JH, Briner JP, Thomas EK, Sepúlveda J, Shapiro B, Bunce M, Miller GH (2021) Ancient plant DNA reveals high Arctic greening during the last interglacial. Proc Natl Acad Sci 118(13):e2019069118. https://doi.org/10.1073/pnas.2019069118

Curd EE, Gold Z, Kandlikar GS, Gomer J, Ogden M, O'Connell T, Pipes L, Schweizer TM, Rabichow L, Lin M, Shi B, Barber PH, Kraft N, Wayne R, Meyer RS (2019) Anacapa toolkit: an environmental DNA toolkit for processing multilocus metabarcode datasets. Methods Ecol Evol 10(9):1469–1475. https://doi.org/10.1111/2041-210X.13214

Dabney J, Meyer M, Pääbo S (2013) Ancient DNA damage. Cold Spring Harb Perspect Biol 5(7):a012567. https://doi.org/10.1101/cshperspect.a012567

de Filippo C, Meyer M, Prüfer K (2018) Quantifying and reducing spurious alignments for the analysis of ultra-short ancient DNA sequences. BMC Biol 16(1):121. https://doi.org/10.1186/s12915-018-0581-9

Domaizon I, Savichtcheva O, Debroas D, Arnaud F, Villar C, Pignol C, Alric B, Perga ME (2013) DNA from lake sediments reveals the long-term dynamics and diversity of *Synechococcus* assemblages. Biogeosciences 10(6):3817–3838. https://doi.org/10.5194/bg-10-3817-2013

Dussex N, Bergfeldt N, de Anca PV, Dehasque M, Díez-del-Molino D, Ersmark E, Kanellidou F, Larsson P, Lemež Š, Lord E, Mármol-Sánchez E, Meleg IN, Måsviken J, Naidoo T, Studerus J, Vicente M, von Seth J, Götherström A, Dalén L, Heintzman PD (2021) Integrating multi-taxon palaeogenomes and sedimentary ancient DNA to study past ecosystem dynamics. Proc R Soc B Biol Sci 288(1957):20211252. https://doi.org/10.1098/rspb.2021.1252

Eisenstein M (2023) Illumina faces short-read rivals. Nat Biotechnol 41(1):3–5. https://doi.org/10.1038/s41587-022-01632-4

Epp LS, Zimmermann HH, Stoof-Leichsenring KR (2019) Sampling and extraction of ancient DNA from sediments. In: Ancient DNA: methods and protocols. Springer, pp 31–44

Essel E, Zavala EI, Schulz-Kornas E, Kozlikin MB, Fewlass H, Vernot B, Shunkov MV, Derevianko AP, Douka K, Barnes I, Soulier M-C, Schmidt A, Szymanski M, Tsanova T, Sirakov N, Endarova E, McPherron SP, Hublin J-J, Kelso J, Pääbo S, Hajdinjak M, Soressi M, Meyer M (2023) Ancient human DNA recovered from a Palaeolithic pendant. Nature 618:328–332. https://doi.org/10.1038/s41586-023-06035-2

Everett R, Cribdon B (2023) MetaDamage tool: examining post-mortem damage in sedaDNA on a metagenomic scale. Front Ecol Evol 10. https://doi.org/10.3389/fevo.2022.888421

Fellows Yates JA, Andrades Valtueña A, Vågene ÅJ, Cribdon B, Velsko IM, Borry M, Bravo-Lopez MJ, Fernandez-Guerra A, Green EJ, Ramachandran SL, Heintzman PD, Spyrou MA, Hübner A, Gancz AS, Hider J, Allshouse AF, Zaro V, Warinner C (2021) Community-curated and standardised metadata of published ancient metagenomic samples with AncientMetagenomeDir. Sci Data 8(1):31. https://doi.org/10.1038/s41597-021-00816-y

Ficetola GF, Taberlet P (2023) Towards exhaustive community ecology via DNA metabarcoding. Mol Ecol. https://doi.org/10.1111/mec.16881

Ficetola GF, Pansu J, Bonin A, Coissac E, Giguet-Covex C, De Barba M, Gielly L, Lopes CM, Boyer F, Pompanon F, Rayé G, Taberlet P (2015) Replication levels, false presences and the estimation of the presence/absence from eDNA metabarcoding data. Mol Ecol Resour 15(3):543–556. https://doi.org/10.1111/1755-0998.12338

Ficetola GF, Poulenard J, Sabatier P, Messager E, Gielly L, Leloup A, Etienne D, Bakke J, Malet E, Fanget B, Støren E, Reyss J-L, Taberlet P, Arnaud F (2018) DNA from lake sediments reveals long-term ecosystem changes after a biological invasion. Sci Adv 4(5):eaar4292. https://doi.org/10.1126/sciadv.aar4292

Frøslev TG, Kjøller R, Bruun HH, Ejrnæs R, Brunbjerg AK, Pietroni C, Hansen AJ (2017) Algorithm for post-clustering curation of DNA amplicon data yields reliable biodiversity estimates. Nat Commun 8(1):1188. https://doi.org/10.1038/s41467-017-01312-x

Fulton TL, Shapiro B (2019) Setting up an ancient DNA laboratory. Methods Mol Biol 1963:1–13. https://doi.org/10.1007/978-1-4939-9176-1_1

Gansauge M-T, Gerber T, Glocke I, Korlevic P, Lippik L, Nagel S, Riehl LM, Schmidt A, Meyer M (2017) Single-stranded DNA library preparation from highly degraded DNA using T4 DNA ligase. Nucleic Acids Res 45(10):e79. https://doi.org/10.1093/nar/gkx033

Gansauge M-T, Aximu-Petri A, Nagel S, Meyer M (2020) Manual and automated preparation of single-stranded DNA libraries for the sequencing of DNA from ancient biological remains and other sources of highly degraded DNA. Nat Protoc 15(8):2279–2300. https://doi.org/10.1038/s41596-020-0338-0

Garcés-Pastor S, Coissac E, Lavergne S, Schwörer C, Theurillat J-P, Heintzman PD, Wangensteen OS, Tinner W, Rey F, Heer M, Rutzer A, Walsh K, Lammers Y, Brown AG, Goslar T, Rijal DP, Karger DN, Pellissier L, Heiri O, Alsos IG (2022) High resolution ancient sedimentary DNA shows that alpine plant diversity is associated with human land use and climate change. Nat Commun 13(1):6559. https://doi.org/10.1038/s41467-022-34010-4

Gelabert P, Sawyer S, Bergström A, Margaryan A, Collin TC, Meshveliani T, Belfer-Cohen A, Lordkipanidze D, Jakeli N, Matskevich Z, Bar-Oz G, Fernandes DM, Cheronet O, Özdoğan KT, Oberreiter V, Feeney RNM, Stahlschmidt MC, Skoglund P, Pinhasi R (2021) Genome-scale sequencing and analysis of human, wolf, and bison DNA from 25,000-year-old sediment. Curr Biol 31(16):3564–3574.e9. https://doi.org/10.1016/j.cub.2021.06.023

Graham RW, Belmecheri S, Choy K, Culleton BJ, Davies LJ, Froese D, Heintzman PD, Hritz C, Kapp JD, Newsom LA, Rawcliffe R, Saulnier-Talbot É, Shapiro B, Wang Y, Williams JW, Wooller MJ (2016) Timing and causes of mid-Holocene mammoth extinction on St. Paul Island, Alaska. Proc Natl Acad Sci 113(33):9310–9314. https://doi.org/10.1073/pnas.1604903113

Haile J, Holdaway R, Oliver K, Bunce M, Gilbert MTP, Nielsen R, Munch K, Ho SYW, Shapiro B, Willerslev E (2007) Ancient DNA chronology within sediment deposits: are paleobiological reconstructions possible and is DNA leaching a factor? Mol Biol Evol 24(4):982–989. https://doi.org/10.1093/molbev/msm016

Handsley-Davis M, Kowal E, Russell L, Weyrich LS (2021) Researchers using environmental DNA must engage ethically with Indigenous communities. Nat Ecol Evol 5(2):146–148. https://doi.org/10.1038/s41559-020-01351-6

Hofreiter M, Mead JI, Martin P, Poinar HN (2003) Molecular caving. Curr Biol 13(18):R693–R695. https://doi.org/10.1016/j.cub.2003.08.039

Huson DH, Auch AF, Qi J, Schuster SC (2007) MEGAN analysis of metagenomic data. Genome Res 17(3):377–386. https://doi.org/10.1101/gr.5969107

Immel A, Le Cabec A, Bonazzi M, Herbig A, Temming H, Schuenemann VJ, Bos KI, Langbein F, Harvati K, Bridault A, Pion G, Julien M-A, Krotova O, Conard NJ, Münzel SC, Drucker DG, Viola B, Hublin J-J, Tafforeau P, Krause J (2016) Effect of X-ray irradiation on ancient DNA in sub-fossil bones – guidelines for safe X-ray imaging. Sci Rep 6(1):32969. https://doi.org/10.1038/srep32969

Jelavic S, Lanson B, Findling N, Lanson M (2023) A route for long-term DNA preservation through nanoconfinement in smectites

Kapp JD, Green RE, Shapiro B (2021) A fast and efficient single-stranded genomic library preparation method optimized for ancient DNA. J Hered 112(3):241–249. https://doi.org/10.1093/jhered/esab012

Keck F, Millet L, Debroas D, Etienne D, Galop D, Rius D, Domaizon I (2020) Assessing the response of micro-eukaryotic diversity to the great acceleration using lake sedimentary DNA. Nat Commun 11(1):3831. https://doi.org/10.1038/s41467-020-17682-8

Kircher M, Sawyer S, Meyer M (2012) Double indexing overcomes inaccuracies in multiplex sequencing on the Illumina platform. Nucleic Acids Res 40(1):e3. https://doi.org/10.1093/nar/gkr771

Kistler L, Smith O, Ware R, Momber G, Bates R, Garwood P, Fitch S, Pallen M, Gaffney V, Allaby RG (2015) Thermal age, cytosine deamination and the veracity of 8,000 year old wheat DNA from sediments. 032060

Kjær KH, Winther Pedersen M, De Sanctis B, De Cahsan B, Korneliussen TS, Michelsen CS, Sand KK, Jelavić S, Ruter AH, Schmidt AMA, Kjeldsen KK, Tesakov AS, Snowball I, Gosse JC, Alsos IG, Wang Y, Dockter C, Rasmussen M, Jørgensen ME, Skadhauge B, Prohaska A, Kristensen JÅ, Bjerager M, Allentoft ME, Coissac E, Rouillard A, Simakova A, Fernandez-Guerra A, Bowler C, Macias-Fauria M, Vinner L, Welch JJ, Hidy AJ, Sikora M, Collins MJ, Durbin R, Larsen NK, Willerslev E (2022) A 2-million-year-old ecosystem in Greenland uncovered by environmental DNA. Nature 612(7939):283–291. https://doi.org/10.1038/s41586-022-05453-y

Kowal E, Weyrich LS, Argüelles JM, Bader AC, Colwell C, Cortez AD, Davis JL, Figueiro G, Fox K, Malhi RS, Matisoo-Smith E, Nayak A, Nelson EA, Nicholas G, Nieves-Colón MA, Russell L, Ulm S, Vergara-Silva F, Villanea FA, Wagner JK, Yracheta JM, Tsosie KS (2023) Community partnerships are fundamental to ethical ancient DNA research. Hum Genet Genom Adv 4(2):100161. https://doi.org/10.1016/j.xhgg.2022.100161

Lammers Y, Heintzman PD, Alsos IG (2021) Environmental palaeogenomic reconstruction of an ice age algal population. Commun Biol 4(1):1–11. https://doi.org/10.1038/s42003-021-01710-4

Langmead B, Salzberg SL (2012) Fast gapped-read alignment with Bowtie 2. Nat Methods 9(4):357–359. https://doi.org/10.1038/nmeth.1923

Lewin HA, Robinson GE, Kress WJ, Baker WJ, Coddington J, Crandall KA, Durbin R, Edwards SV, Forest F, Gilbert MTP, Goldstein MM, Grigoriev IV, Hackett KJ, Haussler D, Jarvis ED, Johnson WE, Patrinos A, Richards S, Castilla-Rubio JC, van Sluys M-A, Soltis PS, Xu X, Yang

H, Zhang G (2018) Earth BioGenome project: sequencing life for the future of life. Proc Natl Acad Sci 115(17):4325–4333. https://doi.org/10.1073/pnas.1720115115

Li H, Durbin R (2009) Fast and accurate short read alignment with burrows-wheeler transform. Bioinformatics 25(14):1754–1760. https://doi.org/10.1093/bioinformatics/btp324

Liew M, Pryor R, Palais R, Meadows C, Erali M, Lyon E, Wittwer C (2004) Genotyping of single-nucleotide polymorphisms by high-resolution melting of small amplicons. Clin Chem 50(7):1156–1164. https://doi.org/10.1373/clinchem.2004.032136

MacConaill LE, Burns RT, Nag A, Coleman HA, Slevin MK, Giorda K, Light M, Lai K, Jarosz M, McNeill MS, Ducar MD, Meyerson M, Thorner AR (2018) Unique, dual-indexed sequencing adapters with UMIs effectively eliminate index cross-talk and significantly improve sensitivity of massively parallel sequencing. BMC Genomics 19(1):30. https://doi.org/10.1186/s12864-017-4428-5

Maricic T, Whitten M, Pääbo S (2010) Multiplexed DNA sequence capture of mitochondrial genomes using PCR products. PLoS One 5(11):e14004. https://doi.org/10.1371/journal.pone.0014004

Massilani D, Morley MW, Mentzer SM, Aldeias V, Vernot B, Miller C, Stahlschmidt M, Kozlikin MB, Shunkov MV, Derevianko AP, Conard NJ, Wurz S, Henshilwood CS, Vasquez J, Essel E, Nagel S, Richter J, Nickel B, Roberts RG, Pääbo S, Slon V, Goldberg P, Meyer M (2022) Microstratigraphic preservation of ancient faunal and hominin DNA in Pleistocene cave sediments. Proc Natl Acad Sci 119(1):e2113666118. https://doi.org/10.1073/pnas.2113666118

Meyer M, Kircher M (2010) Illumina sequencing library preparation for highly multiplexed target capture and sequencing. Cold Spring Harb Protoc 2010(6):pdb.prot5448. https://doi.org/10.1101/pdb.prot5448

Michelsen C, Pedersen MW, Fernandez-Guerra A, Zhao L, Petersen TC, Korneliussen TS (2022) metaDMG – a fast and accurate ancient DNA damage toolkit for metagenomic data. 2022.12.06.519264

Murchie TJ, Kuch M, Duggan AT, Ledger ML, Roche K, Klunk J, Karpinski E, Hackenberger D, Sadoway T, MacPhee R, Froese D, Poinar H (2021) Optimizing extraction and targeted capture of ancient environmental DNA for reconstructing past environments using the PalaeoChip Arctic-1.0 bait-set. Quat Res 99:305–328. https://doi.org/10.1017/qua.2020.59

Murchie TJ, Karpinski E, Eaton K, Duggan AT, Baleka S, Zazula G, MacPhee RDE, Froese D, Poinar HN (2022) Pleistocene mitogenomes reconstructed from the environmental DNA of permafrost sediments. Curr Biol 32(4):851–860.e7. https://doi.org/10.1016/j.cub.2021.12.023

Nesje A (1992) A piston corer for lacustrine and marine sediments. Arct Alp Res 24(3):257–259. https://doi.org/10.1080/00040851.1992.12002956

Nichols RV, Vollmers C, Newsom LA, Wang Y, Heintzman PD, Leighton M, Green RE, Shapiro B (2018) Minimizing polymerase biases in metabarcoding. Mol Ecol Resour 18(5):927–939. https://doi.org/10.1111/1755-0998.12895

Nichols RV, Curd E, Heintzman PD, Shapiro B (2019) Targeted amplification and sequencing of ancient environmental and sedimentary DNA. Methods Mol Biol 1963:149–161. https://doi.org/10.1007/978-1-4939-9176-1_16

Nota K, Klaminder J, Milesi P, Bindler R, Nobile A, van Steijn T, Bertilsson S, Svensson B, Hirota SK, Matsuo A, Gunnarsson U, Seppä H, Väliranta MM, Wohlfarth B, Suyama Y, Parducci L (2022) Norway spruce postglacial recolonization of Fennoscandia. Nat Commun 13(1):1333. https://doi.org/10.1038/s41467-022-28976-4

Nwosu EC, Brauer A, Monchamp M-E, Pinkerneil S, Bartholomäus A, Theuerkauf M, Schmidt J-P, Stoof-Leichsenring KR, Wietelmann T, Kaiser J, Wagner D, Liebner S (2023) Early human impact on lake cyanobacteria revealed by a Holocene record of sedimentary ancient DNA. Commun Biol 6(1):1–12. https://doi.org/10.1038/s42003-023-04430-z

O'Leary NA, Wright MW, Brister JR, Ciufo S, Haddad D, McVeigh R, Rajput B, Robbertse B, Smith-White B, Ako-Adjei D, Astashyn A, Badretdin A, Bao Y, Blinkova O, Brover V, Chetvernin V, Choi J, Cox E, Ermolaeva O, Farrell CM, Goldfarb T, Gupta T, Haft D, Hatcher E, Hlavina W, Joardar VS, Kodali VK, Li W, Maglott D, Masterson P, McGarvey KM, Murphy

MR, O'Neill K, Pujar S, Rangwala SH, Rausch D, Riddick LD, Schoch C, Shkeda A, Storz SS, Sun H, Thibaud-Nissen F, Tolstoy I, Tully RE, Vatsan AR, Wallin C, Webb D, Wu W, Landrum MJ, Kimchi A, Tatusova T, DiCuccio M, Kitts P, Murphy TD, Pruitt KD (2016) Reference sequence (RefSeq) database at NCBI: current status, taxonomic expansion, and functional annotation. Nucleic Acids Res 44(D1):D733–D745. https://doi.org/10.1093/nar/gkv1189

Pal S, Gregory-Eaves I, Pick FR (2015) Temporal trends in cyanobacteria revealed through DNA and pigment analyses of temperate lake sediment cores. J Paleolimnol 54(1):87–101. https://doi.org/10.1007/s10933-015-9839-1

Parducci L, Väliranta M, Salonen JS, Ronkainen T, Matetovici I, Fontana SL, Eskola T, Sarala P, Suyama Y (2015) Proxy comparison in ancient peat sediments: pollen, macrofossil and plant DNA. Philos Trans R Soc Lond Ser B Biol Sci 370(1660):20130382. https://doi.org/10.1098/rstb.2013.0382

Parducci L, Bennett KD, Ficetola GF, Alsos IG, Suyama Y, Wood JR, Pedersen MW (2017) Ancient plant DNA in lake sediments. New Phytol 214(3):924–942. https://doi.org/10.1111/nph.14470

Parks DH, Chuvochina M, Rinke C, Mussig AJ, Chaumeil P-A, Hugenholtz P (2022) GTDB: an ongoing census of bacterial and archaeal diversity through a phylogenetically consistent, rank normalized and complete genome-based taxonomy. Nucleic Acids Res 50(D1):D785–D794. https://doi.org/10.1093/nar/gkab776

Pawlowski J, Bruce K, Panksep K, Aguirre FI, Amalfitano S, Apothéloz-Perret-Gentil L, Baussant T, Bouchez A, Carugati L, Cermakova K, Cordier T, Corinaldesi C, Costa FO, Danovaro R, Dell'Anno A, Duarte S, Eisendle U, Ferrari BJD, Frontalini F, Frühe L, Fazi S (2022) Environmental DNA metabarcoding for benthic monitoring: A review of sediment sampling and DNA extraction methods. Science of the Total Environment 818:151783. https://doi.org/10.1016/j.scitotenv.2021.151783

Pedersen MW, Ruter A, Schweger C, Friebe H, Staff RA, Kjeldsen KK, Mendoza MLZ, Beaudoin AB, Zutter C, Larsen NK, Potter BA, Nielsen R, Rainville RA, Orlando L, Meltzer DJ, Kjær KH, Willerslev E (2016) Postglacial viability and colonization in North America's ice-free corridor. Nature 537(7618):45–49. https://doi.org/10.1038/nature19085

Pedersen MW, De Sanctis B, Saremi NF, Sikora M, Puckett EE, Gu Z, Moon KL, Kapp JD, Vinner L, Vardanyan Z, Ardelean CF, Arroyo-Cabrales J, Cahill JA, Heintzman PD, Zazula G, MacPhee RDE, Shapiro B, Durbin R, Willerslev E (2021) Environmental genomics of late Pleistocene black bears and giant short-faced bears. Curr Biol 31(12):2728–2736.e8. https://doi.org/10.1016/j.cub.2021.04.027

Picard M, Pochon X, Atalah J, Pearman JK, Rees A, Howarth JD, Moy CM, Vandergoes MJ, Hawes I, Khan S, Wood SA (2022) Using metabarcoding and droplet digital PCR to investigate drivers of historical shifts in cyanobacteria from six contrasting lakes. Sci Rep 12(1):12810. https://doi.org/10.1038/s41598-022-14216-8

Pilon S, Zastepa A, Taranu ZE, Gregory-Eaves I, Racine M, Blais JM, Poulain AJ, Pick FR (2019) Contrasting histories of microcystin-producing cyanobacteria in two temperate lakes as inferred from quantitative sediment DNA analyses. Lake Reserv Manag 35(1):102–117. https://doi.org/10.1080/10402381.2018.1549625

Poulain AJ, Aris-Brosou S, Blais JM, Brazeau M, Keller WB, Paterson AM (2015) Microbial DNA records historical delivery of anthropogenic mercury. ISME J 9(12):2541–2550. https://doi.org/10.1038/ismej.2015.86

Ratnasingham S, Hebert PDN (2007) Bold: the barcode of life data system. Mol Ecol Notes 7(3):355–364. https://doi.org/10.1111/j.1471-8286.2007.01678.x

Revéret A, Rijal DP, Heintzman PD, Brown AG, Stoof-Leichsenring KR, Alsos IG (2023) Environmental DNA of aquatic macrophytes: The potential for reconstructing past and present vegetation and environments. Freshw Biol. https://doi.org/10.1111/fwb.14158

Rijal DP, Heintzman PD, Lammers Y, Yoccoz NG, Lorberau KE, Pitelkova I, Goslar T, Murguzur FJA, Salonen JS, Helmens KF, Bakke J, Edwards ME, Alm T, Bråthen KA, Brown AG, Alsos IG (2021) Sedimentary ancient DNA shows terrestrial plant richness continuously increased over the Holocene in northern Fennoscandia. Sci Adv 7(31):eabf9557. https://doi.org/10.1126/sciadv.abf9557

Rodriguez-Martinez S, Klaminder J, Morlock MA, Dalén L, Huang DY-T (2023) The topological nature of tag jumping in environmental DNA metabarcoding studies. Mol Ecol Resour 23(3):621–631. https://doi.org/10.1111/1755-0998.13745

Ross KS, Haites NE, Kelly KF (1990) Repeated freezing and thawing of peripheral blood and DNA in suspension: effects on DNA yield and integrity. J Med Genet 27(9):569–570. https://doi.org/10.1136/jmg.27.9.569

Safarikova M, Kubena AA, Frankova V, Zima T, Kalousova M (2021) The effects of different storage conditions and repeated freeze/thaw cycles on the concentration, purity and integrity of genomic DNA. Folia Biol (Praha) 67(1):10–15

Sand KK, Jelavić S, Kjær KH, Prohaska A (2023) Importance of eDNA taphonomy and provenance for robust ecological inference: insights from interfacial geochemistry. 2023.01.24.525431

Savichtcheva O, Debroas D, Kurmayer R, Villar C, Jenny JP, Arnaud F, Perga ME, Domaizon I (2011) Quantitative PCR enumeration of Total/toxic Planktothrix rubescens and Total cyanobacteria in preserved DNA isolated from Lake sediments. Appl Environ Microbiol 77(24):8744–8753. https://doi.org/10.1128/AEM.06106-11

Schnurrenberger DW, Kelts KR, Johnson TC, Shane LCK, Ito E (2001) National lacustrine core repository (LacCore). J Paleolimnol 25(1):123–127. https://doi.org/10.1023/A:1008171027125

Schubert M, Lindgreen S, Orlando L (2016) AdapterRemoval v2: rapid adapter trimming, identification, and read merging. BMC Res Notes 9(1):88. https://doi.org/10.1186/s13104-016-1900-2

Schulte L, Bernhardt N, Stoof-Leichsenring K, Zimmermann HH, Pestryakova LA, Epp LS, Herzschuh U (2021) Hybridization capture of larch (Larix mill.) chloroplast genomes from sedimentary ancient DNA reveals past changes of Siberian forest. Mol Ecol Resour 21(3):801–815. https://doi.org/10.1111/1755-0998.13311

Schulte L, Meucci S, Stoof-Leichsenring KR, Heitkam T, Schmidt N, von Hippel B, Andreev AA, Diekmann B, Biskaborn BK, Wagner B, Melles M, Pestryakova LA, Alsos IG, Clarke C, Krutovsky KV, Herzschuh U (2022) Larix species range dynamics in Siberia since the last glacial captured from sedimentary ancient DNA. Commun Biol 5(1):1–11. https://doi.org/10.1038/s42003-022-03455-0

Seeber PA, von Hippel B, Kauserud H, Löber U, Stoof-Leichsenring KR, Herzschuh U, Epp LS (2022) Evaluation of lake sedimentary ancient DNA metabarcoding to assess fungal biodiversity in Arctic paleoecosystems. Environmental DNA 4(5):1150–1163. https://doi.org/10.1002/edn3.315

Seersholm FV, Pedersen MW, Søe MJ, Shokry H, Mak SST, Ruter A, Raghavan M, Fitzhugh W, Kjær KH, Willerslev E, Meldgaard M, Kapel CMO, Hansen AJ (2016) DNA evidence of bowhead whale exploitation by Greenlandic Paleo-Inuit 4,000 years ago. Nat Commun 7(1):13389. https://doi.org/10.1038/ncomms13389

Selway CA, Armbrecht L, Thornalley D (2022) An outlook for the Acquisition of Marine Sedimentary Ancient DNA (sedaDNA) from North Atlantic Ocean archive material. Paleoceanogr Paleoclimatol 37(5):e2021PA004372. https://doi.org/10.1029/2021PA004372

Shao W, Khin S, Kopp WC (2012) Characterization of effect of repeated freeze and thaw cycles on stability of genomic DNA using pulsed field gel electrophoresis. Biopreserv Biobank 10(1):4–11. https://doi.org/10.1089/bio.2011.0016

Shapiro B, Ho SYW, Drummond AJ, Suchard MA, Pybus OG, Rambaut A (2011) A Bayesian phylogenetic method to estimate unknown sequence ages. Mol Biol Evol 28(2):879–887. https://doi.org/10.1093/molbev/msq262

Sjögren P, Edwards ME, Gielly L, Langdon CT, Croudace IW, Merkel MKF, Fonville T, Alsos IG (2017) Lake sedimentary DNA accurately records 20th century introductions of exotic conifers in Scotland. New Phytol 213(2):929–941. https://doi.org/10.1111/nph.14199

Slon V, Hopfe C, Weiß CL, Mafessoni F, de la Rasilla M, Lalueza-Fox C, Rosas A, Soressi M, Knul MV, Miller R, Stewart JR, Derevianko AP, Jacobs Z, Li B, Roberts RG, Shunkov MV, de Lumley H, Perrenoud C, Gušić I, Kućan Ž, Rudan P, Aximu-Petri A, Essel E, Nagel S, Nickel B, Schmidt A, Prüfer K, Kelso J, Burbano HA, Pääbo S, Meyer M (2017) Neandertal and Denisovan DNA from Pleistocene sediments. Science 356(6338):605–608. https://doi.org/10.1126/science.aam9695

Smith O, Momber G, Bates R, Garwood P, Fitch S, Pallen M, Gaffney V, Allaby RG (2015) Archaeology. Sedimentary DNA from a submerged site reveals wheat in the British Isles 8000 years ago. Science 347(6225):998–1001. https://doi.org/10.1126/science.1261278

Soares AER (2019) Hybridization capture of ancient DNA using RNA baits. In: Shapiro B, Barlow A, Heintzman PD, Hofreiter M, Paijmans JLA, Soares AER (eds) Ancient DNA: methods and protocols. Springer, New York, pp 121–128

Taberlet P, Coissac E, Pompanon F, Gielly L, Miquel C, Valentini A, Vermat T, Corthier G, Brochmann C, Willerslev E (2007) Power and limitations of the chloroplast trnL (UAA) intron for plant DNA barcoding. Nucleic Acids Res 35(3):e14. https://doi.org/10.1093/nar/gkl938

Taberlet P, Bonin A, Zinger L, Coissac E, Taberlet P, Bonin A, Zinger L, Coissac E (2018) Environmental DNA: for biodiversity research and monitoring. Oxford University Press, Oxford, New York

ter Schure ATM, Bajard M, Loftsgarden K, Høeg HI, Ballo E, Bakke J, Støren EWN, Iversen F, Kool A, Brysting AK, Krüger K, Boessenkool S (2021) Anthropogenic and environmental drivers of vegetation change in southeastern Norway during the Holocene. Quat Sci Rev 270:107175. https://doi.org/10.1016/j.quascirev.2021.107175

van Vugt L, Garcés-Pastor S, Gobet E, Brechbühl S, Knetge A, Lammers Y, Stengele K, Alsos IG, Tinner W, Schwörer C (2022) Pollen, macrofossils and sedaDNA reveal climate and land use impacts on Holocene mountain vegetation of the Lepontine Alps, Italy. Quat Sci Rev 296:107749. https://doi.org/10.1016/j.quascirev.2022.107749

Vernot B, Zavala EI, Gómez-Olivencia A, Jacobs Z, Slon V, Mafessoni F, Romagné F, Pearson A, Petr M, Sala N, Pablos A, Aranbur A, De Castro JMB, Carbonell E, Li B, Krajcarz MT, Krivoshapkin AI, Kolobova KA, Kozlikin MB, Shunkov MV, Derevianko AP, Viola B, Grote S, Essel E, Herraéz DL, Nagel S, Nickel B, Richter J, Schmidt A, Peter B, Kelso J, Roberts RG, Arsuaga JL, Meyer M (2021) Unearthing Neanderthal population history using nuclear and mitochondrial DNA from cave sediments. Science 372(6542). https://doi.org/10.1126/science.abf1667

Von Eggers J, Monchamp M-E, Capo E, Giguet-Covex C, Spanbauer T, Heintzman PD (2022) Inventory of ancient environmental DNA from sedimentary archives: locations, methods, and target taxa. Zenodo. https://doi.org/10.5281/ZENODO.6847522

Wang Y, Pedersen MW, Alsos IG, De Sanctis B, Racimo F, Prohaska A, Coissac E, Owens HL, Merkel MKF, Fernandez-Guerra A, Rouillard A, Lammers Y, Alberti A, Denoeud F, Money D, Ruter AH, McColl H, Larsen NK, Cherezova AA, Edwards ME, Fedorov GB, Haile J, Orlando L, Vinner L, Korneliussen TS, Beilman DW, Bjørk AA, Cao J, Dockter C, Esdale J, Gusarova G, Kjeldsen KK, Mangerud J, Rasic JT, Skadhauge B, Svendsen JI, Tikhonov A, Wincker P, Xing Y, Zhang Y, Froese DG, Rahbek C, Bravo DN, Holden PB, Edwards NR, Durbin R, Meltzer DJ, Kjær KH, Möller P, Willerslev E (2021) Late quaternary dynamics of Arctic biota from ancient environmental genomics. Nature 600(7887):86–92. https://doi.org/10.1038/s41586-021-04016-x

Wang Y, Korneliussen TS, Holman LE, Manica A, Pedersen MW (2022) ngsLCA – A toolkit for fast and flexible lowest common ancestor inference and taxonomic profiling of metagenomic data. Methods Ecol Evol 13(12):2699–2708. https://doi.org/10.1111/2041-210X.14006

Warinner C, Rodrigues JFM, Vyas R, Trachsel C, Shved N, Grossmann J, Radini A, Hancock Y, Tito RY, Fiddyment S, Speller C, Hendy J, Charlton S, Luder HU, Salazar-García DC, Eppler E, Seiler R, Hansen LH, Castruita JAS, Barkow-Oesterreicher S, Teoh KY, Kelstrup CD, Olsen JV, Nanni P, Kawai T, Willerslev E, von Mering C, Lewis CM, Collins MJ, Gilbert MTP, Rühli F, Cappellini E (2014) Pathogens and host immunity in the ancient human oral cavity. Nat Genet 46(4):336–344. https://doi.org/10.1038/ng.2906

Weber S, Junk I, Brink L, Wörner M, Künzel S, Veith M, Teubner D, Klein R, Paulus M, Krehenwinkel H (2023) Molecular diet analysis in mussels and other metazoan filter feeders and an assessment of their utility as natural eDNA samplers. Mol Ecol Resour 23(2):471–485. https://doi.org/10.1111/1755-0998.13710

Weiß CL, Dannemann M, Prüfer K, Burbano HA (2015) Contesting the presence of wheat in the British Isles 8,000 years ago by assessing ancient DNA authenticity from low-coverage data. elife 4:e10005. https://doi.org/10.7554/eLife.10005

Wilkinson MD, Dumontier M, Aalbersberg IJ, Appleton G, Axton M, Baak A et al (2016) The FAIR guiding principles for scientific data management and stewardship. Sci Data 3(1):160018. https://doi.org/10.1038/sdata.2016.18

Willerslev E, Hansen AJ, Binladen J, Brand TB, Gilbert MTP, Shapiro B, Bunce M, Wiuf C, Gilichinsky DA, Cooper A (2003) Diverse plant and animal genetic records from Holocene and Pleistocene sediments. Science 300(5620):791–795. https://doi.org/10.1126/science.1084114

Williams JW, Spanbauer TL, Heintzman PD, Blois J, Capo E, Goring SJ, Monchamp M-E, Parducci L, Eggers JMV, Alsos IG, Bowler C, Coolen MJL, Cullen N, Crump S, Epp LS, Fernandez-Guerra A, Grimm E, Herzschuh U, Mereghetti A, Meyer RS, Nota K, Pedersen MW, Pérez V, Shapiro B, Stoof-Leichsenring KR, Wood J (2023) Strengthening global-change science by integrating aeDNA with paleoecoinformatics. Trends Ecol Evol. 38(10):P946-960. https://doi.org/10.1016/j.tree.2023.04.016

Wood DE, Lu J, Langmead B (2019) Improved metagenomic analysis with kraken 2. Genome Biol 20(1):257. https://doi.org/10.1186/s13059-019-1891-0

Zavala EI, Jacobs Z, Vernot B, Shunkov MV, Kozlikin MB, Derevianko AP, Essel E, de Fillipo C, Nagel S, Richter J, Romagné F, Schmidt A, Li B, O'Gorman K, Slon V, Kelso J, Pääbo S, Roberts RG, Meyer M (2021) Pleistocene sediment DNA reveals hominin and faunal turnovers at Denisova cave. Nature 595(7867):399–403. https://doi.org/10.1038/s41586-021-03675-0

Ziesemer KA, Mann AE, Sankaranarayanan K, Schroeder H, Ozga AT, Brandt BW, Zaura E, Waters-Rist A, Hoogland M, Salazar-García DC, Aldenderfer M, Speller C, Hendy J, Weston DA, MacDonald SJ, Thomas GH, Collins MJ, Lewis CM, Hofman C, Warinner C (2015) Intrinsic challenges in ancient microbiome reconstruction using 16S rRNA gene amplification. Sci Rep 5:16498. https://doi.org/10.1038/srep16498

Zinger L, Bonin A, Alsos IG, Bálint M, Bik H, Boyer F, Chariton AA, Creer S, Coissac E, Deagle BE, De Barba M, Dickie IA, Dumbrell AJ, Ficetola GF, Fierer N, Fumagalli L, Gilbert MTP, Jarman S, Jumpponen A, Kauserud H, Orlando L, Pansu J, Pawlowski J, Tedersoo L, Thomsen PF, Willerslev E, Taberlet P (2019) DNA metabarcoding – need for robust experimental designs to draw sound ecological conclusions. Mol Ecol 28(8):1857–1862. https://doi.org/10.1111/mec.15060

Chapter 4
Bacterial and Archaeal DNA from Lake Sediments

Aurèle Vuillemin, Marco J. L. Coolen, Jens Kallmeyer, Susanne Liebner, and Stefan Bertilsson

Keywords Aquatic microbiology · Limnogeology · Bacteria · Archaea · Subsurface biosphere · Biogeochemical cycles · Terminal electron acceptors · Fermentation · Sediment stratigraphy · Early diagenesis

Introduction

Lake Sediments as Microbial Niches and Prokaryotic DNA Repositories

Microorganisms are abundant and diverse in inland waters (Newton et al. 2011; Magnabosco et al. 2018; Bertilsson and Mehrshad 2021), where they contribute significantly to global element fluxes through regulation of biogeochemical cycles. The distribution, density, and composition of microbial communities in aquatic sediments arise from multiple environmental processes that interactively create microbial niches. In contrast to the global ocean, where microbial taxa distributions are mainly affected by large-scale or global environmental factors, continental habitats are patchier and local processes such as geological and ecological events exert

A. Vuillemin (✉) · J. Kallmeyer · S. Liebner
GFZ German Research Centre for Geosciences, Section Geomicrobiology, Potsdam, Germany
e-mail: aurele.vuillemin@gfz-potsdam.de; jens.kallmeyer@gfz-potsdam.de; susanne.liebner@gfz-potsdam.de

M. J. L. Coolen
School of Earth and Planetary Sciences, Curtin University, Bentley, Australia
e-mail: marco.coolen@curtin.edu.au

S. Bertilsson
Department of Aquatic Sciences and Assessment, Swedish University of Agricultural Sciences, Uppsala, Sweden
e-mail: stefan.bertilsson@slu.se

E. Capo et al. (eds.), *Tracking Environmental Change Using Lake Sediments*, Developments in Paleoenvironmental Research 21,
https://doi.org/10.1007/978-3-031-43799-1_4

specific constraints on the assembly of microbial communities. Microorganisms recovered in lacustrine deposits are broadly assumed to reflect open aquatic ecosystems with typically high phylogenetic diversity and active metabolisms at the time of deposition, whereas at the pelagic-benthic transition zone, also called sediment-water interface, geochemical conditions drastically shift towards anoxia in the sediments. As sediments accumulate, microorganisms undergo growth-based selection paralleled by depletion of substrates. Thereby certain populations persist while communities are increasingly becoming confined, isolated and inactive. Because of this, the distribution of sedimentary bacterial and archaeal DNA is influenced and controlled by variable accumulation, turnover rates and preservation processes that act over short (seasonal, decadal) and longer (geologic) timescales.

Lacustrine sediments that accumulate over geologic timescales tend to form stratified archives that can reflect past regional climatic or local trophic conditions and biogeochemical processes, where specific microorganisms have thrived in response to, for example, resources, ecological processes and environmental constraints. Microbial responses to short-term environmental changes are currently a topic of increasing attention as anthropogenic disturbances have been found to alter aquatic conditions on a large scale (Jane et al. 2021). Because microbial communities drive key processes in biogeochemical cycles, they stand at the base of ecosystem functioning (Rousk and Bengtson 2014). Therefore, their physiological acclimation to environmental stress and feedback to higher trophic levels can help restore or, conversely, further degrade aquatic conditions and ecosystems. For instance, lakes act as significant components of the carbon pump as they accumulate large amounts of organic carbon in their sediments. Even if most lacustrine organic matter (OM) is terrestrial in origin and thereby assumed to be recalcitrant, sediment-dwelling prokaryotes have the potential to convert a significant but variable portion of this buried OM into methane and carbon dioxide that can be emitted to the atmosphere. Geochemical composition, physical regime, temperature, and sedimentation rate are some of the variables that exert significant influence on the pathways and rates of OM burial and degradation in lakes (Dean and Gorham 1998; Mendonça et al. 2017). This significant role of microorganisms as OM degraders also pertains to more variable, sometimes extreme, lacustrine environments, such as soda lakes (Zorz et al. 2019), salt lakes (Vavourakis et al. 2018), volcanic lakes (İnceoğlu et al. 2015), thermokarst lakes (in‘t Zandt et al. 2020), as well as periglacial (Anesio et al. 2017), seasonally (Thomas et al. 2019) and permanently frozen (Li and Morgan-Kiss 2019) lakes, or anthropogenically impacted systems, such as acidic mining lakes (Quatrini and Johnson 2018).

Currently, there is no universal answer to the actual relationships between the presence of microorganisms in lacustrine systems and the physical and geochemical processes that characterize their sediments as microbial habitats. A thorough and spatially resolved characterization of sediment-living microorganisms in terms of abundance, diversity and metabolic traits can accordingly shed light on nutrient cycling, OM degradation and redox processes during burial and many other biogeochemical processes. Compiling sedimentary DNA (sedDNA) data on lacustrine sediment microbiomes from a wide range of environmental and geochemical conditions will provide information on microbial ecosystem functioning in terms of core taxa, in response to anthropogenic perturbations and changing environmental conditions over recent geologic timescales.

In an effort to disentangle the significance and origin of prokaryotic sedDNA, this chapter presents taxonomic assemblages and metabolic traits of lacustrine microbiomes from source to sink and describes characteristic shifts in microbial functional guilds that develop during burial and subsequent sediment aging that shape the geochemical zonation. We further attempt to trace processes of microbial assembly and selection with increasing sediment age that foster an extant and persistently active, though less diverse, microbial biosphere even at great depths where thermodynamic conditions approach the limits for life. In order to correlate past physico-chemical processes in lakes to the stratigraphic distribution of microorganisms, we will also discuss specific methodological aspects that make best use of this prokaryotic sedDNA. We will briefly present an aseptic sampling strategy tailored for field operations and discuss how these can be efficiently adopted and modified to characterize extended sedDNA records. The chapter serves as an introduction for how prokaryotic sedDNA in lacustrine sediment archives can inform about biogeochemical cycles and their dynamic change over variable timescales.

Assembly of Lake Sediment Microbiomes

Biogeographic patterns of microbial distribution in terms of presence, abundance, taxonomic diversity and functions principally arise from four basic environmental processes that shape microbial consortia (Hanson et al. 2012; Hanson 2017). This includes adaptive emergence of new genetic variation in the environment under scrutiny by evolutionary processes (diversification); passive transport of viable microorganisms during watershed runoff or river flooding (dispersal); species interactions related to, for example, competition for resources such as nutrients, organic and inorganic substrates (selection); and stochastic changes in community composition resulting from isolation, turnover, and replacement of the active community, with reduction of genetic richness during burial (drift). The development and increased availability of high-throughput sequencing technologies over the last decade have broadened the field of ecology to now encompass the composition, phylogeny, biogeography as well as functional and metabolic attributes of the subsurface biosphere (Orsi et al. 2013; Salazar et al. 2019; Vuillemin et al. 2019a, 2020a; Pearman et al. 2022a). Advancing sequencing technologies now allow to build global catalogs of microbiomes and their communities across Earth, and to investigate in particular the subsurface biosphere and microbial dark matter (Thompson et al. 2017; Nayfach et al. 2021).

In the global ocean, microbial niches and biogeographical patterns are strongly influenced and controlled by large-scale hydrodynamic factors and isolation of water masses (DeLong and Pace 2001; Sunagawa et al. 2015) and this also applies to the sediments in terms of distance from the shore, surface ocean productivity, and depositional rates and redox conditions (Kallmeyer et al. 2012; Hoshino et al. 2020). In comparison, continental habitats rather reflect the local history of geological events, resulting in a much more geographically constrained and dynamic biogeographical distribution patterns (Pavoine and Bonsall 2011; Nemergut et al. 2011). In theory, sedDNA originates from multiple sources contributed from interconnected aquatic habitats, e.g. outflows, rivers, water column, shallow sediment (Zwart et al. 2002; De Oliveira and Margis 2015; Tang et al. 2020). Considering a series of climatic, hydrological, geochemical and depositional events, the combination of these four basic assembly processes exerts constraint on the composition and functions of microbial communities (Niño-García et al. 2016; Paver et al. 2020), with selective environmental pressures determining the subsequent local growth of microorganisms (Lacap et al. 2011). Dispersed taxa that are either passively buried or actively grow in lake sediments thereby represent a potential record of past source-to-sink processes, suggesting that the distribution of microbial features along lake stratigraphic deposits reflects broad ecological processes (Gugliandolo et al. 2016) and environmental changes (Pearman et al. 2020, 2023). By mapping the variability of lacustrine microbiomes across a wide range of environmental conditions, environmental "omics" can advance our understanding of microbial ecosystem functioning at both the local and global scale (Escalas et al. 2019). Ultimately, the systematic investigation of lacustrine microbiomes at the time of sampling and in their stratigraphic context (Hall et al. 2018) makes it possible to define core community subsets across very different lacustrine ecosystems (Gibbons and Gilbert 2015; Ruiz-González et al. 2017).

Lacustrine Stratigraphy and Preservation of Sedimentary DNA

In lacustrine ecosystems, sediments that accumulate over time can, at least in theory, reflect trophic conditions and biogeochemical processes at the time of deposition, i.e. conditions where specific microorganisms thrived in response to available resources and environmental constraints (Fig. 4.1a). Lacustrine sediments thereby record and reflect past and recent climatic conditions while providing at the same time a wide range of ecological niches for actively growing contemporary microorganisms. Geomicrobiological investigations of extensive lacustrine records recovered through scientific drilling supported the hypothesis that microbial features correlate with contrasting climatic and sedimentological context (Glombitza et al. 2013; Vuillemin et al. 2013a, 2016a; Holm et al. 2020). Because substrates accumulating in sediments exert selective pressure during the initial phase after deposition (Fig. 4.1a), in-lake processes underscoring the deposition of such substrates in essence define the degree to which the lake floor can be colonized by prokaryotes (Vuillemin et al. 2014a), thereby potentially defining microbial community shifts across different lacustrine environments and climatic periods (Ariztegui et al. 2015). Prokaryotic sedDNA can thus be developed into a novel biomarker for linking

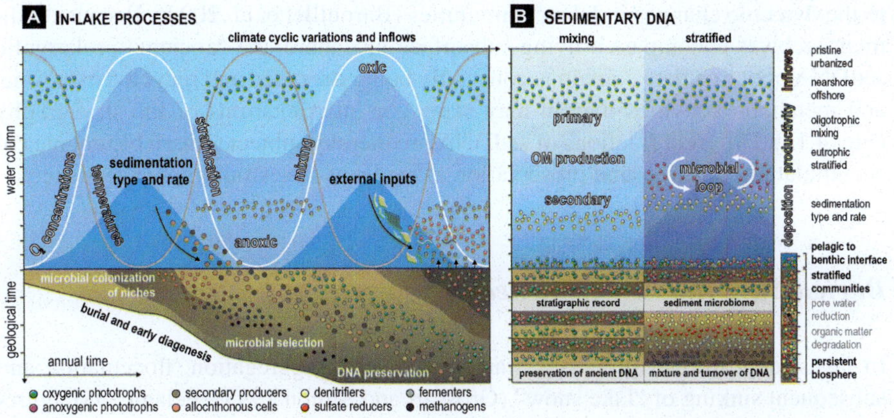

Fig. 4.1 Biotic and abiotic in-lake factors shaping sedimentary niches and their colonization by microbial communities over time. (**a**) In-lake processes, such as seasonal variations in temperature, oxygen level, as well as external inflows, OM production, sedimentation rate and OM content will result in sediment strata that are differently colonized by microorganisms. (**b**) SedDNA accumulates over time from multiple environmental sources that, once identified, can be indicative of catchment, in-lake and sediment processes. External sources of DNA preserved in the sediment can be considered ancient and thereby of the same stratigraphic age as the sediment, whereas the degree of sediment colonization by resident microbial populations varies according to the substrate initially deposited in the corresponding layers (i.e. pore water ions, OM types). While substrate depletion results in selective survival of taxa, it also induces a replacement of living cells and turnover of sedDNA whose composition then gradually departs from the time of deposition

microbial communities and biogeochemical cycles in sediments to past ecological processes recorded by lacustrine archives (Garner et al. 2020).

For Holocene time series, lake ecosystem function and characteristics have been successfully reconstructed using ancient sources of eukaryotic planktonic DNA (Coolen et al. 2004; Capo et al. 2021a; Kisand et al. 2018), enabling backtracking of vegetation in the watershed (Parducci et al. 2015, 2017; Alsos et al. 2018), lake colonization by freshwater fishes (Nelson-Chorney et al. 2019), primary production in the water column (Stoof-Leichsenring et al. 2015) and recent anthropogenic influences on food webs (Li et al. 2019a; Barouillet et al. 2023). In comparison, prokaryotic sedDNA essentially originates from a combination of inputs from the catchment, primary production and microbial secondary production in the water column whereas sediments as a whole feature a combination of such imported allochthonous and autochthonous DNA alongside actively growing sediment microbiota (Fig. 4.1b). The direct impact of environmental change on the distribution and diversity of microbial communities in the lacustrine subsurface has been documented for the Late Glacial (Vuillemin et al. 2018a; Thomas et al. 2020), Holocene transition (Coolen et al. 2008; Møller et al. 2020) and in modern times, e.g. Medieval Climate Anomaly, Little Ice Age (Vuillemin et al. 2013a; Garcia-Lopez et al. 2021), with a special focus on the Anthropocene (Cai et al. 2018; Han et al. 2020). However, there are still challenges in accurately interpreting prokaryotic sedDNA in relation to the dynamic changes in lake communities (Barouillet et al. 2023). Because sediment archives encompass a living subsurface biosphere, the "ancient" prokaryotic sedDNA pool that they contain is constantly being degraded and recycled over time, and metabolic activities therein may skew our interpretations of the stratigraphy (Fig. 4.1b). Different fractions of sedDNA are thereby subject to variable degrees of accumulation, turnover and preservation along lacustrine stratigraphic sequences.

Buried Alive in Lacustrine Sediment Archives

In lakes, depositional processes start with particle aggregation, flocculation and subsequent sinking of "lake snow" (Grossart and Simon 1998). Pelagic OM aggregates are rapidly colonized and partially degraded by heterotrophic bacterial communities (Schweitzer et al. 2001) which tends to decrease the efficiency of carbon burial in lake sediments (Bižić-Ionescu et al. 2018). Depending on the regime, depth, and geochemical conditions of the water column, a succession of aerobic and anaerobic redox processes takes place during particle settling (Grossart et al. 1997; Lu et al. 2013; Li et al. 2021), resulting in early replacement of cells (Reiche et al. 2011) and DNA turnover (Torti et al. 2015). Prior knowledge of microbial diversity in the water column is thus key to discriminating allochthonous, and possibly ancient sources of sedDNA for environmental reconstructions, from those that are autochthonous to the sediment (Garner et al. 2020). Upon deposition at the sediment-water interface, local geochemical conditions exert selective pressure on the growth of microorganisms and thereby shape the composition of the viable microbial communities emerging and accumulating in the sediments. From the sediment-water interface down into the first layers of the sediment, microbial biomass production experiences an exponential reduction in cell density and generation times (Haglund et al. 2003). As sediments accumulate, microbial metabolism slows down along with a concomitant depletion of electron acceptors and decreasing reactivity of organic substrates (Jørgensen and Marshall 2016). This gradual decrease in bioavailable substrates and cell generation times results in a "depauperate horizon" (Wurzbacher et al. 2017) reflecting a selective change in community composition towards a so-called "deep biosphere assembly" (Walsh et al. 2016; Starnawski et al. 2017) that can still sustain metabolic activity over geologic timescales (Vuillemin et al. 2020a, b).

Because prokaryotic sedDNA is a combination of ancient sources and active cells, uncertainties remain about the signal extractable from molecular inventories and the contribution of past and active communities to the sedimentary record (Stoeva et al. 2014; Pearman et al. 2022a). Depending on the changing metabolic constraints after sediment deposition and burial, prokaryotic sedDNA experiences different degrees of preservation depending on whether it persists intracellularly in living, albeit metabolically less active, and dormant cells (Ellegaard et al. 2020), or extracellularly after cell lysis as a free fraction or adsorbed to mineral surfaces (Torti et al. 2015; Vuillemin et al. 2017). Few studies have so far quantified the relative influence of initial environmental factors in regulating the final composition of total prokaryotic sedDNA in the sediment (Pearman et al. 2022b; Thorpe et al. 2022), and for lacustrine systems there is limited access to data for establishing a robust relationship between their environmental context and resident microbes that prevail over extended stratigraphic records (Ariztegui et al. 2015; Vuillemin et al. 2018b; Thomas et al. 2020).

Environmental fluctuations in lacustrine systems generate microheterogeneity and patchiness along the sediment stratigraphy, potentially contributing allochthonous sources of sedDNA and terrestrial OM in sporadic strata, or massively disrupting microbial structure at the lake floor (Brasell et al. 2021), possibly also modifying diffusion processes over sediment depth profiles. Extensive comparisons of community composition across sediment strata are thus required to benchmark the approach and evaluate whether sedDNA derived from prokaryotic organisms can be used to infer climatic and environmental conditions at the time of deposition (Bischoff et al. 2013; Holm et al. 2020; More et al. 2019). The parallel use of alternative (non-DNA) paleoenvironmental proxies is still recommended to identify and attribute prokaryotic sedDNA sources shaped by catchment or water body conditions at the time of deposition (Orsi et al. 2017; Vuillemin et al. 2016a, b; More et al. 2019, 2021).

Taxonomic Assemblages and Metabolic Guilds from Source to Sink

Allochthonous Versus Autochthonous Microorganisms in Lake Sediments

Environmental Processes: Dispersal, Dormancy and Resuscitation of Microbial Seed Banks

Prokaryotic cells derived from terrestrial and aquatic sources are subject to different degrees of preservation and dormancy (Lennon and Jones 2011) and often constitute a dominant fraction of sedDNA in aquatic systems (Aanderud et al. 2016; Vuillemin et al. 2017). Such microbial seed banks include active, viable but dormant cells, and resting stages (e.g. cysts, endospores, exospores, myxospores, akinetes), either transported as particulate OM from the catchment (e.g. soil bacteria) or emerging as planktonic aggregates (e.g. primary and secondary producers) that sink to the lake floor, that can be considered allochthonous from a sediment perspective (Tang et al. 2020). Dormant cells displaying sufficient metabolic flexibility can potentially be resuscitated outside of their preferred habitats and become active in the water column or sediment. The remaining dormant cells that are not resuscitated in the sediment environment only experience weak or no selection during burial (Lennon et al. 2021). Thus, microbial resting stages are ubiquitous in anoxic sediments (Wunderlin et al. 2014a), whereas dead cells and those that do not produce persistent resting stages typically lyse, senesce and in the process provide labile substrates for growth and expansion of active communities responsible for replacement processes (Wurzbacher et al. 2017; Pearman et al. 2021). For this reason, microbiomes at the sediment-water interface are expected to initially reflect an open geochemical system that displays high phylogenetic diversity and metabolic activity. As the sediments become isolated and undergo aging, this diversity is purged and depleted (Martiny et al. 2006; Walsh et al. 2016).

To accurately link diverse microbial communities to their corresponding habitats and geochemical niches, a suite of functional marker genes (Fig. 4.2) reflective of metabolic traits and processes inherent to specific sediment settings or geochemical conditions can be investigated in parallel to taxonomic ribosomal RNA (rRNA) genes, enabling analysis of microbial guilds and ecological functional groups (Green et al. 2008). Since microbial guilds are never monophyletic (Boucher et al. 2003; Falkowski et al. 2008), taxonomic and functional genes are studied using a combination of amplicon sequencing, quantitative PCR (qPCR) assays, metagenomics (i.e. total DNA) and metatranscriptomics (i.e. total RNA) to characterize and quantify their functional diversity and ensure that the corresponding metabolic potential is present and realized (Perez-Garcia et al. 2016).

Starting with the watershed, members of the phyla Acidobacteriota, Actinobacteriota, Firmicutes and Verrucomicrobiota are all major components of the soil microbiome known to produce resting stages (Janssen et al. 2002; Paul et al. 2019). These persisting cells can be mobilized from terrestrial soils and hydrologically imported to lacustrine systems, especially during periods of high precipitation or floods (Crump et al. 2012; Ruiz-González et al. 2015). Although aerobic heterotrophs among remobilized resting stages often actively colonize planktonic aggregates in the water column, once deposited at the lake floor they undergo rapid replacement below the sediment-water interface to persist only as dormant cells, thus constituting rare taxa (Cardman et al. 2014; Ren et al. 2022). In contrast, allochthonous exospores (Actinobacteriota) and endospores (Firmicutes) from anaerobes imported and sedimented in lakes can only resuscitate in the deeper anoxic layers (Aüllo et al. 2013; Cupit et al. 2019). Furthermore, depending on OM productivity, light penetration, depth and redox zonation of the water column (Pearman et al. 2021), prokaryotic photoautotrophs (i.e. primary producers) can evolve both as planktonic and benthic colonies and settle actively at the lake floor (Pearman et al. 2022a). Akinete-forming Cyanobacteria (Ellegaard and Ribeiro 2018) and obligate anaerobic Chlorobia (Gregersen et al. 2009) preserved in the sediment can later act as indicators of past productivity (Pearman et al. 2022b) and stratification events (Coolen and Overmann 2007). It has been suggested that members of Planctomycetota genera (e.g. *Phycisphaera*, *Pirellula*) and other heterotrophs (e.g. Chloroflexota, Bacteroidia, Actinobacteriota) in the microbial loop act as secondary producers and scavengers of collapsing or senescent algal blooms (Lage and Bondoso 2014), resulting in the overrepresentation of mixotrophs over phototrophs in the sedDNA pool (Kurilkina et al. 2016; Vuillemin et al. 2017). Moreover, oxic conditions at the lake floor allow benthic macro- and microfauna to

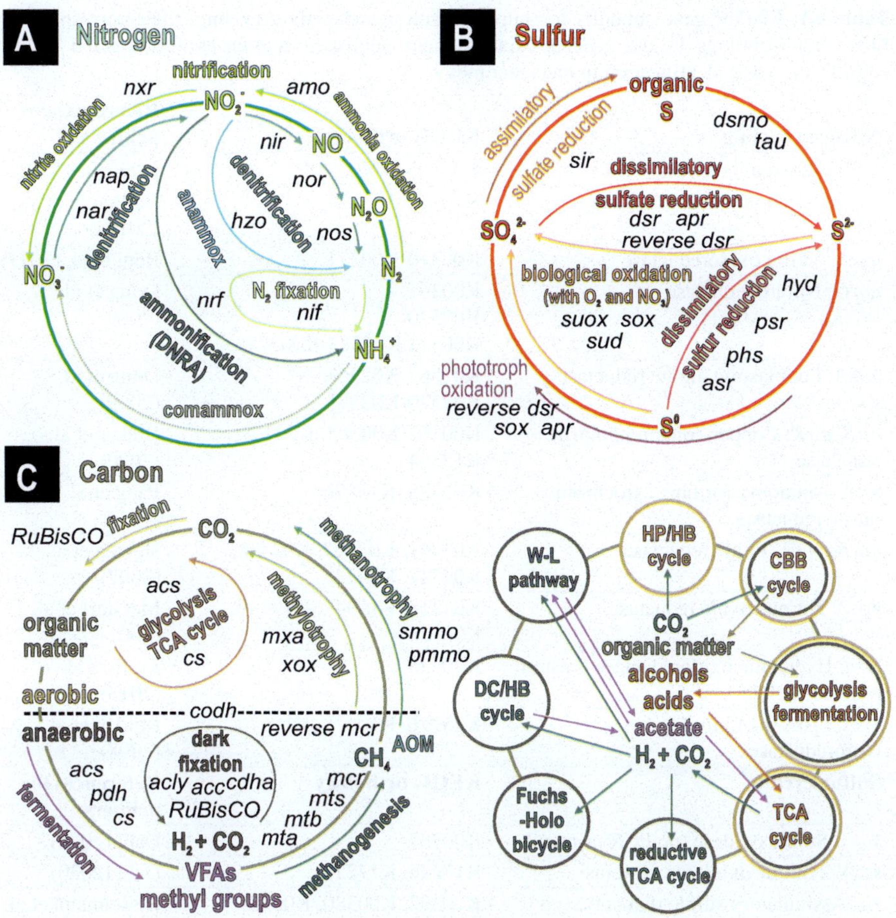

Fig. 4.2 Biogeochemical cycling of nitrogen, sulfur and carbon and the main corresponding functional marker genes and pathways. (**a**) Nitrogen cycling functional genes related to processes of nitrogen fixation (*nif*), nitrification (*amo*, *nxr*), anaerobic ammonia oxidation or anammox (*hzo*), dissimilatory nitrate reduction to ammonium (DNRA, *nrf*), and denitrification (*nar*, *nap*, *nir*, *nor*, *nos*). (**b**) Sulfur cycling functional genes related to processes of sulfur (*sox*, *suox*, *sud*, reverse *dsr*, *apr*) and organic sulfur (*dsm*o, *tau*) oxidation, assimilatory sulfite reduction (*sir*), dissimilatory sulfate (*dsr*, *apr*) and intermediate sulfur species reduction (*asr*, *phs*, *psr*, *hyd*). (**c**) Carbon cycling functional genes related to aerobic and anaerobic processes of carbon fixation (*RuBisCo*, *acly*, *acc*, *codh*, *cdha*), glycolysis and fermentation (*acs*, *cs*, *pdh*), methylotrophy (*mxa*, *xox*), methanotrophy (*smmo*, *pmmo*, reverse *mcr)* and methanogenesis (*mta*, *mtb*, *mts*, *mcr*), followed by metabolic pathways of autotrophic carbon fixation (Berg 2011) including the light and dark Calvin-Benson-Bassham (CBB) cycle, the aerobic 3-hydroxypropionate/4-hydroxybutyrate (HP/HB) and anaerobic dicarboxylate/4-hydroxybutyrate (DC/HB) cycle, the reductive tricarboxylic acid (TCA) cycle, the Wood-Ljungdahl (W-L) pathway (i.e. acetogenesis, methanogenesis) and Fuchs-Holo (or 3-hydroxypropionate) bicycle that can be coupled to glycolysis, fermentation and TCA cycle via production and uptake of metabolic intermediates. For gene abbreviations and corresponding KEGG orthology, see Table 4.1

Table 4.1 Marker gene subunits for nitrogen, sulfur and carbon cycling, their corresponding functional orthologs to the KEGG database, and published references for targeted primer amplification and identification in metagenomes

Nitrogen cycling	**KEGG orthology**	**References for primers**
nifA: Nitrogen fixation	K02585 to K02588	Pang et al. (2019)
amoA: Ammonia monooxygenase	K10944	Vuillemin et al. (2019a)
nxrA: Nitrite oxidoreductase	K00370-K00371	Rani et al. (2017)
narG: Nitrate reductase	K00367, K00370-K00371,K00373-K00374	Dong et al. (2009)
napA: Periplasmic nitrate reductase	K02567-K02568, K02570-K02571	Dong et al. (2009)
nirS, nirK: Copper-containing nitrite reductase	K00362-K00363, K00368, K15864	Braker et al. (2000)
nrfA: Ammonia-forming cytochrome nitrite reductase	K03385, K15876	Pang et al. (2019)
norB: Nitric oxide reductase	K04561, K02305, K02448, K04747, K04748	Heylen et al. (2007)
nosZ: Nitrous oxide reductase	K07218, K19342, K19339, K00376	McGuirl et al. (2001)
Hzo: Hydrazine oxidoreductase (custom primers F1-R1)	K20935	Kong et al. (2013)
nfsA, nfsB, snrA, nbzA, pnrA: Nitroreductases	K15976, K10678, K10679	Roldán et al. (2008)
Sulfur cycling	**KEGG orthology**	**References for primers**
suox: Sulfite oxidase (no PCR primers)	K00387	Dahl (2020)
soxYZ: Sulfur-oxidizing proteins	K17226, K17227	Dahl (2020)
sir: Assimilatory sulfite reductase (no PCR primers)	K00392, K00380, K00381	Gisselmann et al. (1993)
dsrA, dsrB: Dissimilatory sulfite reductase	K11179 to K11181, K07236, K07237	Loy et al. (2009)
aprA, aprB: Adenylylsulfate reductase	K00394, K00395	Biderre-Petit et al. (2011)
asrA: Anaerobic sulfite reductase	K16950, K16951, K00385	Vuillemin et al. (2018b)
phsA, psrA: Thiosulfate/polysulfide reductase	K08352 to K08354	Wasmund et al. (2017)
hydB: Sulfhydrogenase	K17993 to K17995-K17996	Wasmund et al. (2017)
dsmo: Dimethylsulfoxide reductase (no PCR primers)	K07306 to K07308	Baltes et al. (2003)
tauA: Taurine dioxygenase	K02975, K19245	Wetzl et al. (2016)

(continued)

Table 4.1 (continued)

Nitrogen cycling	**KEGG orthology**	**References for primers**
Methane cycling	**KEGG orthology**	**References for primers**
mxaF: Methanol dehydrogenase	K23995	Heyer et al. (2002)
xoxF: Methanol dehydrogenase homologue	K23995	Taubert et al. (2015)
pmmoA: Particulate methane monooxygenase	K10944 to K10946	Inagaki et al. (2004)
smmoX: Soluble methane monooxygenase	K16157 to K16162	Heyer et al. (2002)
mtaB: Methanol CoM methyltransferase	K14080	Bose et al. (2008)
mtbA: Methylamine-specific CoM methyltransferase	K14082	Veit et al. (2005)
mtsA: Methylthiol-specific CoM methyltransferase	K16954	Tallant et al. (2001)
mcrA: Methyl-CoM reductase	K00399, K00401, K03421, K03422, K00402	Vuillemin et al. (2018b)
Carbon assimilation pathway	**KEGG orthology**	**References for primers**
accA, accD: Acetyl-CoA carboxylase	K01962, K01963	Freiberg et al. (2004)
acsA: Acetyl-CoA synthetase (glycolysis)	K01895	Gardner et al. (2006)
cs: Citrate synthase (TCA cycle) (*gtlA* primers)	K01647	Castro et al. (2012)
acly: ATP-citrate lyase (reductive TCA cycle) (*aclA, AclB* primers)	K15230, K15231	Voordeckers et al. (2008)
pdhA, pdhB: Pyruvate dehydrogenase (fermentation)	K00161, K00162	Bao et al. (2021)
codH: Carbon monoxide dehydrogenase (W-L pathway) (CODHech primers)	K00198, K00193	Omae et al. (2021)
cdhA, cdhB: Acetyl-CoA synthase (W-L pathway)	K00192, K00194, K00197	Eggen et al. (1991)
RuBisCO: Ribulose-1,5-diphosphate oxidase (CBB cycle) (*cbbL primers*)	K01601-K01602	Chi et al. (2018)

bioturbate the sediments which highly increases replacement and turnover rates of sedDNA (Zeng et al. 2014). Although subsequent sediment anoxia selects against the main microbial populations seen in the water column, microaerobic and mixotrophs can nevertheless persist during the initial stage of burial and dwell over longer timescales, i.e. tens (Vuillemin et al. 2018a) or even hundreds of thousands of years (Thomas et al. 2020).

Taxonomy: Primary Versus Secondary Producers in the Microbial Loop

Primary producers: In aquatic systems, Cyanobacteria are among the main prokaryotic primary producers that can be the cause of toxic blooms and create negative feedbacks on the water column, with excess ammonia release and oxygen depletion during bloom collapse and senescence (Huisman et al. 2018). Some important features among Cyanobacteria identified in sedDNA assemblages (see Chap. 5 of this volume) relate to their lifestyle and ecology, including the capacity to fix nitrogen (i.e. diazotrophy), produce neurotoxins, perform both oxygenic and anoxygenic photosynthesis (i.e. with and without oxygen production), persist and grow under microoxic to sulfidic conditions (Hamilton et al. 2018; Block et al. 2021), and form colonies or akinetes (i.e. resting stages) (Legrand et al. 2016; Ellegaard and Ribeiro 2018). Taxonomic analysis of cyanobacterial sedDNA has been used to determine shifts in lake salinity, such as the postglacial re-connection of the paleo-freshwater Antarctic Ace Lake and subsequent development of modern-day brackish conditions based on the first occurrence of the pelagic cyanobacterium *Synechococcus* strain Ace (Coolen et al. 2008). In stratified aquatic systems where light penetrates down to the chemocline (Gregersen et al. 2009; Storelli et al. 2013), anoxygenic aerobic and anaerobic phototrophs (purple sulfur, purple non-sulfur, green sulfur, green non-sulfur bacteria, Heliobacteriaceae) can actively grow under oxygen minimum, ferruginous and sulfidic conditions, oxidizing either dissolved iron or sulfur in close proximity to the chemocline (Madigan 2003; Crowe et al. 2008). Accordingly, increased representation of sedDNA sequences preserved from taxa among green sulfur Chlorobia, green non-sulfur Chloroflexales, purple alphaproteobacterial Rhodobacteraceae or purple sulfur gammaproteobacterial Chromatiaceae can be indicative of water column stratification and shifts in salinity (Coolen and Overmann 1998, 2007; Asao et al. 2011). Such assemblages of phototrophic taxa are known to constitute the upper layers of stromatolites and to form biofilms (Wong et al. 2015). However, the mere presence of pigments involved in phototrophy does not necessarily imply a significant role in primary production as some pigment-producing bacteria (e.g. among Planctomycetota, Bacteroidia, Chloroflexota, Alphaproteobacteria) may be active as secondary producers in the microbial loop via photoheterotrophy and mixotrophy (Eiler 2006). Phototrophy among prokaryotes rely on specific pigments used to harvest light at different wavelengths (Scheer 2006), such as chlorophyll variants (*Chl-a*, *Chl-d*, *Chl-f*) and bacteriochlorophyll (*Bchl-a* to *Bchl-e*) (Suzuki et al. 1997). Some other pigments are associated exclusively with anoxygenic photosynthesis (*puf*M genes) (Pfennig and Trüper 1992) and some of the genes encoding for these pigments can be compared to the corresponding fossil pigments along stratigraphic records (Leavitt and Hodgson 2006; Vuillemin et al. 2016a). Carbon fixation among phototrophs proceeds mostly via the Calvin-Benson-Bassham (CBB) cycle, with genes encoding the ribulose-1, 5-diphosphate carboxylase (*RuBisCO*) and some specific cytochromes (e.g. *cbb*L, *cbb*S) (Gibson et al. 2002). However, *RuBisCO* genes can also operate in chemoautotrophic dark carbon fixation in the absence of light (Canfield et al. 2005). To interpret the detection of such genes, one has to place observations in an environmental context and also consider taxonomic assignments (Hügler and Sievert 2011). Additionally, while mixotrophs produce photoactive pigments, such as bacteriorhodopsins, proteorhodopsins and various carotenoids (Béjà et al. 2001; Eiler 2006), they cannot use CO_2 as their sole carbon source.

Secondary producers: As OM aggregates from pelagic primary production sink through the water column, a microbial loop is established via mixotrophic and heterotrophic interactions that fuel a complex network of prokaryotes and protists (Ryszard et al. 2009; Xing et al. 2020). Depending on seasonal variations and trophic levels (Lewis 2009; Li and Morgan-Kiss 2019), the microbial loop can drive intense nutrient cycling and heterotrophic processes leading, in turn, to rapid oxygen depletion in the hypolimnion. Subsequently, some taxonomic groups known to evolve as micro-aerobes can thrive in the oxycline – where oxygen concentrations are low – and settle at the sediment-water interface (Vuillemin et al. 2017) where they feed on sinking OM and reduced forms of nutrients diffusing out of the underlying sediment. Some Planctomycetota (e.g. *Phycisphaera*, *Pirellula*, *Rhodopirellula*) are ubiquitous across a broad range of geochemical conditions (Storesund et al. 2020) and produce carotenoids that place them at the intersection between primary and secondary producers (Kallscheuer et al. 2019). These populations can anchor their cells to OM aggregates or form biofilms at the sediment-water interface (Lage and Bondoso 2014), produce sulfatases involved in the degradation of complex sugars or derive energy from ammonia, either aerobically (i.e. nitrification) or anaerobically (i.e anammox). Finally, the Planctomycetota can form cysts in shallow sediments that persist long after burial (Wiegand et al. 2018). Similar traits are reported for filamentous non-phototrophic Chloroflexota, including aerobic and facultative anaerobic representatives (e.g. *Nitrolancea*, *Caldilinea*) and obligate anaerobic taxa (e.g. *Anaerolinea*), displaying the ability to form biofilms, degrade biopolymers, and ferment carbohydrates and drive various nitrogen transformation processes (Speirs et al. 2019). These fermenters are known to thrive in shallow sediments. The class Bacteroidia contains pelagic and benthic members, some of which producing photoactive pigments, degrading algal matter and assimilating carbohydrates via aerobic, microaerobic and anaerobic glycolysis (González et al. 2008; Dalcin Martins et al. 2021) – that play a key role in the turnover of sinking OM after algal blooms, potentially contributing to oxygen depletion in the water column (Vuillemin et al. 2022). As micro-oxic conditions develop (100–60 μM), they can shift to micro-aerobic respiration with metabolic capacity for uptake of dissolved OM and anaplerotic reactions (i.e. non-autotrophic CO_2 fixation) replenishing the tricarboxylic acid (TCA) cycle (Fernández-Gómez et al. 2013; Krüger et al. 2019). Although they do not produce resting stages, some taxa can sustain activity and growth under anoxic conditions (Phillips et al. 2017) and persist in the sediment long after burial (e.g. *Pontibacter*, *Tenacibaculum*, *Imtechella*) (Vuillemin et al. 2022). The protein-encoding genes superoxide dismutase (*sod*, i.e. H_2O disproportionation), and cytochrome-c oxidase (*cox*, i.e. aerobic respiration) are expressed during aerobic respiration, whereas those involved in cytochrome bd2 ubiquinol (*bd2*) and cytochrome cbb3-type (*cbb3*) oxidases are indicative of microaerobic respiration (Kalvelage et al. 2015).

Diazotrophs and nitrifiers: The ability to fix atmospheric nitrogen is carried out, mostly by Cyanobacteria, Actinobacteriota, Firmicutes and Alphaproteobacteria encoding by a complex of nitrogen fixation genes (*nif*) including homologous heterometal (Mo, V and Fe) nitrogenases responsible for the reduction of N_2 to NH_3 (Hu and Ribbe 2015). Nitrifiers, which are known to thrive in soils, in the vicinity of oxyclines and oxic sediments (Vuillemin 2023), constitute a metabolic guild involved in the recycling of inorganic fixed nitrogen (i.e. NH_4^+, NO_3^-, NO_2^-). By preventing a decrease in TN:TP ratio, this metabolic guild can adjust lake trophic status (Downing and McCauley 1992), especially in systems with long residence time (They et al. 2017). Aerobic ammonia oxidizers are often considered to be chemolithoautotrophs although heterotrophy on dissolved organic carbon has also been observed (Aylward and Santoro 2020). This metabolic guild includes several archaeal candidates among the class Nitrososphaeria, such as *Nitrosopumilus*, *Nitrososphaera*, *Nitrosotalea*, *Nitrosarchaeum* and *Nitrosomarinus* sp. (Jung et al. 2022), as well as genera among the Gammaproteobacteria, including *Nitrosomonas*, *Nitrosospira*, *Nitrosolobus*, *Nitrosovibrio* and *Nitrosococcus* sp. (Soliman and Eldyasti 2018). Some well known nitrite oxidizers further include *Nitrospina*, *Nitrospira*, *Nitrococcus*, and *Nitrobacter* sp. among Nitrospinota, Nitrospirota, Gamma- and Alphaproteobacteria, respectively (Ge et al. 2015). Ammonia monooxygenase (*amo*) and nitrite oxidoreductase (*nxr*) (Rani et al. 2017; Kitzinger et al. 2020) are involved in aerobic nitrification (Fig. 4.2a).

Stratified Microbial Communities in Shallow Subsurface Sediments

Environmental Processes: Imprints of Successive Respiration Modes and Fermentation

Although geochemical and depositional conditions exert initial control over the sedimentation and colonization of the lake bottom sediments by microorganisms (Fig. 4.1), selective growth of subpopulations on bioavailable substrates rapidly results in a mismatch between microbial characteristics and past lacustrine conditions during the early burial phase (Lacap et al. 2011). Pore water concentrations and availability of respiratory electron acceptors and other dissolved constituents, as well as the reactivity of sedimentary OM, are regarded as the main factors controlling the distribution and growth of microbial communities in aquatic subsurface sediments. As sediment accumulates and microorganisms are buried, the potential for exergonic metabolic reactions (i.e. Gibbs free energy) decreases with the sequential depletion of pore water oxidants during respiration (i.e. O_2, NO_3^-, Mn^{4+}, Fe^{3+}, SO_4^{2-}, CO_3^{2-}) and the stepwise decomposition of bioavailable OM (McInerney et al. 2009). Expecting that microorganisms utilize the most energetically favorable respiratory electron acceptor available (Fig. 4.3a) and preferentially degrade the most labile OM sources (Fig. 4.3b), microbial communities stratify with respect to

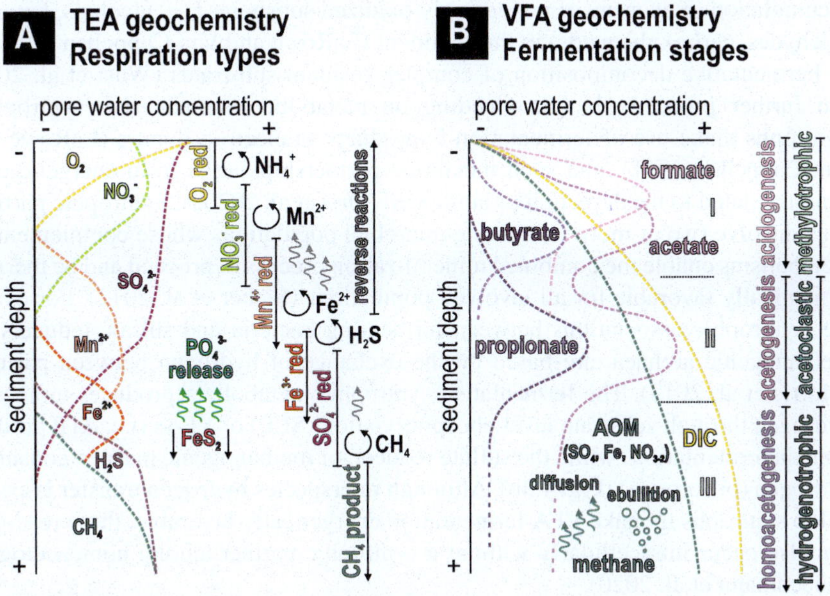

Fig. 4.3 Idealized model of stratified sediment pore water geochemistry related to the respiration of terminal electron acceptors and anaerobic remineralization of organic matter. (**a**) Successive respiratory use of terminal electron acceptors (TEA) according to available Gibbs free energy (i.e. O_2, NO_3^-, Mn^{4+}, Fe^{3+}, SO_4^{2-}, CO_3^{2-}) leading to their gradual depletion in pore waters and increase in their reduced counterparts. Upward diffusion allows for reverse redox reactions (e.g. cryptic sulfur cycling, anaerobic oxidation of methane). (**b**) Fermentation stages involved in the production of volatile fatty acids (VFAs) as metabolic intermediates (I-III: acidogenesis, acetogenesis, homoacetogenesis), and paralleled methanogenic pathways (I-III: methylotrophic, acetoclastic, hydrogenotrophic) leading to stepwise remineralization of OM that results in buildup of methane and dissolved inorganic carbon (DIC) over depth. The increase and subsequent decrease of VFA concentrations with depth reflect their fermentative production and turnover mostly via syntrophy

the gradual depletion of deposited substrates by "layering" metabolic guilds that experience selective growth over sediment depth profiles. The successive use of electron acceptors during microbial respiration as well as their cryptic recycling (Fig. 4.3a) typically result in the strong vertical redox zonation within the sediment column (Frindte et al. 2015).

In anoxic habitats, stepwise degradation of labile OM can also proceed, but then by means of fermentation without the need for external electron acceptors. Such fermentative degradation of OM (McInerney et al. 2009) begins with hydrolysis (i.e. breakdown into soluble compounds), and goes on with acidogenesis (i.e. conversion into organic acids) and acetogenesis (i.e. conversion into acetate) coupled to methylotrophic and acetoclastic methanogenesis (i.e. substrate conversion into methane). Advanced stages of fermentation include heterotrophic dark carbon fixation via homoacetogenesis (Ryan et al. 2008) and hydrogenotrophic methanogenesis (Fig. 4.3b). These reactions lead to formation and buildup of volatile fatty acids (VFAs), dissolved inorganic carbon (DIC) and methane with concomitant accumulation of an array of increasingly oxidized substrates (i.e. alcohols, ketones, aldehydes, carboxylic acids) in the sediment (LaRowe and Van Cappellen 2011).

Fermentative decomposition of complex residual substrates (Wüst et al. 2009) can further proceed via cross-feeding on metabolic products (i.e. syntrophy). Syntrophs make use of fermentation byproducts as electron donors (LaRowe and Van Cappellen 2011), and such metabolic partnerships arise from energetic constraints related to the depletion of labile OM (Orsi et al. 2020a). Syntrophic partnerships involve two or more interacting microbial populations whose complementary metabolisms enable the combined mineralization process to proceed and be thermodynamically favorable for all involved populations (Sieber et al. 2012). For example, syntrophic associations between fermenting bacteria and sulfate reducers, or methanogenic archaea, are based on the exchange of hydrogen between partners (Morris et al. 2013). The fermentative syntrophic metabolizer produces molecular hydrogen through substrate level phosphorylation (SLP) of VFAs (i.e. ATP production via fermentation) while the sulfate reducer, or methanogenic partner, consumes hydrogen (or formate) (Fig. 4.3b). Although interspecies hydrogen transfer is essential to syntrophs to make VFA fermentation energetically favorable, these reactions can also occur intracellularly without a syntrophic partner among homoacetogens (Wiechmann et al. 2020).

Charismatic Microbial Guilds in Shallow Sediments: Denitrifiers, Sulfate Reducers, Syntrophs, Methanogens

Denitrifiers: Denitrification starts in the oxycline and parallels the onset of anoxia. After oxygen has been depleted, denitrification is the second-most thermodynamically favorable respiratory process (Fig. 4.3a). This metabolism involves a stepwise reduction of NO_3^- to gaseous N_2O and N_2 as end products (Pessi et al. 2022). It is mediated by the respiratory (*nar*) and periplasmic (*nap*) nitrate reductase, nitrite reductase (*nir*), nitric oxide reductase (*nor*), and nitrous oxide reductase (*nos*) (Braker et al. 2000; Dong et al. 2009; Dalsgaard et al. 2014). The more reduced forms of inorganic nitrogen are respired in the final steps of denitrification (*nor, nos*) and result in a gaseous loss of fixed nitrogen (Einsle and Kroneck 2004; Lam and Kuypers 2011). Denitrifiers are common in soils, aquatic oxygen minimum zones, surface sediments and other oxygen depleted ecosystems (Tiedje 1994). Many taxa among the Alphaproteobacteria can perform denitrification (e.g. *Paracoccus*, *Agrobacterium*, *Rhizobium*, *Azospirillum*, *Nitrobacter*), as well as within Gammaproteobacteria (e.g. *Nitrosomonas*, *Alcaligenes*, *Pseudomonas*, *Thiobacillus*, *Acidovorax*, *Limnobacter*) and Bacteroidia (e.g. *Flavobacterium*, *Cytophaga*, *Flexibacter*). It is important to note that this metabolic trait is not taxonomically conserved and can therefore not readily be inferred from taxonomic information. Dissimilatory nitrate reduction to ammonium (DNRA), or heterotrophic ammonification, often occurs concurrently with denitrification (Fig. 4.2a) and is encoded by the gene ammonia-forming cytochrome nitrite reductase (*nrf*) (Pang et al. 2019), while nitroreductases (*ntr*) mediate reduction of nitro-compounds (Roldán et al. 2008).

Sulfate reducers: Although the ability to use sulfate as a respiratory electron acceptor (sulfate reduction) is restricted to certain phylogenetic groups (Castro et al. 2000), many apparent sulfate reducers display metabolic flexibility and are also capable of fermentation to produce VFAs and hydrogen, thereby acting as important regulators of OM turnover (Plugge et al. 2011). Sulfate reducers are mostly affiliated with the phylum Desulfobacterota (e.g. *Desulfobacca*, *Desulfomonas*, *Desulfobacter*, *Desulfovibrio*, *Desulfococcus*, *Thermodesulfobacterium*), but also include some Firmicutes (e.g. *Desulfotomaculum*, *Desulfosporosinus*) and specific groups within the Nitrospirota (i.e. *Thermodesulfovibrio*), Thermodesulfobiota (i.e. *Thermodesulfobium*) and Halobacterota (i.e. *Archaeoglobus*) (Muyzer and Stams 2008; Anantharaman et al. 2018). The canonical sequence of genes encoding proteins for respiratory reduction of sulfate to hydrogen sulfide (Fig. 4.2b; Table 4.1) is dissimilatory sulfite reductase (*dsr*), adenylylsulfate reductase (*apr*), anaerobic sulfite reductase (*asr*), thiosulfate/polysulfide reductase (*phs*/*psr*) and sulfhydrogenase (*hyd*) (Wasmund et al. 2017).

Concomitant reactions of cryptic sulfur cycling and redox transformation of intermediate sulfur species (i.e. SO_3^{2-}, $S_2O_3^{2-}$, S_n^{2-}, S^0) are carried out by taxonomically and functionally diverse microorganisms, including denitrifiers and anoxygenic anaerobic phototrophs (van Vliet et al. 2021; Vigneron et al. 2021). Cryptic sulfur cycling is an enigmatic process in which intermediate sulfur species can be re-oxidized anaerobically, in theory in the presence of ferric iron (Holmkvist et al. 2011), via disproportion or "dismutase" (Fig. 4.2b). Some genes involved in the respiratory reduction of sulfur intermediates include the aforementioned genes: *asr* (i.e. SO_3^{2-}, sulfite), *phs* (i.e. $S_2O_3^{2-}$, thiosulfate), *psr* (i.e. S_n^{2-}, polysulfide chains), sulfhydrogenase (*hyd*) (i.e. S_n^{2-} and S^0, elemental sulfur). The metabolic functions driving sulfur dismutase in cryptic cycling remain unclear. Some genes involved in sulfur cycling have demonstrated reverse types, namely reverse sulfide dehydrogenase (*sud*), also known as bifurcating ferredoxin: NADP oxidoreductase or NADH-dependent reduced ferredoxin:NADP+ oxidoreductase (*nfn*) (Pereira et al. 2011), reverse dissimilatory sulfite reductase (*rdsr*), and reverse-acting adenylylsulfate reductase (*apr*) although they are mostly known to be expressed under (micro)aerobic conditions (Loy et al. 2009; Wasmund et al. 2017). The reduction of intermediate sulfur species with molecular hydrogen as electron donor is channeled via heterodisulfide reductase (*hdr*) as a hydrogenase with heme groups and iron-sulfur clusters (Hedderich et al. 1998). By being reversible (Bell et al. 2020), the confurcating-bifurcating electron transport chain involved in cryptic sulfur cycling can prevent the accumulation and precipitation of sulfide (Vuillemin et al. 2018b; Berg et al. 2019). However, because several genes involved in cryptic sulfur reactions are reversible and relate to aerobic or anaerobic functions, they are difficult to assign to a specific environmental context.

Syntrophs: Syntrophic partnerships consist of the "degraders" with fermenting capacity and the "consumers" that remove the excreted fermentation products (e.g. VFAs) so that the first part of the process can proceed with marginal gain of energy.

The fermentative degraders are mostly found among Desulfobacterota (e.g. *Syntrophus*, *Geobacter*, *Smithella*, *Desulfovibrio*, *Syntrophothermus*, *Syntrophobacter*), Firmicutes (e.g. *Pelotomaculum*, *Sporotomaculum*, *Thermosyntropha*, *Ruminococcus*, *Syntrophomonas*, *Clostridium*) and some specific Crenarchaeota (i.e. *Ignicoccus*, *Pyrococcus*), anaerobic methanotrophic archaea (i.e. ANME I and II, *Candidatus* Methanoperedens) and Chloroflexota (i.e. *Bellilinea*, *Longilinea*). The consumers that make use of hydrogen (or formate) produced in the first fermentation step are commonly sulfate reducers, methanogens and (homo)acetogens (McInerney et al. 2008; Morris et al. 2013). The consumption of VFAs (Fig. 4.3b) proceeds via acetate (i.e. $C_2H_3O_2^-$) kinase, propionate (i.e. $C_3H_5O_2^-$) kinase, butyrate (i.e. $C_4H_7O_2^-$) kinase and phosphotransacetylase which are all genes involved in the SLP of VFAs. Formate (i.e. CHO_2^-) is anaerobically oxidized via formate dehydrogenase (*fdh*). The redox complexes involved in interspecies electron transport are composed of several subunits and include, among others, the heterodisulfide reductase (*hdr*), NAD^+-reducing Ni-Fe hydrogenase (*hox*), with the Nuo and Rnf membrane complexes (McInerney et al. 2009; Yan and Ferry 2018). However, the presence of these genes alone cannot robustly be used as markers for syntrophic metabolism.

In addition, hydrogenases catalyze the reversible oxidation of molecular hydrogen as energy and electron source that characterize many syntrophic associations. Metal-based and metal-free hydrogenases display specific sensitivity to the partial pressure of molecular hydrogen, and show different efficiencies in fermentative environments (Vignais et al. 2001) as increased hydrogen concentrations compromise certain hydrogenase activities (Skidmore et al. 2013). In order of decreasing sensitivity to hydrogen partial pressure, such hydrogenases include Fe-Fe hydrogenase (*ddh*), Ni-Fe hydrogenase (*nfh*) with NAD-reducing hydrogenase (*hox*) used in oxidation of molecular hydrogen, and formate hydrogenlyase (*fhl*), coenzyme F420 reducing and non-reducing hydrogenase (*frh*), methyl-viologen hydrogenase (*mvh*) and heterodisulfide reductase (*hdr*) used in the production of molecular hydrogen.

Methanogens: Methanogens in lake sediments typically display succession patterns that reflect the type of carbon substrates that they assimilate (Fig. 4.3b) as well as their tolerance to pore water sulfate and chloride concentrations (Vuillemin et al. 2014b). In the presence of pore water sulfate, sulfate reducers can outcompete methanogens for hydrogen. This causes hydrogenotrophic methanogenesis (detailed below) to be depressed under saline and brackish conditions (Liu and Whitman 2008), that can only prevail in sulfur-poor freshwater lakes or below the sulfate-methane transition zone (Borrel et al. 2011). Twenty species of archaea have been identified as methylotrophic methanogens (e.g. *Methanohalophilus*, *Methanolobus*, *Methanococcus*). This group relies on methyl compounds such as methanol, methylated amines or methylated sulfides as substrates to produce methane. Thirteen of these are obligate methylotrophs (e.g. *Methanosphaera*, *Methanomicrococcus*, *Methanococcoides*, *Methanomassilicoccus*, *Methanomethylophilus*). There are furthermore 9 genera of acetoclastic methanogens (e.g. *Methanoregula*, *Methanosarcina*, *Methanoculleus*) that utilize acetate to produce methane, only one of which is an obligate acetotroph (i.e. *Methanothrix*). Syntrophic acetate oxidation can sometimes replace acetoclastic methanogenesis (Fig. 4.3b) followed by hydrogenotrophic methanogenesis or homoacetogenesis (Dyksma et al. 2020). Hydrogenotrophic methanogens (e.g. *Methanospirillum*, *Methanoregula*, *Methanoplanus*, *Methanopyrus*, *Methanocalculus*, *Methanothermobacter*) oxidize H_2 and reduce CO_3^{2-} to produce methane. Although often found in syntrophic partnerships, they do not depend on them. Among the hydrogenotrophic methanogens, 38 genera can also assimilate formate to form methane, i.e. formatotrophs (e.g. *Methanolinea*, *Methanobacterium*, *Methanococcus*). Some methanogens are capable of performing multiple alternative methanogenic pathways and hence cannot be placed into a single category (Garcia et al. 2000). Finally, some uncultivated clades display an unexpected potential for methane production (Evans et al. 2015), these include the Verstraetearcheota (Vanwonterghem et al. 2016) and certain clades of Bathyarchaeia (Berghuis et al. 2019), which are the only cases of methane metabolism identified outside the phyla Euryarchaeota and Halobacterota (Söllinger and Urich 2019). Functional genes involved in methane production target specific carbon sources (e.g. methanol, methylamine, methylthiol) and can be used to discriminate fermentation stages (Figs. 4.2c and 4.3b), for instance the methanol coenzyme M methyltransferase (*mta*), methylamine-specific coenzyme M methyltransferase (*mtb*), and methylthiol-specific coenzyme M methyltransferase (*mts*) that are involved in methylotrophy, or the methyl-tetrahydromethanopterin coenzyme M methyltransferase (*mtr*) in hydrogenotrophy (Table 4.1). The final step leading to methane production and release from the cell corresponds to methyl-coenzyme M reductase (*mcr*). Because certain archaea hold a partial methanogenic Wood-Ljungdahl (W-L) pathway (Baker et al. 2016; Orsi et al. 2020b), the *mcr* genes are the only valid marker for microbial methane production (Morris et al. 2014).

Vertical Distribution of Microbial Taxa in Recent Sediments from Two Geochemically Distinct Lakes

To reconstruct past lacustrine environmental conditions, the taxonomic composition of prokaryotic communities was investigated by producing 16S rRNA gene amplicon data from recently deposited sediment (down to 1.5 m) from two lakes with contrasting hydrological regimes and geochemical conditions (Fig. 4.4).

Lago di Cadagno is a crenogenic meromictic lake located in the Swiss Alps at 1921 m elevation above sea level (asl) (Del Don et al. 2001). Submerged saline springs flowing through the evaporites of the fractured bedrock supply the lake with high concentrations of SO_4^{2-}, which is reduced by sulfate reducers creating anoxic, sulfidic conditions in the monimolimnion (Berg et al. 2016) and underlying sediments (Berg et al. 2022). The upper 1.5 m sediment profile represents the last 3500 years (0.4 mm × $year^{-1}$) and is composed of pelagic muds and mixed terrestrial deposits (Wirth et al. 2013). The prokaryotic assemblage of Lago di Cadagno sediments (Fig. 4.4a) can be interpreted as taxa derived from soil runoff (e.g. Actinobacteriota, Acidobacteriota); the oxic photic overlying zone (e.g.

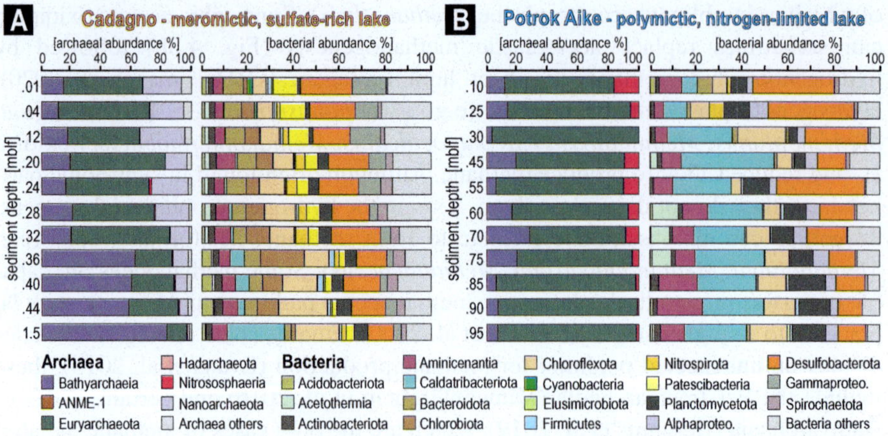

Fig. 4.4 Layered microbial communities in surface sediments corresponding to contrasted lacustrine conditions. (**a**) 16S rRNA gene diversity [relative %] for Archaea and Bacteria in the shallow sediment of Lago di Cadagno, a sulfate-rich permanently stratified alpine lake. (Modified after Berg et al. 2022). (**b**) 16S rRNA gene diversity [relative %] for Archaea and Bacteria in the shallow sediment of Laguna Potrok Aike, an endorheic, polymictic and nitrogen-limited maar lake. (Modified after Vuillemin et al. 2018a)

Cyanobacteria) and the anoxic sulfidic photic zone (e.g. Chlorobia). In addition, the assemblage includes aerobic (e.g. Nitrospirota), microaerobic (e.g. Bacteroidia) and anaerobic heterotrophs (e.g. Nanoarchaeota, Patescibacteria, Planctomycetota). In the sediment, stratified microbial guilds related to contrasting respiratory types (Fig. 4.3a) are predominant, with a succession from denitrifiers (e.g. Alphaproteobacteria) to sulfate reducers, (e.g. Desulfobacterota) and a few methanogens that are tolerant to sulfate (e.g. Euryarchaeota) combined with fermenters providing molecular hydrogen for methanogenesis (e.g. Firmicutes). The vertical succession of fermenters implies a stepwise degradation of organic substrates through primary (e.g. Acetothermia, Aminicenantia) and secondary (e.g. Caldatribacteriota, Chloroflexota, Bathyarchaeia) fermentation (Fig. 4.4a).

Laguna Potrok Aike is a maar lake located in Southern Patagonia, Argentina within the Pali Aike volcanic field. The steep morphology of this 100 m deep maar results in a pelagic-hemipelagic regime with frequent slumps and turbidites (Zolitschka et al. 2013). Due to the persistent influence of the Westerly winds on the study site, the lake is polymictic and the water column does not stratify in any season (Mayr et al. 2007) with mean annual temperatures between 4 °C and 10 °C. The basin is endorheic with nitrogen-limited oxic conditions down to the sediment-water interface, but oxygen penetration is restricted to the uppermost few millimeters (Vuillemin et al. 2013a). The top meter of sediment was assessed to cover a period of 1500 years (i.e 0.6 mm × year^{-1}). The prokaryotic assemblage of Laguna Potrok Aike (Fig. 4.4b) reflect the oligotrophic conditions of the water column, with taxa related to nitrogen recycling by nitrifiers under oxic (e.g. Nitrososphaeria, Nitrospirota) and microoxic (e.g. Planctomycetota) conditions and only a few aerobic heterotrophs from the water column (e.g. Bacteroidia). Apart from some terrigenous inputs (e.g. Acidobacteriota, Actinobacteriota), stratified guilds related to various respiratory guilds (Fig. 4.3a) are present, i.e. denitrifiers (e.g. Alphaproteobacteria), sulfate reducers, (e.g. Desulfobacterota). Due to the low salinity of this lake, the more thermodynamically favorable electron acceptors are depleted at shallow depths (ca. 40 cm) and methanogenesis prevails (e.g. Euryarchaeota) as a dominant process (Fig. 4.4b). The concomitant prokaryotic assemblage is mainly composed of primary (e.g. Acetothermia, Aminicenantia) and secondary (e.g. Chloroflexota, Bathyarchaeia) fermenters, some of which may be involved in syntrophic associations (e.g. Caldatribacteriota, Firmicutes) with hydrogenotrophic methanogens (Fig. 4.3b).

Together, the limited water depth and redox zonation in Lago di Cadagno promote the preservation of "ancient" sedDNA exogenous to the sediment, which can readily be discriminated from the active resident microbial populations. The related 16S rRNA gene abundances are indicative of specific subpopulations that allow inferences about past regime, redox stratification and type of sedimentation in the lake. Notable replacements and turnover of taxa exogenous to the sediment still occur within the upper meter of sediment. Finally, the lake salinity and related availability of favorable electron acceptors in the pore water allow sulfate reduction to be maintained deeper into the sediment profile, delaying the onset of fermentation and methanogenesis. In contrast, Laguna Potrok Aike's deep and constantly mixing oxic water column leaves little room for preservation of "ancient" sedDNA derived from planktonic sources. Fully oxic conditions allows for colonization of the sediment-water interface by aerobic heterotrophs, whereas the low salinity results in rapid depletion of pore water electron acceptors and promotes rapid transition to fermentation and hydrogenotrophic methanogenesis as the main processes in the sediment.

Slow Growers and Persistent Prokaryotes in Deep Lake Sediments

Environmental Processes: Fermentation of Residual Substrates, Necromass and DNA Turnover

Microbial communities in deep lacustrine sediments are structured according to energy-conserving fermentative metabolisms along with the gradual depletion of pore water electron acceptors (Berg et al. 2022) and labile OM (Glombitza et al. 2013). Microbial cell densities experience an exponential decrease into the fermentation zone (Orsi et al. 2020a), resulting in a "depauperate horizon" (Wurzbacher et al. 2017) corresponding to a sharp decrease in microbial diversity, followed by a steady increase in taxonomic assemblages representing "persister" populations (Vuillemin et al. 2018a; Thomas et al. 2020) where starvation induces dormancy (Ellegaard et al. 2020). Metabolic features that exert selection of pre-adapted persisters seem to relate to their ability to drive secondary redox reactions for OM fermentation in substrate-limited sediment (Friese et al. 2021). These processes include hydrogen production (i.e hydrogenases) and electron bifurcation (i.e. cytochromes, flavins) used in heterotrophic dark carbon fixation (Santoro et al. 2013), such as in the W-L pathway, anaerobic dicarboxylate/4-hydroxybutyrate (DC/HB) cycle, or reductive TCA cycle (Fig. 4.2c). Although these metabolisms are similar to those involved in syntrophic associations (Conrad 2020), it is not clear whether deep biosphere taxa actively collaborate in stepwise degradation of refractory OM via non-methanogenic interactions (Bertolet et al. 2019; Conrad et al. 2020) or alternatively persist opportunistically by making use of hydrolysed products (Orsi 2018). Deep sediments accumulate diagenetically altered and recalcitrant necromass over time (LaRowe and Van Cappellen 2011; Lever et al. 2015). The pool of OM is at least partially degraded and fermented by (homo)acetogenic bacteria (Vuillemin et al. 2020a, b; Zemskaya et al. 2021), and in theory via fermentative Stickland reactions between amino acids that can serve as either electron donors or acceptors (de Vladar 2012). While the energy yield from such reactions remains insufficient to produce a net increase in cellular biomass (Orsi et al. 2020a), such microbial activity sustained at the limit-to-life will result in turnover of microbial populations and sediment nucleic acids (Mhatre et al. 2019).

Depth profiles of 16S rRNA genes retrieved along lacustrine sequences spanning beyond the Last Glacial Maximum (i.e. sediment older than ca. 21 ka) are predominantly composed of yet-uncultivated fermenters and (homo)acetogenic degraders of complex organic carbon substrates (Fig. 4.5). Metabolic potential for heterotrophic dark carbon fixation (i.e homoacetogenesis, methanogenesis) is found in bacterial genomes from Aminicenantia, Elusimicrobiota, Chloroflexota (class Dehalococcoidia), and Caldatribacteriota, and archaeal genomes of Hadarchaeia and Bathyarchaeia which both display predicted metabolisms revolving around C_1 compounds (Colman et al. 2017). The distribution of metabolic traits across yet-uncultivated candidate clades and the evolutionary processes underlying them are incompletely resolved as even some core metabolic pathways (e.g. glycolysis, pentose phosphate pathway, pyruvate metabolism) are only partially predicted by phylogenetic relationships (Jaffe et al. 2020).

Populations that Live and Persist at the Thermodynamic Limit of Life

We proceed to present candidate clades reported to be "persisters" in deep lake sediments, acknowledging that there could be metabolic diversification within each of these candidate phyla and classes whose metabolisms are only inferred from cultivation-independent metagenome-assembled genomes (Castelle and Banfield 2018).

Aminicenantia: This class (phylum Acidobacteriota) features a metabolism that allows its taxa to only ferment proteinaceous substrates and sugars, albeit they can access a wide range of short and long-chain carbohydrates including chitin and cellulose via respiration of nitrite (Kadnikov et al. 2019). As anaerobic organotrophs, they act as initial degraders of buried biomass and OM, producing hydrogen and

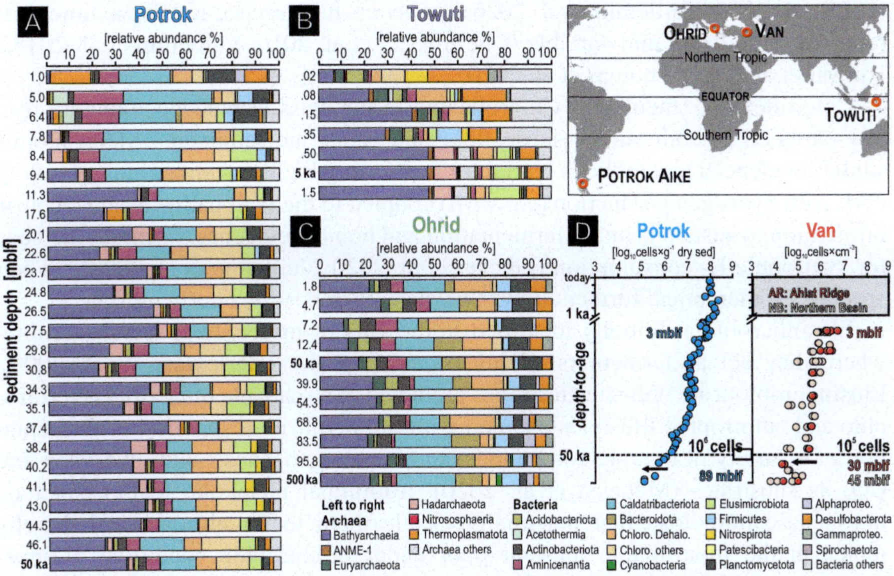

Fig. 4.5 16S rRNA gene taxonomic diversity and cell density in long-term sedimentary sediment archives from four different lakes retrieved by the International Continent Scientific Drilling Program (ICDP). (**a**) 16S rRNA-based microbial community composition (relative abundance [%] at phylum level) in sediments of Laguna Potrok Aike, an endorheic nitrogen-limited and polymictic 100 m deep maar lake. (After Vuillemin et al. 2018a). (**b**) 16S rRNA-based microbial community composition (relative abundance [%] at phylum level) in sediments of Lake Towuti, a ferruginous phosphorus-limited and stratified 200 m deep tectonic lake. (After Vuillemin et al. 2018b). (**c**) 16S rRNA-based microbial community composition (relative abundance [%] at phylum level) in sediments of Lake Ohrid, an endorheic 288 m deep tectonic basin fed by underground karstic channels. (After Thomas et al. 2020). (**d**) DAPI-stained total cell counts [$\log_{10}$ cells × g^{-1} dry sed] in sediments of Laguna Potrok Aike. (After Vuillemin et al. 2014a) and SYBR Green-stained total cell counts [$\log_{10}$ cells × cm^{-3} wet sed] in sediments of Lake Van. (After Kallmeyer et al. 2015)

acetate as metabolic products. This clade tends to decrease in the microbial assemblage as it transits from fermentative stage I to II (Fig. 4.3b).

Caldatribacteriota: Members of this phylum perform complete anaerobic fermentative glycolysis (i.e. glycolysis, TCA cycle, SLP) and are putatively involved in syntrophic acetate oxidation in hydrocarbon-enriched environments (Liu et al. 2019). Its complex double membrane hinders diffusive loss of metabolites (e.g. VFAs, alcohols, aldehydes) out of the cell (Katayama et al. 2020). Its metabolic machinery includes a microcompartment that assists in secondary fermentation of aldehydes and conversion to pentose (Nobu et al. 2016) and a membrane complex for electron bifurcation that harnesses additional redox energy channeled through the W-L pathway (Vuillemin et al. 2020a). Altogether, this indicates that candidate clades among Caldatribacteriota can grow and sustain active populations in anoxic environments without a syntrophic partner. It is noteworthy that the Caldatribacteriota clade JS1 is the most abundant and widespread clade in oceanic anoxic sediments

(Lee et al. 2018; Vuillemin et al. 2020a), whereas its presence in lake sediments is reported to be lower and variable (Chernitsyna et al. 2016; Vuillemin et al. 2018a; Panwar et al. 2020; Thomas et al. 2020).

Chloroflexota: Uncultured Chloroflexota in the class Dehalococcoidia are usually enriched in both anoxic subseafloor and lacustrine sediments. Metagenomic studies evidenced metabolic potential for fermenting highly oxidized and refractory OM, with hydrogen production (e.g. *mvh*) coupled to the W-L pathway and electron bifurcation to sustain primary fermentation and homoacetogenesis without the need for syntrophic H_2 consumption (Fincker et al. 2020; Vuillemin et al. 2020b). Their special cell membrane further allows this clade to persist in a dormant state.

Elusimicrobiota: Initially identified in the guts of insects (e.g. scarab, termite) where they act as fermenters of sugars and peptides (Herlemann et al. 2009), Elusimicrobiota are obligate anaerobes predicted to be capable of flexible heterotrophic and autotrophic lifestyles coupled with nitrate/nitrite respiration or fermentation of a variety of organic compounds, and homoacetogenesis with hydrogen and CO_2 as substrates (Méheust et al. 2020). Additional metabolic features include nitrogenase homologs that potentially enable them to fix nitrogen and thereby replenish the bioavailable nutrient pool under oligotrophic conditions (Lopez-Fernandez et al. 2018).

Hadarchaeota: Firstly discovered in deep South-African gold mines (Takai et al. 2001), Hadarchaeota display genome streamlining to mediate key geochemical processes in subsurface environments. They have genes putatively involved in degradation of short-chain alkanes, CO^- and H_2 oxidation, oxidation of VFAs and H_2 production associated with a partial W-L pathway (Wang et al. 2019). In the complex interactions of OM remineralization that characterize the fermentative zone (Orsi et al. 2020a), C_1 and C_2 compounds produced as fermentative by-products, such as formate, CO^-, and acetate, can be readily utilized by strictly anaerobic Hadarchaeota (Baker et al. 2016), potentially marking the lower boundary of syntrophic associations in the sediment.

Bathyarchaeia: As a generalistic class (phylum Crenarchaeota) that includes 17 subgroups, they are obligately found in anoxic sediments and are expected to anaerobically utilize a wide range of organic compounds, such as detrital proteins, carbohydrates, VFAs, aromatic compounds, short chain alkanes, methylated compounds (Zhou et al. 2018), and other refractory organic compounds such as lignin (Yu et al. 2018). In addition, they have inferred metabolic potential for hydrogen production, methanogenesis (Evans et al. 2015; Borrel et al. 2016; Berghuis et al. 2019) and homoacetogenesis (He et al. 2016) via the W-L pathway (Fig. 4.2c). This apparently allows them to dominate OM remineralization in deep sediment (Fig. 4.5) and efficiently drive heterotrophic dark carbon fixation (Feng et al. 2019). Consistently with their postulated hydrothermal origin (Feng et al. 2019), Bathyarchaeia also display metabolic capacity to use reduced sulfur.

Microorganisms involved in autotrophic and heterotrophic dark carbon fixation can make a considerable contribution to inorganic carbon fixation in lake sediments (Zhao et al. 2020). Genes involved in dark carbon fixation (Fig. 4.2c) include ATP-citrate lyase (*acly*) as the first step of the reductive TCA cycle (Kanao et al. 2001), ribulose-1,5-diphosphate carboxylase gene (RuBisCO) in the non-phototrophic CBB cycle, different carboxylases (Erb 2011) used in mixotrophic DC/HP cycle and Fuchs-Holo bicycle, and several functional genes involved in the (homo)acetogenic W-L pathway from formyl-tetrahydrofolate synthetase (*fthfs*) to carbon monoxide dehydrogenase (*codh*). Metabolic mechanisms at the thermodynamic limits for life, such as secondary fermentations in the (homo)acetogenic W-L pathway, implement redox complexes (e.g. NADH, ferredoxin, flavins) for energy conservation via electron bifurcation (Schuchmann and Müller 2016; Buckel and Thauer 2018). Electron bifurcation, combined with the reversible oxidation of molecular hydrogen (McInerney et al. 2009), is an indirect way of conserving redox energy by reducing protons to molecular hydrogen and makes use of it in heterotrophic dark carbon fixation (Lemaire et al. 2020). Some of the main hydrogenases coupled with electron bifurcation (Vignais et al. 2001) are the heterodisulfide reductase (*hdr*), methyl-viologen reductase (*mvh*), coenzyme F420 reducing and non-reducing hydrogenase (*frh*), and in theory different L-amino acid dehydrogenases involved in fermentative Stickland reactions (de Vladar 2012).

Prokaryotic Assemblages of the Deep Biosphere Persisting in Lake Sediments Older than 50 Ka

Geomicrobiological investigations of lacustrine sediment archives from ancient lakes has enabled characterization of the long and deeply buried biosphere in terms of cell concentrations (Vuillemin et al. 2014a; Kallmeyer et al. 2015), community composition (Vuillemin et al. 2018a; Thomas et al. 2020), and metabolic potential and activity (Glombitza et al. 2013; Vuillemin et al. 2018b; Friese et al. 2021). To illustrate the exploration of the lacustrine deep biosphere, we compared 16S rRNA gene diversities and cell densities in up to 500 ka old sediment profiles from four different lakes.

The stratigraphic sediment profile recovered from the 100 m deep maar Laguna Potrok Aike represents 51 ka years of sediment deposition (Zolitschka et al. 2013). Predicting functions based on identified taxa representative of respiration types, the 16S rRNA assemblages in the upper 10 m below lake floor (mblf) reflect successive replacement of denitrifiers (i.e. Alphaproteobacteria) with sulfate reducers (i.e. Desulfobacterota), and methanogens (i.e. Thermoplasmatota, Euryarchaeota) in syntrophic association with primary (i.e. Acetothermia, Aminicenantia) and secondary (e.g. Caldatribacteriota, Firmicutes) fermenters, generally trending towards homoacetogenesis (e.g. Chloroflexota). The Late Glacial transition around 10 mblf corresponds to a drastic shift in sedimentation types that resulted in less abundant and more refractory organic substrates in sediments of the glacial period (Vuillemin et al. 2014a, 2018a). This transition is accompanied by the increasing presence of a different microbial community predominantly composed of Bathyarchaeia, Hadarchaeota, and Elusimicrobiota with Aminicenantia, Caldatribacteriota and Dehalococcoidia persisting at slow metabolic rates throughout the sediment archive (Fig. 4.5a). Although intervals corresponding to gravity flows were systematically discarded, putative terrigenous inputs are still observed as sequences assigned to Acidobacteriota, Actinobacteriota and Bacteroidia (Fig. 4.5a).

Seated in ultramafic rocks on the Indonesian island of Sulawesi (Fig. 4.5b), Lake Towuti is a long-term stratified, 200 m deep tectonic basin. Lateritic weathering in the catchment delivers abundant iron oxides but very little sulfate to the lake, leading to ferruginous conditions that scavenge phosphorus from the water column. Its sedimentary column is estimated to span 1 Ma of depositional history with variable sedimentary rates (Russell et al. 2020), trophic state and redox conditions across wet and dry periods of the Quaternary (Vuillemin et al. 2023). Given the lake's ferruginous chemistry and oxygen depletion below 130 m in the water column, Lake Towuti has an overall shortage of electron acceptors (i.e. oxygen, nitrate, sulfate) in its bottom waters (Vuillemin et al. 2016b). Thus, pore water electron acceptors are depleted even at shallow sediment depth (Vuillemin et al. 2016b, 2018b), leading to the emergence of taxonomic assemblages with fermentation as prominent features and methane as the final degradation product (Friese et al. 2021). This observation is surprising, as there are abundant minerals containing ferric iron. However, these minerals appear to be poorly reactive towards microbial reduction, contradicting the canonical sequence of electron acceptor use. The main initial microbial constituents reflecting metabolic guilds include sulfate reducers (e.g. Desulfobacterota, Nitrospirota) and methanogens (e.g. Euryarchaeota) in concomitance with primary fermenters (e.g. Firmicutes, Aminicenantia). Below 0.5 mblf, the taxonomic assemblage is predominantly composed of Bathyarchaeia, Hadarchaeota, Elusimicrobiota and Chloroflexota, with predicted metabolisms revolving around H_2, CO^-, formate and CH_4 (Colman et al. 2017). We propose that the main active metabolic processes in sediments deeper than 2 mblf putatively include fermentation of proteinaceous substrates by Bathyarchaeia and recycling of excreted ammonium by Elusimicrobiota (Fig. 4.5b).

Lake Ohrid is an endorheic oligotrophic 288 m deep tectonic basin located in the Balkans mostly fed by karstic groundwater channels (Fig. 4.5c). The upper 200 m of the water column experience seasonal overturn, whereas the oxic bottom waters only mix on decadal scales and typically remain stratified due to strong salinity-induced density gradients. Its history is estimated to cover 1.2 Ma, with sedimentation rates that highly fluctuated across glacial-interglacial periods (Francke et al. 2016). The deep sediment record of Lake Ohrid displays a 16S rRNA gene diversity dominated by Caldatribacteriota, Bathyarchaeia and Dehalococcoidia, which is in accordance with processes expected to shape communities in substrate-limited and hundreds ka old anoxic sediments (Fig. 4.5c). The sporadic occurrence of gammaproteobacterial, cyanobacterial and putative soil representatives further suggests that some DNA sources that originated from the catchment and water column were preserved hundreds ka after burial, particularly in sediments corresponding to sudden deposits during dry glacial intervals (Thomas et al. 2020).

Seated on the high plateau of far east Turkey, Lake Van is an endorheic 171 m deep soda lake. Due to hydrothermal activity in the catchment, the water column is strongly alkaline (pH 9.7). Although annual temperatures fluctuate between 25° and − 12 °C, the high salinity of the lake prevents freezing in winter. The stratigraphic sequence of the deeper of the two tectonic sub-basins is estimated to span 600 ka of climatic history, with a succession of finely laminated sediment, chemical varves and frequent tephra layers (Stockhecke et al. 2014). The differences in sedimentation rates across Lake Potrok Aike (1 mm $year^{-1}$), Towuti (0.22 mm $year^{-1}$), Ohrid (0.43 mm $year^{-1}$) and Van (0.4 mm $year^{-1}$) can account for discrepancies between sediment depth and age. Still in the case of Lake Potrok Aike and Lake Van (Kallmeyer et al. 2015), population cell densities experience a continuous decrease with depth and stabilize around 10^6 to 10^5 cells × cm^{-3} after some 50 ka years of burial (Fig. 4.5d).

These four case studies also show that sudden sediment deposition (e.g. turbidites, slumps, floods), especially during glacial periods, can result in enhanced preservation of DNA sources derived from the catchment and in-lake processes as increased sedimentation rates tend to disrupt stratified microbial communities and abort, or at least, hinder early diagenetic processes.

Anthropogenic Expansion of Adaptive and Versatile Ubiquists

The biogeochemical feedbacks of microbial communities on their respective ecosystems (Hall et al. 2018) are currently gaining attention, as pristine lakes are increasingly exposed to anthropogenic disturbances (Han et al. 2020; Jiang et al. 2017). Such perturbations may include land use (Kraemer et al. 2020; Pearman et al. 2022b) with increased nutrient or OM inputs (Wu et al. 2019) resulting in lake browning and eutrophication (Fiskal et al. 2019, 2021), as well as wastewater outfalls (McLellan et al. 2015) and oil spills (Zemskaya et al. 2015) delivering persistent organic pollutants (Morasch et al. 2010), heavy metals (Bravo et al. 2018) and potential pathogens (Chandran et al. 2011) across the watershed.

Microbial activities can have significant and variable impacts in pristine and human-impacted ecosystems (Hall et al. 2018), for instance by creating resilience to environmental perturbations (Shade et al. 2012a, b), enabling on-site bioremediation (Aalto et al. 2018), or accelerating nutrient and metal release from sediments into the overlying waters (Sinkko et al. 2013). As eutrophication progresses and increases the flux of OM to the sediment, microbial processes produce specific metabolic solutes (e.g. N_2O, NH_4^+, PO_4^{3-}, H_2S, Fe^{2+}, CO_2, CH_4) that can potentially escape the sediment or diffuse across the sediment-water interface (Fig. 4.3). If such metabolites reach overlying waters, they could cause negative feedback that would further increase lake productivity (van Helmond et al. 2020) or enhance production of greenhouse gasses. In contrast, pore water saturation can trigger precipitation of authigenic minerals that can help maintain stable water column conditions by promoting long-term sequestration of nutrients and metals in the sediment (Taylor and Boult 2007; Vuillemin et al. 2013b, 2020b, 2023; Berg et al. 2019). Since these effects are influenced by microbial community composition, it is necessary to study the link between lacustrine environmental conditions (Lewis 2009), species distributions and mechanisms of selection (Bryanskaya et al. 2016). Thus, sedDNA surveys are necessary in order to understand microbial community assembly in emergent ecological niches (Mandakovic et al. 2018) and assess environmental resilience as pristine ecosystems transit to human-dominated ecosystems of the Anthropocene (Sun et al. 2017; Blanchette and Lund 2021; Fiskal et al. 2021).

Environmental Processes: Pollutant Loading, Eutrophication, and Excess Nutrient

Sediment microbiomes downstream of urbanized catchments (Jiang et al. 2017; Han et al. 2020) are increasingly being used as models for elucidating how anthropogenic disturbances affect lacustrine habitats (Barouillet et al. 2023). Excessive nutrient inputs to lacustrine systems typically cause enhanced productivity in the water column, thereby increasing organic carbon sedimentation and promoting OM mineralization via organotrophic sulfate reduction, ammonification and methanogenesis at shallow sediment depths (Fiskal et al. 2019). Such redox conditions could invoke negative feedback in the overlying waters (Liikanen and Martikainen 2003). The identification of core microbiomes, typically defined as the suite of populations consistently shared across variable microbial communities from similar habitats (i.e. adaptive ubiquists) (Shade and Handelsman 2012; Ruiz-González et al. 2017), would facilitate a systematic assessment and monitoring of microbial assemblies characteristic for either pristine or resilient conditions (Yadav et al. 2018; Ji et al. 2019). This would help pinpoint populations and taxa that may be responsible for *in situ* bioremediation (i.e. bioremediators) or carry signals of human-impacted conditions (Gibbons and Gilbert 2015; Kato et al. 2015; Escalas et al. 2019; Wu et al. 2019).

Microbial Mitigation of Anthropogenic Pressures: Sulfur and Iron Oxidizers, Methylotrophs, Methanotrophs, Anammox Bacteria, Opportunists

Sulfur oxidizers: In organic-rich sediments, excess organotrophic sulfate reduction associated with limited iron reduction leads to an accumulation of H_2S in pore water with sediment sulfidization that can have potentially toxic effects. Specific taxa affiliated with Gammaproteobacteria (e.g. *Thioploca*, *Thiothrix*, *Beggiatoa*, *Thioglobus*, *Sulfuriflexus*, *Acidiferrobacter*, *Thiobacillus*) play important ecological roles in sulfide detoxification via sulfur oxidation coupled to autotrophic dark carbon fixation (i.e. chemolithoautotrophy) in the water column and at the sediment-water interface (Kojima et al. 2003; Lavik et al. 2009; Pjevac et al. 2015; Vuillemin et al. 2022). Although sulfur oxidation is the prevalent process, the use of reduced forms of sulfur as electron donors by these taxa can also be coupled to DNRA, or ammonification. DNRA is a process in competition with canonical denitrification (Fig. 4.2a, b) that can potentially cause accumulation of excess ammonia in the system (Fernández et al. 2009). In acidic mining lakes (Lu et al. 2016; Quatrini and Johnson 2018), some phototrophic sulfur bacteria (e.g. *Allochromatium*,

Ectothiorhodospira) and bacterial candidates (e.g. *Acidithiobacillus, Ferrimicrobium*) are also capable of sustaining growth via sulfide, sulfur and iron oxidation to potentially drive in-lake neutralization and remediate groundwater pollution (Zak et al. 2021). Increased OM inputs to the sediment will stimulate mineralization coupled with sulfate reduction and methanogenesis. Excess sulfide buildup in the system can be mitigated under aerobic conditions by microbial expression of genes encoding the sulfite oxidase (*suox*) and periplasmic sulfur-oxidizing proteins (*sox*) as part of the sulfur oxidative pathway, as well as reverse dissimilatory sulfite reductase (*rdsr*), reversible adenylylsulfate reductase (*apr*) and reversible sulfide dehydrogenase (*sud*) (Dahl 2020) also known as bifurcating ferredoxin: NADP oxidoreductase (*nfn*) (Pereira et al. 2011). However, it is not clearly established whether these genes are actively expressed in anaerobic sulfur oxidation under anoxic conditions (Orsi et al. 2016). Oxidation of organic forms of sulfur can aerobically take place by means of genes encoding taurine dioxygenase (*tauD*) and potentially via a heterodisulfide-like pathway. Under anoxic conditions, the dimethyl sulfoxide reductase (*dsmo*) encoding gene can potentially increase the production of toxic dimethyl sulfides, byproducts which can serve as substrates in producing methane via the methylthiol-specific coenzyme M methyltransferase (*mts*).

Iron oxidizers: Reduction of iron can cause excessive release of phosphorus from the sediment to the overlying waters (Gächter et al. 1988; Taylor and Boult 2007) in association with solubilization of heavy metal ions accelerated by either reduction (e.g. Hg^{2+}, As^{3+}) or oxidation (e.g. U^{6+}, Se^{6+}, Cu^{2+}, Cr^{4+}). This is because many metals and nutrients tend to form complexes with ferric iron, and accordingly microbial iron oxidizers have the potential to mitigate pollutant releases from reduced sediments (Bravo et al. 2018; Kappler et al. 2021). Iron oxidizers are phylogenetically diverse and include scattered genera across diverse lineages of Proteobacteria (Hedrich et al. 2011) that thrive under variable environmental conditions (Singh et al. 2018). They display a certain degree of metabolic versatility, some being aerobic acidophiles (e.g. *Acidithiobacillus*, *Sideroxydans*), neutrophiles (e.g. *Leptothrix*, *Ferritrophicum*, *Gallionella*), phototrophs (e.g. *Rhodovulum*, *Rhodobacter*, *Rhodomicrobium*), or nitrate-dependent facultative anaerobes (e.g. *Paracoccus*, *Thermomonas*, *Azospira*). Iron oxidizers as a metabolic guild also include members of several other phyla (Emerson et al. 2010, 2013), including Actinobacteriota (e.g. *Acidimicrobium*, *Ferrimicrobium*, *Ferrithrix*), Firmicutes (e.g. *Alicyclobacillus*), Nitrospirota (e.g. *Leptospirillum*), and some archaeal clades that are mainly relevant to acid mining lakes. There is no clear functional marker gene for iron oxidation or reduction, which are metabolic processes involving complex redox cofactors, e.g. cytochromes, ferredoxins, iron-sulfur proteins and flavins (e.g. rubredoxins, riboflavins, rusticyanin), involved in electron transfer (Kappler et al. 2021). Some operons (i.e. *rus*, *petI*, *fox*) are clearly involved in iron oxidation and the corresponding iron respiratory transport chain, which include several subunits encoding the high potential iron sulfur (i.e. Fe_4S_4) protein (*Iro*) with different cytochrome oxidases (a, b, c) and variations of the related *cyc* and *cox* genes (Hedrich et al. 2011). Microbial reduction of metalloids is traceable via functional marker genes, such as arsenate reductase (i.e. glutaredoxin; *arsABC*)

(Saltikov and Newman 2003), selenite reductase (i.e. thioredoxin; *srrA-F*) (Wen et al. 2016) and mercury methylation via a corrinoid binding domain (*hgcA*) and a coupled ferredoxin (*hgcB*) (Parks et al. 2013).

Methylotrophs and methanotrophs: Canonical methylotrophs are obligate aerobes whose isolates have originally been classified based on membrane characteristics into gammaproteobacterial type I (e.g. *Methylococcus*, *Methylomonas*, *Methylobacter*, *Methylosarcina*, *Methylomicrobium*) or alphaproteobacterial type II (e.g. *Methylosinus*, *Methylocystis*, *Methylocella*). However, most methylotrophs are also facultative methanotrophs as they can use methane as a substrate in addition to a range of other reduced C_1 compounds (e.g. methanol, formate, alcohol, formaldehyde) (Higgins et al. 1981; McDonald et al. 2008). These facultative methane oxidizers can grow and survive under a broad range of environmental conditions with regards to, e.g. temperature, pH, and salinity, including the extremes (Trotsenko and Khmelenina 2002). Substantial presence of this guild is expected to reflect high methane availability under oxic conditions, and thus potentially relate to increased productivity and stable methanogenic conditions with diffusive release from underlying sediments (Yang et al. 2016). Processes of anaerobic oxidation of methane (AOM) are performed with electron acceptors other than O_2. For instance, the anaerobic methanotrophic archaea (i.e. ANME groups) among the phylum Euryarchaeota can use SO_4^{2-} (Schubert et al. 2011), Fe^{3+} and Mn^{4+} (Yang et al. 2021) as electron acceptors. *Candidatus* Methanoperedens can also couple AOM processes to NO_3^- reduction (Haroon et al. 2013), while the bacterial candidate phylum Methylomirabilota mediates this process with NO_2^- (i.e. *Candidatus* Methylomirabilis) (Welte et al. 2016). Both aerobic and anaerobic methanotrophs are found in freshwater lake sediments but it is still not clear whether they play a significant role in mitigating emissions of the potent greenhouse gas methane to the atmosphere (Borrel et al. 2011; Norði et al. 2013; He et al. 2018). Methane escapes from anoxic sediments and can be oxidized aerobically (Fig. 4.2c) in the water column or at the sediment-water interface by means of particulate (i.e. membrane bound) methane monooxygenase (*pmmo*) and soluble (i.e. cytoplasmic) methane monooxygenase (*smmo*) which both catalyze the oxidation of methane to methanol (Iguchi et al. 2010). Methane can also be oxidized anaerobically in the sediment via reverse methyl-coenzyme M reductase (*mcr*) which is the same enzyme as the one involved in methane production. Methylotrophic oxidation of other reduced C_1 compounds proceeds via genes encoding for instance methanol dehydrogenase (*mxa*), methanol dehydrogenase homologue (*xox*) and formate dehydrogenase (*fdh*).

Anammox bacteria: Microbial fermentation of proteins is the main source of the NH_4^+ released to pore water. Eleven candidates among the phylum Planctomycetota are known to perform anammox, a process where NO_2^- derived from either aerobic NH_4^+ oxidation or partial NO_3^- reduction is required. Under microoxic conditions, these candidates can play a significant role in removing excess NH_4^+ diffusing to the sediment-water interface (Kuenen 2008), but can also result in production of nitrous oxide (N_2O), another potent greenhouse gas (Kuypers et al. 2005). They are affiliated with the genera *Kuenenia*, *Brocardia*, *Anammoxoglobus*, *Jettenia*, *Brasilis*, and *Scalindula* and display metabolic features, such as a membrane-bound organelle (i.e. anammoxosome), resistance to desiccation and long-term dormancy (Pereira et al. 2017; Zhu et al. 2019). Whereas nitrification (*amo*, *nxr*) only operates under aerobic conditions, anammox processes take place under microaerophilic conditions (Kong et al. 2013) via hydrazine oxidoreductase (*hzo*). The oxygen-sensitive conversion of organic and reduced nitrogen species in eutrophic systems is most effective under concomitant anaerobic oxidation and partial nitrification-denitrification (Hosseinlou et al. 2019), i.e. conditions where anammox can be combined with nitrite-nitrate dependent AOM (van Kessel et al. 2018). In contrast, aerobic chemolithoautotrophic sulfur oxidizers hold genes encoding ammonia-forming cytochrome nitrite reductase (*nrf*), which leads to the production and release of ammonia (i.e. DNRA), and thereby compete with the aforementioned processes. Otherwise, excess organic forms of nitrogen are commonly converted to ammonia via different aerobic ureases, cyanases, nitrilases and oxygen-insensitive nitroreductases (Vuillemin 2023).

Opportunists: Opportunists are typically microbes that can sustain life from a number of resources and immediately take advantage of the opportunity to thrive when favorable conditions arise. Some opportunistic microaerophilic aerobes, or facultative anaerobes, found mostly within the Rhodobacterales among Alphaproteobacteria, Burkholderiales and Rhodocyclales in the class Gammaproteobacteria, or Pseudomonadales among Gammaproteobacteria, are metabolically versatile performing H_2 oxidation with oxalate (i.e. $C_2O_4^{2-}$) as a carbon source. They can make use of reduced metabolites diffusing across the sediment-water interface from underlying eutrophic sediments (e.g. HS^-, NH_4^+, CH_4, CO_3^-, H_2), remain viable long after burial and grow rapidly after even minor sediment exposure to oxygen (Vuillemin et al. 2010; Capo et al. 2021b). Many pathogens can survive in a dormant stage and be transported far downstream from e.g. wastewater treatment plants (Fisher et al. 2015; McLellan et al. 2015). Fecal matter from terrestrial and aquatic animals is also an important source of pathogenic fermenters to the sediment, including some Bacteroidia (e.g. *Prevotella*, *Porphyromonas*, *Rickenella*), Firmicutes (e.g. *Clostridium*, *Lachnospira*, *Trichococcus*, *Bacillus*), Gammaproteobacteria (e.g. *Aeromonas*, *Klebsiella*) and Campylobacterota (e.g. *Arcobacter*, *Campylobacter*). Depending on the features of the ecosystems in which these potentially pathogenic populations are imported, some of them can persist in impacted sediments where they contribute to enhanced mineralization (Haller et al. 2011; Han et al. 2020) or represent indicators of pollution gradients (Li et al. 2019b) or a potential health hazard (Chandran et al. 2011; McLellan et al. 2015; Bai et al. 2022). Dissemination of sedDNA from feces also has specific applications in the field of microbial archaeology (Rampelli et al. 2021; Weyrich and Pérez 2023).

Indicators of Excess Nutrient in Lake Sediments Downstream Urbanized Catchments

Anthropogenic eutrophication downstream of urbanized catchments increases the amount of organic carbon deposited in lacustrine sediments (Anderson et al. 2014), which in turn shapes the redox chemistry and the degree of OM remineralization during shallow burial. Excessive microbial turnover of OM in the sediment can thereby result in increased sulfidization, ammonification and methanogenesis in the sediment, whereas metal reduction triggers nutrient releases from organic-rich sediments.

Five Swiss lakes displaying different degrees of increasing eutrophication over time (Fig. 4.5a) were investigated to trace the fate of aquatic OM loads (Fiskal et al. 2019, 2021) and decipher the role of microbial respiratory activities during shallow burial on increasingly OM-rich sediments over recent times (Han et al. 2020). Phylogenetic assemblages were analyzed statistically (i.e. similarity percentages breakdown, geochemistry and phylogeny correlation matrix, co-occurrence patterns, permutational multivariate analysis of variance) to sort out the taxa associated with permanent changes in microbial community structure (Fig. 4.6b).

Metabolic guilds that seem to thrive in increasingly eutrophic sediments (Fig. 4.5b) are nitrifiers (i.e. Nitrososphaeria, Gammaproteobacteria), methylotrophic methanogens (e.g. Euryarchaeota, Thermoplasmatota), sulfate reducers (i.e. Desulfobacterota) and ammonifying sulfur oxidizers (e.g. Gammaproteobacteria). Increasing preservation of putative planktonic sequences from the water column was observed along the eutrophication gradient, mostly affiliated with Cyanobacteria and Bacteroidia as well as anaerobic fermenters at shallow depth (e.g. Acetothermia, Aminicenantia). The main microbial drivers are found among the class Gammaproteobacteria that can make use of reduced metabolites diffusing from underlying anoxic sediments, e.g. oxidation of NH_4^+ and H_2S by *Nitrosomonadaceae*, and oxidation of both H_2 and CH_4 by *Hydrogenophilaceae*. The *Xanthomonadaceae* may include phytopathogenic populations feeding on algal blooms and nitrate reducers in the sediment. The bacterial assemblages that emerge from eutrophic conditions are apparently composed of several fast-growing opportunists which can disrupt syntrophic associations to short-cut the community core assemblage. This example raises questions on the actual resilience of sediment core microbiomes to anthropogenic disturbances over the long term.

Research Perspectives and Methodological Considerations

Prokaryotic Sedimentary DNA as an Indicator of Past and Current Environmental Shifts

Whether prokaryotic DNA in the sediment can be used as a stratigraphic recording tool to inform us about past environmental changes and depositional conditions is still a matter of debate. A selection of proxies from sedDNA sequence data and gravity/piston/hydraulic coring techniques can foster novel multidisciplinary exploration of information preserved in sedimentary archives (Fig. 4.7). Such an approach would comprehensively integrate past climate and biogeochemical cycles and enable reconstruction of the dynamic roles of microbial activity in the water column and sediment, for instance in the climatic context of the last glacial-interglacial

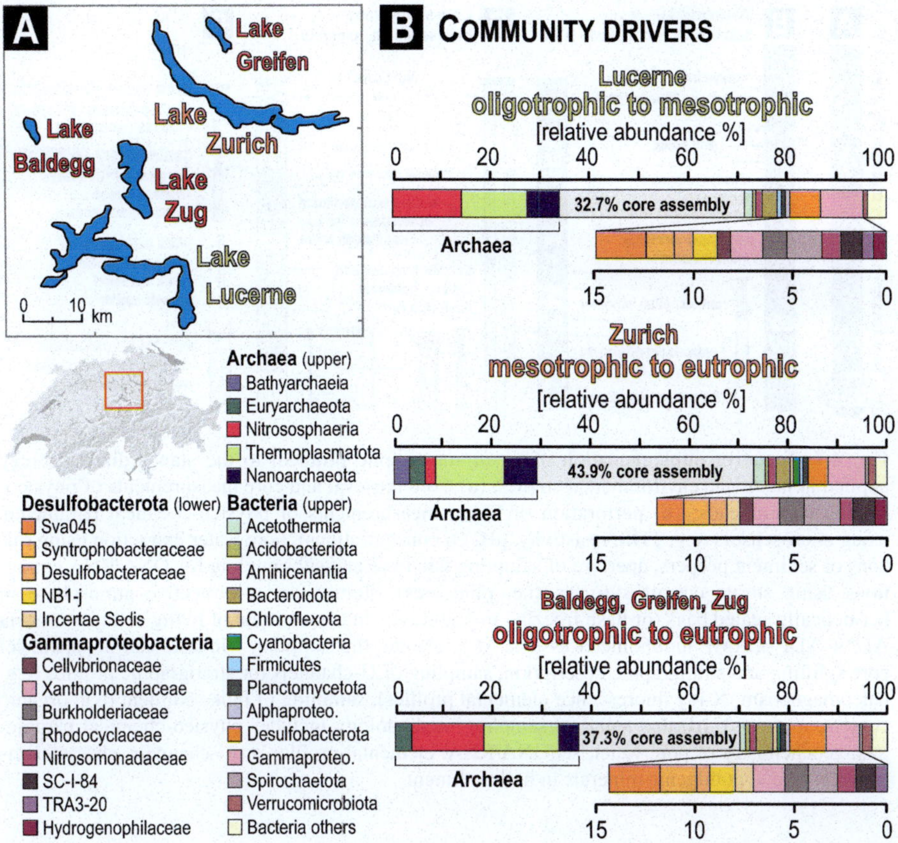

Fig. 4.6 Microbial indicators of excess nutrients in sediments of five Swiss lakes displaying increasing degrees of eutrophication. (**a**) Geographic location of Lake Lucerne, Zurich, Zug, Greifen, and Baldegg (**b**) 16S rRNA gene diversity of statistically relevant taxa putatively useful as proxies for tracing lake eutrophication over time (data from Han et al. 2020), with Desulfobacterota and Gammaproteobacteria enlarged in the lower bar charts. The nutrient gradient increases bottom to top in the sediment cores, reflecting an increase in sedimentary organic carbon related to anthropogenic influence over time. The most influential bacterial and archaeal groups in driving community differences between eutrophic Lake Greifen, Lake Baldegg and Lake Zug, mesotrophic Lake Zurich, and oligotrophic Lake Lucerne were identified via similarity percentages breakdown, correlation matrix of geochemistry and phylogeny, co-occurrence patterns and permutational multivariate variance. (Han et al. 2020)

transition and the onset of the Anthropocene. A particularly exciting opportunity is the implementation of metagenomic, metatranscriptomic and metaproteomic approaches to the very extensive sedimentary archives retrieved from scientific drilling campaigns taking place over the last decades (Anselmetti et al. 2020). So far, such drilling has encompassed a collection of material from lakes displaying different geochemical conditions (Ariztegui et al. 2015; Kallmeyer et al. 2015), intensity

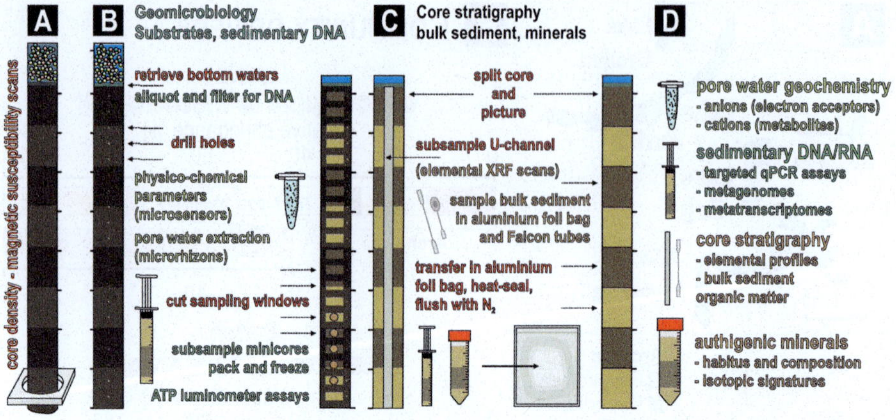

Fig. 4.7 Tentative implementation of an *in situ* aseptic protocol to the standardized strategy applied in lacustrine environmental studies. (**a**) Core retrieval and core measurements of physico-chemical parameters; (**b**) perforation of holes, measurements of physico-chemical parameters using microsensors (e.g. TDS, resistivity, pH, O_2 concentrations), pore water extraction using rhizons or sediment peepers, aperture of sampling windows and subsampling for DNA/RNA extractions using sterile cut-off syringes (i.e. minicores), storage under protective atmosphere in hermetically sealed bags (or flash freezing in liquid N_2), and assessment of living cell biomass via ATP + ADP + AMP luminometer assays. At this point, the sampling windows can be taped; (**c**) core splitting and stratigraphic description, sampling of U-channels for stratigraphic profiles (e.g. paleomagnetism, X-ray fluorescence elemental profiles), sampling of bulk sediment (e.g. organic content, minerals); (**d**) this tentative sampling methodology includes physico-chemical parameters, geochemistry of pore water, sedDNA/RNA, elemental profiles on U-channels, characterization of OM and authigenic minerals in bulk sediment

of anthropogenic disturbances (Michas et al. 2017; Herold et al. 2020) and environmental management regimes along urbanization gradients (Blaser et al. 2016).

The implementation of high-throughput sequencing tools to paleolimnology can thus constitute a significant advance for stratigraphic studies and past environmental reconstructions. In depth characterization of planktonic and benthic microbial communities via a more complete inventory of taxonomic and functional marker genes presents a powerful and novel avenue for exploration of climatic time series along sedimentary records. The identification of microbial taxa and their related metabolic activities can potentially shed light on both in-lake and sediment nutrient cycling, OM degradation and redox processes in relation to the deposited substrates. By tracing microbially mediated biogeochemical cycling in relation to physico-chemical processes and lacustrine endogenic conditions, one can shed new light on past climatic transitions and validate the reliability of parallel proxies used in environmental reconstructions of lake archives (e.g. lipids, minerals, isotope signals). Finally, metagenomic sequence data (i.e. total DNA) and metatranscriptomes (i.e. total RNA sequences) enable a broader assessment of metabolic potential and activities of sediment microbiomes at stratigraphic transitions and identification of key sequence signatures.

In the Anthropocene, lacustrine ecosystems are continuously exposed to increasing urbanization of their catchments and other disturbances, calling for efforts to monitor anthropogenic impacts (Picard et al. 2022; Barouillet et al. 2023) and the resilience of lacustrine microbiomes to such disturbances (Sun et al. 2017; Blanchette and Lund 2021). Because microbial communities constitute the base of aquatic food webs, nutrient disequilibrium constitutes a major threat to the state of lacustrine ecosystems. By driving OM remineralization and redox processes in the shallow subsurface, microbial activities impact the sediment sorption capacities and thereby the long-term sequestration of pollutants. Compiled sedDNA studies targeting prokaryotic communities have shown the potential of integrating meta-omics methodology in limnogeology for *in situ* investigations of microbial processes. Such a combined approach could help us to study how global change brought by human-related impacts will influence lacustrine systems and what mitigation measures could be effective.

Meta-Omics Adapted to Limnogeology and Stratigraphic Studies

Parallel Field Sampling for Limnogeology and Microbiology

Sediment sampling of gravity/piston/hydraulic cores is commonly processed for multidisciplinary approach of lacustrine archives integrating a series of climatic proxies (Ohlendorf et al. 2011). Here, we propose to routinely implement *in situ* aseptic protocols (e.g. Vuillemin et al. 2013a; Friese et al. 2017) that allows for combined studies of microbial systems and past climatic events (Fig. 4.7). Establishing controlled aseptic conditions on site is mandatory for any downstream analytical protocol that involves techniques based on PCR amplification and high-throughput sequencing of prokaryotic sedDNA (Capo et al. 2021b and references therein).

Using primer pairs that target specific marker genes for taxa or functions (Fig. 4.2; Table 4.1), qPCR assays can effectively profile lacustrine microbial communities in terms of depth distribution and abundance of these targets. Subsequent high-throughput sequencing of barcoded amplicons could subsequently validate the assays, whereas in depth sequencing of metagenomes and metatranscriptomes more quantitatively and broadly unravels the metabolic potential and gene expression patterns of entire prokaryotic communities. A multi-pronged characterization of prokaryotic genetic material via meta-omics can thereby provide a holistic understanding of aquatic ecosystems along stratigraphic records, and in this way revolutionize the conventional field of limnogeology. Here, we review some state-of-the-art techniques that allow for the characterization of the subsurface biosphere as well as ancient DNA sources, and share some opportunities and pitfalls with regards to the feasibility and accuracy of meta-omics targeted approaches applied to limnogeology.

Selective Extractions of Nucleic Acid Fractions

Applying some modifications to commercially available DNA extraction and purification kits can initially allow for the separation of specific fractions from the total sedDNA pool prior to PCR amplification.

Intra- and extracellular DNA extractions: sedDNA not enclosed in living cells may in some instances account for the largest fraction of the total pool of sedDNA, and includes molecules locked within dead cells, organic aggregates, adsorbed to mineral matrices, and viral DNA (Torti et al. 2015). Specific protocols allow for separate extraction of the genetic material enclosed in intact cells from the pool contained extracellularly (Corinaldesi et al. 2005; Alawi et al. 2014). Applying such protocols, one can refine the interpretation of prokaryotic sedDNA signals and their significance along stratigraphic sequences (Vuillemin et al. 2017). The principles for such DNA-partitioning methods is to retrieve the soluble extracellular DNA fraction from the extraction buffer, while cells are desorbed, put into flotation and concentrated into a pellet by centrifugation.

Ancient DNA: Ancient DNA preserved in sediments partially degrades and breaks into shorter DNA strands over time. Ancient DNA represents a relatively low percentage of the total extractable prokaryotic sedDNA. Furthermore, prokaryotic sedDNA experiences a constant turnover due to metabolic processes by the viable and active part of the sediment microbial community, making it hard to discriminate the recent from the ancient fraction (Capo et al. 2022), or assign a consistent stratigraphic age to taxa persisting in dormancy compared to those initially deposited as resting stages (Coolen and Overmann 2007). Because of the typically very low biomass in ancient sediment samples, it is recommended to work in ultraclean lab facilities to minimize the potential for contamination and increase recovery of the true targets. DNA extraction replicates from low biomass samples can for example be pooled, or the volume of sediments and buffers can be scaled up and extracts can be concentrated using Amicon filters (Vuillemin et al. 2019a, b, 2020a, b, c). Hybridization capture techniques allow to focus metagenomic sequencing on specific biological targets (Beaudry et al. 2021), such as taxonomic and functional marker genes or metagenome-assembled genomes, with the possibility to bioinformatically identify DNA damage signals characteristic for ancient DNA (Armbrecht et al. 2021). This hybridization capture is performed by adding biotinylated baits that are complementary to the DNA sequences of interest prior to library synthesis to increase the proportion of the selected genomic regions in metagenomic libraries. Similarly, library preparation can be achieved from short fragments of single-stranded DNA using a biotinylated adapter oligonucleotide that have been tethered to magnetic beads for later recovery and amplification (Gansauge and Meyer 2013). Cytosine deamination (observed as C-to-T transitions) is typically used to authenticate ancient sequences bearing characteristic DNA damage but it is important to keep in mind that computational filtering of ancient DNA leads to significant data loss (Borry et al. 2021; Pérez et al. 2022).

Endospore extractions: Endospore formation is a survival strategy found among Firmicutes. Although Firmicutes are one of the most frequently enriched and isolated groups in cultivation-based studies, they are often absent in molecular biodiversity inventories as their sedDNA remains locked in these resistant structures. Physical isolation of endospores from sediment samples can be used to specifically target endospore-forming bacteria, which can account for 5 to 20% of the combined community composition in total sedDNA extracts (Wunderlin et al. 2014b). The procedure includes a step where vegetative cells are lysed, followed by DNAse treatment that allow for subsequent filtration and enrichment of endospores from environmental sediment samples (Wunderlin et al. 2016). DNA is then extracted from the endospore enrichment by performing an enzymatic digestion directly on the filter. The relative abundance of bacteria with the potential to form endospores can be assessed via qPCR assays with primer pairs targeting the endosporulation-specific *spo0A* gene (Wunderlin et al. 2014a), and endospores can also be quantified based on the diagnostic biomarker dipicolinic acid (Wörmer et al. 2019).

RNA extractions: There are different commercially available extraction kits to recover RNA from frozen sediment samples, the critical step being the precipitation and efficient recovery of the RNA separate from inhibiting substances (Coolen and Orsi 2015; Vuillemin et al. 2020a; Orsi et al. 2020b). In low biomass samples, RNA extracts often remain below detection, even using high-sensitivity kits (Vuillemin et al. 2020b). To circumvent this issue, the initial amount of sample material and volume of extraction reagents can be scaled up. We discourage the use of RNAlater® solution to preserve samples in the field, as it is not adapted to sediment samples and often results in low nucleic acid yield and inhibition in later amplification stages (Rissanen et al. 2010). Instead, we recommend retrieving RNA from sediment minicores kept frozen at −80 °C after sectioning them into single use aliquots as this will minimize biases caused by repeated freezing-thawing. Once released from the cells, RNA molecules hydrolyse quickly and it may be prudent to perform transcription to synthesize the more stable complementary DNA (cDNA) prior to storage (Rosenow et al. 2001) or directly proceed with the preparation of metatranscriptomic libraries.

Activity rates: The incorporation of ^{3}H-thymidine into DNA (Kirschner and Velimirov 1999) and ^{14}C-leucine into proteins (Fischer and Pusch 1999) is a technique that allows monitoring bacterial secondary production in the aerobic zone. For anoxic sediments, specific process rates can be measured via sediment incubations with radiolabeled substrates, such as potential rates of sulfate reduction (Kallmeyer et al. 2004) or anaerobic oxidation of methane (Norði and Thamdrup 2014). In deep sediments, modeling of D:L amino acid racemization rates allows inferring the distribution and turnover time of the deep biosphere in terms of living and dormant microbial biomass (Lomstein et al. 2012; Braun et al. 2017).

Negative control: In addition to *in situ* contamination controls (Kallmeyer 2017), blank samples and mock communities are indispensable tools to identify and, if needed, also subtract DNA signals from ambient contamination in laboratories (e.g. airborne endospores, exospores, myxospores) (O'Sullivan et al. 2021) or extraction reagents (Glassing et al. 2016). These controls allow for later curation of the sequencing results (Pichler et al. 2018) as well as standardization and quality assurance of microbiome analysis (Tourlousse et al. 2022).

Targeted PCR Assays for Linking Taxonomy to Functions, Fluorescence Techniques

With the development of molecular probes based on the 16S rRNA and other functional marker genes (Table 4.1), qPCR and multiplex droplet digital PCR (Weisbrod et al. 2020) can be used to quantify and compare key features and populations within microbial communities, both with regards to taxonomic groups and functions (e.g. Fish et al. 2013; Alexander et al. 2015). A general technical difficulty in qPCR assays is the preparation and amplification of accurate and precise standard curves for all the primers used in targeting specific taxa and functional genes, as well as internal controls on amplification efficiency to compensate for coextracted inhibitors in different samples. Limits of detection are usually around 10^2 to 10^3 gene copies, depending on the dilution factor used (Boulter et al. 2016; Bender et al. 2018).

Taxonomic genes: Total sedDNA, which is comprised of both ancient and recent DNA from both eukaryotic and dormant/active prokaryotic material, can be superficially analyzed and compared by amplicon sequencing using primers that target universal 18S rRNA (eukaryotic) and 16S rRNA (prokaryotic) genes (e.g. Nichols et al. 2019; Capo et al. 2021b). The density and relative abundance of prokaryotic populations are primarily investigated using primers targeting the V4–V6 regions of the 16S rRNA gene, assuming a minor bias inherent to extraction and amplification efficiencies. However, due to the variable number of 16S rRNA gene copies in genomes across phyla (Větrovský and Baldrian 2013), the use of such "universal" primer pairs can result in potential quantitative biases between the different clades studied (Starke et al. 2021). Primer pairs can also be designed *in silico* to target less conserved regions for a more selective identification of taxonomic groups (Morales and Holben 2009). For example, this can be done at the level of domains to compare the relative contribution of Bacteria and Archaea (Ye et al. 2012) with primers such as Arch519F-915R and Eub341F-534R, but also for more highly resolved populations (Narrowe et al. 2017). Taxonomic assignment of 16S rRNA gene amplicon sequences is commonly performed by grouping of recovered sequences into operational taxonomic units (OTUs) sharing an arbitrary level of sequence homology or unique amplicon sequence variants (ASVs). Representative sequences are then taxonomically assigned performing BLASTn searches against databases such as SILVA 16S rRNA small subunit (SSU) reference database (Pruesse et al. 2007; Quast et al. 2013).

Although cultivation-independent 16S rRNA gene surveys can be used to assess and compare the taxonomically diverse communities typical for sediments, they do not provide reliable information on the ecological functions of yet-uncultivated microorganisms or taxa with unresolved or variable functions (Macalady et al. 2013; Bier et al. 2015). Specific software (e.g. PICRUSt2, paprica) does at least in theory allow for reasonable predictions to be made about metabolism inferred from phylogeny and thereby describe the functionality of microbes in lacustrine ecosystems merely using taxonomic data (Douglas et al. 2020; Erazo et al. 2021). It should however be noted that such predictions critically depend on the availability of metabolically annotated reference genomes for functionally coherent target groups, and may thus be less well suited for sediments and other environments with a paucity of such reference data (i.e. where many organisms are uncultured and functionally undescribed).

After assessment of the community composition and as an alternative to the above mentioned approach, additional primer pairs targeting functional genes (Table 4.1) can be used to quantify the potential for specific processes related to biogeochemical cycling, and thereby indirectly associate taxonomy to functional traits. For instance, primers targeting different respiration types (Fig. 4.2), such as those involved in denitrification (i.e. *narG*, *napA*, *nirK*), sulfate reduction (i.e. *dsrAB*, *aprAB*), and methanogenesis (i.e. *mcrA*) can be amplified from the same DNA extracts to track the vertical distribution of microbial guilds and correlate them with the taxonomic assignment of 16S rRNA genes while avoiding inferences based on information obtained from their sometimes quite distantly related cultivated relatives.

Comparison with cell count techniques: Classic quantification of microbial biomass is usually achieved via cell counts by means of fluorescence microscopy, using different DNA stains, such as Acridine Orange (Parkes et al. 1994), 4′,6′-Diamidin-2-phenylindol (DAPI) (Suzuki et al. 1993), and SYBR Green I (Kallmeyer et al. 2008). Optimizations of fluorescent microscopic cell enumeration protocols may include LIVE/DEAD cell counts to discriminate living from dead cells (Haglund et al. 2003), fluorescent *in situ* hybridization (FISH) with molecular probes targeting specific taxonomic groups (Moter and Göbel 2000) and related protocols used to enhance the signal, e.g. catalyzed reporter deposition or CARD (Yamaguchi et al. 2015), or detect metabolic activity, i.e. bioorthogonal non-canonical amino acid tagging or BONCAT (Hatzenpichler et al. 2014). Enumeration of microorganisms can be further streamlined using flow cytometry (Morono et al. 2013) and fluorescence-activated cell sorting (FACS) (Kalyuzhnaya et al. 2006). Although the respective shape of the downcore profiles is usually reliable, qPCR values and cell counts can easily be offset by orders of magnitude in the sediment (Lloyd et al. 2013).

Metagenomics and Metatranscriptomics

High-throughput shotgun sequencing techniques can provide extensive information on the taxonomic identities and patterns of environmental occurrence of bacteria, archaea as well as viruses and eukaryotes (Rinke et al. 2013; Hug et al. 2016). Assigning metabolic functions to the phylogenetic diversity documented from the lacustrine subsurface does not always unravel cryptic interactions across metabolic guilds (Macalady et al. 2013), nor their actual metabolic activities in link to past trophic levels in the corresponding stratigraphic sequences (Ellegaard et al. 2020; Garner et al. 2020). Metagenomes and metatranscriptomes provide detailed insights into the genomic structure and metabolic potential and even activity of complex microbial communities (Vavourakis et al. 2018). The genetic potential of hitherto uncultivated microbial populations can be retrieved from shotgun metagenomes by reconstruction and analysis of metagenomic-assembled genomes (MAGs) (Solden et al. 2016; Jiao et al. 2021). The systematic and cumulative recovery of MAGs from different locations will in the longer run allow for pangenomic analyses of lacustrine microbiomes (Thompson et al. 2017; Nayfach et al. 2021), and such genomic resources will also facilitate and enable a population-centric interpretation of functional gene expression patterns via metatranscriptomics for a holistic understanding of the subsurface biosphere.

The critical role of read length: High-throughput sequencing can provide either short reads (2 × 150 bps, i.e. base pairs) that allow for rapid screening of functional genes in diverse microbial communities, or reads of intermediate length (2 × 300 bps) that are more effective for contig assembly (Knight et al. 2018). Contrary to ancient DNA that represents short fragments, longer reads are more likely to derive from intact cells for which a combination of DNA size selection long read sequencing technology (Oxford Nanopore, Pacific Biosciences) would allow the researcher to specifically target viable cells and rule out long-dead cells. The latter long-read technology is mainly feasible for low diversity samples where the aim could be to obtain reads that are long enough for MAG completion (Moss et al. 2020). Metagenomic sequencing of an environmental sample displaying a high taxonomic diversity may for example produce multiple fragmented bins, with potential chimeric edges and inclusions of false positive reads across bins. In searching gene expression, most of the initial RNA extracts are composed of non-coding ribosomal RNA (He et al. 2010). The reads assembled from metatranscriptomic libraries are then only analyzed in terms of messenger RNA, which is indicative of active protein synthesis and represents transcripts. A read length of <300 bp is typically sufficient to identify such transcripts and a separate bioinformatic pipeline is needed to taxonomically assign them or map the transcripts back to the corresponding MAGs and SAGs. Similarly, in targeting ancient DNA, metagenomic assembly usually results in partial and less well supported bins due to the overall short read lengths of the fragmented DNA (Der Sarkissian et al. 2021). Because of the intrinsic nature of sedDNA and a necessity to optimize sequencing costs to the usefulness of the data, short reads are best suited for sediments, for which an alternative approach to the binning step can be applied (Key et al. 2017). In this case, the focus is rather on mining specific marker genes relevant to the studied environment, for instance by performing BlastP and taxonomic assignments against the corresponding functional gene database, e.g. FunGene (Fish et al. 2013).

On-Site Contamination Control and Core Sampling

Gravity coring techniques allow for retrieval of undisturbed sediment-water interface and underlying fine grained sediment sequences from lakes (Weaver and Schultheiss 1990). Given the lack of drilling fluids, the core samples are uncontaminated, except for possibly the outer few mm where some surface material might be dragged down by the core barrel (Kallmeyer 2017). These rims have to be scraped off under a protective atmosphere (e.g. glove bag, anaerobic chamber) using a sterile spatula.

In hydraulic drilling operations, it is almost impossible to maintain sterile conditions on site as the drill fluid itself contains surface microbes and other contaminants. Moreover, lake surface waters are usually pumped in the hydraulic system used to shoot core liners into the sediment, which can potentially result in a trans-contamination that is hardly detectable at the time of sampling and carried on in downstream analysis of the samples. The addition of tracer compounds or fluorescent tracer particles can then be used for identification and rejection of the parts of the hydraulic cores that were infiltrated by the drill fluid (Friese et al. 2017; Kallmeyer 2017). Once retrieved, the sediment core liners from drilling operations can be subsampled on site according to a strategy that allows for successive core measurements of physico-chemical properties (Fig. 4.7a), retrieval of bottom waters and pore water samples with minimal exposure to ambient conditions (Fig. 4.7b), sediment material for sedDNA and sedRNA using sterile end-cut syringe minicores (Coolen et al. 2013) and ATP + ADP + AMP luminometer assays as an indication of living cell biomass (Vuillemin et al. 2010; Orsi 2023). The stratigraphic description and subsampling for bulk parameters (Fig. 4.7c) can be performed according to standardized procedures (e.g. Ohlendorf et al. 2011). The sampling resolution across the sediment stratigraphy has to take into consideration factors, such as sedimentation rates, sample heterogeneity, and degree of sedDNA degradation with depth, and biological replicates are required (Pearman et al. 2021).

Summary

General microbial patterns of biogeography can be established based on sedDNA retrieved from diverse inland aquatic ecosystems, provided that certain variable environmental factors are taken into consideration in order to trace the admixture of prokaryotic sedDNA preserved in lacustrine sediments from source to sink. These include several watershed characteristics, such as climate and mineralogy in the catchment and degree of urbanization, which in turn determine the lake regime, productivity and geochemistry of the water body, redox conditions at the sediment-water interface, sedimentation rates and OM content over time.

Records of environmental conditions through the preservation of prokaryotic sedDNA will partially be obscured by the presence of a subsurface biosphere that remains active during burial. Sediment colonization and selective growth of specific sub-populations of the resident microbiomes are inherent to the substrates available in lacustrine deposits. Microbial guilds reflective of respiration types and OM decomposition usually go hand-in-hand with the geochemical gradient that develops in the sedimentary column during early diagenetic processes, but may also prevail in specific stratigraphic intervals. Thus, the correct attribution of prokaryotic sedDNA sources to an environmental context is essential for resolving past and modern ecosystem functioning. Because the succession of microbial taxa pre-adapted to gradual depletion of electron donors and acceptors is relatively well known, sources of sedDNA inherent to resident microbial communities can be discriminated and provide evidence of diagenetic processes during burial. In contrast, ancient sources of sedDNA that persist in microbially active sediments are exposed to a constant turnover, leading to preferential preservation of dormant stages over free sedDNA.

Bacteria and archaea are tremendously diverse, and therefore there is the need to rigorously assign their metabolic functions and assess their degree of activity along sediment archives. Beyond taxonomic surveys and geochemical profiling of the subsurface biosphere, novel procedures are being customized to sort, target, sequence and assemble sedDNA fragments of interest to unravel specific biogeochemical interactions in lacustrine ecosystems pertaining past and actual ecological changes at the local and global scale.

Acknowledgements We thanked John Pearman and Christian Wurzbacher for their reviews, corrections, and comments on the present chapter.

References

Aalto SL, Saarenheimo J, Mikkonen A, Rissanen AJ, Tiirola M (2018) Resistant ammonia-oxidizing archaea endure, but adapting ammonia-oxidizing bacteria thrive in boreal lake sediments receiving nutrient-rich effluents. Environ Microbiol 20(10):3616–3628. https://doi.org/10.1111/1462-2920.14354

Aanderud ZT, Vert JC, Lennon JT, Magnusson TW, Breakwell DP, Harker AR (2016) Bacterial dormancy is more prevalent in freshwater than hypersaline lakes. Front Microbiol 7

Alawi M, Schneider B, Kallmeyer J (2014) A procedure for separate recovery of extra- and intracellular DNA from a single marine sediment sample. J Microbiol Methods 104:36–42. https://doi.org/10.1016/j.mimet.2014.06.009

Alexander J, Bollmann A, Seitz W, Schwartz T (2015) Microbiological characterization of aquatic microbiomes targeting taxonomical marker genes and antibiotic resistance genes of opportunistic bacteria. Sci Total Environ 512–513:316–325. https://doi.org/10.1016/j.scitotenv.2015.01.046

Alsos IG, Lammers Y, Yoccoz NG, Jørgensen T, Sjögren P, Gielly L, Edwards ME (2018) Plant DNA metabarcoding of lake sediments: how does it represent the contemporary vegetation. PLoS One 13(4):e0195403. https://doi.org/10.1371/journal.pone.0195403

Anantharaman K, Hausmann B, Jungbluth SP, Kantor RS, Lavy A, Warren LA, Rappé MS, Pester M, Loy A, Thomas BC, Banfield JF (2018) Expanded diversity of microbial groups that shape the dissimilatory sulfur cycle. ISME J 12(7):1715–1728. https://doi.org/10.1038/s41396-018-0078-0

Anderson NJ, Bennion H, Lotter AF (2014) Lake eutrophication and its implications for organic carbon sequestration in Europe. Glob Chang Biol 20(9):2741–2751. https://doi.org/10.1111/gcb.12584

Anesio AM, Lutz S, Chrismas NAM, Benning LG (2017) The microbiome of glaciers and ice sheets. NPJ Biofilms Microbiomes 3(1):1–11. https://doi.org/10.1038/s41522-017-0019-0

Anselmetti F, Ashwal L, Ariztegui D, Bohnhoff M, Bomberg M, Claeys P, Eichelberger J, Ellsworth WL, Goodenough K, Heubeck C, Kelemen P, Kipfer R, Koeberl C, Kopf A, Miller K, Nordgulen O, Noren A, Onstott T, Pease V, Philippot P, Russell J, Soreghan G, Stein M, Verschuren D, Yamada Y (2020) ICDP science plan: 2020–2030. https://doi.org/10.2312/icdp.2020.001

Ariztegui D, Thomas C, Vuillemin A (2015) Present and future of subsurface biosphere studies in lacustrine sediments through scientific drilling. Int J Earth Sci (Geol Rundsch) 104(6):1655–1665. https://doi.org/10.1007/s00531-015-1148-4

Armbrecht L, Hallegraeff G, Bolch CJS, Woodward C, Cooper A (2021) Hybridisation capture allows DNA damage analysis of ancient marine eukaryotes. Sci Rep 11:3220. https://doi.org/10.1038/s41598-021-82578-6

Asao M, Pinkart HC, Madigan MT (2011) Diversity of extremophilic purple phototrophic bacteria in soap Lake, a Central Washington (USA) Soda Lake. Environ Microbiol 13(8):2146–2157. https://doi.org/10.1111/j.1462-2920.2011.02449.x

Aüllo T, Ranchou-Peyruse A, Ollivier B, Magot M (2013) Desulfotomaculum spp. and related gram-positive sulfate-reducing bacteria in deep subsurface environments. Front Microbiol 4

Aylward FO, Santoro AE (2020) Heterotrophic Thaumarchaea with small genomes are widespread in the Dark Ocean. mSystems 5(3):e00415-20. https://doi.org/10.1128/mSystems.00415-20

Bai Y, Ruan X, Li R, Zhang Y, Wang Z (2022) Metagenomics-based antibiotic resistance genes diversity and prevalence risk revealed by pathogenic bacterial host in Taihu Lake, China. Environ Geochem Health 44(8):2531–2543. https://doi.org/10.1007/s10653-021-01021-x

Baker BJ, Saw JH, Lind AE, Lazar CS, Hinrichs K-U, Teske AP, Ettema TJG (2016) Genomic inference of the metabolism of cosmopolitan subsurface Archaea. Hadesarchaea Nat Microbiol 1:16002. https://doi.org/10.1038/nmicrobiol.2016.2

Baltes N, Hennig-Pauka I, Jacobsen I, Gruber AD, Gerlach GF (2003) Identification of dimethyl sulfoxide reductase in Actinobacillus pleuropneumoniae and its role in infection. Infect Immun 71(12):6784–6792. https://doi.org/10.1128/IAI.71.12.6784-6792.2003

Bao S, Ding X, Yu S, Xing X, Ding C (2021) Characterization of pyruvate dehydrogenase complex E1 alpha and beta subunits of mycoplasma synoviae. Microb Pathog 155:104851. https://doi.org/10.1016/j.micpath.2021.104851

Barouillet C, Monchamp M-E, Bertilsson S, Brasell K, Domaizon I, Epp LS, Ibrahim A, Mejbel H, Nwosu EC, Pearman JK, Picard M, Thomson-Laing G, Tsugeki N, Von Eggers J, Gregory-Eaves I, Pick FR, Wood SA, Capo E (2023) Investigating the effects of anthropogenic stressors on lake biota using sedimentary DNA. Freshw Biol. https://doi.org/10.1111/fwb.14027

Beaudry MS, Wang J, Kieran TJ, Thomas J, Bayona-Vásquez NJ, Gao B, Devault A, Brunelle B, Lu K, Wang J-S, Rhodes OE, Glenn TC (2021) Improved microbial community characterization of 16S rRNA via metagenome hybridization capture enrichment. Front Microbiol 12

Béjà O, Spudich EN, Spudich JL, Leclerc M, DeLong EF (2001) Proteorhodopsin phototrophy in the ocean. Nature 411(6839):786–789. https://doi.org/10.1038/35081051

Bell E, Lamminmäki T, Alneberg J, Andersson AF, Qian C, Xiong W, Hettich RL, Frutschi M, Bernier-Latmani R (2020) Active sulfur cycling in the terrestrial deep subsurface. ISME J 14(5):1260–1272. https://doi.org/10.1038/s41396-020-0602-x

Bender JM, Li F, Adisetiyo H, Lee D, Zabih S, Hung L, Wilkinson TA, Pannaraj PS, She RC, Bard JD, Tobin NH, Aldrovandi GM (2018) Quantification of variation and the impact of biomass in targeted 16S rRNA gene sequencing studies. Microbiome 6(1):155. https://doi.org/10.1186/s40168-018-0543-z

Berg IA (2011) Ecological aspects of the distribution of different autotrophic CO_2 fixation pathways. Appl Environ Microbiol 77(6):1925–1936. https://doi.org/10.1128/AEM.02473-10

Berg JS, Michellod D, Pjevac P, Martinez-Perez C, Buckner CRT, Hach PF, Schubert CJ, Milucka J, Kuypers MMM (2016) Intensive cryptic microbial iron cycling in the low iron water column of the meromictic Lake Cadagno. Environ Microbiol 18(12):5288–5302. https://doi.org/10.1111/1462-2920.13587

Berg JS, Jézéquel D, Duverger A, Lamy D, Laberty-Robert C, Miot J (2019) Microbial diversity involved in iron and cryptic sulfur cycling in the ferruginous, low-sulfate waters of Lake Pavin. PLoS One 14(2):e0212787. https://doi.org/10.1371/journal.pone.0212787

Berg JS, Lepine M, Laymand E, Han X, Vogel H, Morlock MA, Gajendra N, Gilli A, Bernasconi SM, Schubert CJ, Su G, Lever MA (2022) Ancient and modern geochemical signatures in the 13,500-year sedimentary record of Lake Cadagno. Front Earth Sci 9

Berghuis BA, Yu FB, Schulz F, Blainey PC, Woyke T, Quake SR (2019) Hydrogenotrophic methanogenesis in archaeal phylum Verstraetearchaeota reveals the shared ancestry of all methanogens. Proc Natl Acad Sci USA 116(11):5037–5044. https://doi.org/10.1073/pnas.1815631116

Bertilsson S, Mehrshad M (2021) Diversity and dynamics of bacterial communities in freshwater lakes. In: Mehner T, Tockner K (eds) Encyclopedia of inland Waters, vol 1. Elsevier, pp 601–615

Bertolet BL, West WE, Armitage DW, Jones SE (2019) Organic matter supply and bacterial community composition predict methanogenesis rates in temperate lake sediments. Limnol Oceanogr Lett 4(5):164–172. https://doi.org/10.1002/lol2.10114

Biderre-Petit C, Boucher D, Kuever J, Alberic P, Jézéquel D, Chebance B, Borrel G, Fonty G, Peyret P (2011) Identification of sulfur-cycle prokaryotes in a low-sulfate Lake (Lake Pavin) using aprA and 16S rRNA gene markers. Microb Ecol 61(2):313–327. https://doi.org/10.1007/s00248-010-9769-4

Bier RL, Bernhardt ES, Boot CM, Graham EB, Hall EK, Lennon JT, Nemergut DR, Osborne BB, Ruiz-González C, Schimel JP, Waldrop MP, Wallenstein MD (2015) Linking microbial community structure and microbial processes: an empirical and conceptual overview. FEMS Microbiol Ecol 91(10):fiv113. https://doi.org/10.1093/femsec/fiv113

Bischoff J, Mangelsdorf K, Gattinger A, Schloter M, Kurchatova AN, Herzschuh U, Wagner D (2013) Response of methanogenic archaea to late Pleistocene and Holocene climate changes in the Siberian Arctic. Glob Biogeochem Cycles 27:305–317. https://doi.org/10.1029/2011GB004238

Bižić-Ionescu M, Ionescu D, Grossart H-P (2018) Organic particles: heterogeneous hubs for microbial interactions in aquatic ecosystems. Front Microbiol 9

Blanchette ML, Lund MA (2021) Aquatic ecosystems of the Anthropocene: limnology and microbial ecology of mine pit lakes. Microorganisms 9(6):1207. https://doi.org/10.3390/microorganisms9061207

Blaser MJ, Cardon ZG, Cho MK, Dangl JL, Donohue TJ, Green JL, Knight R, Maxon ME, Northen TR, Pollard KS, Brodie EL (2016) Toward a predictive understanding of Earth's microbiomes to address 21st century challenges. MBio 7(3):e00714-16. https://doi.org/10.1128/mBio.00714-16

Block KR, O'Brien JM, Edwards WJ, Marnocha CL (2021) Vertical structure of the bacterial diversity in meromictic Fayetteville Green Lake. MicrobiologyOpen 10(4):e1228. https://doi.org/10.1002/mbo3.1228

Borrel G, Jézéquel D, Biderre-Petit C, Morel-Desrosiers N, Morel J-P, Peyret P, Fonty G, Lehours A-C (2011) Production and consumption of methane in freshwater lake ecosystems. Res Microbiol 162(9):832–847. https://doi.org/10.1016/j.resmic.2011.06.004

Borrel G, Adam PS, Gribaldo S (2016) Methanogenesis and the Wood–Ljungdahl pathway: an ancient, versatile, and fragile association. Genome Biol Evol 8(6):1706–1711. https://doi.org/10.1093/gbe/evw114

Borry M, Hübner A, Rohrlach AB, Warinner C (2021) PyDamage: automated ancient damage identification and estimation for contigs in ancient DNA de novo assembly. PeerJ 9:e11845. https://doi.org/10.7717/peerj.11845

Bose A, Pritchett MA, Metcalf WW (2008) Genetic analysis of the methanol- and methylamine-specific methyltransferase 2 genes of Methanosarcina acetivorans C2A. J Bacteriol 190(11):4017–4026. https://doi.org/10.1128/JB.00117-08

Boucher Y, Douady CJ, Papke RT, Walsh DA, Boudreau MER, Nesbø CL, Case RJ, Doolittle WF (2003) Lateral gene transfer and the origins of prokaryotic groups. Annu Rev Genet 37:283–328. https://doi.org/10.1146/annurev.genet.37.050503.084247

Boulter N, Suarez FG, Schibeci S, Sunderland T, Tolhurst O, Hunter T, Hodge G, Handelsman D, Simanainen U, Hendriks E, Duggan K (2016) A simple, accurate and universal method for quantification of PCR. BMC Biotechnol 16(1):27. https://doi.org/10.1186/s12896-016-0256-y

Braker G, Zhou J, Wu L, Devol AH, Tiedje JM (2000) Nitrite reductase genes (nirK and nirS) as functional markers to investigate diversity of denitrifying bacteria in pacific northwest marine sediment communities. Appl Environ Microbiol 66(5):2096–2104. https://doi.org/10.1128/AEM.66.5.2096-2104.2000

Brasell KA, Howarth J, Pearman JK, Fitzsimons SJ, Zaiko A, Pochon X, Vandergoes MJ, Simon KS, Wood SA (2021) Lake microbial communities are not resistant or resilient to repeated large-scale natural pulse disturbances. Mol Ecol 30(20):5137–5150. https://doi.org/10.1111/mec.16110

Braun S, Mhatre SS, Jaussi M, Røy H, Kjeldsen KU, Pearce C, Seidenkrantz M-S, Jørgensen BB, Lomstein BA (2017) Microbial turnover times in the deep seabed studied by amino acid racemization modelling. Sci Rep 7(1):5680. https://doi.org/10.1038/s41598-017-05972-z

Bravo AG, Zopfi J, Buck M, Xu J, Bertilsson S, Schaefer JK, Poté J, Cosio C (2018) Geobacteraceae are important members of mercury-methylating microbial communities of sediments impacted by waste water releases. ISME J 12(3):802–812. https://doi.org/10.1038/s41396-017-0007-7

Bryanskaya AV, Malup TK, Lazareva EV, Taran OP, Rozanov AS, Efimov VM, Peltek SE (2016) The role of environmental factors for the composition of microbial communities of saline lakes in the Novosibirsk region (Russia). BMC Microbiol 16(1):S4. https://doi.org/10.1186/s12866-015-0618-y

Buckel W, Thauer RK (2018) Flavin-based electron bifurcation, ferredoxin, Flavodoxin, and anaerobic respiration with protons (Ech) or NAD+ (Rnf) as electron acceptors: a historical review. Front Microbiol 9

Cai J, Bai C, Tang X, Dai J, Gong Y, Hu Y, Shao K, Zhou L, Gao G (2018) Characterization of bacterial and microbial eukaryotic communities associated with an ephemeral hypoxia event in Taihu Lake, a shallow eutrophic Chinese lake. Environ Sci Pollut Res 25(31):31543–31557. https://doi.org/10.1007/s11356-018-2987-x

Canfield DE, Kristensen E, Thamdrup B (2005) Carbon fixation and phototrophy. Adv Mar Biol 48:95–127. https://doi.org/10.1016/s0065-2881(05)48004-9

Capo E, Ninnes S, Domaizon I, Bertilsson S, Bigler C, Wang X-R, Bindler R, Rydberg J (2021a) Landscape setting drives the microbial eukaryotic community structure in four Swedish Mountain lakes over the Holocene. Microorganisms 9(2):355. https://doi.org/10.3390/microorganisms9020355

Capo E, Giguet-Covex C, Rouillard A, Nota K, Heintzman PD, Vuillemin A, Ariztegui D, Arnaud F, Belle S, Bertilsson S, Bigler C, Bindler R, Brown AG, Clarke CL, Crump SE, Debroas D, Englund G, Ficetola GF, Garner RE, Gauthier J, Gregory-Eaves I, Heinecke L, Herzschuh U, Ibrahim A, Kisand V, Kjær KH, Lammers Y, Littlefair J, Messager E, Monchamp M-E, Olajos F, Orsi W, Pedersen MW, Rijal DP, Rydberg J, Spanbauer T, Stoof-Leichsenring KR, Taberlet P, Talas L, Thomas C, Walsh DA, Wang Y, Willerslev E, van Woerkom A, Zimmermann HH, Coolen MJL, Epp LS, Domaizon IG, Alsos I, Parducci L (2021b) Lake sedimentary DNA research on past terrestrial and aquatic biodiversity: overview and recommendations. Quaternary 4(1):6. https://doi.org/10.3390/quat4010006

Capo E, Monchamp M-E, MJL C, Domaizon I, Armbrecht L, Bertilsson S (2022) Environmental paleomicrobiology: using DNA preserved in aquatic sediments to its full potential. Environ Microbiol 24(5):2201–2209. https://doi.org/10.1111/1462-2920.15913

Cardman Z, Arnosti C, Durbin A, Ziervogel K, Cox C, Steen AD, Teske A (2014) Verrucomicrobia are candidates for polysaccharide-degrading Bacterioplankton in an Arctic Fjord of Svalbard. Appl Environ Microbiol 80(12):3749–3756. https://doi.org/10.1128/AEM.00899-14

Castelle CJ, Banfield JF (2018) Major new microbial groups expand diversity and Alter our understanding of the Tree of life. Cell 172(6):1181–1197. https://doi.org/10.1016/j.cell.2018.02.016

Castro HF, Williams NH, Ogram A (2000) Phylogeny of sulfate-reducing bacteria1. FEMS Microbiol Ecol 31(1):1–9. https://doi.org/10.1111/j.1574-6941.2000.tb00665.x

Castro HF, Classen AT, Austin EE, Crawford KM, Schadt CW (2012) Development and validation of a citrate synthase directed quantitative PCR marker for soil bacterial communities. Appl Soil Ecol 61:69–75. https://doi.org/10.1016/j.apsoil.2012.05.007

Chandran A, Varghese S, Kandeler E, Thomas A, Hatha M, Mazumder A (2011) An assessment of potential public health risk associated with the extended survival of indicator and pathogenic bacteria in freshwater lake sediments. Int J Hyg Environ Health 214(3):258–264. https://doi.org/10.1016/j.ijheh.2011.01.002

Chernitsyna SM, Mamaeva EV, Lomakina AV, Pogodaeva TV, Galach'yants YP, Bukin SV, Pimenov NV, Khlystov OM, Zemskaya TI (2016) Phylogenetic diversity of microbial communities of the Posolsk Bank bottom sediments, Lake Baikal. Microbiology 85(6):672–680. https://doi.org/10.1134/S0026261716060060

Chi X-Q, Wang L, Guo R, Zhao D, Li J, Zhang Y, Jiao N (2018) RuBisCO large subunit gene primers for assessing the CO2-assimilating planktonic community structure in Jiaozhou Bay, China. FEMS Microbiol Lett 365(14):fny140. https://doi.org/10.1093/femsle/fny140

Colman DR, Poudel S, Stamps BW, Boyd ES, Spear JR (2017) The deep, hot biosphere: twenty-five years of retrospection. Proc Natl Acad Sci U S A 114(27):6895–6903. https://doi.org/10.1073/pnas.1701266114

Conrad R (2020) Importance of hydrogenotrophic, aceticlastic and methylotrophic methanogenesis for methane production in terrestrial, aquatic and other anoxic environments: a mini review. Pedosphere 30(1):25–39. https://doi.org/10.1016/S1002-0160(18)60052-9

Conrad R, Klose M, Enrich-Prast A (2020) Acetate turnover and methanogenic pathways in Amazonian lake sediments. Biogeosciences 17(4):1063–1069. https://doi.org/10.5194/bg-17-1063-2020

Coolen MJL, Orsi WD (2015) The transcriptional response of microbial communities in thawing Alaskan permafrost soils. Front Microbiol 6

Coolen MJL, Overmann J (1998) Analysis of subfossil molecular remains of purple sulfur bacteria in a Lake sediment. Appl Environ Microbiol 64(11):4513–4521. https://doi.org/10.1128/AEM.64.11.4513-4521.1998

Coolen MJL, Overmann J (2007) 217 000-year-old DNA sequences of green sulfur bacteria in Mediterranean sapropels and their implications for the reconstruction of the paleoenvironment. Environ Microbiol 9(1):238–249. https://doi.org/10.1111/j.1462-2920.2006.01134.x

Coolen MJL, Muyzer G, Rijpstra WIC, Schouten S, Volkman JK, Sinninghe Damsté JS (2004) Combined DNA and lipid analyses of sediments reveal changes in Holocene haptophyte and diatom populations in an Antarctic lake. Earth Planet Sci Lett 223:225–239. https://doi.org/10.1016/j.epsl.2004.04.014

Coolen MJL, Talbot HM, Abbas BA, Ward C, Schouten S, Volkman JK, Damsté JSS (2008) Sources for sedimentary bacteriohopanepolyols as revealed by 16S rDNA stratigraphy. Environ Microbiol 10(7):1783–1803. https://doi.org/10.1111/j.1462-2920.2008.01601.x

Coolen MJL, Orsi WD, Balkema C, Quince C, Harris K, Sylva SP, Filipova-Marinova M, Giosan L (2013) Evolution of the plankton paleome in the Black Sea from the Deglacial to Anthropocene. Proc Natl Acad Sci 110(21):8609–8614. https://doi.org/10.1073/pnas.1219283110

Corinaldesi C, Danovaro R, Dell'Anno A (2005) Simultaneous recovery of extracellular and intracellular DNA suitable for molecular studies from marine sediments. Appl Environ Microbiol 71(1):46–50. https://doi.org/10.1128/AEM.71.1.46-50.2005

Crowe S, Jones C, Katsev S, Magen C, O'Neill A, Sturm A, Canfield D, Haffner G, Mucci A, Sundby B, Fowle D (2008) Photoferrotrophs thrive in an Archean Ocean analogue. Proc Natl Acad Sci U S A 105(41):15938–15943. https://doi.org/10.1073/pnas.0805313105

Crump BC, Amaral-Zettler LA, Kling GW (2012) Microbial diversity in arctic freshwaters is structured by inoculation of microbes from soils. ISME J 6(9):1629–1639. https://doi.org/10.1038/ismej.2012.9

Cupit C, Lomstein BA, Kjeldsen KU (2019) Contrasting community composition of endospores and vegetative Firmicutes in a marine sediment suggests both endogenous and exogenous sources of endospore accumulation. Environ Microbiol Rep 11(3):352–360. https://doi.org/10.1111/1758-2229.12679

Dahl C (2020) A biochemical view on the biological sulfur cycle. Enviro Technol Treat Sulfur Pollu Princ Eng 2:55–96. https://doi.org/10.2166/9781789060966_0055

Dalcin Martins P, de Jong A, Lenstra WK, van Helmond NAGM, Slomp CP, Jetten MSM, Welte CU, Rasigraf O (2021) Enrichment of novel Verrucomicrobia, Bacteroidetes, and Krumholzibacteria in an oxygen-limited methane- and iron-fed bioreactor inoculated with Bothnian Sea sediments. MicrobiologyOpen 10(1):e1175. https://doi.org/10.1002/mbo3.1175

Dalsgaard T, Stewart FJ, Thamdrup B, De Brabandere L, Revsbech NP, Ulloa O, Canfield DE, De Long EF (2014) Oxygen at nanomolar levels reversibly suppresses process rates and gene expression in anammox and denitrification in the oxygen minimum zone off northern Chile. MBio 5(6):e01966. https://doi.org/10.1128/mBio.01966-14

De Oliveira LFV, Margis R (2015) The source of the river as a nursery for microbial diversity. PLoS One 10(3):e0120608. https://doi.org/10.1371/journal.pone.0120608

de Vladar HP (2012) Amino acid fermentation at the origin of the genetic code. Biol Direct 7(1):6. https://doi.org/10.1186/1745-6150-7-6

Dean WE, Gorham E (1998) Magnitude and significance of carbon burial in lakes, reservoirs, and peatlands. Geology 26(6):535–538

Del Don C, Hanselmann KW, Peduzzi R, Bachofen R (2001) The meromictic alpine Lake Cadagno: Orographical and biogeochemical description. Aquat Sci 63(1):70–90. https://doi.org/10.1007/PL00001345

DeLong EF, Pace NR (2001) Environmental diversity of bacteria and archaea. Syst Biol 50(4):470–478

Der Sarkissian C, Velsko IM, Fotakis AK, Vågene ÅJ, Hübner A, Fellows Yates JA (2021) Ancient metagenomic studies: considerations for the wider scientific community. mSystems 6(6):e01315-21. https://doi.org/10.1128/msystems.01315-21

Dong LF, Smith CJ, Papaspyrou S, Stott A, Osborn AM, Nedwell DB (2009) Changes in benthic denitrification, nitrate ammonification, and Anammox process rates and nitrate and nitrite reductase gene abundances along an estuarine nutrient gradient (the Colne estuary, United Kingdom). Appl Environ Microbiol 75(10):3171–3179. https://doi.org/10.1128/AEM.02511-08

Douglas GM, Maffei VJ, Zaneveld JR, Yurgel SN, Brown JR, Taylor CM, Huttenhower C, Langille MGI (2020) PICRUSt2 for prediction of metagenome functions. Nat Biotechnol 38(6):685–688. https://doi.org/10.1038/s41587-020-0548-6

Downing JA, McCauley E (1992) The nitrogen: phosphorus relationship in lakes. Limnol Oceanogr 37(5):936–945. https://doi.org/10.4319/lo.1992.37.5.0936

Dyksma S, Jansen L, Gallert C (2020) Syntrophic acetate oxidation replaces acetoclastic methanogenesis during thermophilic digestion of biowaste. Microbiome 8(1):105. https://doi.org/10.1186/s40168-020-00862-5

Eggen RI, Geerling AC, Jetten MS, de Vos WM (1991) Cloning, expression, and sequence analysis of the genes for carbon monoxide dehydrogenase of Methanothrix soehngenii. J Biol Chem 266(11):6883–6887

Eiler A (2006) Evidence for the ubiquity of mixotrophic bacteria in the Upper Ocean: implications and consequences. Appl Environ Microbiol 72(12):7431–7437. https://doi.org/10.1128/AEM.01559-06

Einsle O, Kroneck PMH (2004) Structural basis of denitrification. Biol Chem 385(10):875–883. https://doi.org/10.1515/BC.2004.115

Ellegaard M, Ribeiro S (2018) The long-term persistence of phytoplankton resting stages in aquatic "seed banks". Biol Rev Camb Philos Soc 93(1):166–183. https://doi.org/10.1111/brv.12338

Ellegaard M, Clokie MRJ, Czypionka T, Frisch D, Godhe A, Kremp A, Letarov A, McGenity TJ, Ribeiro S, John Anderson N (2020) Dead or alive: sediment DNA archives as tools for tracking aquatic evolution and adaptation. Commun Biol 3(1):1–11. https://doi.org/10.1038/s42003-020-0899-z

Emerson D, Fleming EJ, McBeth JM (2010) Iron-oxidizing bacteria: an environmental and genomic perspective. Annu Rev Microbiol 64:561–583. https://doi.org/10.1146/annurev.micro.112408.134208

Emerson D, Field E, Chertkov O, Davenport K, Goodwin L, Munk C, Nolan M, Woyke T (2013) Comparative genomics of freshwater Fe-oxidizing bacteria: implications for physiology, ecology, and systematics. Front Microbiol 4

Erazo NG, Dutta A, Bowman JS (2021) From microbial community structure to metabolic inference using paprica. STAR Protocols 2(4):101005. https://doi.org/10.1016/j.xpro.2021.101005

Erb TJ (2011) Carboxylases in natural and synthetic microbial pathways. Appl Environ Microbiol 77(24):8466–8477. https://doi.org/10.1128/AEM.05702-11

Escalas A, Hale L, Voordeckers JW, Yang Y, Firestone MK, Alvarez-Cohen L, Zhou J (2019) Microbial functional diversity: from concepts to applications. Ecol Evol 9(20):12000–12016. https://doi.org/10.1002/ece3.5670

Evans PN, Parks DH, Chadwick GL, Robbins SJ, Orphan VJ, Golding SD, Tyson GW (2015) Methane metabolism in the archaeal phylum Bathyarchaeota revealed by genome-centric metagenomics. Science 350(6259):434–438. https://doi.org/10.1126/science.aac7745

Falkowski PG, Fenchel T, Delong EF (2008) The microbial engines that drive earth's biogeochemical cycles. Science 320(5879):1034–1039. https://doi.org/10.1126/science.1153213

Feng X, Wang Y, Zubin R, Wang F (2019) Core metabolic features and hot origin of Bathyarchaeota. Engineering 5(3):498–504. https://doi.org/10.1016/j.eng.2019.01.011

Fernández N, Sierra-Alvarez R, Amils R, Field JA, Sanz JL (2009) Compared microbiology of granular sludge under autotrophic, mixotrophic and heterotrophic denitrification conditions. Water Sci Technol 59(6):1227–1236. https://doi.org/10.2166/wst.2009.092

Fernández-Gómez B, Richter M, Schüler M, Pinhassi J, Acinas SG, González JM, Pedrós-Alió C (2013) Ecology of marine Bacteroidetes: a comparative genomics approach. ISME J 7(5):1026–1037. https://doi.org/10.1038/ismej.2012.169

Fincker M, Huber JA, Orphan VJ, Rappé MS, Teske A, Spormann AM (2020) Metabolic strategies of marine subseafloor Chloroflexi inferred from genome reconstructions. Environ Microbiol 22(8):3188–3204. https://doi.org/10.1111/1462-2920.15061

Fischer H, Pusch M (1999) Use of the [14C]leucine incorporation technique to measure bacterial production in river sediments and the Epiphyton. Appl Environ Microbiol 65(10):4411–4418

Fish J, Chai B, Wang Q, Sun Y, Brown CT, Tiedje J, Cole J (2013) FunGene: the functional gene pipeline and repository. Front Microbiol 4

Fisher JC, Eren AM, Green HC, Shanks OC, Morrison HG, Vineis JH, Sogin ML, McLellan SL (2015) Comparison of sewage and animal fecal microbiomes by using Oligotyping reveals potential human fecal indicators in multiple taxonomic groups. Appl Environ Microbiol 81(20):7023–7033. https://doi.org/10.1128/AEM.01524-15

Fiskal A, Deng L, Michel A, Eickenbusch P, Han X, Lagostina L, Zhu R, Sander M, Schroth MH, Bernasconi SM, Dubois N, Lever MA (2019) Effects of eutrophication on sedimentary organic carbon cycling in five temperate lakes. Biogeosciences 16(19):3725–3746. https://doi.org/10.5194/bg-16-3725-2019

Fiskal A, Anthamatten E, Deng L, Han X, Lagostina L, Michel A, Zhu R, Dubois N, Schubert CJ, Bernasconi SM, Lever MA (2021) Carbon sources of benthic fauna in temperate lakes across multiple trophic states. Biogeosciences 18(14):4369–4388. https://doi.org/10.5194/bg-18-4369-2021

Francke A, Wagner B, Just J, Leicher N, Gromig R, Baumgarten H, Vogel H, Lacey JH, Sadori L, Wonik T, Leng MJ, Zanchetta G, Sulpizio R, Giaccio B (2016) Sedimentological processes and environmental variability at Lake Ohrid (Macedonia, Albania) between 637 ka and the present. Biogeosciences 13(4):1179–1196. https://doi.org/10.5194/bg-13-1179-2016

Freiberg C, Brunner NA, Schiffer G, Lampe T, Pohlmann J, Brands M, Raabe M, Häbich D, Ziegelbauer K (2004) Identification and characterization of the first class of potent bacterial acetyl-CoA carboxylase inhibitors with antibacterial activity. J Biol Chem 279(25):26066–26073. https://doi.org/10.1074/jbc.M402989200

Friese A, Kallmeyer J, Axel Kitte J, Montaño Martínez I, Bijaksana S, Wagner D, The ICDP Lake Chalco Drilling Science Team and the ICDP Towuti Drilling Science Team (2017) A simple and inexpensive technique for assessing contamination during drilling operations. Limnol Oceanogr Methods 15(2):200–211. https://doi.org/10.1002/lom3.10159

Friese A, Bauer K, Glombitza C, Ordoñez L, Ariztegui D, Heuer VB, Vuillemin A, Henny C, Nomosatryo S, Simister R, Wagner D, Bijaksana S, Vogel H, Melles M, Russell JM, Crowe SA, Kallmeyer J (2021) Organic matter mineralization in modern and ancient ferruginous sediments. Nat Commun 12(1):2216. https://doi.org/10.1038/s41467-021-22453-0

Frindte K, Allgaier M, Grossart H-P, Eckert W (2015) Microbial response to experimentally controlled redox transitions at the sediment water interface. PLoS One 10(11):e0143428. https://doi.org/10.1371/journal.pone.0143428

Gächter R, Meyer JS, Mares A (1988) Contribution of bacteria to release and fixation of phosphorus in lake sediments. Limnol Oceanogr 33(6 part2):1542–1558. https://doi.org/10.4319/lo.1988.33.6part2.1542

Gansauge M-T, Meyer M (2013) Single-stranded DNA library preparation for the sequencing of ancient or damaged DNA. Nat Protoc 8(4):737–748. https://doi.org/10.1038/nprot.2013.038

Garcia JL, Patel BK, Ollivier B (2000) Taxonomic, phylogenetic, and ecological diversity of methanogenic archaea. Anaerobe 6(4):205–226. https://doi.org/10.1006/anae.2000.0345

Garcia-Lopez E, Moreno A, Bartolomé M, Leunda M, Sancho C, Cid C (2021) Glacial ice age shapes microbiome composition in a receding southern European glacier. Front Microbiol 12

Gardner JG, Grundy FJ, Henkin TM, Escalante-Semerena JC (2006) Control of acetyl-coenzyme a Synthetase (AcsA) activity by acetylation/deacetylation without NAD+ involvement in Bacillus subtilis. J Bacteriol 188(15):5460–5468. https://doi.org/10.1128/JB.00215-06

Garner RE, Gregory-Eaves I, Walsh DA (2020) Sediment metagenomes as time capsules of Lake microbiomes. mSphere 5(6):e00512-20. https://doi.org/10.1128/mSphere.00512-20

Ge S, Wang S, Yang X, Qiu S, Li B, Peng Y (2015) Detection of nitrifiers and evaluation of partial nitrification for wastewater treatment: a review. Chemosphere 140:85–98. https://doi.org/10.1016/j.chemosphere.2015.02.004

Gibbons SM, Gilbert JA (2015) Microbial diversity – exploration of natural ecosystems and microbiomes. Curr Opin Genet Dev 35:66–72. https://doi.org/10.1016/j.gde.2015.10.003

Gibson JL, Dubbs JM, Tabita FR (2002) Differential expression of the CO2 fixation operons of Rhodobacter sphaeroides by the Prr/Reg two-component system during chemoautotrophic growth. J Bacteriol 184(23):6654–6664. https://doi.org/10.1128/JB.184.23.6654-6664.2002

Gisselmann G, Klausmeier P, Schwenn JD (1993) The ferredoxin:sulphite reductase gene from Synechococcus PCC7942. Biochim Biophys Acta 1144(1):102–106. https://doi.org/10.1016/0005-2728(93)90037-g

Glassing A, Dowd SE, Galandiuk S, Davis B, Chiodini RJ (2016) Inherent bacterial DNA contamination of extraction and sequencing reagents may affect interpretation of microbiota in low bacterial biomass samples. Gut Pathogens 8(1):24. https://doi.org/10.1186/s13099-016-0103-7

Glombitza C, Stockhecke M, Schubert C, Vetter A, Kallmeyer J (2013) Sulfate reduction controlled by organic matter availability in deep sediment cores from the saline, alkaline Lake Van (Eastern Anatolia, Turkey). Front Microbiol 4

González JM, Fernández-Gómez B, Fernàndez-Guerra A, Gómez-Consarnau L, Sánchez O, Coll-Lladó M, del Campo J, Escudero L, Rodríguez-Martínez R, Alonso-Sáez L, Latasa M, Paulsen I, Nedashkovskaya O, Lekunberri I, Pinhassi J, Pedrós-Alió C (2008) Genome analysis of the proteorhodopsin-containing marine bacterium Polaribacter sp. MED152 (Flavobacteria). Proc Natl Acad Sci 105(25):8724–8729. https://doi.org/10.1073/pnas.0712027105

Green JL, Bohannan BJM, Whitaker RJ (2008) Microbial biogeography: from taxonomy to traits. Sci 320(5879):1039–1043. https://www.science.org/doi/10.1126/science.1153475

Gregersen LH, Habicht KS, Peduzzi S, Tonolla M, Canfield DE, Miller M, Cox RP, Frigaard N-U (2009) Dominance of a clonal green sulfur bacterial population in a stratified lake. FEMS Microbiol Ecol 70(1):30–41. https://doi.org/10.1111/j.1574-6941.2009.00737.x

Grossart HP, Simon M (1998) Significance of limnetic organic aggregates (lake snow) for the sinking flux of particulate organic matter in a large lake. Aquat Microb Ecol 15:115

Grossart H-P, Simon M, Logan BE (1997) Formation of macroscopic organic aggregates (lake snow) in a large lake: the significance of transparent exopolymer particles, plankton, and zooplankton. Limnol Oceanogr 42(8):1651–1659. https://doi.org/10.4319/lo.1997.42.8.1651

Gugliandolo C, Michaud L, Lo Giudice A, Lentini V, Rochera C, Camacho A, Maugeri TL (2016) Prokaryotic community in Lacustrine Sediments of byers peninsula (Livingston Island, Maritime Antarctica). Microb Ecol 71(2):387–400. https://doi.org/10.1007/s00248-015-0666-8
Haglund A-L, Lantz P, Törnblom E, Tranvik L (2003) Depth distribution of active bacteria and bacterial activity in lake sediment. FEMS Microbiol Ecol 46(1):31–38. https://doi.org/10.1016/S0168-6496(03)00190-9
Hall E, Bernhardt E, Bier R, Bradford M, Boot C, Cotner J, del Giorgio P, Evans S, Graham E, Jones S, Lennon J, Locey K, Nemergut D, Osborne B, Rocca J, Schimel J, Waldrop M, Wallenstein M (2018) Understanding how microbiomes influence the systems they inhabit. Nat Microbiol 3:977–982. https://doi.org/10.1038/s41564-018-0201-z
Haller L, Tonolla M, Zopfi J, Peduzzi R, Wildi W, Poté J (2011) Composition of bacterial and archaeal communities in freshwater sediments with different contamination levels (Lake Geneva, Switzerland). Water Res 45(3):1213–1228. https://doi.org/10.1016/j.watres.2010.11.018
Hamilton TL, Klatt JM, de Beer D, Macalady JL (2018) Cyanobacterial photosynthesis under sulfidic conditions: insights from the isolate Leptolyngbya sp. strain hensonii. ISME J 12(2):568–584. https://doi.org/10.1038/ismej.2017.193
Han X, Schubert CJ, Fiskal A, Dubois N, Lever MA (2020) Eutrophication as a driver of microbial community structure in lake sediments. Environ Microbiol 22(8):3446–3462. https://doi.org/10.1111/1462-2920.15115
Hanson CA (2017) Microbial biogeography. In: International encyclopedia of geography. Wiley, pp 1–6. https://doi.org/10.1002/9781118786352.wbieg0231
Hanson CA, Fuhrman JA, Horner-Devine MC, Martiny JBH (2012) Beyond biogeographic patterns: processes shaping the microbial landscape. Nat Rev Microbiol 10(7):497–506. https://doi.org/10.1038/nrmicro2795
Haroon MF, Hu S, Shi Y, Imelfort M, Keller J, Hugenholtz P, Yuan Z, Tyson GW (2013) Anaerobic oxidation of methane coupled to nitrate reduction in a novel archaeal lineage. Nature 500(7464):567–570. https://doi.org/10.1038/nature12375
Hatzenpichler R, Scheller S, Tavormina PL, Babin BM, Tirrell DA, Orphan VJ (2014) In situ visualization of newly synthesized proteins in environmental microbes using amino acid tagging and click chemistry. Environ Microbiol 16(8):2568–2590. https://doi.org/10.1111/1462-2920.12436
He S, Wurtzel O, Singh K, Froula JL, Yilmaz S, Tringe SG, Wang Z, Chen F, Lindquist EA, Sorek R, Hugenholtz P (2010) Validation of two ribosomal RNA removal methods for microbial metatranscriptomics. Nat Methods 7(10):807–812. https://doi.org/10.1038/nmeth.1507
He Y, Li M, Perumal V, Feng X, Fang J, Xie J, Sievert SM, Wang F (2016) Genomic and enzymatic evidence for acetogenesis among multiple lineages of the archaeal phylum Bathyarchaeota widespread in marine sediments. Nat Microbiol 1(6):16035. https://doi.org/10.1038/nmicrobiol.2016.35
He Z, Zhang Q, Feng Y, Luo H, Pan X, Gadd GM (2018) Microbiological and environmental significance of metal-dependent anaerobic oxidation of methane. Sci Total Environ 610–611:759–768. https://doi.org/10.1016/j.scitotenv.2017.08.140
Hedderich R, Klimmek O, Kröger A, Dirmeier R, Keller M, Stetter KO (1998) Anaerobic respiration with elemental sulfur and with disulfides. FEMS Microbiol Rev 22(5):353–381. https://doi.org/10.1111/j.1574-6976.1998.tb00376.x
Hedrich S, Schlömann M, Johnson DB (2011) The iron-oxidizing proteobacteria. Microbiology (Reading) 157(Pt 6):1551–1564. https://doi.org/10.1099/mic.0.045344-0
Herlemann DPR, Geissinger O, Ikeda-Ohtsubo W, Kunin V, Sun H, Lapidus A, Hugenholtz P, Brune A (2009) Genomic analysis of "Elusimicrobium minutum," the first cultivated representative of the phylum "Elusimicrobia" (formerly termite group 1). Appl Environ Microbiol 75(9):2841–2849. https://doi.org/10.1128/AEM.02698-08
Herold M, Martínez Arbas S, Narayanasamy S, Sheik AR, Kleine-Borgmann LAK, Lebrun LA, Kunath BJ, Roume H, Bessarab I, Williams RBH, Gillece JD, Schupp JM, Keim PS, Jäger C, Hoopmann MR, Moritz RL, Ye Y, Li S, Tang H, Heintz-Buschart A, May P, Muller EEL, Laczny CC, Wilmes P (2020) Integration of time-series meta-omics data reveals how microbial

ecosystems respond to disturbance. Nature. Communications 11(1). https://doi.org/10.1038/s41467-020-19006-2

Heyer J, Galchenko VF, Dunfield PF (2002) Molecular phylogeny of type II methane-oxidizing bacteria isolated from various environments. Microbiology (Reading) 148(Pt 9):2831–2846. https://doi.org/10.1099/00221287-148-9-2831

Heylen K, Vanparys B, Gevers D, Wittebolle L, Boon N, De Vos P (2007) Nitric oxide reductase (norB) gene sequence analysis reveals discrepancies with nitrite reductase (nir) gene phylogeny in cultivated denitrifiers. Environ Microbiol 9(4):1072–1077. https://doi.org/10.1111/j.1462-2920.2006.01194.x

Higgins IJ, Best DJ, Hammond RC, Scott D (1981) Methane-oxidizing microorganisms. Microbiol Rev 45(4):556–590

Holm S, Walz J, Horn F, Yang S, Grigoriev MN, Wagner D, Knoblauch C, Liebner S (2020) Methanogenic response to long-term permafrost thaw is determined by paleoenvironment. FEMS Microbiol Ecol 96(3):fiaa021. https://doi.org/10.1093/femsec/fiaa021

Holmkvist L, Ferdelman TG, Jørgensen BB (2011) A cryptic sulfur cycle driven by iron in the methane zone of marine sediment (Aarhus Bay, Denmark). Geochim Cosmochim Acta 75(12):3581–3599. https://doi.org/10.1016/j.gca.2011.03.033

Hoshino T, Doi H, Uramoto G-I, Wörmer L, Adhikari RR, Xiao N, Morono Y, D'Hondt S, Hinrichs K-U, Inagaki F (2020) Global diversity of microbial communities in marine sediment. Proc Natl Acad Sci 117(44):27587–27597. https://doi.org/10.1073/pnas.1919139117

Hosseinlou D, Sartaj M, Delatolla R (2019) Simultaneous anaerobic oxidation/partial nitrification-denitrification for cost-effective and efficient removal of organic carbon and nitrogen from highly polluted streams. Environ Technol 40(16):2114–2126. https://doi.org/10.1080/09593330.2018.1438522

Hu Y, Ribbe MW (2015) Nitrogenase and homologs. J Biol Inorg Chem 20(2):435–445. https://doi.org/10.1007/s00775-014-1225-3

Hug LA, Baker BJ, Anantharaman K, Brown CT, Probst AJ, Castelle CJ, Butterfield CN, Hernsdorf AW, Amano Y, Ise K, Suzuki Y, Dudek N, Relman DA, Finstad KM, Amundson R, Thomas BC, Banfield JF (2016) A new view of the tree of life. Nat Microbiol 1:16048. https://doi.org/10.1038/nmicrobiol.2016.48

Hügler M, Sievert SM (2011) Beyond the Calvin cycle: autotrophic carbon fixation in the ocean. Annu Rev Mar Sci 3:261–289. https://doi.org/10.1146/annurev-marine-120709-142712

Huisman J, Codd GA, Paerl HW, Ibelings BW, Verspagen JMH, Visser PM (2018) Cyanobacterial blooms. Nat Rev Microbiol 16(8):471–483. https://doi.org/10.1038/s41579-018-0040-1

Iguchi H, Yurimoto H, Sakai Y (2010) Soluble and particulate methane monooxygenase gene clusters of the type I methanotroph Methylovulum miyakonense HT12. FEMS Microbiol Lett 312(1):71–76. https://doi.org/10.1111/j.1574-6968.2010.02101.x

Inagaki F, Tsunogai U, Suzuki M, Kosaka A, Machiyama H, Takai K, Nunoura T, Nealson KH, Horikoshi K (2004) Characterization of C1-metabolizing prokaryotic communities in methane seep habitats at the Kuroshima Knoll, southern Ryukyu arc, by analyzing pmoA, mmoX, mxaF, mcrA, and 16S rRNA genes. Appl Environ Microbiol 70(12):7445–7455. https://doi.org/10.1128/AEM.70.12.7445-7455.2004

İnceoğlu Ö, Llirós M, Crowe SA, García-Armisen T, Morana C, Darchambeau F, Borges AV, Descy J-P, Servais P (2015) Vertical distribution of functional potential and active microbial communities in meromictic Lake Kivu. Microb Ecol 70(3):596–611. https://doi.org/10.1007/s00248-015-0612-9

in 't Zandt MH , Frank J, Yilmaz P, Cremers G, Jetten MSM, Welte CU (2020) Long-term enriched methanogenic communities from thermokarst lake sediments show species-specific responses to warming, FEMS Microb 1(1), xtaa008. https://doi.org/10.1093/femsmc/xtaa008

Jaffe AL, Castelle CJ, Matheus Carnevali PB, Gribaldo S, Banfield JF (2020) The rise of diversity in metabolic platforms across the candidate phyla radiation. BMC Biol 18(1):69. https://doi.org/10.1186/s12915-020-00804-5

Jane SF, Hansen GJA, Kraemer BM, Leavitt PR, Mincer JL, North RL, Pilla RM, Stetler JT, Williamson CE, Woolway RI, Arvola L, Chandra S, DeGasperi CL, Diemer L, Dunalska J, Erina O, Flaim G, Grossart H-P, Hambright KD, Hein C, Hejzlar J, Janus LL, Jenny J-P, Jones JR, Knoll LB, Leoni B, Mackay E, Matsuzaki S-IS, McBride C, Müller-Navarra DC, Paterson AM, Pierson D, Rogora M, Rusak JA, Sadro S, Saulnier-Talbot E, Schmid M, Sommaruga R, Thiery W, Verburg P, Weathers KC, Weyhenmeyer GA, Yokota K, Rose KC (2021) Widespread deoxygenation of temperate lakes. Nature 594(7861):66–70. https://doi.org/10.1038/s41586-021-03550-y

Janssen PH, Yates PS, Grinton BE, Taylor PM, Sait M (2002) Improved culturability of soil bacteria and isolation in pure culture of novel members of the divisions Acidobacteria, Actinobacteria, Proteobacteria, and Verrucomicrobia. Appl Environ Microbiol 68(5):2391–2396. https://doi.org/10.1128/AEM.68.5.2391-2396.2002

Ji B, Liang J, Ma Y, Zhu L, Liu Y (2019) Bacterial community and eutrophic index analysis of the East Lake. Environ Pollut 252:682–688. https://doi.org/10.1016/j.envpol.2019.05.138

Jiang H, Liu Y, King G (2017) Editorial: special thematic issue on inland aquatic ecosystems. FEMS Microbiol Ecol 93(12):fix148. https://doi.org/10.1093/femsec/fix148

Jiao J-Y, Liu L, Hua Z-S, Fang B-Z, Zhou E-M, Salam N, Hedlund BP, Li W-J (2021) Microbial dark matter coming to light: challenges and opportunities. Natl Sci Rev 8(3):nwaa280. https://doi.org/10.1093/nsr/nwaa280

Jørgensen BB, Marshall IPG (2016) Slow microbial life in the seabed. Annu Rev Mar Sci 8:311–332. https://doi.org/10.1146/annurev-marine-010814-015535

Jung M-Y, Sedlacek CJ, Kits KD, Mueller AJ, Rhee S-K, Hink L, Nicol GW, Bayer B, Lehtovirta-Morley L, Wright C, de la Torre JR, Herbold CW, Pjevac P, Daims H, Wagner M (2022) Ammonia-oxidizing archaea possess a wide range of cellular ammonia affinities. ISME J 16(1):272–283. https://doi.org/10.1038/s41396-021-01064-z

Kadnikov VV, Mardanov AV, Beletsky AV, Karnachuk OV, Ravin NV (2019) Genome of the candidate phylum Aminicenantes bacterium from a deep subsurface thermal aquifer revealed its fermentative saccharolytic lifestyle. Extremophiles 23(2):189–200. https://doi.org/10.1007/s00792-018-01073-5

Kallmeyer J (2017) Chapter three – contamination control for scientific drilling operations. In: Sariaslani S, Gadd GM (eds) Advances in applied microbiology. Academic, pp 61–91

Kallmeyer J, Ferdelman TG, Weber A, Fossing H, Jørgensen BB (2004) A cold chromium distillation procedure for radiolabeled sulfide applied to sulfate reduction measurements. Limnol Oceanogr Methods 2(6):171–180. https://doi.org/10.4319/lom.2004.2.171

Kallmeyer J, Smith DC, Spivack AJ, D'Hondt S (2008) New cell extraction procedure applied to deep subsurface sediments. Limnol Oceanogr Methods 6(6):236–245. https://doi.org/10.4319/lom.2008.6.236

Kallmeyer J, Pockalny R, Adhikari RR, Smith DC, D'Hondt S (2012) Global distribution of microbial abundance and biomass in subseafloor sediment. Proc Natl Acad Sci 109(40):16213–16216. https://doi.org/10.1073/pnas.1203849109

Kallmeyer J, Grewe S, Glombitza C, Kitte JA (2015) Microbial abundance in lacustrine sediments: a case study from Lake Van, Turkey. Int J Earth Sci (Geol Rundsch) 104(6):1667–1677. https://doi.org/10.1007/s00531-015-1219-6

Kallscheuer N, Moreira C, Airs R, Llewellyn CA, Wiegand S, Jogler C, Lage OM (2019) Pink- and orange-pigmented Planctomycetes produce saproxanthin-type carotenoids including a rare C45 carotenoid. Environ Microbiol Rep 11(6):741–748. https://doi.org/10.1111/1758-2229.12796

Kalvelage T, Lavik G, Jensen MM, Revsbech NP, Löscher C, Schunck H, Desai DK, Hauss H, Kiko R, Holtappels M, LaRoche J, Schmitz RA, Graco MI, Kuypers MMM (2015) Aerobic microbial respiration in oceanic oxygen minimum zones. PLoS One 10(7). https://doi.org/10.1371/journal.pone.0133526

Kalyuzhnaya MG, Zabinsky R, Bowerman S, Baker DR, Lidstrom ME, Chistoserdova L (2006) Fluorescence in situ hybridization-flow cytometry-cell sorting-based method for separation and enrichment of type I and type II methanotroph populations. Appl Environ Microbiol 72(6):4293–4301. https://doi.org/10.1128/AEM.00161-06

Kanao T, Fukui T, Atomi H, Imanaka T (2001) ATP-citrate lyase from the green sulfur bacterium Chlorobium limicola is a heteromeric enzyme composed of two distinct gene products. Eur J Biochem 268(6):1670–1678

Kappler A, Bryce C, Mansor M, Lueder U, Byrne JM, Swanner ED (2021) An evolving view on biogeochemical cycling of iron. Nat Rev Microbiol 19(6):360–374. https://doi.org/10.1038/s41579-020-00502-7

Katayama T, Nobu MK, Kusada H, Meng X-Y, Hosogi N, Uematsu K, Yoshioka H, Kamagata Y, Tamaki H (2020) Isolation of a member of the candidate phylum 'Atribacteria' reveals a unique cell membrane structure. Nat Commun 11(1):6381. https://doi.org/10.1038/s41467-020-20149-5

Kato H, Mori H, Maruyama F, Toyoda A, Oshima K, Endo R, Fuchu G, Miyakoshi M, Dozono A, Ohtsubo Y, Nagata Y, Hattori M, Fujiyama A, Kurokawa K, Tsuda M (2015) Time-series metagenomic analysis reveals robustness of soil microbiome against chemical disturbance. DNA Res 22(6):413–424. https://doi.org/10.1093/dnares/dsv023

Key FM, Posth C, Krause J, Herbig A, Bos KI (2017) Mining metagenomic data sets for ancient DNA: recommended protocols for authentication. Trends Genet 33(8):508–520. https://doi.org/10.1016/j.tig.2017.05.005

Kirschner AKT, Velimirov B (1999) Benthic bacterial secondary production measured via simultaneous 3H-thymidine and 14C-leucine incorporation, and its implication for the carbon cycle of a shallow macrophyte-dominated backwater system. Limnol Oceanogr 44(8):1871–1881. https://doi.org/10.4319/lo.1999.44.8.1871

Kisand V, Talas L, Kisand A, Stivrins N, Reitalu T, Alliksaar T, Vassiljev J, Liiv M, Heinsalu A, Seppä H, Veski S (2018) From microbial eukaryotes to metazoan vertebrates: wide spectrum paleo-diversity in sedimentary ancient DNA over the last ~14,500 years. Geobiology 16(6):628–639. https://doi.org/10.1111/gbi.12307

Kitzinger K, Marchant HK, Bristow LA, Herbold CW, Padilla CC, Kidane AT, Littmann S, Daims H, Pjevac P, Stewart FJ, Wagner M, Kuypers MMM (2020) Single cell analyses reveal contrasting life strategies of the two main nitrifiers in the ocean. Nat Commun 11(1):767. https://doi.org/10.1038/s41467-020-14542-3

Knight R, Vrbanac A, Taylor B, Aksenov A, Callewaert C, Debelius J, Gonzalez A, Kosciolek T, McCall L-I, McDonald D, Melnik A, Morton J, Navas J, Quinn R, Sanders J, Swafford A, Thompson L, Tripathi A, Xu Z, Zaneveld J, Zhu Q, Caporaso JG, Dorrestein P (2018) Best practices for Analysing microbiomes. Nat Rev Microbiol 16:410–422

Kojima H, Teske A, Fukui M (2003) Morphological and phylogenetic characterizations of freshwater Thioploca species from Lake Biwa, Japan, and Lake Constance, Germany. Appl Environ Microbiol 69(1):390–398. https://doi.org/10.1128/AEM.69.1.390-398.2003

Kong L, Jing H, Kataoka T, Buchwald C, Liu H (2013) Diversity and spatial distribution of hydrazine oxidoreductase (hzo) gene in the oxygen minimum zone off Costa Rica. PLoS One 8(10):e78275. https://doi.org/10.1371/journal.pone.0078275

Kraemer SA, Barbosa da Costa N, Shapiro BJ, Fradette M, Huot Y, Walsh DA (2020) A large-scale assessment of lakes reveals a pervasive signal of land use on bacterial communities. ISME J 14(12):3011–3023. https://doi.org/10.1038/s41396-020-0733-0

Krüger K, Chafee M, Ben Francis T, Glavina del Rio T, Becher D, Schweder T, Amann RI, Teeling H (2019) In marine Bacteroidetes the bulk of glycan degradation during algae blooms is mediated by few clades using a restricted set of genes. ISME J 13(11):2800–2816. https://doi.org/10.1038/s41396-019-0476-y

Kuenen J (2008) Anammox bacteria: from discovery to application. Nat Rev Microbiol 6:320–326. https://doi.org/10.1038/nrmicro1857

Kurilkina MI, Zakharova YR, Galachyants YP, Petrova DP, Bukin YS, Domysheva VM, Blinov VV, Likhoshway YV (2016) Bacterial community composition in the water column of the deepest freshwater Lake Baikal as determined by next-generation sequencing. FEMS Microbiol Ecol 92(7):fiw094. https://doi.org/10.1093/femsec/fiw094

Kuypers MMM, Lavik G, Woebken D, Schmid M, Fuchs BM, Amann R, Jørgensen BB, Jetten MSM (2005) Massive nitrogen loss from the Benguela upwelling system through anaerobic ammonium oxidation. Proc Natl Acad Sci 102(18):6478–6483. https://doi.org/10.1073/pnas.0502088102

Lacap DC, Lau MCY, Pointing SB (2011) Biogeography of prokaryotes. In: Fontaneto D (ed) Biogeography of microscopic organisms, is everything small everywhere? Cambridge University Press, pp 35–42

Lage OM, Bondoso J (2014) Planctomycetes and macroalgae, a striking association. Front Microbiol 5

Lam P, Kuypers MMM (2011) Microbial nitrogen cycling processes in oxygen minimum zones. Annu Rev Mar Sci 3(1):317–345. https://doi.org/10.1146/annurev-marine-120709-142814

LaRowe DE, Van Cappellen P (2011) Degradation of natural organic matter: a thermodynamic analysis. Geochim Cosmochim Acta 75:2030–2042. https://doi.org/10.1016/j.gca.2011.01.020

Lavik G, Stührmann T, Brüchert V, Van der Plas A, Mohrholz V, Lam P, Mussmann M, Fuchs BM, Amann R, Lass U, Kuypers MMM (2009) Detoxification of sulphidic African shelf waters by blooming chemolithotrophs. Nature 457(7229):581–U86

Leavitt P, Hodgson D (2006) Sedimentary pigments. In: Developments in paleoenvironmental research, pp 295–325

Lee YM, Hwang K, Lee JI, Kim M, Hwang CY, Noh H-J, Choi H, Lee HK, Chun J, Hong SG, Shin SC (2018) Genomic insight into the predominance of candidate phylum Atribacteria JS1 lineage in marine sediments. Front Microbiol 9

Legrand B, Lamarque A, Sabart M, Latour D (2016) Characterization of akinetes from cyanobacterial strains and lake sediment: a study of their resistance and toxic potential. Harmful Algae 59:42–50. https://doi.org/10.1016/j.hal.2016.09.003

Lemaire ON, Jespersen M, Wagner T (2020) CO2-fixation strategies in energy extremophiles: what can we learn from Acetogens? Front Microbiol 11

Lennon JT, Jones SE (2011) Microbial seed banks: the ecological and evolutionary implications of dormancy. Nat Rev Microbiol 9(2):119–130. https://doi.org/10.1038/nrmicro2504

Lennon JT, den Hollander F, Wilke-Berenguer M, Blath J (2021) Principles of seed banks and the emergence of complexity from dormancy. Nat Commun 12(1):4807. https://doi.org/10.1038/s41467-021-24733-1

Lever MA, Rogers KL, Lloyd KG, Overmann J, Schink B, Thauer RK, Hoehler TM, Jørgensen BB (2015) Life under extreme energy limitation: a synthesis of laboratory- and field-based investigations. FEMS Microbiol Rev 39(5):688–728. https://doi.org/10.1093/femsre/fuv020

Lewis WM (2009) Ecological zonation in lakes. In: Likens GE (ed) Encyclopedia of inland waters. Academic, Oxford, pp 416–422

Li W, Morgan-Kiss RM (2019) Influence of environmental drivers and potential interactions on the distribution of microbial communities from three permanently stratified Antarctic Lakes. Front Microbiol 10

Li F, Zhang X, Xie Y, Wang J (2019a) Sedimentary DNA reveals over 150 years of ecosystem change by human activities in Lake Chao. China Environ Int 133:105214. https://doi.org/10.1016/j.envint.2019.105214

Li Y, Wu H, Shen Y, Wang C, Wang P, Zhang W, Gao Y, Niu L (2019b) Statistical determination of crucial taxa indicative of pollution gradients in sediments of Lake Taihu, China. Environ Pollut 246:753–762. https://doi.org/10.1016/j.envpol.2018.12.087

Li Q, Cooper RE, Wegner C-E, Taubert M, Jehmlich N, von Bergen M, Küsel K (2021) Insights into autotrophic activities and carbon flow in iron-rich pelagic aggregates (iron snow). Microorganisms 9(7):1368. https://doi.org/10.3390/microorganisms9071368

Liikanen A, Martikainen PJ (2003) Effect of ammonium and oxygen on methane and nitrous oxide fluxes across sediment-water interface in a eutrophic lake. Chemosphere 52(8):1287–1293. https://doi.org/10.1016/S0045-6535(03)00224-8

Liu Y, Whitman WB (2008) Metabolic, phylogenetic, and ecological diversity of the methanogenic archaea. Ann N Y Acad Sci 1125:171–189. https://doi.org/10.1196/annals.1419.019

Liu Y-F, Qi Z-Z, Shou L-B, Liu J-F, Yang S-Z, Gu J-D, Mu B-Z (2019) Anaerobic hydrocarbon degradation in candidate phylum 'Atribacteria' (JS1) inferred from genomics. ISME J 13(9):2377–2390. https://doi.org/10.1038/s41396-019-0448-2

Lloyd KG, May MK, Kevorkian RT, Steen AD (2013) Meta-analysis of quantification methods shows that archaea and bacteria have similar abundances in the subseafloor. Appl Environ Microbiol 79(24):7790–7799. https://doi.org/10.1128/AEM.02090-13

Lomstein BA, Langerhuus AT, D'Hondt S, Jørgensen BB, Spivack AJ (2012) Endospore abundance, microbial growth and necromass turnover in deep sub-seafloor sediment. Nature 484(7392):101–104. https://doi.org/10.1038/nature10905

Lopez-Fernandez M, Simone D, Wu X, Soler L, Nilsson E, Holmfeldt K, Lantz H, Bertilsson S, Dopson M (2018) Metatranscriptomes reveal that all three domains of life are active but are dominated by bacteria in the Fennoscandian crystalline granitic continental deep biosphere. MBio 9(6):e01792-18. https://doi.org/10.1128/mBio.01792-18

Loy A, Duller S, Baranyi C, Mußmann M, Ott J, Sharon I, Béjà O, Le Paslier D, Dahl C, Wagner M (2009) Reverse dissimilatory sulfite reductase as phylogenetic marker for a subgroup of sulfur-oxidizing prokaryotes. Environ Microbiol 11(2):289–299. https://doi.org/10.1111/j.1462-2920.2008.01760.x

Lu S, Chourey K, Reiche M, Nietzsche S, Shah MB, Neu TR, Hettich RL, Küsel K (2013) Insights into the structure and metabolic function of microbes that shape pelagic iron-rich aggregates ("iron snow"). Appl Environ Microbiol 79(14):4272–4281. https://doi.org/10.1128/AEM.00467-13

Lu S, Peiffer S, Lazar CS, Oldham C, Neu TR, Ciobota V, Näb O, Lillicrap A, Rösch P, Popp J, Küsel K (2016) Extremophile microbiomes in acidic and hypersaline river sediments of Western Australia. Environ Microbiol Rep 8(1):58–67. https://doi.org/10.1111/1758-2229.12351

Macalady JL, Hamilton TL, Grettenberger CL, Jones DS, Tsao LE, Burgos WD (2013) Energy, ecology and the distribution of microbial life. Philos Trans R Soc Lond Ser B Biol Sci 368(1622):20120383. https://doi.org/10.1098/rstb.2012.0383

Madigan MT (2003) Anoxygenic phototrophic bacteria from extreme environments. Photosynth Res 76(1–3):157–171. https://doi.org/10.1023/A:1024998212684

Magnabosco C, Lin L-H, Dong H, Bomberg M, Ghiorse W, Stan-Lotter H, Pedersen K, Kieft TL, van Heerden E, Onstott TC (2018) The biomass and biodiversity of the continental subsurface. Nat Geosci 11(10):707–717. https://doi.org/10.1038/s41561-018-0221-6

Mandakovic D, Rojas C, Maldonado J, Latorre M, Travisany D, Delage E, Bihouée A, Jean G, Díaz FP, Fernández-Gómez B, Cabrera P, Gaete A, Latorre C, Gutiérrez RA, Maass A, Cambiazo V, Navarrete SA, Eveillard D, González M (2018) Structure and co-occurrence patterns in microbial communities under acute environmental stress reveal ecological factors fostering resilience. Sci Rep 8:5875. https://doi.org/10.1038/s41598-018-23931-0

Martiny JBH, Bohannan BJM, Brown JH, Colwell RK, Fuhrman JA, Green JL, Horner-Devine MC, Kane M, Krumins JA, Kuske CR, Morin PJ, Naeem S, Øvreås L, Reysenbach A-L, Smith VH, Staley JT (2006) Microbial biogeography: putting microorganisms on the map. Nat Rev Microbiol 4(2):102–112. https://doi.org/10.1038/nrmicro1341

Mayr C, Wille M, Haberzettl T, Fey M, Janssen S, Lücke A, Ohlendorf C, Oliva G, Schäbitz F, Schleser GH, Zolitschka B (2007) Holocene variability of the southern hemisphere westerlies in Argentinean Patagonia (52°S). Quat Sci Rev 26:579–584. https://doi.org/10.1016/j.quascirev.2006.11.013

McDonald IR, Bodrossy L, Chen Y, Murrell JC (2008) Molecular ecology techniques for the study of aerobic methanotrophs. Appl Environ Microbiol 74(5):1305–1315. https://doi.org/10.1128/AEM.02233-07

McGuirl MA, Bollinger JA, Cosper N, Scott RA, Dooley DM (2001) Expression, purification, and characterization of NosL, a novel Cu(I) protein of the nitrous oxide reductase (nos) gene cluster. J Biol Inorg Chem 6(2):189–195. https://doi.org/10.1007/s007750000190

McInerney MJ, Struchtemeyer CG, Sieber J, Mouttaki H, Stams AJM, Schink B, Rohlin L, Gunsalus RP (2008) Physiology, ecology, phylogeny, and genomics of microorganisms capable of syntrophic metabolism. Ann N Y Acad Sci 1125:58–72. https://doi.org/10.1196/annals.1419.005

McInerney MJ, Sieber JR, Gunsalus RP (2009) Syntrophy in anaerobic global carbon cycles. Curr Opin Biotechnol 20(6):623–632. https://doi.org/10.1016/j.copbio.2009.10.001

McLellan SL, Fisher JC, Newton RJ (2015) The microbiome of urban waters. Int Microbiol 18(3):141–149. https://doi.org/10.2436/20.1501.01.244

Méheust R, Castelle CJ, Matheus Carnevali PB, Farag IF, He C, Chen L-X, Amano Y, Hug LA, Banfield JF (2020) Groundwater Elusimicrobia are metabolically diverse compared to gut microbiome Elusimicrobia and some have a novel nitrogenase paralog. ISME J 14(12):2907–2922. https://doi.org/10.1038/s41396-020-0716-1

Mendonça R, Müller RA, Clow D, Verpoorter C, Raymond P, Tranvik LJ, Sobek S (2017) Organic carbon burial in global lakes and reservoirs. Nat Commun 8(1):1694. https://doi.org/10.1038/s41467-017-01789-6

Mhatre SS, Kaufmann S, Marshall IPG, Obrochta S, Andrèn T, Jørgensen BB, Lomstein BA (2019) Microbial biomass turnover times and clues to cellular protein repair in energy-limited deep Baltic Sea sediments. FEMS Microbiol Ecol 95(6):fiz068. https://doi.org/10.1093/femsec/fiz068

Michas A, Vestergaard G, Trautwein K, Avramidis P, Hatzinikolaou DG, Vorgias CE, Wilkes H, Rabus R, Schloter M, Schöler A (2017) More than 2500 years of oil exposure shape sediment microbiomes with the potential for syntrophic degradation of hydrocarbons linked to methanogenesis. Microbiome 5:118. https://doi.org/10.1186/s40168-017-0337-8

Møller TE, van der Bilt WGM, Roerdink DL, Jørgensen SL (2020) Microbial community structure in Arctic Lake sediments reflect variations in Holocene climate conditions. Front Microbiol 11

Morales SE, Holben WE (2009) Empirical testing of 16S rRNA gene PCR primer pairs reveals variance in target specificity and efficacy not suggested by in silico analysis. Appl Environ Microbiol 75(9):2677–2683. https://doi.org/10.1128/AEM.02166-08

Morasch B, Bonvin F, Reiser H, Grandjean D, de Alencastro LF, Perazzolo C, Chèvre N, Kohn T (2010) Occurrence and fate of micropollutants in the Vidy Bay of Lake Geneva, Switzerland. Part II: micropollutant removal between wastewater and raw drinking water. Environ Toxicol Chem 29(8):1658–1668. https://doi.org/10.1002/etc.222

More KD, Giosan L, Grice K, Coolen MJL (2019) Holocene paleodepositional changes reflected in the sedimentary microbiome of the Black Sea. Geobiology 17(4):436–448. https://doi.org/10.1111/gbi.12338

More KD, Wuchter C, Irigoien X, Tierney JE, Giosan L, Grice K, Coolen MJL (2021) Subseafloor archaea reflect 139 kyrs of paleodepositional changes in the northern Red Sea. Geobiology 19(2):162–172. https://doi.org/10.1111/gbi.12421

Morono Y, Terada T, Kallmeyer J, Inagaki F (2013) An improved cell separation technique for marine subsurface sediments: applications for high-throughput analysis using flow cytometry and cell sorting. Environ Microbiol 15(10):2841–2849. https://doi.org/10.1111/1462-2920.12153

Morris BEL, Henneberger R, Huber H, Moissl-Eichinger C (2013) Microbial syntrophy: interaction for the common good. FEMS Microbiol Rev 37(3):384–406. https://doi.org/10.1111/1574-6976.12019

Morris R, Schauer-Gimenez A, Bhattad U, Kearney C, Struble CA, Zitomer D, Maki JS (2014) Methyl coenzyme M reductase (mcrA) gene abundance correlates with activity measurements of methanogenic H2/CO2-enriched anaerobic biomass. Microb Biotechnol 7(1):77–84. https://doi.org/10.1111/1751-7915.12094

Moss EL, Maghini DG, Bhatt AS (2020) Complete, closed bacterial genomes from microbiomes using nanopore sequencing. Nat Biotechnol 38(6):701–707. https://doi.org/10.1038/s41587-020-0422-6

Moter A, Göbel UB (2000) Fluorescence in situ hybridization (FISH) for direct visualization of microorganisms. J Microbiol Methods 41(2):85–112. https://doi.org/10.1016/s0167-7012(00)00152-4

Muyzer G, Stams AJM (2008) The ecology and biotechnology of sulphate-reducing bacteria. Nat Rev Microbiol 6(6):441–454. https://doi.org/10.1038/nrmicro1892

Narrowe AB, Angle JC, Daly RA, Stefanik KC, Wrighton KC, Miller CS (2017) High-resolution sequencing reveals unexplored archaeal diversity in freshwater wetland soils. Environ Microbiol 19(6):2192–2209. https://doi.org/10.1111/1462-2920.13703

Nayfach S, Roux S, Seshadri R, Udwary D, Varghese N, Schulz F, Wu D, Paez-Espino D, Chen I-M, Huntemann M, Palaniappan K, Ladau J, Mukherjee S, Reddy TBK, Nielsen T, Kirton E, Faria JP, Edirisinghe JN, Henry CS, Jungbluth SP, Chivian D, Dehal P, Wood-Charlson EM, Arkin AP, Tringe SG, Visel A, Woyke T, Mouncey NJ, Ivanova NN, Kyrpides NC, Eloe-Fadrosh EA (2021) A genomic catalog of Earth's microbiomes. Nat Biotechnol 39(4):499–509. https://doi.org/10.1038/s41587-020-0718-6

Nelson-Chorney HT, Davis CS, Poesch MS, Vinebrooke RD, Carli CM, Taylor MK (2019) Environmental DNA in lake sediment reveals biogeography of native genetic diversity. Front Ecol Environ 17(6):313–318. https://doi.org/10.1002/fee.2073

Nemergut DR, Costello EK, Hamady M, Lozupone C, Jiang L, Schmidt SK, Fierer N, Townsend AR, Cleveland CC, Stanish L, Knight R (2011) Global patterns in the biogeography of bacterial taxa. Environ Microbiol 13(1):135–144. https://doi.org/10.1111/j.1462-2920.2010.02315.x

Newton RJ, Jones SE, Eiler A, McMahon KD, Bertilsson S (2011) A guide to the natural history of freshwater lake bacteria. Microbiol Mol Biol Rev 75(1):14–49. https://doi.org/10.1128/MMBR.00028-10

Nichols RV, Curd E, Heintzman PD, Shapiro B (2019) Targeted amplification and sequencing of ancient environmental and sedimentary DNA. Methods Mol Biol 1963:149–161. https://doi.org/10.1007/978-1-4939-9176-1_16

Niño-García JP, Ruiz-González C, del Giorgio PA (2016) Interactions between hydrology and water chemistry shape bacterioplankton biogeography across boreal freshwater networks. ISME J 10(7):1755–1766. https://doi.org/10.1038/ismej.2015.226

Nobu MK, Dodsworth JA, Murugapiran SK, Rinke C, Gies EA, Webster G, Schwientek P, Kille P, Parkes RJ, Sass H, Jørgensen BB, Weightman AJ, Liu W-T, Hallam SJ, Tsiamis G, Woyke T, Hedlund BP (2016) Phylogeny and physiology of candidate phylum 'Atribacteria' (OP9/JS1) inferred from cultivation-independent genomics. ISME J 10(2):273–286. https://doi.org/10.1038/ismej.2015.97

Norði K, Thamdrup B (2014) Nitrate-dependent anaerobic methane oxidation in a freshwater sediment. Geochim Cosmochim Acta 132:141–150. https://doi.org/10.1016/j.gca.2014.01.032

Norði K, Thamdrup B, Schubert CJ (2013) Anaerobic oxidation of methane in an iron-rich Danish freshwater lake sediment. Limnol Oceanogr:546–554. https://doi.org/10.4319/lo.2013.58.2.0546

O'Sullivan DM, Doyle RM, Temisak S, Redshaw N, Whale AS, Logan G, Huang J, Fischer N, Amos GCA, Preston MD, Marchesi JR, Wagner J, Parkhill J, Motro Y, Denise H, Finn RD, Harris KA, Kay GL, O'Grady J, Ransom-Jones E, Wu H, Laing E, Studholme DJ, Benavente ED, Phelan J, Clark TG, Moran-Gilad J, Huggett JF (2021) An inter-laboratory study to investigate the impact of the bioinformatics component on microbiome analysis using mock communities. Sci Rep 11(1):10590. https://doi.org/10.1038/s41598-021-89881-2

Ohlendorf C, Gebhardt C, Hahn A, Kliem P, Zolitschka B, PASADO Science Team (2011) The PASADO core processing strategy – a proposed new protocol for sediment core treatment in multidisciplinary lake drilling projects. Sediment Geol 239(1):104–115. https://doi.org/10.1016/j.sedgeo.2011.06.007

Omae K, Oguro T, Inoue M, Fukuyama Y, Yoshida T, Sako Y (2021) Diversity analysis of thermophilic hydrogenogenic carboxydotrophs by carbon monoxide dehydrogenase amplicon sequencing using new primers. Extremophiles 25(1):61–76. https://doi.org/10.1007/s00792-020-01211-y

Orsi WD (2018) Ecology and evolution of seafloor and subseafloor microbial communities. Nat Rev Microbiol 16(11):671–683. https://doi.org/10.1038/s41579-018-0046-8

Orsi WD (2023) A rapid method for measuring ATP + ADP + AMP in marine sediment. Environ Microbiol. https://doi.org/10.1111/1462-2920.16359

Orsi W, Edgcomb VP, Christman GD, Biddle JF (2013) Gene expression in the deep biosphere. Nature 499:205–208

Orsi WD, Barker Jørgensen B, Biddle JF (2016) Transcriptional analysis of sulfate reducing and chemolithoautotrophic sulfur oxidizing bacteria in the deep subseafloor. Environ Microbiol Rep 8(4):452–460. https://doi.org/10.1111/1758-2229.12387

Orsi WD, Coolen MJL, Wuchter C, He L, More KD, Irigoien X, Chust G, Johnson C, Hemingway JD, Lee M, Galy V, Giosan L (2017) Climate oscillations reflected within the microbiome of Arabian Sea sediments. Sci Rep 7(1):6040. https://doi.org/10.1038/s41598-017-05590-9

Orsi WD, Schink B, Buckel W, Martin WF (2020a) Physiological limits to life in anoxic subseafloor sediment. FEMS Microbiol Rev 44(2):219–231. https://doi.org/10.1093/femsre/fuaa004

Orsi WD, Vuillemin A, Rodriguez P, Coskun ÖK, Gomez-Saez GV, Lavik G, Mohrholz V, Ferdelman TG (2020b) Metabolic activity analyses demonstrate that Lokiarchaeon exhibits homoacetogenesis in sulfidic marine sediments. Nat Microbiol 5(2):248–255. https://doi.org/10.1038/s41564-019-0630-3

Pang J, Yamato M, Soda S, Inoue D, Ike M (2019) Nitrogen-cycling functional genes in brackish and freshwater sediments in Yodo River in Japan. J Water Environ Technol 17(2):109–116. https://doi.org/10.2965/jwet.18-074

Panwar P, Allen MA, Williams TJ, Hancock AM, Brazendale S, Bevington J, Roux S, Páez-Espino D, Nayfach S, Berg M, Schulz F, Chen I-MA, Huntemann M, Shapiro N, Kyrpides NC, Woyke T, Eloe-Fadrosh EA, Cavicchioli R (2020) Influence of the polar light cycle on seasonal dynamics of an Antarctic lake microbial community. Microbiome 8(1):116. https://doi.org/10.1186/s40168-020-00889-8

Parducci L, Väliranta M, Salonen JS, Ronkainen T, Matetovici I, Fontana SL, Eskola T, Sarala P, Suyama Y (2015) Proxy comparison in ancient peat sediments: pollen, macrofossil and plant DNA. Philos Trans R Soc Lond Ser B Biol Sci 370(1660):20130382. https://doi.org/10.1098/rstb.2013.0382

Parducci L, Bennett KD, Ficetola GF, Alsos IG, Suyama Y, Wood JR, Pedersen MW (2017) Ancient plant DNA in lake sediments. New Phytol 214(3):924–942. https://doi.org/10.1111/nph.14470

Parkes RJ, Cragg BA, Bale SJ, Getlifff JM, Goodman K, Rochelle PA, Fry JC, Weightman AJ, Harvey SM (1994) Deep bacterial biosphere in Pacific Ocean sediments. Nature 371(6496):410–413. https://doi.org/10.1038/371410a0

Parks JM, Johs A, Podar M, Bridou R, Hurt RA, Smith SD, Tomanicek SJ, Qian Y, Brown SD, Brandt CC, Palumbo AV, Smith JC, Wall JD, Elias DA, Liang L (2013) The genetic basis for bacterial mercury methylation. Science 339(6125):1332–1335. https://doi.org/10.1126/science.1230667

Paul C, Filippidou S, Jamil I, Kooli W, House GL, Estoppey A, Hayoz M, Junier T, Palmieri F, Wunderlin T, Lehmann A, Bindschedler S, Vennemann T, Chain PSG, Junier P (2019) Bacterial spores, from ecology to biotechnology. Adv Appl Microbiol 106:79–111. https://doi.org/10.1016/bs.aambs.2018.10.002

Paver SF, Newton RJ, Coleman ML (2020) Microbial communities of the Laurentian Great Lakes reflect connectivity and local biogeochemistry. Environ Microbiol 22(1):433–446. https://doi.org/10.1111/1462-2920.14862

Pavoine S, Bonsall MB (2011) Measuring biodiversity to explain community assembly: a unified approach. Biol Rev 86(4):792–812. https://doi.org/10.1111/j.1469-185X.2010.00171.x

Pearman JK, Biessy L, Thomson-Laing G, Waters S, Vandergoes MJ, Howarth JD, Rees A, Moy C, Pochon X, Wood SA (2020) Local factors drive bacterial and microeukaryotic community composition in lake surface sediment collected across an altitudinal gradient. FEMS Microbiol Ecol 96(6):fiaa070. https://doi.org/10.1093/femsec/fiaa070

Pearman JK, Thomson-Laing G, Howarth JD, Vandergoes MJ, Thompson L, Rees A, Wood SA (2021) Investigating variability in microbial community composition in replicate environmental DNA samples down lake sediment cores. PLoS One 16(5):e0250783. https://doi.org/10.1371/journal.pone.0250783

Pearman JK, Biessy L, Howarth JD, Vandergoes MJ, Rees A, Wood SA (2022a) Deciphering the molecular signal from past and alive bacterial communities in aquatic sedimentary archives. Mol Ecol Resour 22(3):877–890. https://doi.org/10.1111/1755-0998.13515

Pearman JK, Wood SA, Vandergoes MJ, Atalah J, Waters S, Adamson J, Thomson-Laing G, Thompson L, Howarth JD, Hamilton DP, Pochon X, Biessy L, Brasell KA, Dahl J, Ellison R, Fitzsimons SJ, Gard H, Gerrard T, Gregersen R, Holloway M, Li X, Kelly DJ, Martin R, McFarlane K, McKay NP, Moody A, Moy CM, Naeher S, Newnham R, Parai R, Picard M, Puddick J, Rees ABH, Reyes L, Schallenberg M, Shepherd C, Short J, Simon KS, Steiner K, Šunde C, Terezow M, Tibby J (2022b) A bacterial index to estimate lake trophic level: national scale validation. Sci Total Environ 812:152385. https://doi.org/10.1016/j.scitotenv.2021.152385

Pearman JK, Adamson J, Thomson-Laing G, Thompson L, Waters S, Vandergoes MJ, Howarth JD, Wood SA (2023) Deterministic processes drive national-scale patterns in lake surface sediment bacteria and eukaryotic assemblage composition. Limnol Oceanogr 68(1):40–55. https://doi.org/10.1002/lno.12247

Pereira IAC, Ramos AR, Grein F, Marques MC, da Silva SM, Venceslau SS (2011) A comparative genomic analysis of energy metabolism in sulfate reducing bacteria and archaea. Front Microbiol 2:69. https://doi.org/10.3389/fmicb.2011.00069

Pereira AD, Cabezas A, Etchebehere C, Chernicharo CA, de Araújo JC (2017) Microbial communities in anammox reactors: a review. Environ Technol Rev 6(1):74–93. https://doi.org/10.1080/21622515.2017.1304457

Pérez V, Liu Y, Hengst MB, Weyrich LS (2022) A case study for the recovery of authentic microbial ancient DNA from soil samples. Microorganisms 10(8):1623. https://doi.org/10.3390/microorganisms10081623

Perez-Garcia O, Lear G, Singhal N (2016) Metabolic network modeling of microbial interactions in natural and engineered environmental systems. Front Microbiol 7

Pessi IS, Viitamäki S, Virkkala AM, Eronen-Rasimus E, Delmont TO, Marushchak ME, Hultman J (2022) In-depth characterization of denitrifier communities across different soil ecosystems in the tundra. Environ Microbiome 17(1):30

Pfennig N, Trüper HG (1992) The family chromatiaceae. In: Balows A, Trüper HG, Dworkin M, Harder W, Schleifer K-H (eds) The prokaryotes: a handbook on the biology of bacteria: ecophysiology, isolation, identification, applications. Springer, New York, pp 3200–3221

Phillips AA, Speth DR, Miller LG, Wang XT, Wu F, Medeiros PM, Monteverde DR, Osburn MR, Berelson WM, Betts HL, Wijker RS, Mullin SW, Johnson HA, Orphan VJ, Fischer WW, Course (2017) Microbial succession and dynamics in meromictic Mono Lake, California. Geobiology 19(4):376–393. https://doi.org/10.1111/gbi.12437

Picard M, Pochon X, Atalah J, Pearman JK, Rees A, Howarth JD, Moy CM, Vandergoes MJ, Hawes I, Khan S, Wood SA (2022) Using metabarcoding and droplet digital PCR to investigate drivers of historical shifts in cyanobacteria from six contrasting lakes. Sci Rep 12(1):12810. https://doi.org/10.1038/s41598-022-14216-8

Pichler M, Coskun ÖK, Ortega-Arbulú A-S, Conci N, Wörheide G, Vargas S, Orsi WD (2018) A 16S rRNA gene sequencing and analysis protocol for the Illumina MiniSeq platform. MicrobiologyOpen 7(6):e00611. https://doi.org/10.1002/mbo3.611

Pjevac P, Korlević M, Berg JS, Bura-Nakić E, Ciglenečki I, Amann R, Orlić S (2015) Community shift from phototrophic to chemotrophic sulfide oxidation following anoxic Holomixis in a stratified seawater Lake. Appl Environ Microbiol 81(1):298–308. https://doi.org/10.1128/AEM.02435-14

Plugge C, Zhang W, Scholten J, Stams A (2011) Metabolic flexibility of sulfate-reducing bacteria. Front Microbiol 2

Pruesse E, Quast C, Knittel K, Fuchs BM, Ludwig W, Peplies J, Glöckner FO (2007) SILVA: a comprehensive online resource for quality checked and aligned ribosomal RNA sequence data compatible with ARB. Nucleic Acids Res 35(21):7188–7196. https://doi.org/10.1093/nar/gkm864

Quast C, Pruesse E, Yilmaz P, Gerken J, Schweer T, Yarza P, Peplies J, Glöckner FO (2013) The SILVA ribosomal RNA gene database project: improved data processing and web-based tools. Nucleic Acids Res 41:D590–D596. https://doi.org/10.1093/nar/gks1219

Quatrini R, Johnson DB (2018) Microbiomes in extremely acidic environments: functionalities and interactions that allow survival and growth of prokaryotes at low pH. Curr Opin Microbiol 43:139–147. https://doi.org/10.1016/j.mib.2018.01.011

Rampelli S, Turroni S, Mallol C, Hernandez C, Galván B, Sistiaga A, Biagi E, Astolfi A, Brigidi P, Benazzi S, Lewis CM, Warinner C, Hofman CA, Schnorr SL, Candela M (2021) Components of a Neanderthal gut microbiome recovered from fecal sediments from El Salt. Commun Biol 4(1):1–10. https://doi.org/10.1038/s42003-021-01689-y

Rani S, Koh H-W, Rhee S-K, Fujitani H, Park S-J (2017) Detection and diversity of the nitrite oxidoreductase alpha subunit (nxrA) gene of Nitrospina in marine sediments. Microb Ecol 73(1):111–122. https://doi.org/10.1007/s00248-016-0897-3

Reiche M, Lu S, Ciobotă V, Neu TR, Nietzsche S, Rösch P, Popp J, Küsel K (2011) Pelagic boundary conditions affect the biological formation of iron-rich particles (iron snow) and their microbial communities. Limnol Oceanogr 56(4):1386–1398. https://doi.org/10.4319/lo.2011.56.4.1386

Ren Z, Zhang C, Li X, Ma K, Cui B (2022) Abundant and rare bacterial taxa structuring differently in sediment and water in Thermokarst Lakes in the Yellow River source area, Qinghai-Tibet plateau. Front Microbiol 13:774514. https://doi.org/10.3389/fmicb.2022.774514

Rinke C, Schwientek P, Sczyrba A, Ivanova NN, Anderson IJ, Cheng J-F, Darling A, Malfatti S, Swan BK, Gies EA, Dodsworth JA, Hedlund BP, Tsiamis G, Sievert SM, Liu W-T, Eisen JA, Hallam SJ, Kyrpides NC, Stepanauskas R, Rubin EM, Hugenholtz P, Woyke T (2013) Insights into the phylogeny and coding potential of microbial dark matter. Nature 499(7459)

Rissanen AJ, Kurhela E, Aho T, Oittinen T, Tiirola M (2010) Storage of environmental samples for guaranteeing nucleic acid yields for molecular microbiological studies. Appl Microbiol Biotechnol 88(4):977–984. https://doi.org/10.1007/s00253-010-2838-2

Roldán MD, Pérez-Reinado E, Castillo F, Moreno-Vivián C (2008) Reduction of polynitroaromatic compounds: the bacterial nitroreductases. FEMS Microbiol Rev 32(3):474–500. https://doi.org/10.1111/j.1574-6976.2008.00107.x

Rosenow C, Saxena RM, Durst M, Gingeras TR (2001) Prokaryotic RNA preparation methods useful for high density array analysis: comparison of two approaches. Nucleic Acids Res 29(22):E112. https://doi.org/10.1093/nar/29.22.e112

Rousk J, Bengtson P (2014) Microbial regulation of global biogeochemical cycles. Front Microbiol 5:103. https://doi.org/10.3389/fmicb.2014.00103

Ruiz-González C, Niño-García JP, del Giorgio PA (2015) Terrestrial origin of bacterial communities in complex boreal freshwater networks. Ecol Lett 18(11):1198–1206. https://doi.org/10.1111/ele.12499

Ruiz-González C, Niño-García JP, Kembel SW, del Giorgio PA (2017) Identifying the core seed bank of a complex boreal bacterial metacommunity. ISME J 11(9):2012–2021. https://doi.org/10.1038/ismej.2017.67

Russell JM, Vogel H, Bijaksana S, Melles M, Deino AL, Hafidz A, Haffner D, Hasberg A, Morlock M, von Rintelen T, Sheppard R, Stelbrink B, Stevenson J (2020) The late quaternary tectonic, biogeochemical, and environmental evolution of ferruginous Lake Towuti, Indonesia. Palaeogeogr Palaeoclimatol Palaeoecol 556:1–15. https://doi.org/10.1016/j.palaeo.2020.109905

Ryan P, Forbes C, Colleran E (2008) Investigation of the diversity of homoacetogenic bacteria in mesophilic and thermophilic anaerobic sludges using the formyltetrahydrofolate synthetase gene. Water Sci Technol 57(5):675–680. https://doi.org/10.2166/wst.2008.059

Ryszard CJ, Tomasz A, Kalinowska K, Skowronska A (2009) Abundance and structure of microbial loop components (bacteria and protists) in lakes of different trophic status. J Microbiol Biotechnol 19(9):858–868

Salazar G, Paoli L, Alberti A, Huerta-Cepas J, Ruscheweyh H-J, Cuenca M, Field CM, Coelho LP, Cruaud C, Engelen S, Gregory AC, Labadie K, Marec C, Pelletier E, Royo-Llonch M, Roux S, Sánchez P, Uehara H, Zayed AA, Zeller G, Carmichael M, Dimier C, Ferland J, Kandels S, Picheral M, Pisarev S, Poulain J, Acinas SG, Babin M, Bork P, Boss E, Bowler C, Cochrane G, de Vargas C, Follows M, Gorsky G, Grimsley N, Guidi L, Hingamp P, Iudicone D, Jaillon O, Kandels-Lewis S, Karp-Boss L, Karsenti E, Not F, Ogata H, Pesant S, Poulton N, Raes J, Sardet C, Speich S, Stemmann L, Sullivan MB, Sunagawa S, Wincker P, Acinas SG, Babin M, Bork P, Bowler C, de Vargas C, Guidi L, Hingamp P, Iudicone D, Karp-Boss L, Karsenti E, Ogata H, Pesant S, Speich S, Sullivan MB, Wincker P, Sunagawa S (2019) Gene expression changes and community turnover differentially shape the global ocean metatranscriptome. Cell 179(5):1068–1083.e21. https://doi.org/10.1016/j.cell.2019.10.014

Saltikov CW, Newman DK (2003) Genetic identification of a respiratory arsenate reductase. Proc Natl Acad Sci 100(19):10983–10988. https://doi.org/10.1073/pnas.1834303100

Santoro AL, Bastviken D, Gudasz C, Tranvik L, Enrich-Prast A (2013) Dark carbon fixation: an important process in lake sediments. PLoS One 8(6):e65813. https://doi.org/10.1371/journal.pone.0065813

Scheer H (2006) An overview of chlorophylls and bacteriochlorophylls: biochemistry, biophysics, functions and applications. In: Grimm B, Porra RJ, Rüdiger W, Scheer H (eds) Chlorophylls and bacteriochlorophylls: biochemistry, biophysics, functions and applications. Springer, Dordrecht, pp 1–26

Schuchmann K, Müller V (2016) Energetics and application of heterotrophy in acetogenic bacteria. Appl Environ Microbiol 82(14):4056–4069. https://doi.org/10.1128/AEM.00882-16

Schubert CJ, Vazquez F, Lösekann-Behrens T, Knittel K, Tonolla M, Boetius A (2011) Evidence for anaerobic oxidation of methane in sediments of a freshwater system (Lago di Cadagno). FEMS Microbiol Ecol 76(1):26–38. https://doi.org/10.1111/j.1574-6941.2010.01036.x

Schweitzer B, Huber I, Amann R, Ludwig W, Simon M (2001) α- and β-Proteobacteria control the consumption and release of amino acids on lake snow aggregates. Appl Environ Microbiol 67(2):632–645. https://doi.org/10.1128/AEM.67.2.632-645.2001

Shade A, Handelsman J (2012) Beyond the Venn diagram: the hunt for a core microbiome. Environ Microbiol 14(1):4–12. https://doi.org/10.1111/j.1462-2920.2011.02585.x

Shade A, Peter H, Allison S, Baho D, Berga M, Buergmann H, Huber D, Langenheder S, Lennon J, Martiny J, Matulich K, Schmidt T, Handelsman J (2012a) Fundamentals of microbial community resistance and resilience. Front Microbiol 3

Shade A, Read JS, Youngblut ND, Fierer N, Knight R, Kratz TK, Lottig NR, Roden EE, Stanley EH, Stombaugh J, Whitaker RJ, Wu CH, McMahon KD (2012b) Lake microbial communities are resilient after a whole-ecosystem disturbance. ISME J 6(12):2153–2167. https://doi.org/10.1038/ismej.2012.56

Sieber JR, McInerney MJ, Gunsalus RP (2012) Genomic insights into syntrophy: the paradigm for anaerobic metabolic cooperation. Annu Rev Microbiol 66:429–452. https://doi.org/10.1146/annurev-micro-090110-102844

Singh VK, Singh AL, Singh R, Kumar A (2018) Iron oxidizing bacteria: insights on diversity, mechanism of iron oxidation and role in management of metal pollution. Environ Sustain 1(3):221–231. https://doi.org/10.1007/s42398-018-0024-0

Sinkko H, Lukkari K, Sihvonen LM, Sivonen K, Leivuori M, Rantanen M, Paulin L, Lyra C (2013) Bacteria contribute to sediment nutrient release and reflect progressed eutrophication-driven hypoxia in an organic-rich continental sea. PLoS One 8(6):e67061. https://doi.org/10.1371/journal.pone.0067061

Skidmore BE, Baker RA, Banjade DR, Bray JM, Tree DR, Lewis RS (2013) Syngas fermentation to biofuels: effects of hydrogen partial pressure on hydrogenase efficiency. Biomass Bioenergy 55:156–162. https://doi.org/10.1016/j.biombioe.2013.01.034

Solden L, Lloyd K, Wrighton K (2016) The bright side of microbial dark matter: lessons learned from the uncultivated majority. Curr Opin Microbiol 31:217–226. https://doi.org/10.1016/j.mib.2016.04.020

Soliman M, Eldyasti A (2018) Ammonia-oxidizing bacteria (AOB): opportunities and applications – a review. Rev Environ Sci Biotechnol 2(17):285–321. https://doi.org/10.1007/s11157-018-9463-4

Söllinger A, Urich T (2019) Methylotrophic methanogens everywhere – physiology and ecology of novel players in global methane cycling. Biochem Soc Trans 47(6):1895–1907. https://doi.org/10.1042/BST20180565

Speirs LBM, Rice DTF, Petrovski S, Seviour RJ (2019) The phylogeny, biodiversity, and ecology of the chloroflexi in activated sludge. Front Microbiol 10

Starke R, Pylro VS, Morais DK (2021) 16S rRNA gene copy number normalization does not provide more reliable conclusions in metataxonomic surveys. Microb Ecol 81(2):535–539. https://doi.org/10.1007/s00248-020-01586-7

Starnawski P, Bataillon T, Ettema TJG, Jochum LM, Schreiber L, Chen X, Lever MA, Polz MF, Jørgensen BB, Schramm A, Kjeldsen KU (2017) Microbial community assembly and evolution in subseafloor sediment. Proc Natl Acad Sci U S A 114(11):2940–2945. https://doi.org/10.1073/pnas.1614190114

Stockhecke M, Kwiecien O, Vigliotti L, Anselmetti FS, Beer J, Çağatay MN, Channell JET, Kipfer R, Lachner J, Litt T, Pickarski N, Sturm M (2014) Chronostratigraphy of the 600,000 year old continental record of Lake Van (Turkey). Quat Sci Rev 104:8–17. https://doi.org/10.1016/j.quascirev.2014.04.008

Stoeva MK, Aris-Brosou S, Chételat J, Hintelmann H, Pelletier P, Poulain AJ (2014) Microbial community structure in lake and wetland sediments from a high Arctic polar desert revealed by targeted transcriptomics. PLoS One 9(3):e89531. https://doi.org/10.1371/journal.pone.0089531

Stoof-Leichsenring KR, Herzschuh U, Pestryakova LA, Klemm J, Epp LS, Tiedemann R (2015) Genetic data from algae sedimentary DNA reflect the influence of environment over geography. Sci Rep 5(1):12924. https://doi.org/10.1038/srep12924

Storelli N, Peduzzi S, Saad MM, Frigaard N-U, Perret X, Tonolla M (2013) CO_2 assimilation in the chemocline of Lake Cadagno is dominated by a few types of phototrophic purple sulfur bacteria. FEMS Microbiol Ecol 84(2):421–432. https://doi.org/10.1111/1574-6941.12074

Storesund JE, Lanzèn A, Nordmann E-L, Armo HR, Lage OM, Øvreås L (2020) Planctomycetes as a vital constituent of the microbial communities inhabiting different layers of the meromictic Lake Sælenvannet (Norway). Microorganisms 8(8):1150. https://doi.org/10.3390/microorganisms8081150

Sun G, Hallema D, Asbjornsen H (2017) Ecohydrological processes and ecosystem services in the Anthropocene: a review. Ecol Process 6(1):35. https://doi.org/10.1186/s13717-017-0104-6

Sunagawa S, Coelho LP, Chaffron S, Kultima JR, Labadie K, Salazar G, Djahanschiri B, Zeller G, Mende DR, Alberti A, Cornejo-Castillo FM (2015) Structure and function of the global ocean microbiome. Science 348(6237):1261359. https://doi.org/10.1126/science.1261359

Suzuki MT, Sherr EB, Sherr BF (1993) DAPI direct counting underestimates bacterial abundances and average cell size compared to AO direct counting. Limnol Oceanogr 38(7):1566–1570. https://doi.org/10.4319/lo.1993.38.7.1566

Suzuki JY, Bollivar DW, Bauer CE (1997) Genetic analysis of chlorophyll biosynthesis. Ann Rev Genet 31(1):61–89. https://doi.org/10.1146/annurev.genet.31.1.61

Takai K, Moser DP, DeFlaun M, Onstott TC, Fredrickson JK (2001) Archaeal diversity in waters from deep south African gold mines. Appl Environ Microbiol 67(12):5750–5760. https://doi.org/10.1128/AEM.67.21.5750-5760.2001

Tallant TC, Paul L, Krzycki JA (2001) The MtsA subunit of the methylthiol:coenzyme M methyltransferase of Methanosarcina barkeri catalyses both half-reactions of corrinoid-dependent dimethylsulfide: coenzyme M methyl transfer. J Biol Chem 276(6):4485–4493. https://doi.org/10.1074/jbc.M007514200

Tang X, Xie G, Shao K, Hu Y, Cai J, Bai C, Gong Y, Gao G (2020) Contrast diversity patterns and processes of microbial community assembly in a river-lake continuum across a catchment scale in northwestern China. Environ Microbiome 15(1):10. https://doi.org/10.1186/s40793-020-00356-9

Taubert M, Grob C, Howat AM, Burns OJ, Dixon JL, Chen Y, Murrell JC (2015) XoxF encoding an alternative methanol dehydrogenase is widespread in coastal marine environments. Environ Microbiol 17(10):3937–3948. https://doi.org/10.1111/1462-2920.12896

Taylor KG, Boult S (2007) The role of grain dissolution and diagenetic mineral precipitation in the cycling of metals and phosphorus: a study of a contaminated urban freshwater sediment. Appl Geochem 22(7):1344–1358. https://doi.org/10.1016/j.apgeochem.2007.01.008

They NH, Amado AM, Cotner JB (2017) Redfield ratios in inland waters: higher biological control of C:N:P ratios in tropical semi-arid high water residence time lakes. Front Microbiol 8

Thomas C, Frossard V, Perga M-E, Tofield-Pasche N, Hofmann H, Dubois N, Belkina N, Zobkova M, Robert S, Lyautey E (2019) Lateral variations and vertical structure of the microbial methane cycle in the sediment of Lake Onego (Russia). Inland Waters 9(2):205–226. https://doi.org/10.1080/20442041.2018.1500227

Thomas C, Francke A, Vogel H, Wagner B, Ariztegui D (2020) Weak influence of paleoenvironmental conditions on the subsurface biosphere of Lake Ohrid over the last 515 ka. Microorganisms 8(11):1736. https://doi.org/10.3390/microorganisms8111736

Thompson LR, Sanders JG, McDonald D, Amir A, Ladau J, Locey KJ, Prill RJ, Tripathi A, Gibbons SM, Ackermann G, Navas-Molina JA, Janssen S, Kopylova E, Vázquez-Baeza Y, González A, Morton JT, Mirarab S, Zech Xu Z, Jiang L, Haroon MF, Kanbar J, Zhu Q, Jin Song S, Kosciolek T, Bokulich NA, Lefler J, Brislawn CJ, Humphrey G, Owens SM, Hampton-Marcell J, Berg-Lyons D, McKenzie V, Fierer N, Fuhrman JA, Clauset A, Stevens RL, Shade A, Pollard KS, Goodwin KD, Jansson JK, Gilbert JA, Knight R, Earth Microbiome Project Consortium (2017) A communal catalogue reveals earth's multiscale microbial diversity. Nature 551(7681):457–463. https://doi.org/10.1038/nature24621

Thorpe AC, Anderson A, Goodall T, Thackeray SJ, Maberly SC, Bendle JA, Gweon HS, Read DS (2022) Sedimentary DNA records long-term changes in a lake bacterial community in response to varying nutrient availability. Environmental DNA 4(6):1340–1355. https://doi.org/10.1002/edn3.344

Tiedje JM (1994) Denitrifiers. In: Methods of soil analysis. Wiley, pp 245–267

Torti A, Lever MA, Jørgensen BB (2015) Origin, dynamics, and implications of extracellular DNA pools in marine sediments. Mar Genomics 24:185–196. https://doi.org/10.1016/j.margen.2015.08.007

Tourlousse DM, Narita K, Miura T, Ohashi A, Matsuda M, Ohyama Y, Shimamura M, Furukawa M, Kasahara K, Kameyama K, Saito S, Goto M, Shimizu R, Mishima R, Nakayama J, Hosomi K, Kunisawa J, Terauchi J, Sekiguchi Y, Kawasaki H (2022) Characterization and demonstration of mock communities as control reagents for accurate human microbiome community measurements. Microbiol Spectr 10(2):e01915-21. https://doi.org/10.1128/spectrum.01915-21

Trotsenko YA, Khmelenina VN (2002) Biology of extremophilic and extremotolerant methanotrophs. Arch Microbiol 177(2):123–131. https://doi.org/10.1007/s00203-001-0368-0

van Helmond NAGM, Robertson EK, Conley DJ, Hermans M, Humborg C, Kubeneck LJ, Lenstra WK, Slomp CP (2020) Removal of phosphorus and nitrogen in sediments of the eutrophic Stockholm archipelago. Baltic Sea Biogeosciences 17(10):2745–2766. https://doi.org/10.5194/bg-17-2745-2020

van Kessel MA, Stultiens K, Slegers MF, Guerrero Cruz S, Jetten MS, Kartal B, Op den Camp HJ (2018) Current perspectives on the application of N-damo and anammox in wastewater treatment. Curr Opin Biotechnol 50:222–227. https://doi.org/10.1016/j.copbio.2018.01.031

van Vliet DM, von Meijenfeldt FAB, Dutilh BE, Villanueva L, Sinninghe Damsté JS, Stams AJM, Sánchez-Andrea I (2021) The bacterial sulfur cycle in expanding dysoxic and euxinic marine waters. Environ Microbiol 23(6):2834–2857. https://doi.org/10.1111/1462-2920.15265

Vanwonterghem I, Evans PN, Parks DH, Jensen PD, Woodcroft BJ, Hugenholtz P, Tyson GW (2016) Methylotrophic methanogenesis discovered in the archaeal phylum Verstraetearchaeota. Nat Microbiol 1(12):1–9. https://doi.org/10.1038/nmicrobiol.2016.170

Vavourakis CD, Andrei A-S, Mehrshad M, Ghai R, Sorokin DY, Muyzer G (2018) A metagenomics roadmap to the uncultured genome diversity in hypersaline soda lake sediments. Microbiome 6(1):168. https://doi.org/10.1186/s40168-018-0548-7

Veit K, Ehlers C, Schmitz RA (2005) Effects of nitrogen and carbon sources on transcription of soluble methyltransferases in Methanosarcina mazei strain Gö1. J Bacteriol 187(17):6147–6154. https://doi.org/10.1128/JB.187.17.6147-6154.2005

Větrovský T, Baldrian P (2013) The variability of the 16S rRNA gene in bacterial genomes and its consequences for bacterial community analyses. PLoS One 8(2):e57923. https://doi.org/10.1371/journal.pone.0057923

Vignais PM, Billoud B, Meyer J (2001) Classification and phylogeny of hydrogenases. FEMS Microbiol Rev 25(4):455–501. https://doi.org/10.1111/j.1574-6976.2001.tb00587.x

Vigneron A, Cruaud P, Culley AI, Couture R-M, Lovejoy C, Vincent WF (2021) Genomic evidence for sulfur intermediates as new biogeochemical hubs in a model aquatic microbial ecosystem. Microbiome 9(1):46. https://doi.org/10.1186/s40168-021-00999-x

Voordeckers JW, Do MH, Hügler M, Ko V, Sievert SM, Vetriani C (2008) Culture dependent and independent analyses of 16S rRNA and ATP citrate lyase genes: a comparison of microbial communities from different black smoker chimneys on the Mid-Atlantic Ridge. Extremophiles 12:627–640. https://doi.org/10.1007/s00792-008-0167-5

Vuillemin A (2023) Nitrogen cycling activities during decreased stratification in the coastal oxygen minimum zone off Namibia. Front Microbiol 14

Vuillemin A, Ariztegui D, Vasconcelos C, The PASADO Scientific Drilling Party (2010) Establishing sampling procedures in lake cores for subsurface biosphere studies: assessing in situ microbial activity. Sci Drill 10:35–39. https://doi.org/10.2204/iodp.sd.10.04.2010

Vuillemin A, Ariztegui D, The PASADO Science Team (2013a) Geomicrobiological investigations in subsaline maar lake sediments over the last 1500 years. Quat Sci Rev 71:119–130. https://doi.org/10.1016/j.quascirev.2012.04.011

Vuillemin A, Ariztegui D, De Coninck AS, Lücke A, Mayr C, Schubert CJ, The PASADO Scientific Team (2013b) Origin and significance of diagenetic concretions in sediments of Laguna Potrok Aike, southern Argentina. J Paleolimnol 50(3):275–291. https://doi.org/10.1007/s10933-013-9723-9

Vuillemin A, Ariztegui D, Lücke A, Mayr C, The PASADO Science Team (2014a) Paleoenvironmental conditions define current sustainability of microbial populations in Laguna Potrok Aike sediments. Argentina Aquat Sci 76(1):101–114. https://doi.org/10.1007/s00027-013-0317-4

Vuillemin A, Ariztegui D, Nobbe G, Schubert CJ (2014b) Influence of methanogenic populations in Holocene lacustrine sediments revealed by clone libraries and fatty acid biogeochemistry. Geomicrobiol J:285–298. https://doi.org/10.1080/01490451.2013.824050

Vuillemin A, Ariztegui D, Leavitt PR, Bunting L, The PASADO Science Team (2016a) Recording of climate and diagenesis through sedimentary DNA and fossil pigments at Laguna Potrok Aike. Argentina Biogeosciences 13(8):2475–2492. https://doi.org/10.5194/bg-13-2475-2016

Vuillemin A, Friese A, Alawi M, Henny C, Nomosatryo S, Wagner D, Crowe SA, Kallmeyer J (2016b) Geomicrobiological features of ferruginous sediments from Lake Towuti. Indonesia Front Microbiol 7:1007. https://doi.org/10.3389/fmicb.2016.01007

Vuillemin A, Horn F, Alawi M, Henny C, Wagner D, Crowe SA, Kallmeyer J (2017) Preservation and significance of extracellular DNA in ferruginous sediments from Lake Towuti, Indonesia. Front Microbiol 8

Vuillemin A, Ariztegui D, Horn F, Kallmeyer J, Orsi WD, The PASADO Science Team (2018a) Microbial community composition along a 50 000-year lacustrine sediment sequence. FEMS Microbiol Ecol 94(4):fiy029. https://doi.org/10.1093/femsec/fiy029

Vuillemin A, Horn F, Friese A, Winkel M, Alawi M, Wagner D, Henny C, Orsi WD, Crowe SA, Kallmeyer J (2018b) Metabolic potential of microbial communities from ferruginous sediments. Environ Microbiol 20(12):4297–4313. https://doi.org/10.1111/1462-2920.14343

Vuillemin A, Wankel SD, Coskun ÖK, Magritsch T, Vargas S, Estes ER, Spivack AJ, Smith DC, Pockalny R, Murray RW, D'Hondt S, Orsi WD (2019a) Archaea dominate oxic subseafloor communities over multimillion-year time scales. Sci Adv 5(6):eaaw4108. https://doi.org/10.1126/sciadv.aaw4108

Vuillemin A, Wirth R, Kemnitz H, Schleicher AM, Friese A, Bauer KW, Simister R, Nomosatryo S, Ordoñez L, Ariztegui D, Henny C, Crowe SA, Benning LG, Kallmeyer J, Russell JM, Bijaksana S, Vogel H, Towuti Drilling Project Science Team T (2019b) Formation of diagenetic siderite in modern ferruginous sediments. Geology 47(6):540–544. https://doi.org/10.1130/G46100.1

Vuillemin A, Vargas S, Coskun ÖK, Pockalny R, Murray RW, Smith DC, D'Hondt S, Orsi WD (2020a) Atribacteria reproducing over millions of years in the Atlantic abyssal subseafloor. MBio 11(5):e01937-20. https://doi.org/10.1128/mBio.01937-20

Vuillemin A, Friese A, Wirth R, Schuessler JA, Schleicher AM, Kemnitz H, Lücke A, Bauer KW, Nomosatryo S, von Blanckenburg F, Simister R, Ordoñez LG, Ariztegui D, Henny C, Russell JM, Bijaksana S, Vogel H, Crowe SA, Kallmeyer J (2020b) Vivianite formation in ferruginous sediments from Lake Towuti, Indonesia. Biogeosciences 17:1955–1973. https://doi.org/10.5194/bg-17-1955-2020

Vuillemin A, Kerrigan Z, D'Hondt S, Orsi WD (2020c) Exploring the abundance, metabolic potential and gene expression of subseafloor chloroflexi in million-year-old oxic and anoxic abyssal clay. FEMS Microbiol Ecol 96(12):fiaa223. https://doi.org/10.1093/femsec/fiaa223

Vuillemin A, Coskun ÖK, Orsi WD (2022) Microbial activities and selection from surface ocean to subseafloor on the Namibian continental shelf. Appl Environ Microbiol 88(9):e00216–e00222. https://doi.org/10.1128/aem.00216-22

Vuillemin A, Mayr C, Schuessler JA, Friese A, Bauer KW, Lücke A, Heuer VB, Glombitza C, Henny C, von Blanckenburg F, Russell JM, Bijaksana S, Vogel H, Crowe SA, Kallmeyer J (2023) A one-million-year isotope record from siderites formed in modern ferruginous sediments. GSA Bull 135(1–2):504–522. https://doi.org/10.1130/B36211.1

Walsh EA, Kirkpatrick JB, Rutherford SD, Smith DC, Sogin M, D'Hondt S (2016) Bacterial diversity and community composition from seasurface to subseafloor. ISME J 10(4):979–989. https://doi.org/10.1038/ismej.2015.175

Wang Y, Wegener G, Hou J, Wang F, Xiao X (2019) Expanding anaerobic alkane metabolism in the domain of archaea. Nat Microbiol 4(4):595–602. https://doi.org/10.1038/s41564-019-0364-2

Wasmund K, Mußmann M, Loy A (2017) The life sulfuric: microbial ecology of sulfur cycling in marine sediments. Environ Microbiol Rep 9(4):323–344. https://doi.org/10.1111/1758-2229.12538

Weaver PPE, Schultheiss PJ (1990) Current methods for obtaining, logging and splitting marine sediment cores. Mar Geophys Res 12:85–100. https://doi.org/10.1007/BF00310565

Weisbrod B, Wood SA, Steiner K, Whyte-Wilding R, Puddick J, Laroche O, Dietrich DR (2020) Is a central sediment sample sufficient? Exploring spatial and temporal microbial diversity in a small lake. Toxins 12(9):580. https://doi.org/10.3390/toxins12090580

Welte CU, Rasigraf O, Vaksmaa A, Versantvoort W, Arshad A, Op den Camp HJM, Jetten MSM, Lüke C, Reimann J (2016) Nitrate- and nitrite-dependent anaerobic oxidation of methane. Environ Microbiol Rep 8(6):941–955. https://doi.org/10.1111/1758-2229.12487

Wen L-L, Lai C-Y, Yang Q, Chen J-X, Zhang Y, Ontiveros-Valencia A, Zhao H-P (2016) Quantitative detection of selenate-reducing bacteria by real-time PCR targeting the selenate reductase gene. Enzym Microb Technol 85:19–24. https://doi.org/10.1016/j.enzmictec.2016.01.001

Wetzl D, Bolsinger J, Nestl BM, Hauer B (2016) α-Hydroxylation of carboxylic acids catalyzed by taurine dioxygenase. ChemCatChem 8(7):1361–1366. https://doi.org/10.1002/cctc.201501244

Weyrich LS, Pérez V (2023) Archaeological microbiology. In: Handbook of archaeological sciences. Wiley, pp 557–574

Wiechmann A, Ciurus S, Oswald F, Seiler VN, Müller V (2020) It does not always take two to tango: "Syntrophy" via hydrogen cycling in one bacterial cell. ISME J 14(6):1561–1570. https://doi.org/10.1038/s41396-020-0627-1

Wiegand S, Jogler M, Jogler C (2018) On the maverick planctomycetes. FEMS Microbiol Rev 42(6):739–760. https://doi.org/10.1093/femsre/fuy029

Wirth SB, Gilli A, Niemann H, Dahl TW, Ravasi D, Sax N, Hamann Y, Peduzzi R, Peduzzi S, Tonolla M, Lehmann MF, Anselmetti FS (2013) Combining sedimentological, trace metal (Mn, Mo) and molecular evidence for reconstructing past water-column redox conditions: the example of meromictic Lake Cadagno (Swiss Alps). Geochim Cosmochim Acta 120:220–238. https://doi.org/10.1016/j.gca.2013.06.017

Wong HL, Smith D-L, Visscher PT, Burns BP (2015) Niche differentiation of bacterial communities at a millimeter scale in Shark Bay microbial mats. Sci Rep 5(1):15607. https://doi.org/10.1038/srep15607

Wörmer L, Hoshino T, Bowles MW, Viehweger B, Adhikari RR, Xiao N, Uramoto G, Könneke M, Lazar CS, Morono Y, Inagaki F, Hinrichs K-U (2019) Microbial dormancy in the marine subsurface: global endospore abundance and response to burial. Sci Adv 5(2):eaav1024. https://doi.org/10.1126/sciadv.aav1024

Wu L, Ning D, Zhang B, Li Y, Zhang P, Shan X, Zhang Q, Brown MR, Li Z, Van Nostrand JD, Ling F, Xiao N, Zhang Y, Vierheilig J, Wells GF, Yang Y, Deng Y, Tu Q, Wang A, Global Water Microbiome Consortium, Zhang T, He Z, Keller J, Nielsen PH, PJJ A, Criddle CS, Wagner M, Tiedje JM, He Q, Curtis TP, Stahl DA, Alvarez-Cohen L, Rittmann BE, Wen X, Zhou J (2019) Global diversity and biogeography of bacterial communities in wastewater treatment plants. Nat Microbiol 4(7):1183–1195. https://doi.org/10.1038/s41564-019-0426-5

Wunderlin T, Corella JP, Junier T, Bueche M, Loizeau J-L, Girardclos S, Junier P (2014a) Endospore-forming bacteria as new proxies to assess impact of eutrophication in Lake Geneva (Switzerland–France). Aquat Sci 76(1):103–116. https://doi.org/10.1007/s00027-013-0329-0

Wunderlin T, Junier T, Roussel-Delif L, Jeanneret N, Junier P (2014b) Endospore-enriched sequencing approach reveals unprecedented diversity of firmicutes in sediments. Environ Microbiol Rep 6(6):631–639. https://doi.org/10.1111/1758-2229.12179

Wunderlin T, Junier T, Paul C, Jeanneret N, Junier P (2016) Physical isolation of endospores from environmental samples by targeted lysis of vegetative cells. J Vis Exp 107:53411. https://doi.org/10.3791/53411

Wurzbacher C, Fuchs A, Attermeyer K, Frindte K, Grossart H-P, Hupfer M, Casper P, Monaghan MT (2017) Shifts among eukaryota, bacteria, and archaea define the vertical organization of a lake sediment. Microbiome 5(1):41. https://doi.org/10.1186/s40168-017-0255-9

Wüst PK, Horn MA, Drake HL (2009) Trophic links between fermenters and methanogens in a moderately acidic fen soil. Environ Microbiol 11(6):1395–1409. https://doi.org/10.1111/j.1462-2920.2009.01867.x

Xing P, Tao Y, Luo J, Wang L, Li B, Li H, Wu QL (2020) Stratification of microbiomes during the holomictic period of lake fuxian, an alpine monomictic lake. Limnol Oceanogr 65(S1):S134–S148. https://doi.org/10.1002/lno.11346

Yadav N, Kour D, Yadav A (2018) Microbiomes of freshwater lake ecosystems. J Microbiol Exp 6(6). https://doi.org/10.15406/jmen.2018.06.00223

Yamaguchi T, Fuchs BM, Amann R, Kawakami S, Kubota K, Hatamoto M, Yamaguchi T (2015) Rapid and sensitive identification of marine bacteria by an improved in situ DNA hybridization chain reaction (quickHCR-FISH). Syst Appl Microbiol 38(6):400–405. https://doi.org/10.1016/j.syapm.2015.06.007

Yan Z, Ferry JG (2018) Electron bifurcation and confurcation in methanogenesis and reverse methanogenesis. Front Microbiol 9

Yang Y, Zhao Q, Cui Y, Wang Y, Xie S, Liu Y (2016) Spatio-temporal variation of sediment methanotrophic microorganisms in a large eutrophic lake. Microb Ecol 71(1):9–17. https://doi.org/10.1007/s00248-015-0667-7

Yang H, Yu S, Lu H (2021) Iron-coupled anaerobic oxidation of methane in marine sediments: a review. J Mar Sci Eng 9(8):875. https://doi.org/10.3390/jmse9080875

Ye J, Coulouris G, Zaretskaya I, Cutcutache I, Rozen S, Madden TL (2012) Primer-BLAST: a tool to design target-specific primers for polymerase chain reaction. BMC Bioinform 13(1):134. https://doi.org/10.1186/1471-2105-13-134

Yu T, Wu W, Liang W, Lever MA, Hinrichs K-U, Wang F (2018) Growth of sedimentary bathyarchaeota on lignin as an energy source. Proc Natl Acad Sci 115(23):6022–6027. https://doi.org/10.1073/pnas.1718854115

Zak D, Hupfer M, Cabezas A, Jurasinski G, Audet J, Kleeberg A, McInnes R, Kristiansen SM, Petersen RJ, Liu H, Goldhammer T (2021) Sulphate in freshwater ecosystems: a review of sources, biogeochemical cycles, ecotoxicological effects and bioremediation. Earth Sci Rev 212:103446. https://doi.org/10.1016/j.earscirev.2020.103446

Zemskaya TI, Lomakina AV, Mamaeva EV, Zakharenko AS, Pogodaeva TV, Petrova DP, Galachyants YP (2015) Bacterial communities in sediments of Lake Baikal from areas with oil and gas discharge. Aquat Microb Ecol 76(2):95–109. https://doi.org/10.3354/ame01773

Zemskaya TI, Bukin SV, Lomakina AV, Pavlova ON (2021) Microorganisms in the sediments of Lake Baikal, the deepest and oldest lake in the world. Microbiology 90(3):298–313. https://doi.org/10.1134/S0026261721030140

Zeng J, Zhao D-Y, Liu P, Yu Z-B, Huang R, Wu QL (2014) Effects of benthic macrofauna bioturbation on the bacterial community composition in lake sediments. Can J Microbiol 60(8):517–524. https://doi.org/10.1139/cjm-2014-0132

Zhao Y, Liu P, Rui J, Cheng L, Wang Q, Liu X, Yuan Q (2020) Dark carbon fixation and chemolithotrophic microbial community in surface sediments of the cascade reservoirs, Southwest China. Sci Total Environ 698:134316. https://doi.org/10.1016/j.scitotenv.2019.134316

Zhou Z, Pan J, Wang F, Gu J-D, Li M (2018) Bathyarchaeota: globally distributed metabolic generalists in anoxic environments. FEMS Microbiol Rev 42(5):639–655. https://doi.org/10.1093/femsre/fuy023

Zhu G, Wang S, Wang C, Zhou L, Zhao S, Li Y, Li F, Jetten MSM, Lu Y, Schwark L (2019) Resuscitation of anammox bacteria after >10,000 years of dormancy. ISME J 13(4):1098–1109. https://doi.org/10.1038/s41396-018-0316-5

Zolitschka B, Anselmetti F, Ariztegui D, Corbella H, Francus P, Lücke A, Maidana NI, Ohlendorf C, Schäbitz F, Wastegård S (2013) Environment and climate of the last 51,000 years – new insights from the Potrok Aike maar lake sediment archive drilling prOject (PASADO). Quat Sci Rev 71:1–12. https://doi.org/10.1016/j.quascirev.2012.11.024

Zorz JK, Sharp C, Kleiner M, Gordon PMK, Pon RT, Dong X, Strous M (2019) A shared core microbiome in soda lakes separated by large distances. Nat Commun 10(1):4230. https://doi.org/10.1038/s41467-019-12195-5

Zwart G, Crump BC, Kamst-van Agterveld MP, Hagen F, Han SK (2002) Typical freshwater bacteria: an analysis of available 16S rRNA gene sequences from plankton of lakes and rivers. Aquat Microb Ecol 28(2):141–155. https://doi.org/10.3354/ame028141

Chapter 5
Cyanobacterial DNA from Lake Sediments

Marie-Eve Monchamp and Frances R. Pick

Keywords Sedimentary DNA · Environmental DNA · PCR · High-throughput sequencing · Metagenomics · Eutrophication · Climate change · Biotic homogenization · Cyanobacteria · Oxyphotobacteria · Lakes · Paleolimnology · Paleo-indicators

Introduction

Cyanobacteria have become infamous due to their impacts on aquatic ecosystems, in particular negative consequences on ecosystem functioning and services to human populations. Observations of cyanobacterial dominance have been reported since the end of the 1800s and the global rise in industrialisation characterized by enhanced human perturbations and a warming climate. We have limited information on past cyanobacteria dynamics and geographic distribution due to the paucity of long-term monitoring data. In addition, challenges remain surrounding the detection and the identification of cyanobacterial taxa deposited in sedimentary archives.

In this chapter, we synthesized the scientific knowledge gained via studies using environmental DNA archived in lake sediments (i.e., sedimentary DNA; sedDNA; Crump 2021) to reconstruct the temporal changes in the ecology, habitat, and function of cyanobacteria in relation to environmental change. We begin with an overview of the general ecology and distribution of cyanobacteria, followed by a characterization of the nature of the sedDNA signal and relevant taphonomic

M.-E. Monchamp (✉)
Department of Biology, McGill University, Montreal, QC, Canada

Groupe de Recherche Interuniversitaire en Limnologie, Montreal, QC, Canada

F. R. Pick
Department of Biology, University of Ottawa, Ottawa, ON, Canada
e-mail: frances.pick@uottawa.ca

E. Capo et al. (eds.), *Tracking Environmental Change Using Lake Sediments*, Developments in Paleoenvironmental Research 21,
https://doi.org/10.1007/978-3-031-43799-1_5

processes, with respect to cyanobacteria. We then report the current state of knowledge of cyanobacterial past dynamics in lacustrine environments through a review of the published cyanobacteria sedDNA literature. Last, we discuss how the field of sedDNA has the potential to shed light on the timing and specific causes of environmental change in addition to guiding ecosystem restoration and management goals.

Diversity and Distribution of Cyanobacteria

Cyanobacteria (a.k.a. Oxyphotobacteria) are photosynthetic bacteria (Whitton 2012). They are ancient organisms that were early colonizers of the planet and are arguably one of the most important microbial groups on Earth because of their contribution to the oxygenation of the atmosphere for the last ca. three billion years. Cyanobacteria are ubiquitous, thriving in terrestrial or aquatic (marine and freshwater) environments ranging from cold to tropical regions. They can survive under the harshest environmental conditions (e.g., geothermal springs, cold and hot deserts). The phylum Cyanobacteria exhibits a high diversity, with an estimated 6000 to 8000 species across more than 300 genera (Guiry 2012; Nabout et al. 2013; Hauer and Komárek 2022) with many species yet to be discovered, particularly in poorly studied regions (e.g., the tropics, extreme habitats). While this diversity can be seen in the range of sizes from unicellular forms below one micron to macroscopic colonies of various shapes (Bonilla and Pick 2017; Fig. 5.1), molecular analyses are leading to the discovery of more species and re-writing the phylogeny and taxonomy of phylum Cyanobacteria (Komárek 2016a, b). In addition to taxonomic diversity, cyanobacteria exhibit a range of ecological and physiological traits that enable adaptation to many environmental extremes (e.g., temperature, oxygen, nutrients, UV, salinity). Changes in the surrounding environment induce rapid physiological changes in cyanobacteria, as such they are highly adaptable and competitive.

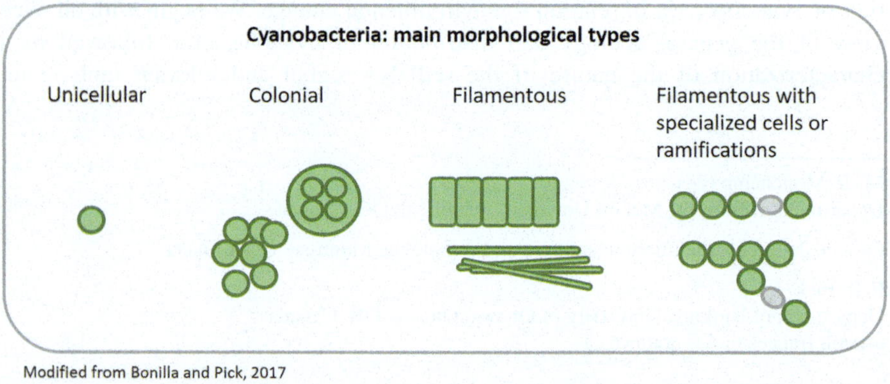

Fig. 5.1 Main morphological types of cyanobacteria. (Modified from Bonilla and Pick (2017))

Physiological and Morphological Traits of Cyanobacteria Favoring Their Dominance

Cyanobacteria have several eco-physiological traits providing a competitive advantage under diverse and changing environments. For example, in terrestrial ecosystems, tolerance to desiccation is a key trait for survival (Whitton and Potts 2014). Nitrogen control and the capacity of some species to fix nitrogen gas (N_2), typically inside thick-walled specialized cells (heterocytes or heterocysts), is also an important trait in several aquatic and terrestrial habitats, providing a unique competitive advantage under low dissolved nitrogen availability (Herrero et al., 2001). In aquatic environments, key adaptations also include buoyancy regulation through gas vesicles, efficient light harvesting, tolerance to extreme solar radiation, ability to grow in warm temperatures, high affinity and high storage capacity for both phosphorus and nitrogen, akinete (resting stage) production, and toxin synthesis (Carey et al. 2012; Bonilla and Pick 2017). For example, picocyanobacteria (cells within the picoplankton size range of 0.2–2 μm) have a high nutrient uptake capacity at low nutrient concentrations, which favors their dominance both in terms of biomass and primary production in the oligotrophic ocean (Whitton and Potts 2014) as well as in oligotrophic lakes (Callieri et al. 2013). Other taxa such as the widespread *Microcystis* genus, which thrives in eutrophic lakes, can form large visible colonies (up to 5 mm) that contribute to dominance and bloom formation (i.e., massive accumulations of cyanobacterial cells typically observed at the water surface).

Current Trends in the Frequency and Magnitude of Cyanobacterial Blooms

Cyanobacteria can proliferate in coastal and inland waters, with accumulations characteristic of harmful algal blooms, referred to as cyanoHABs, which degrade the health of aquatic ecosystems and alter ecosystem services, raising concerns among the general public, lake managers, policymakers, and scientists. CyanoHABs can result in deoxygenation of the water column, nuisance odors and tastes, and the production of cyanotoxins, a suite of secondary metabolites potentially harmful to humans, wildlife, and domesticated animals (Chorus and Welker 2021). There are a variety of cyanotoxin types (hepatotoxins, neurotoxins, dermatoxins) with different effects (Carmichael 1994; Chorus and Batram 1999; Chorus and Welker 2021). Many species can produce multiple toxins, and the same toxin can be synthesized by many species. Overall, the factors triggering and mediating toxin synthesis in cyanobacteria, as well as the role(s) of these secondary metabolites, remain poorly understood (Chorus et al. 2021).

The incidence of cyanobacterial bloom events in lakes has risen considerably in recent decades (Ho et al. 2019). Based on sedimentary pigment analysis, the dominance of cyanobacteria relative to other phytoplankton in temperate lakes appears to have increased globally since the mid twentieth century (Taranu et al. 2015) in parallel with human population growth, urbanization and the rise in use of chemical fertilizers. The intensity and frequency of cyanoHABs is expected to continue to increase in response to rising levels of atmospheric CO_2 with enhanced combined (and possible synergistic) effects of anthropogenic-induced eutrophication (Moss et al. 2011; Huisman et al. 2018; Kakouei et al. 2021). Changes in environmental conditions, such as lake trophic status, can trigger community shifts towards taxa better adapted to the new environmental conditions (Watson et al. 1997). Compositional turnover often goes unnoticed in lakes unless they are closely monitored, but they can indicate a major change in ecosystem functioning as new taxa become dominant. An increased nutrient input and/or lake water warming often leads to the seasonal dominance of large colonial and buoyant taxa, in the orders Chroococcales, Nostocales and Oscillatoriales (Bonilla and Pick 2017). These shifts in community composition can inform on the timing of past environmental changes and on whether the system has recovered after nutrient abatement or if a regime shift has occurred (Monchamp et al. 2021).

Sediment Deposition of Cyanobacteria

Lake sediments are formed by the accumulation over time of organic and inorganic matter originating from the lake itself, the catchment, and atmospheric input, creating an archive of past biota, trophic conditions, and biogeochemical processes prevailing at the time of deposition (Smol 2009). Undisturbed lacustrine (and marine) sediments therefore constitute a source of information for paleoecologists and paleoclimatologists to study past life and climatic conditions using a variety of paleolimnological tools and proxies. Living, dead or dormant cyanobacterial cells can accumulate in sediments through several processes such as aggregation on particles or concentration in fecal pellets followed by sedimentation, as well as through active migration. Some cyanobacteria are also able to grow in surficial sediments (e.g., Gaget et al. 2020), particularly in littoral areas and shallow lakes where light reaches the water-sediment interface. Most cyanobacteria have thick gelatinous cell walls making them less prone to cell degradation than other phytoplankton taxa, particularly in the case of the resting stages of Nostocales (akinetes), which have been shown to successfully germinate under laboratory conditions decades to centuries after burial in sediments (Livingstone and Jaworski 1980; Legrand et al. 2019).

Biogeochemical proxies, including lipid biomarkers and pigments (photosynthetic and UV-screening), as well as isotope composition, have been used for many decades to detect and quantify photosynthetic microbes, including cyanobacteria, in aquatic sediments (Leavitt and Hodgson 2006). More recently cyanotoxins have been quantified in sediment cores: these include the more persistent chemical species e.g., several congeners of microcystins (Zastepa et al. 2017) and the cylindrospermopsins (Waters 2016). However, such proxies offer limited information on community composition or the presence of potentially toxic taxa, because they can only provide a coarse taxonomic identification of the main phytoplankton groups. This is a challenge that sedDNA studies are currently addressing.

Taphonomic Processes: Production, Sedimentation, and Preservation of Cyanobacterial DNA in Sediments

In-lake drivers, such as grazing, parasitism, water currents and wind, can influence the rate at which phytoplankton cells are deposited in sediments. How DNA from various organisms reaches bottom sediments, is altered through diagenesis, or becomes preserved over time is still not fully understood. Investigations of the correspondence between pelagic and sediment communities provide new information that can guide the interpretation of sedDNA historical archives. For example, Nwosu et al. (2021a) compared the composition, richness, and evenness of the cyanobacterial DNA signal between pelagic water, sediment trap, and surface sediment samples, and they found no significant difference in detection of cyanobacterial taxonomic composition across sample types. This suggests minimal taxonomic bias between sediments and the water column. However, the authors reported large taxon-specific differences in relative abundances between water and sediment samples, suggesting differential deposition or preservation of the DNA signal between cyanobacterial clades in sediments. For instance, the genus *Planktothrix* was found in relatively low proportion compared to other cyanobacteria in both surface sediment and sediment trap samples, despite this taxon being dominant in the water column, especially near the thermocline. On the other hand, the DNA signal for colonial taxa (*Aphanizomenon*, *Snowella*) was stronger in surface sediment samples. In a comparative study of metagenomes from lake water and both contemporary and pre-industrial sediments, Garner et al. (2020) found that some dominant bacterial taxa from the pelagic zone were not detected in sediment samples, while some of the preserved microorganisms detected had never been reported from limnological surveys. Only a few such *in situ* studies on the efficiency of phytoplankton/microbial eukaryotes deposition and recovery in lake sediments are currently published (see Gauthier et al. 2021). Laboratory experiments suggest that, in addition to differential sedimentation behavior within the water column depending on taxa features (e.g., habitats, morphologies), there may be differences in degradation rates among major cyanobacteria taxa once settled in the sediments (Mejbel et al. 2022) as well as an impact of sediment storage on sedDNA concentration and species recovery (Brasell et al. 2022). Finally, only a few studies have investigated the congruence of the sedDNA signal between sediment cores collected at various locations in a same lake (Weisbrod et al. 2020; Thorpe et al. 2022). These works report good congruence of both cyanobacterial and other prokaryotes composition across sites.

Overall, cold and hypoxic/anoxic conditions are more favorable for DNA preservation in lake sediments (Mejbel et al. 2022 and references therein; see Chap. 2 of this volume) and thus deep cold-water lakes are generally expected to be adequate systems in which to retrieve cyanobacterial archives. However, there have been a handful of paleolimnological studies on subtropical lakes (e.g., Martínez de la Escalera et al. 2014; Huo et al. 2022a) as well as shallow temperate lakes (e.g., Picard et al. 2022a) that have successfully obtained cyanobacterial DNA records. More work that specifically aims at (1) evaluating the extent to which the limnetic biological signal reaches the sediments and (2) understanding the processes behind sinking/sedimentation/degradation rates of various planktonic forms including cyanobacteria on a range of lake types with different sediment geochemical properties remain necessary for improving ecological inferences based on sedDNA archives.

Cyanobacterial Signal in Lacustrine Sediments: Dead or Alive?

One of the challenges in the field of environmental paleomicrobiology is distinguishing between active, dormant, and dead cells in lake sediments (Capo et al. 2022). A further complication is that certain cyanobacteria taxa can overwinter in sediments and recolonize the water column the following spring under favorable environmental conditions (Brunberg and Blomqvist 2002). Nostocales can remain viable in sediments, possibly up to hundreds of years (Legrand et al. 2019) and germinate when environmental conditions improve. It is therefore difficult to disentangle the dead vs. live signal in the cyanobacterial DNA deposited to sedimentary archives, which can constitute a limitation of sedDNA-based reconstructions. Nevertheless, the recovery of cells in sediment layers, whether they are currently active or not, reflect the ecological conditions at the time of deposition and are accepted as a reasonable means of reconstructing past community composition dynamics.

Discriminating the presence of living organisms in the pool of sedDNA is not possible based on DNA analyses alone because the DNA signal can come from both active and inactive (i.e., dead) cells. Even though DNA starts degrading as soon as cell death or cell lysis occurs, it can preserve for an extended period of time in lake water and in sediments (Barnes et al. 2014; Pedersen et al. 2015). However, damage starts accumulating inside DNA molecules deposited hundreds of years ago (Dabney et al. 2013), creating a specific signature that can be authenticated bioinformatically by calculating the ratios of damaged vs. intact DNA fragments sequenced (see Chap. 3 of this volume). Finally, it has been proposed that RNA could be used to distinguish between living and dead cells because of its higher degradation rate compared to DNA. However, a recent investigation based on the recovery of ca. 200 years-old RNA molecules in lake sediments shows that cyanobacterial RNA can also be quite stable over decadal to centennial times (Pearman et al. 2022), which suggests that the potential of RNA for distinguishing between living and dead cells may extend to sediments deposited at least several hundred years ago.

SedDNA-Based Approaches: A Means of Reconstructing Past Cyanobacterial Histories

Over the last two decades, the rapid development of environmental DNA techniques and high-throughput sequencing technologies (HTS) has opened new possibilities to identify taxa within all domains of life, including cyanobacteria, in their natural habitat (Pedersen et al. 2015; Taberlet et al. 2018). Molecular genetic approaches allow for a more refined taxonomic level compared to other proxies like pigments (with some caveats), potentially at the resolution of species and subspecies/strains. Analyses of DNA preserved in aquatic sediments have expanded our knowledge of cyanobacteria temporal dynamics in various aquatic habitats around the globe. In a literature search using *Web of Science* (as of July 2022), and based on the list of sampling sites compiled by Von Eggers et al. (2022), we found 50 research articles exploring the long-term dynamics of cyanobacteria with sedDNA at a total of 71 lacustrine sites (Fig. 5.2). Of the 123 DNA records from aquatic sediments, 104 were from lake sediment cores (time-series), four were from lake surface sediments only, and two were obtained from sediment traps. The majority (31 out of 50) of the published cyanobacterial sedDNA papers are based on amplicon sequencing (metabarcoding; Fig. 5.3) and most studies (90%) target a region of the 16S rRNA

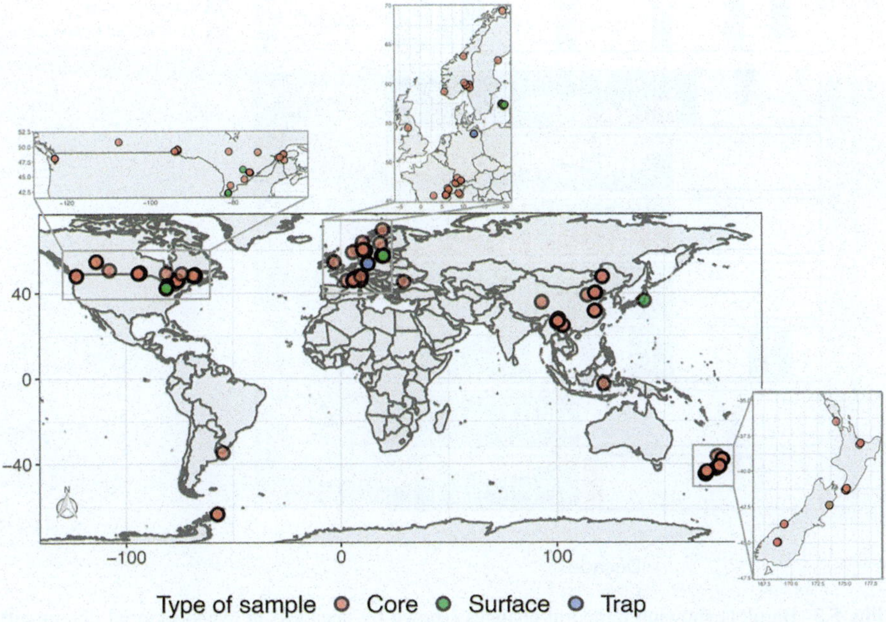

Fig. 5.2 Global map showing the sites where cyanobacteria have been targeted using sedimentary DNA approaches in lakes (n = 71 unique sites). Sampling sites are color-coded by sample type: sediment cores (i.e., time series data) in red, surface sediments only in green, and sediment trap samples in blue. The data were compiled from the dataset published by Von Eggers et al. (2022). Theses and unpublished works were not included in this map

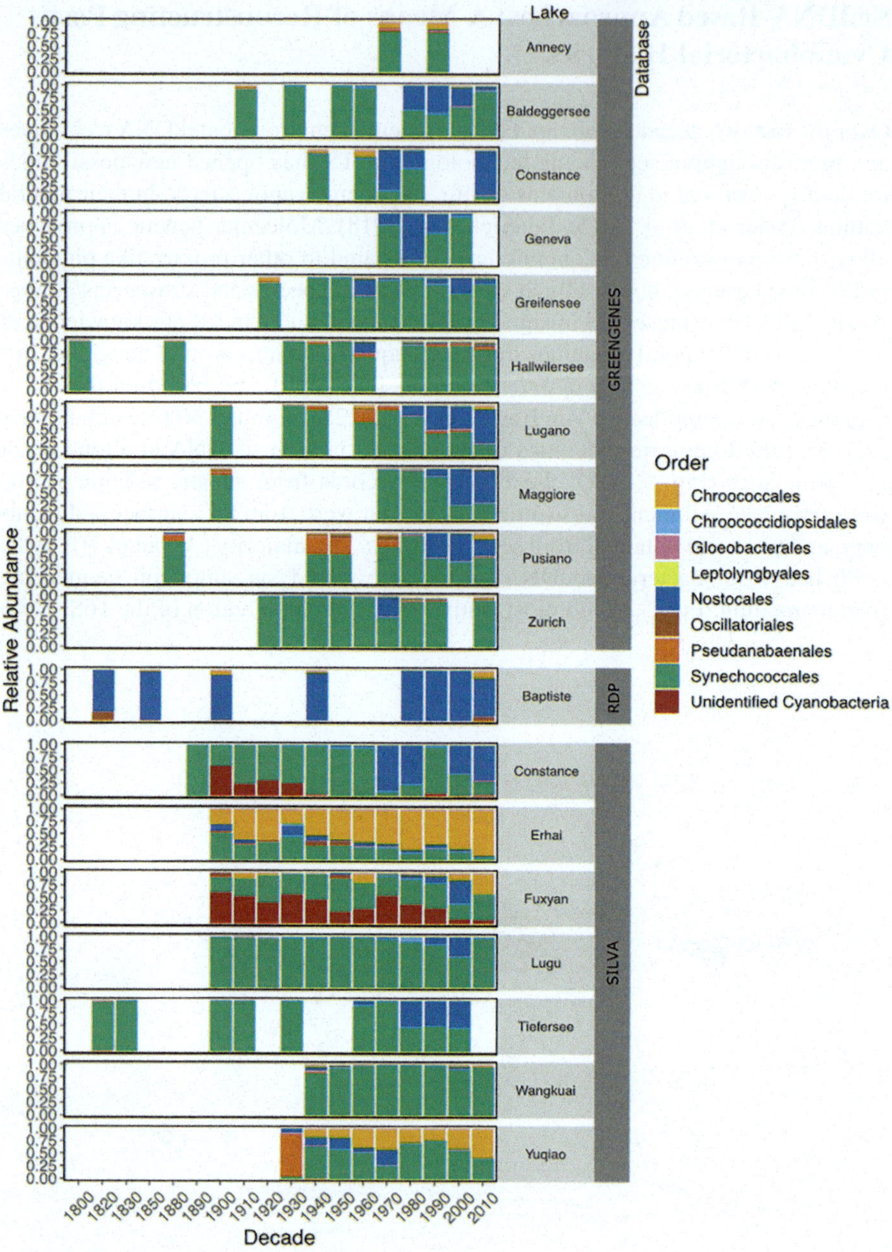

Fig. 5.3 Barplots showing temporal changes (shown by decades) in cyanobacterial community composition based on relative abundances of molecular taxonomic units (Operational taxonomic units (OTUs) or amplicon sequence variants (ASVs)) in various lakes. Data were collected from published literature presenting metabarcoding datasets (Monchamp et al. 2018; Pilon et al. 2019; Nwosu et al. 2021; Huo et al. 2022b). Sampling sites and reference sequence databases used for taxonomic assignment are shown on the right side of the barplots (Greengenes, SILVA, or RDP Classifier)

gene to reconstruct past community diversity or quantify cyanobacterial abundance. Finally, most studies span the last ~100 to ~200 years, i.e., a period of enhanced anthropogenic impacts, but overall cyanobacterial sedDNA investigations extend from recent times (i.e., surface sediments) to 10,000 years BP (Von Eggers et al. 2022).

One of the earliest studies to examine microbial sedDNA was that of Coolen and Overmann (1998) who recovered traces of DNA from purple sulfur bacteria in ~9000-year-old sediments of a Canadian meromictic lake. Subsequently, Rinta-Kanto et al. (2009) analyzed sedDNA specifically for cyanobacteria in surficial sediments collected in 2004 and 2009 in Lake Erie along with an archived surficial sediment sample (3 top centimeters collected with a Shipek grab sampler) from the 1970s. The authors successfully amplified fragments of the *mcy*A gene involved in the synthesis of microcystins, which were identified as produced by *Microcystis* spp. in both recent sediments and those retrieved in the 1970s. Savichtcheva et al. (2011) also applied quantitative PCR (qPCR) assays to study the dynamics of cyanobacterial communities and potentially toxic *Planktothrix* populations over a period of *ca.* 50 to 100 years from sediment cores collected in peri-alpine lakes. Further, Domaizon et al. (2013) analyzed ~100-years of temporal changes in *Synechococcus* populations in Lake Bourget (France) and reported changes in cyanobacterial composition through time as a response to climate warming and changes in lake trophic status. More studies quantifying gene copy numbers in sediments through time for either total cyanobacteria, specific taxa and/or targeted ecological functions have followed. For example, Savichtcheva et al. (2015) investigated cyanobacterial composition using qPCR over a period of ~60 years and discovered that phosphorus was the dominant driver of the relative abundance of *Planktothrix rubescens*, which was favored under mesotrophic conditions. The authors further reported that, despite fluctuating environmental conditions, the same *Planktothrix* species remained present in the lake since the 1920s. In another quantitative study, Pal et al. (2015) compared qPCR and sedimentary pigment data sets downcore in five temperate lakes located within and outside a protected area (Quebec, Canada) and found that the number of cyanobacteria 16S rRNA gene copies (in parallel with cyanobacterial pigments) had significantly increased since the 1980s across the lakes in comparison to the past ~150 years most likely in response to regional climate change and land development. More recently, a digital droplet (dd)PCR approach has been applied to quantify specific gene abundances in sediment samples as the technique has several advantages over qPCR in quantifying low DNA template (Mejbel et al. 2021). Overall, there was coherence in ddPCR and amplicon sequencing approaches, but the authors noted that research aimed at quantifying absolute abundances from complex environments are recommended using ddPCR, due to the higher tolerance of this method to PCR inhibitors when compared to HTS.

Alongside qPCR and ddPCR approaches targeting specific genes, sedDNA-based studies analyzing whole assemblages at the community level started to emerge, often combined with traditional paleo-proxies (e.g., pigments) to complement the sedDNA signal. For example, Fernandez-Carazo et al. (2013) combined denaturing gel gradient electrophoresis (DGGE) followed by cloning and sequencing, and carotenoid and chlorophyll analysis to report changes in cyanobacterial community structure in two maritime Antarctic lakes during the late Holocene. A few years later, Monchamp et al. (2016) applied for the first time sedDNA metabarcoding to reconstruct cyanobacterial communities in two peri-Alpine lakes over a period of ~60 years spanning pre- and post-eutrophication periods. The authors reported coherent compositional and diversity (richness) patterns from cyanobacterial communities sampled in sedDNA archives and in water samples (i.e., microscopy counts of cyanobacterial cells) collected as part of a long-term monitoring program.

Based on sedDNA community composition analyses of lakes published to date, community structure has shifted over the past ~50–100 years either towards dominance of Nostocales or towards greater dominance of Chroococcales (at the expense of Synechococcales, Fig. 5.3). Both these orders contain bloom-forming and toxigenic taxa. Direct comparisons of suborder changes are more difficult, given the use of different bioinformatics tools and reference sequence databases. Suborder and species level changes are also likely more context dependent on specific limnological characteristics of lakes.

Overall, paleolimnological multi-proxy studies published to date analyzing pigments, biomarkers, and hard subfossils data in addition to cyanobacteria sedDNA have collectively helped: (1) evaluate the congruence between sedDNA datasets and other proxies and (2) evaluate the past impact of environmental change on cyanobacteria (e.g., Pal et al. 2015; Tse et al. 2018; Pilon et al. 2019; Monchamp et al. 2021; Nwosu et al. 2021b; Erratt et al. 2022; Picard et al. 2022b; Nwosu et al. 2023). Such research continues to be relevant as the field of sedDNA is in its infancy and there is a need to evaluate methods and tools across a range of lake types (e.g., from shallow to deep, from acidic to alkaline) and different climatic conditions to accurately interpret sedDNA results as a proxy of past environmental and biotic changes.

An Expanded Representation of Phylum Cyanobacteria: Contribution of sedDNA Analyses

The application of HTS technologies to sequence the genomes of uncultured bacteria collected in natural environments has also led to the discovery of a previously unknown group of ancestral cyanobacteria (Di Rienzi et al. 2013; Soo et al. 2014; Soo et al. 2015). These clades are currently comprised of three described classes (determined on the basis of 16S rRNA gene analyses): Oxyphotobacteria (typically called Cyanobacteria), Melainabacteria, and the basal clade ML635J-21 for which the name Sericytochromatia has been proposed (Soo et al. 2017).

Taxa from the Melainabacteria and Sericytochromatia clades have been identified in a wide variety of anoxic and oxic environments, including subsurface groundwater (Di Rienzi et al. 2013), algal biofilms (Soo et al. 2017), animal and human intestinal tracts, and feces/animal guts (Ley et al. 2005; Di Rienzi et al.

2013; Soo et al. 2014). Several recent observations of their presence in freshwater lakes based on sedDNA and water eDNA analyses have been reported (Salmaso et al. 2018; Monchamp et al. 2019; Ibrahim et al. 2021; Nwosu et al. 2021b). Members of these clades have a range of metabolic pathways different from the Oxyphotobacteria (e.g., anoxic respiration, fermentation). However, the factors controlling their physiology, ecology, and distribution in lakes and other natural settings have yet to be explored.

The Use of sedDNA to Track Anthropogenic-Induced Change: Impact of climate and Pollution on Cyanobacteria Abundance, Community Structure and Taxonomic Composition

Growing evidence over the last few decades is confirming the impacts of climate warming and pollution by eutrophication on lake biota (Barouillet et al. 2023). As such, determining cyanobacteria dominance in freshwaters in relation to human activities is a topic that has gained a lot of interest. DNA reconstructions from lacustrine sediments have provided new insights on the timing and specific causes of environmental change, with great potential to help set restoration and management goals. From investigations of individual lakes and lake districts, climate warming, in addition to eutrophication and hydrological modifications, has likely increased the overall abundance of cyanobacteria and of specific taxa based on quantitative sedDNA analyses over the past several hundred years (Savichtcheva et al. 2011, 2015; Pal et al. 2015; Pilon et al. 2019; Yan et al. 2019; Erratt et al. 2022). Importantly, deconvoluting the effect of climate change and eutrophication is often challenging since the two drivers coincide temporally in many regions that have experienced population and land use changes (De Senerpont Domis et al. 2013; Rigosi et al. 2015). In some cases the absence of significant change in independent proxies of nutrient enrichment can point to the importance of climate change alone (Pilon et al. 2019).

One trend that has been observed worldwide is the regional homogenisation of biological communities (e.g., Rahel 2002; Dornelas et al. 2014; Magurran et al. 2015), that is the increased taxonomic similarity between communities in a given region. In freshwaters, this phenomenon is most commonly caused by alterations to the watershed as well as changes in climate and lake productivity (Petsch 2016). Studies based on sedDNA have pointed to the role of climate warming and anthropogenically-induced eutrophication as drivers of biotic homogenisation in various biological groups in freshwaters, including cyanobacteria and other photosynthetic phytoplankton (Monchamp et al. 2018; Cao et al. 2020; Zhang et al. 2021; Huo et al. 2022b), as well as heterotrophic protists (Keck et al. 2020) and ciliates (Barouillet et al. 2022). Monchamp et al. (2018) first reported cyanobacterial homogenization in 10 peri-Alpine lakes since the ~1980s due to the influence of climate warming (affecting the strength and duration of water column stratification) combined with nitrogen and phosphorus enrichment that allow cyanobacterial growth. Huo et al. (2022b) has also reported the recent homogenisation of cyanobacterial communities in north temperate lakes and reservoirs attributed mainly to enhanced nutrient pollution and warming. These findings are important because biotic homogenisation of cyanobacteria tends to be associated with the dominance of colonial and bloom-forming taxa which also include toxigenic taxa. Both observational (Salmaso et al. 2015) and sedDNA datasets (Monchamp et al. 2018) suggest the recent colonization of several large temperate lakes in northern Italy by the potentially toxic *Dolichospermum lemmermannii*, contributing to biotic homogenisation.

The Use of sedDNA in Evaluating the Temporal Dynamics of Potentially Toxic Cyanobacteria

Cyanotoxins are synthesized via the activation of a cluster of genes encoding the non-ribosomal peptide and polyketide synthetases (Kultschar et al. 2018). This suite of genes can be detected in sedDNA archives to inform on the temporal dynamics of potentially toxic cyanobacterial taxa in the past (Henao et al. 2019). However, detecting their presence is not sufficient to confirm the active synthesis of cyanotoxins, but rather informs on the potential toxicity of cyanobacterial species. The ability to detect and quantify these functional genes has high relevance for research aimed at identifying the timing and causal factors behind toxic cyanobacteria blooms and for the purposes of lake management (Pilon et al. 2019). Rinta-Kanto et al. (2009) developed *mcy*D primers allowing for quantification of microcystin synthetase gene via qPCR and applied these to surficial sediment in Lake Erie. Additional primers for microcystins and other cyanotoxins (saxitoxins, anatoxins) have since been developed and used to detect and quantify toxin genes in lake sediments. For example, Martínez de la Escalera et al. (2014) reported a compositional shift in cyanobacterial communities in a freshwater lagune (Laguna Blanca, Uruguay) between the nineteenth and the twenty-first centuries. The appearance of potentially toxic cyanobacteria was attributed to changes in trophic status. Quantitative PCR of the *sxt*U gene (saxitoxin synthesis) further revealed the presence of saxitoxin-producing *Raphidiopsis* (previously *Cylindrospermopsis*) *raciborskii* (Wołoszyńska) in recent sediment samples (e.g., from 2000).

Legrand et al. (2017) investigated the distribution of *Dolichospermum macrosporum* and *Dolichospermum flos-aquae* in sediment cores of the French Lake Aydat over a period of ~220 years by microscopy counts of akinetes PCR detection of 16S rRNA gene and both anatoxin-a (*ana*C) and microcystin (*mcy*A) biosynthetic genes. While *mcy*A was detected throughout the core, *ana*C was only detected in the most recent sediments (since the 1980s). Based on these observations and the distribution of akinetes of the two *Dolichospermum* species downcore, the authors concluded that the presence of *ana*C was most likely linked to the dominance of *D. macrosporum* in Lake Aydat since the mid-1980s related to anthropogenic activities.

Quantitative PCR applied to sedDNA has shown contrasting histories of toxic cyanobacteria in two temperate lakes of central Canada (Pilon et al. 2019). In Lake of the Woods (straddling the USA and Canada), *Microcystis* (genus specific 16S rRNA) and microcystin genes (*myc*D) were both detected in the samples dated to the 1980s, with climate change rather than eutrophication likely responsible for the recent rise in copy numbers of both genes. In contrast, in naturally eutrophic Lake Baptiste (Alberta, Canada) both these gene markers were found throughout the entire core and were present well before European settlement. In both lakes, the trends in sedDNA gene copies and microcystin concentrations measured in sediments were correlated, although in Lake of the Woods the detection of toxic genes preceded the detection of microcystins by ~20 years, because of the greater sensitivity of DNA techniques compared to the analytical chemical methods for microcystins. In another quantitative study tracing back both photosynthetic pigments and anatoxin-a genes in sedimentary archives, Hobbs et al. (2021) found that a change in nutrient cycling due to farming history and possibly fish stocking in Anderson Lake (Washington State, USA) caused a shift in the bacterial community and led to the dominance of anatoxin-a producing *Dolichospermum* sp.

Some species, such as the tropical and potentially toxic filamentous *Raphidiopsis raciborskii* (Wołoszyńska), are thought to have spread to several regions including northern latitudes through the 1980's and 1990's (Neilan et al. 2003; Hamilton et al. 2005). *Dolichospermum lemmermannii* expanding to the Italian peri-Alpine lakes is another example (Salmaso et al. 2015). The invasion histories of toxigenic cyanobacterial taxa remain generally unknown, but paleolimnological work based on sedDNA analyses could help elucidate the timing of invasion and the phylogeography of cyanobacteria of interest both from an ecological and a lake management perspective.

Methodological Considerations

DNA-based paleo-reconstructions remain limited by a number of methodological considerations (Capo et al. 2021). These include the low amount of starting DNA material that can be retrieved from ancient sediments, the limited knowledge of deposition and taphonomic processes, and the bioinformatics treatment of sedDNA sequence data (e.g., raw sequence preprocessing, contaminant removal, taxonomic assignment). Therefore, adhering to strict ancient DNA laboratory protocols to reduce contamination and the use of blanks (negative controls) at every step of the workflow are especially important in order to ascertain the quality of sedDNA datasets (Chap. 3 of this volume). Further, the choice of analytical methods and pipelines have an incidence on the type of ecological inferences that can be made based on these data. For example, PCR-based approaches (including amplicon sequencing and quantitative PCR methods) efficiently enhance the signal of the target organisms, but they introduce abundance biases due to the exponential nature of the PCR amplification. Furthermore, amplicon sequencing does not inform as to the patterns of DNA damage (i.e., size fractionation of DNA molecules and deamination patterns), which can be used to authenticate ancient DNA molecules (Dabney et al. 2013). Copy number variation within genes used to analyzed cyanobacterial community composition is another source of error as it can skew the abundance estimates in favor of taxa that have higher numbers of copies of these genes: presently the number of gene copies has not been determined for all taxa and a major effort of sequencing cyanobacteria whole genomes is required here. Finally, a limitation that remains is related to primer design. Indeed, designing truly cyanobacteria specific (or genus-specific) 16S primers is difficult, and consequently a high number of metabarcoding amplicons belong to other (non-target) bacterial taxa, which translates to a skewed representation of cyanobacteria in sequencing assays.

Despite some limitations, sedDNA metabarcoding has proven to be a valuable approach for reconstructing the taxonomic diversity of lake cyanobacteria and relating changes in community composition and relative abundance with anthropogenic stressors in the watershed of lakes (e.g., water-level fluctuations due to damming, nutrient pollution, climate warming) (Fig. 5.3). However, due to the lack of standardized protocols, databases, and the different molecular units used (e.g., operational taxonomic units; OTUs vs. amplicon sequence variants; ASVs), metabarcoding datasets generated by different research groups can be difficult to compare in a meaningful way.

The current paucity of reference DNA sequence databases (for cyanobacteria as well as for most biological groups, especially microbes) is another major limitation of cyanobacterial DNA studies, both in modern and paleo-environments. In a recent gap-analysis study, it was shown that approximately 50% of the cyanobacterial species identified by light microscopy were not represented in the reference databases (Salmaso et al. 2022). Finally, inconsistencies and mis-identification of cyanobacterial taxa in existing 16S rRNA reference databases have been identified which have led to the development of a curated database for cyanobacteria (Ramos et al. 2017). Cyanobacteria taxonomy is under revision with new molecular data changing nomenclature and phylogenetic assignments.

Future Directions

Similar to the study of genes responsible for the synthesis of cyanotoxins, other functional genes could be examined to monitor cyanobacteria and their specific functioning. One example is pertinent to the proliferation of benthic cyanobacteria, which are problematic because they produce taste and odor compounds as well as toxins. Gaget et al. (2020) developed qPCR assays to monitor taste and odor genes (*geo*A, *MIB* synthetase genes) in benthic biofilms. Paleo-reconstructions aimed at monitoring changes in the abundance of functional genes regulating secondary metabolites in cyanobacteria (such as toxins, taste and odor compounds) have important implications for lake and reservoir management. Other targets could also be relevant to explore, such as functional genes associated with gas vesicle formation, akinete formation, nitrogen fixation and nutrient storage unique to cyanobacteria.

The field of paleogenomics has begun exploring the possibility of recovering the RNA signal in sedimentary archives to study the taxonomic composition and function of microorganisms. The analysis of RNA can also enable the separation and comparison of total vs. "active" bacteria (Pawlowski et al. 2022), even in ancient sediments (Pearman et al. 2022). Although it is assumed that RNA degrades more quickly in the environment compared to DNA, RNA molecules can apparently preserve over long time periods in aquatic sedimentary archives and can be used to study function in past biological communities through genetic and metagenomic analyses (Lindqvist and Rajora 2019). Even more care is required in the analysis of RNA in sediments compared to sedDNA to minimize signal contamination and degradation.

A developing trend in paleo-molecular studies is to use shotgun sequencing (metagenomics) which allows retrieval of the DNA signal of all organisms in a sample in proportion to their original presence. This approach allows for the reconstruction of the entire pool of lake biota and thus to potentially explore relationships between different organisms across trophic levels (see Chap. 12 of this volume). Paleogenomics coupled with DNA enrichment or hybridization capture techniques have also proven efficient to recover planktonic algae and to track ancient DNA damage. Recently, hybridization capture was used on sediment seafloor to target both phytoplankton (including cyanobacteria) and zooplankton, demonstrating the potential of this approach to reconstruct the composition and relative abundance of various planktonic forms (Armbrecht et al. 2021). Recent work has also highlighted the potential of paleogenomic for reconstructing whole microbial time-series diversity from lake sediments metagenomes (Garner et al. 2020) and the diversity of an Ice Age algal population based on mitochondrial genomes reconstructed from lake sediments (Lammers et al. 2021). Collectively, these approaches demonstrate the power of sedDNA for paleogenomics once the potential of sediment nucleic acid analyses is fully unlocked.

Summary

The number of molecular genetic explorations of cyanobacterial past trajectories in lacustrine sediments continues to increase. The main applications of cyanobacterial sedDNA studies so far include evaluating the impact of human activities and eutrophication on community composition and dominance, determining the presence and abundance of toxin biosynthesis genes over time, and evaluating sedimentation and taphonomy processes to better interpret the cyanobacterial DNA signal in lake sedimentary archives. This new knowledge complements other findings from classical paleo-indicators, while also contributing to improving our understanding of the causes of environmental and ecological change in lakes. Additional studies comparing sedDNA-based proxies with other proxies (fossil organisms, biogeochemical markers) or modern surveys are still much needed to validate sedDNA findings in a variety of lacustrine settings and in response to environmental stressors (e.g., eutrophication, climate change, invasive species, changes in hydrology (water residence time or water levels), salinization, acidification, overfishing). In addition to the question of biotic homogenization of communities and environmental selection, sedDNA approaches can address questions of evolutionary change as a result of human activities, as well as historical biogeography distribution.

Acknowledgements We thank E. Nwosu for providing the cyanobacterial OTU table for Lake Tiefer See. We also wish to thank members of the *Seda*DNA Scientific Society who contributed ancient DNA data from lake sampling sites to the database. We thank Paul MacKeigan for feedback on a previous version of the manuscript. Finally, we extend a special thanks to Drs. Susie Wood, Nico Salmaso, Eric Capo, John Smol and Cécilia Barouillet for their constructive reviews of this chapter. MEM was supported by a postdoctoral fellowship from the Natural Sciences and Engineering Research Council of Canada (NSERC) and FRP through an NSERC Discovery Grant.

References

Armbrecht L, Hallegraeff G, Bolch CJS, Woodward C, Cooper A (2021) Hybridisation capture allows DNA damage analysis of ancient marine eukaryotes. Sci Rep 11:3220. https://doi.org/10.1038/s41598-021-82578-6

Barnes MA, Turner CR, Jerde CL, Renshaw MA, Chadderton WL, Lodge DM (2014) Environmental conditions influence eDNA persistence in aquatic systems. Environ Sci Technol 48(3):1819–1827. https://doi.org/10.1021/es404734p

Barouillet C, Monchamp M-E, Bertilsson S, Brasell K, Domaizon I, Epp LS, Ibrahim A, Mejbel H, Nwosu EC, Pearman JK, Picard M, Thomson-Laing G, Tsugeki N, Von Eggers J, Gregory-Eaves I, Pick FR, Wood SA, Capo E (2023) Investigating the effects of anthropogenic stressors on lake biota using sedimentary DNA. Freshwater biology n/a(n/a). https://doi.org/10.1111/fwb.14027

Barouillet C, Vasselon V, Keck F, Millet L, Etienne D, Galop D, Rius D, Domaizon I (2022) Paleoreconstructions of ciliate communities reveal long-term ecological changes in temperate lakes. Sci Rep 12(1):7899. https://doi.org/10.1038/s41598-022-12041-7

Bonilla S, Pick FR (2017) Freshwater bloom-forming cyanobacteria and anthropogenic change. Limnology and Oceanography e-Lectures 7(2):1–62. https://doi.org/10.1002/loe2.10006

Brasell KA, Pochon X, Howarth J, Pearman JK, Zaiko A, Thompson L, Vandergoes MJ, Simon KS, Wood SA (2022) Shifts in DNA yield and biological community composition in stored sediment: implications for paleogenomic studies. Metabarcoding and Metagenomics 6:e78128. https://doi.org/10.3897/mbmg.6.78128

Brunberg A-K, Blomqvist P (2002) Benthic overwintering of Microcystis colonies under different environmental conditions. J Plankton Res 24(11):1247–1252. https://doi.org/10.1093/plankt/24.11.1247

Callieri C, Coci M, Corno G, Macek M, Modenutti B, Balseiro E, Bertoni R (2013) Phylogenetic diversity of nonmarine picocyanobacteria. FEMS Microbiol Ecol 85(2):293–301. https://doi.org/10.1111/1574-6941.12118

Cao X, Xu X, Bian R, Wang Y, Yu H, Xu Y, Duan G, Bi L, Chen P, Gao S, Wang J, Peng J, Qu J (2020) Sedimentary ancient DNA metabarcoding delineates the contrastingly temporal change of lake cyanobacterial communities. Water Res 183:116077. https://doi.org/10.1016/j.watres.2020.116077

Capo E, Giguet-Covex C, Rouillard A, Nota K, Heintzman PD, Vuillemin A, Ariztegui D, Arnaud F, Belle S, Bertilsson S, Bigler C, Bindler R, Brown AG, Clarke CL, Crump SE, Debroas D, Englund G, Ficetola GF, Garner RE, Gauthier J, Gregory-Eaves I, Heinecke L, Herzschuh U, Ibrahim A, Kisand V, Kjær KH, Lammers Y, Littlefair J, Messager E, Monchamp M-E, Olajos

F, Orsi W, Pedersen MW, Rijal DP, Rydberg J, Spanbauer T, Stoof-Leichsenring KR, Taberlet P, Talas L, Thomas C, Walsh DA, Wang Y, Willerslev E, van Woerkom A, Zimmermann HH, Coolen MJL, Epp LS, Domaizon I, G. Alsos I, Parducci L (2021) Lake sedimentary DNA research on past terrestrial and aquatic biodiversity: overview and recommendations. Quaternary 4(1):6. https://doi.org/10.3390/quat4010006

Capo E, Monchamp M-E, Coolen MJL, Domaizon I, Armbrecht L, Bertilsson S (2022) Environmental paleomicrobiology: using DNA preserved in aquatic sediments to its full potential. Environ Microbiol 24(5):2201–2209. https://doi.org/10.1111/1462-2920.15913

Carey CC, Ibelings BW, Hoffmann EP, Hamilton DP, Brookes JD (2012) Eco-physiological adaptations that favour freshwater cyanobacteria in a changing climate. Water Res 46(5):1394–1407. https://doi.org/10.1016/j.watres.2011.12.016

Carmichael WW (1994) The toxins of cyanobacteria. Sci Am 270(1):78–86. https://doi.org/10.1038/scientificamerican0194-78

Chorus I, Batram J (1999) Toxic cyanobacteria in water - a guide to their public health consequences, monitoring and management

Chorus I, Fastner J, Welker M (2021) Cyanobacteria and cyanotoxins in a changing environment: concepts, controversies. Challenges Water 13(18):2463. https://doi.org/10.3390/w13182463

Chorus I, Welker M (eds) (2021) Toxic cyanobacteria in water: a guide to their public health consequences, monitoring and management, 2nd edn. CRC Press, London

Coolen MJL, Overmann J (1998) Analysis of subfossil molecular remains of purple Sulfur bacteria in a Lake sediment. Appl Environ Microbiol 64(11):4513–4521. https://doi.org/10.1128/AEM.64.11.4513-4521.1998

Crump SE (2021) Sedimentary ancient DNA as a tool in paleoecology. Nat Rev Earth Environ 2(4):229–229. https://doi.org/10.1038/s43017-021-00158-8

Dabney J, Meyer M, Pääbo S (2013) Ancient DNA damage. Cold Spring Harb Perspect Biol 5(7):a012567. https://doi.org/10.1101/cshperspect.a012567

De Senerpont Domis LN, Elser JJ, Gsell AS, Huszar VLM, Ibelings BW, Jeppesen E, Kosten S, Mooij WM, Roland F, Sommer U, Van Donk E, Winder M, Lürling M (2013) Plankton dynamics under different climatic conditions in space and time. Freshw Biol 58(3):463–482. https://doi.org/10.1111/fwb.12053

Di Rienzi SC, Sharon I, Wrighton KC, Koren O, Hug LA, Thomas BC, Goodrich JK, Bell JT, Spector TD, Banfield JF, Ley RE (2013) The human gut and groundwater harbor non-photosynthetic bacteria belonging to a new candidate phylum sibling to cyanobacteria. eLife 2:e01102. https://doi.org/10.7554/eLife.01102

Domaizon I, Savichtcheva O, Debroas D, Arnaud F, Villar C, Pignol C, Alric B, Perga ME (2013) DNA from lake sediments reveals the long-term dynamics and diversity of *Synechococcus* assemblages. Biogeosciences 10(6):3817–3838. https://doi.org/10.5194/bg-10-3817-2013

Dornelas M, Gotelli NJ, McGill B, Shimadzu H, Moyes F, Sievers C, Magurran AE (2014) Assemblage time series reveal biodiversity change but not systematic loss. Science 344(6181):296–299. https://doi.org/10.1126/science.1248484

Erratt K, Creed IF, Favot EJ, Todoran I, Tai V, Smol JP, Trick CG (2022) Paleolimnological evidence reveals climate-related preeminence of cyanobacteria in a temperate meromictic lake. https://doi.org/10.1139/cjfas-2021-0095

Fernandez-Carazo R, Verleyen E, Hodgson DA, Roberts SJ, Waleron K, Vyverman W, Wilmotte A (2013) Late Holocene changes in cyanobacterial community structure in maritime Antarctic lakes. J Paleolimnol 50(1):15–31. https://doi.org/10.1007/s10933-013-9700-3

Gaget V, Hobson P, Keulen A, Newton K, Monis P, Humpage AR, Weyrich LS, Brookes JD (2020) Toolbox for the sampling and monitoring of benthic cyanobacteria. Water Res 169:115222. https://doi.org/10.1016/j.watres.2019.115222

Garner RE, Gregory-Eaves I, Walsh DA (2020) Sediment metagenomes as time capsules of Lake microbiomes. mSphere 5(6):e00512–e00520. https://doi.org/10.1128/mSphere.00512-20

Gauthier J, Walsh D, Selbie DT, Bourgeois A, Griffiths K, Domaizon I, Gregory-Eaves I (2021) Evaluating the congruence between DNA-based and morphological taxonomic approaches in water and sediment trap samples: analyses of a 36-month time series from a temperate monomictic lake. Limnol Oceanogr 66(8):3020–3039. https://doi.org/10.1002/lno.11856
Guiry MD (2012) How many species of algae are there? J Phycol 48(5):1057–1063. https://doi.org/10.1111/j.1529-8817.2012.01222.x
Hamilton PB, Ley LM, Dean S, Pick FR (2005) The occurrence of the cyanobacterium Cylindrospermopsis raciborskii in Constance Lake: an exotic cyanoprokaryote new to Canada. Phycologia 44(1):17–25. https://doi.org/10.2216/0031-8884(2005)44[17:TOOTCC]2.0.CO;2
Hauer T, Komárek J (2022) CyanoDB 2.0 - on-line database of cyanobacterial genera. World-Wide Electron Publ
Henao E, Rzymski P, Waters MN (2019) A review on the study of cyanotoxins in Paleolimnological research: current knowledge and future needs. Toxins (Basel) 12(1):6. https://doi.org/10.3390/toxins12010006
Ho JC, Michalak AM, Pahlevan N (2019) Widespread global increase in intense lake phytoplankton blooms since the 1980s. Nature 574(7780):667–670. https://doi.org/10.1038/s41586-019-1648-7
Hobbs WO, Dreher TW, Davis EW, Vinebrooke RD, Wong S, Weissman T, Dawson M (2021) Using a lake sediment record to infer the long-term history of cyanobacteria and the recent rise of an anatoxin producing Dolichospermum sp. Harmful Algae 101:101971. https://doi.org/10.1016/j.hal.2020.101971
Huisman J, Codd GA, Paerl HW, Ibelings BW, Verspagen JMH, Visser PM (2018) Cyanobacterial blooms. Nat Rev Microbiol 16(8):471–483. https://doi.org/10.1038/s41579-018-0040-1
Huo S, Zhang H, Wang J, Chen J, Wu F (2022a) Temperature and precipitation dominates millennium changes of eukaryotic algal communities in Lake Yamzhog Yumco, southern Tibetan plateau. Sci Total Environ 829:154636. https://doi.org/10.1016/j.scitotenv.2022.154636
Huo S, Zhang H, Monchamp M-E, Wang R, Weng N, Zhang J, Zhang H, Wu F (2022b) Century-long homogenization of algal communities is accelerated by nutrient enrichment and climate warming in lakes and reservoirs of the north temperate zone. Environ Sci Technol 56(6):3780–3790. https://doi.org/10.1021/acs.est.1c06958
Ibrahim A, Capo E, Wessels M, Martin I, Meyer A, Schleheck D, Epp LS (2021) Anthropogenic impact on the historical phytoplankton community of Lake Constance reconstructed by multimarker analysis of sediment-core environmental DNA. Mol Ecol 30(13):3040–3056. https://doi.org/10.1111/mec.15696
Kakouei K, Kraemer BM, Anneville O, Carvalho L, Feuchtmayr H, Graham JL, Higgins S, Pomati F, Rudstam LG, Stockwell JD, Thackeray SJ, Vanni MJ, Adrian R (2021) Phytoplankton and cyanobacteria abundances in mid-21st century lakes depend strongly on future land use and climate projections. Glob Chang Biol 27(24):6409–6422. https://doi.org/10.1111/gcb.15866
Keck F, Millet L, Debroas D, Etienne D, Galop D, Rius D, Domaizon I (2020) Assessing the response of micro-eukaryotic diversity to the great acceleration using lake sedimentary DNA. Nat Commun 11(1):3831. https://doi.org/10.1038/s41467-020-17682-8
Komárek J (2016a) A polyphasic approach for the taxonomy of cyanobacteria: principles and applications. Eur J Phycol 51(3):346–353. https://doi.org/10.1080/09670262.2016.1163738
Komárek J (2016b) Review of the cyanobacterial genera implying planktic species after recent taxonomic revisions according to polyphasic methods: state as of 2014. Hydrobiologia 764(1):259–270. https://doi.org/10.1007/s10750-015-2242-0
Kultschar B, Llewellyn C, Kultschar B, Llewellyn C (2018) Secondary metabolites in cyanobacteria. Secondary Metabolites - Sources and Applications. IntechOpen
Lammers Y, Heintzman PD, Alsos IG (2021) Environmental palaeogenomic reconstruction of an ice age algal population. Commun Biol 4(1):1–11. https://doi.org/10.1038/s42003-021-01710-4
Leavitt P, Hodgson D (2006) Sedimentary pigments. Dev. Paleoenviron. Res, In, pp 295–325
Legrand B, Lamarque A, Sabart M, Latour D (2017) Benthic archives reveal recurrence and dominance of toxigenic cyanobacteria in a eutrophic Lake over the last 220 years. Toxins (Basel) 9(9):271. https://doi.org/10.3390/toxins9090271

Legrand B, Miras Y, Beauger A, Dussauze M, Latour D (2019) Akinetes and ancient DNA reveal toxic cyanobacterial recurrences and their potential for resurrection in a 6700-year-old core from a eutrophic lake. Sci Total Environ 687:1369–1380. https://doi.org/10.1016/j.scitotenv.2019.07.100

Ley RE, Bäckhed F, Turnbaugh P, Lozupone CA, Knight RD, Gordon JI (2005) Obesity alters gut microbial ecology. Proc Natl Acad Sci 102(31):11070–11075. https://doi.org/10.1073/pnas.0504978102

Lindqvist C, Rajora OP (2019) Paleogenomics. Springer, Cham, Switzerland

Livingstone D, Jaworski GHM (1980) The viability of akinetes of blue-green algae recovered from the sediments of Rostherne mere. Br Phycol J 15(4):357–364. https://doi.org/10.1080/00071618000650361

Magurran AE, Dornelas M, Moyes F, Gotelli NJ, McGill B (2015) Rapid biotic homogenization of marine fish assemblages. Nat Commun 6(1):8405. https://doi.org/10.1038/ncomms9405

Martínez de la Escalera G, Antoniades D, Bonilla S, Piccini C (2014) Application of ancient DNA to the reconstruction of past microbial assemblages and for the detection of toxic cyanobacteria in subtropical freshwater ecosystems. Mol Ecol 23(23):5791–5802. https://doi.org/10.1111/mec.12979

Mejbel HS, Dodsworth W, Baud A, Gregory-Eaves I, Pick FR (2021) Comparing quantitative methods for Analyzing sediment DNA Records of Cyanobacteria in experimental and Reference Lakes. Front Microbiol 12

Mejbel HS, Dodsworth W, Pick FR (2022) Effects of temperature and oxygen on cyanobacterial DNA preservation in sediments: a comparison study of major taxa. Environ DNA 4(4):717–731. https://doi.org/10.1002/edn3.289

Monchamp M-È, Bruel R, Frossard V, McGowan S, Lavrieux M, Muschick M, Perga M-É, Dubois N (2021) Paleoecological evidence for a multi-trophic regime shift in a perialpine Lake (lake Joux, Switzerland). Anthropocene 35(100301). https://doi.org/10.1016/j.ancene.2021.100301

Monchamp M-E, Spaak P, Domaizon I, Dubois N, Bouffard D, Pomati F (2018) Homogenization of lake cyanobacterial communities over a century of climate change and eutrophication. Nat Ecol Evol 2(2):317–324. https://doi.org/10.1038/s41559-017-0407-0

Monchamp M-E, Spaak P, Pomati F (2019) Long term diversity and distribution of non-photosynthetic cyanobacteria in Peri-Alpine Lakes. Front Microbiol 9

Monchamp M-E, Walser J-C, Pomati F, Spaak P (2016) Sedimentary DNA reveals cyanobacterial community diversity over 200 years in two Perialpine Lakes. Appl Environ Microbiol 82(21):6472–6482. https://doi.org/10.1128/AEM.02174-16

Moss B, Kosten S, Meerhoff M, Battarbee RW, Jeppesen E, Mazzeo N, Havens K, Lacerot G, Liu Z, De Meester L, Paerl H, Scheffer M (2011) Allied attack: climate change and eutrophication. Inland Waters 1(2):101–105. https://doi.org/10.5268/IW-1.2.359

Nabout JC, da Silva Rocha B, Carneiro FM, Sant'Anna CL (2013) How many species of cyanobacteria are there? Using a discovery curve to predict the species number. Biodivers Conserv 22(12):2907–2918. https://doi.org/10.1007/s10531-013-0561-x

Neilan BA, Saker ML, Fastner J, Törökné A, Burns BP (2003) Phylogeography of the invasive cyanobacterium Cylindrospermopsis raciborskii. Mol Ecol 12(1):133–140. https://doi.org/10.1046/j.1365-294x.2003.01709.x

Nwosu EC, Brauer A, Monchamp M-E, Pinkerneil S, Bartholomäus A, Theuerkauf M, Schmidt J-P, Stoof-Leichsenring KR, Wietelmann T, Kaiser J, Wagner D, Liebner S (2023) Early human impact on lake cyanobacteria revealed by a Holocene record of sedimentary ancient DNA. Commun Biol 6(1):1–12. https://doi.org/10.1038/s42003-023-04430-z

Nwosu EC, Roeser P, Yang S, Ganzert L, Dellwig O, Pinkerneil S, Brauer A, Dittmann E, Wagner D, Liebner S (2021a) From water into sediment—tracing freshwater cyanobacteria via DNA analyses. Microorganisms 9(8):1778. https://doi.org/10.3390/microorganisms9081778

Nwosu EC, Brauer A, Kaiser J, Horn F, Wagner D, Liebner S (2021b) Evaluating sedimentary DNA for tracing changes in cyanobacteria dynamics from sediments spanning the last 350 years of Lake Tiefer see. NE Germany J Paleolimnol 66(3):279–296. https://doi.org/10.1007/s10933-021-00206-9

Pal S, Gregory-Eaves I, Pick FR (2015) Temporal trends in cyanobacteria revealed through DNA and pigment analyses of temperate lake sediment cores. J Paleolimnol 54(1):87–101. https://doi.org/10.1007/s10933-015-9839-1

Pawlowski J, Bruce K, Panksep K, Aguirre FI, Amalfitano S, Apothéloz-Perret-Gentil L, Baussant T, Bouchez A, Carugati L, Cermakova K, Cordier T, Corinaldesi C, Costa FO, Danovaro R, Dell'Anno A, Duarte S, Eisendle U, Ferrari BJD, Frontalini F, Frühe L, Haegerbaeumer A, Kisand V, Krolicka A, Lanzén A, Leese F, Lejzerowicz F, Lyautey E, Maček I, Sagova-Marečková M, Pearman JK, Pochon X, Stoeck T, Vivien R, Weigand A, Fazi S (2022) Environmental DNA metabarcoding for benthic monitoring: a review of sediment sampling and DNA extraction methods. Sci Total Environ 818:151783. https://doi.org/10.1016/j.scitotenv.2021.151783

Pearman JK, Biessy L, Howarth JD, Vandergoes MJ, Rees A, Wood SA (2022) Deciphering the molecular signal from past and alive bacterial communities in aquatic sedimentary archives. Mol Ecol Resour 22(3):877–890. https://doi.org/10.1111/1755-0998.13515

Pedersen MW, Overballe-Petersen S, Ermini L, Sarkissian CD, Haile J, Hellstrom M, Spens J, Thomsen PF, Bohmann K, Cappellini E, Schnell IB, Wales NA, Carøe C, Campos PF, Schmidt AMZ, Gilbert MTP, Hansen AJ, Orlando L, Willerslev E (2015) Ancient and modern environmental DNA. Philosop Transact Royal Society B: Biolog Sci 370(1660):20130383. https://doi.org/10.1098/rstb.2013.0383

Petsch DK (2016) Causes and consequences of biotic homogenization in freshwater ecosystems. Int Rev Hydrobiol 101(3–4):113–122. https://doi.org/10.1002/iroh.201601850

Picard M, Pochon X, Atalah J, Pearman JK, Rees A, Howarth JD, Moy CM, Vandergoes MJ, Hawes I, Khan S, Wood SA (2022a) Using metabarcoding and droplet digital PCR to investigate drivers of historical shifts in cyanobacteria from six contrasting lakes. Sci Rep 12(1):12810. https://doi.org/10.1038/s41598-022-14216-8

Picard M, Wood SA, Pochon X, Vandergoes MJ, Reyes L, Howarth JD, Hawes I, Puddick J (2022b) Molecular and pigment analyses provide comparative results when reconstructing historic cyanobacterial abundances from Lake sediment cores. Microorganisms 10(2):279. https://doi.org/10.3390/microorganisms10020279

Pilon S, Zastepa A, Taranu ZE, Gregory-Eaves I, Racine M, Blais JM, Poulain AJ, Pick FR (2019) Contrasting histories of microcystin-producing cyanobacteria in two temperate lakes as inferred from quantitative sediment DNA analyses. Lake Reservoir Manag 35(1):102–117. https://doi.org/10.1080/10402381.2018.1549625

Rahel FJ (2002) Homogenization of freshwater faunas. Annu Rev Ecol Syst 33(1):291–315. https://doi.org/10.1146/annurev.ecolsys.33.010802.150429

Ramos V, Morais J, Vasconcelos VM (2017) A curated database of cyanobacterial strains relevant for modern taxonomy and phylogenetic studies. Sci Data 4(1):170054. https://doi.org/10.1038/sdata.2017.54

Rigosi A, Hanson P, Hamilton DP, Hipsey M, Rusak JA, Bois J, Sparber K, Chorus I, Watkinson AJ, Qin B, Kim B, Brookes JD (2015) Determining the probability of cyanobacterial blooms: the application of Bayesian networks in multiple lake systems. Ecol Appl 25(1):186–199

Rinta-Kanto JM, Saxton MA, DeBruyn JM, Smith JL, Marvin CH, Krieger KA, Sayler GS, Boyer GL, Wilhelm SW (2009) The diversity and distribution of toxigenic Microcystis spp. in present day and archived pelagic and sediment samples from Lake Erie. Harmful Algae 8(3):385–394. https://doi.org/10.1016/j.hal.2008.08.026

Salmaso N, Albanese D, Capelli C, Boscaini A, Pindo M, Donati C (2018) Diversity and cyclical seasonal transitions in the bacterial Community in a Large and Deep Perialpine Lake. Microb Ecol 76(1):125–143. https://doi.org/10.1007/s00248-017-1120-x

Salmaso N, Capelli C, Shams S, Cerasino L (2015) Expansion of bloom-forming Dolichospermum lemmermannii (Nostocales, cyanobacteria) to the deep lakes south of the Alps: colonization patterns, driving forces and implications for water use. Harmful Algae 50:76–87. https://doi.org/10.1016/j.hal.2015.09.008

Salmaso N, Vasselon V, Rimet F, Vautier M, Elersek T, Boscaini A, Donati C, Moretto M, Pindo M, Riccioni G, Stefani E, Capelli C, Lepori F, Kurmayer R, Mischke U, Klemenčič AK, Novak K, Greco C, Franzini G, Fusato G, Giacomazzi F, Lea A, Menegon S, Zampieri C, Macor A, Virgilio D, Zanut E, Zorza R, Buzzi F, Domaizon I (2022) DNA sequence and taxonomic gap analyses to quantify the coverage of aquatic cyanobacteria and eukaryotic microalgae in reference databases: results of a survey in the alpine region. Sci Total Environ 834:155175. https://doi.org/10.1016/j.scitotenv.2022.155175

Savichtcheva O, Debroas D, Kurmayer R, Villar C, Jenny JP, Arnaud F, Perga ME, Domaizon I (2011) Quantitative PCR enumeration of Total/toxic Planktothrix rubescens and Total cyanobacteria in preserved DNA isolated from Lake sediments. Appl Environ Microbiol 77(24):8744–8753. https://doi.org/10.1128/AEM.06106-11

Savichtcheva O, Debroas D, Perga ME, Arnaud F, Villar C, Lyautey E, Kirkham A, Chardon C, Alric B, Domaizon I (2015) Effects of nutrients and warming on Planktothrix dynamics and diversity: a palaeolimnological view based on sedimentary DNA and RNA. Freshw Biol 60(1):31–49. https://doi.org/10.1111/fwb.12465

Smol JP (2009) Pollution of lakes and Rivers: a paleoenvironmental perspective, 2nd edn. Wiley-Blackwell, Malden, MA

Soo RM, Hemp J, Parks DH, Fischer WW, Hugenholtz P (2017) On the origins of oxygenic photosynthesis and aerobic respiration in cyanobacteria. Science 355(6332):1436–1440. https://doi.org/10.1126/science.aal3794

Soo RM, Skennerton CT, Sekiguchi Y, Imelfort M, Paech SJ, Dennis PG, Steen JA, Parks DH, Tyson GW, Hugenholtz P (2014) An expanded genomic representation of the phylum cyanobacteria. Genome Biol Evol 6(5):1031–1045. https://doi.org/10.1093/gbe/evu073

Soo RM, Woodcroft BJ, Parks DH, Tyson GW, Hugenholtz P (2015) Back from the dead; the curious tale of the predatory cyanobacterium Vampirovibrio chlorellavorus. Peer J 3:e968. https://doi.org/10.7717/peerj.968

Taberlet P, Bonin A, Zinger L, Coissac E, Taberlet P, Bonin A, Zinger L, Coissac E (2018) Environmental DNA: for biodiversity research and monitoring. Oxford University Press, Oxford, New York

Taranu ZE, Gregory-Eaves I, Leavitt PR, Bunting L, Buchaca T, Catalan J, Domaizon I, Guilizzoni P, Lami A, McGowan S, Moorhouse H, Morabito G, Pick FR, Stevenson MA, Thompson PL, Vinebrooke RD (2015) Acceleration of cyanobacterial dominance in north temperate-subarctic lakes during the Anthropocene. Ecol Lett 18(4):375–384. https://doi.org/10.1111/ele.12420

Thorpe AC, Anderson A, Goodall T, Thackeray SJ, Maberly SC, Bendle JA, Gweon HS, Read DS (2022) Sedimentary DNA records long-term changes in a lake bacterial community in response to varying nutrient availability. Environmental DNA 4(6):1340–1355. https://doi.org/10.1002/edn3.344

Tse TJ, Doig LE, Tang S, Zhang X, Sun W, Wiseman SB, Feng CX, Liu H, Giesy JP, Hecker M, Jones PD (2018) Combining high-throughput sequencing of sedaDNA and traditional Paleolimnological techniques to infer historical trends in cyanobacterial communities. Environ Sci Technol 52(12):6842–6853. https://doi.org/10.1021/acs.est.7b06386

Von Eggers J, Monchamp M-E, Capo E, Giguet-Covex C, Spanbauer T, Heintzman PD (2022) Inventory of ancient environmental DNA from sedimentary archives: locations, methods, and target taxa. Zenodo. https://doi.org/10.5281/ZENODO.6847522

Waters MN (2016) A 4700-year history of cyanobacteria toxin production in a shallow subtropical Lake. Ecosystems 19(3):426–436. https://doi.org/10.1007/s10021-015-9943-0

Watson SB, McCauley E, Downing JA (1997) Patterns in phytoplankton taxonomic composition across temperate lakes of differing nutrient status. Limnol Oceanogr 42(3):487–495. https://doi.org/10.4319/lo.1997.42.3.0487

Weisbrod B, Wood SA, Steiner K, Whyte-Wilding R, Puddick J, Laroche O, Dietrich DR (2020) Is a central sediment sample sufficient? Exploring spatial and temporal microbial diversity in a small Lake. Toxins 12(9):580. https://doi.org/10.3390/toxins12090580

Whitton BA (ed) (2012) Ecology of cyanobacteria II. Springer, Netherlands, Dordrecht

Whitton BA, Potts M (2014) Summary for policymakers. In: Whitton BA, Potts M (eds) Climate change 2013 – the physical science basis. Cambridge University Press, New York, NY, USA, pp 1–30

Yan D, Xu H, Yang M, Lan J, Hou W, Wang F, Zhang J, Zhou K, An Z, Goldsmith Y (2019) Responses of cyanobacteria to climate and human activities at Lake Chenghai over the past 100 years. Ecol Indic 104:755–763. https://doi.org/10.1016/j.ecolind.2019.03.019

Zastepa A, Taranu ZE, Kimpe LE, Blais JM, Gregory-Eaves I, Zurawell RW, Pick FR (2017) Reconstructing a long-term record of microcystins from the analysis of lake sediments. Sci Total Environ 579:893–901. https://doi.org/10.1016/j.scitotenv.2016.10.211

Zhang H, Huo S, Yeager KM, Wu F (2021) Sedimentary DNA record of eukaryotic algal and cyanobacterial communities in a shallow Lake driven by human activities and climate change. Sci Total Environ 753:141985. https://doi.org/10.1016/j.scitotenv.2020.141985

Chapter 6
Protist DNA from Lake Sediments

Cécilia Barouillet, Isabelle Domaizon, and Eric Capo

Keywords Protists · Unicellular eukaryotes · Lakes · Sedimentary DNA · Environmental DNA · Paleolimnology · Paleo-indicators

Introduction

Protists are unicellular eukaryotes found in almost all biomes on Earth and represent more than 85% of the diversity of the known eukaryotes (Pawlowski et al. 2012; Adl et al. 2019; Burki et al. 2020). They are sometimes referred to as microbial eukaryotes; however, there is no real consensus about the definition of microbial eukaryotes as some authors consider some metazoans to be microbial eukaryotes (e.g., rotifers; (Bik et al. 2012; Finlay and Fenchel 2004), while others exclude all metazoans and embryophytes but include some multicellular fungi (Kammerlander et al. 2015; Triadó-Margarit and Casamayor 2015). Although the first classifications within the eukaryotic domain originally rely on morphological characteristics, recent phylogenetic analysis revealed new protist lineages (Burki 2014; Burki et al. 2020).

Protists represent a diverse and abundant fraction of planktonic and benthic assemblages of aquatic environments (Debroas et al. 2017; Grossmann et al. 2016). In both freshwater and marine ecosystems, they are found in all components of the food web where they play key functional and ecological roles as phototrophs, predators, phagotrophs, osmotrophs, saprotrophs, virovores, mixotrophs, and parasites (Burki et al. 2021; DeLong et al. 2023; Singer et al. 2021). Protists participate in the

C. Barouillet (✉) · I. Domaizon
INRAE, Université Savoie Mont Blanc, CARRTEL, Thonon le bains, France

Pôle R&D ECLA, INRAE, CARRTEL, Thonon-les-bains, France
e-mail: isabelle.domaizon@inrae.fr

E. Capo
Department of Ecology and Environmental Science, Umeå University, Umeå, Sweden
e-mail: eric.capo@umu.se

E. Capo et al. (eds.), *Tracking Environmental Change Using Lake Sediments*, Developments in Paleoenvironmental Research 21,
https://doi.org/10.1007/978-3-031-43799-1_6

biogeochemical cycles through the transformation and recycling of matter and energy, and are involved in complex biological interactions as parasites and endosymbionts (Bjorbækmo et al. 2020). Despite their critical importance in aquatic ecosystems, there is still a lack of information about how and if anthropogenic disturbances impact the habitats of protists and thus their diversity and abundance. Furthermore to what extent protist communities are resilient to natural vs anthropogenic perturbations remains largely unknown.

Prior to advances in the genetic analyses of protists communities, surveys based on the identification of morphological features were limited to only a few groups that included diatoms, chrysophytes, chlorophytes, cryptophytes, ciliates, and heterotrophic flagellates. While these traditional limnological surveys and paleolimnological investigations provided a wealth of information on the ecology and biogeography of some protist communities in freshwaters, a large part of the diversity is still uncovered as a majority of protists do not leave post-burial diagnostic morphological features in sediments (e.g., diatoms frustules, testate amoebae and chrysophytes scales) (Smol et al. 2001). Improvements in molecular ecology techniques now have expanded our knowledge of the diversity of protists in the environment, and improved our understanding of their ecological and functional roles in ecosystems (Burki et al. 2021; Singer et al. 2021).

In this chapter, we focus on the application of molecular biological techniques to sedimentary DNA (sedDNA) to investigate a broad temporal perspective on the biodiversity, community assemblages and populations dynamic of lake protists (Domaizon et al. 2017; Barouillet et al. 2023). We summarize the state of knowledge about: (i) the taxonomic and functional diversity of protists; (ii) their roles in lake food webs; (iii) the methodology applied with sedDNA approaches; (iv) taphonomic processes influencing their past molecular signal in sediments; (v) potential bias due to the presence of living protists in sediments; and (vi) their responses to long-term environmental changes and potential use as bioindicators. While protists are arguably the most diverse group of eukaryotes, for the sake of simplicity and because many of the estimated 20,000–600,000 protist species have yet to be described and could not be summarized in one single chapter, we focus on the protist groups usually predominant in lake systems (Table 6.1).

Protist Ecology

Taxonomic and Functional Diversity of Protists

The concept of morphospecies with respect to this morphologically diverse group of organisms was particularly unstable in the time of microscopic analysis, thereby limiting good estimates of diversity. Estimating the global diversity of protists is challenging because of the uncertainties related to intraspecies diversity (or microdiversity) and discrepancies between studies based on morphological and genetic

Table 6.1 Majors protists groups in aquatic habitats and their associated trophic ecology

Protist group	Nutritional strategy	Trophic role
Apicomplexa	Heterotrophs	Parasites Predators
Diatoms	Phototrophs	
Dinoflagellates	Phototrophs	
	Heterotrophs	Predators
	Mixotrophs	
Cercozoans	Heterotrophs	Predators
Chlorophytes	Phototrophs	
Chrysophytes	Phototrophs[a]	
	Heterotrophs[b]	
	Mixotrophs	
Ciliates	Heterotrophs	Predators Saprotrophs[c] Parasites Some endocommensal forms Hosts for endosymbionts
	Mixotrophs	
Cryptophytes	Phototrophs	
	Mixotrophs	
Oomycetes	Heterotrophs	Parasites
Perkinseans	Heterotrophs	Parasites
Pirsonia	Heterotrophs	Parasites

[a]species from the synuralean lineage are the only (to our knowledge) strict phototrophic chrysophytes
[b]Heterotrophic chrysophytes are color-less species that lost the ability to photosynthesize. These various heterotrophic species evolved many times independently in different lineages. Example of non-photosynthetic species includes the *Spumella*-like genera
[c]some rare freshwater taxa but most are usually associated with terrestrial environment

analyses (Caron and Hu 2019). In the early days of the molecular analysis of protists, geneticists demonstrated that protist diversity was largely underestimated, by one to two orders of magnitude (Nanney et al. 1998). Protists are paraphyletic (Burki et al. 2020; Singer et al. 2021) (Fig. 6.1, Table 6.1), hence, taxonomically, but also functionally diverse organisms. For instance, some protists can be purely autotrophic or heterotrophic, while others lay in-between these two nutritional strategies as mixotrophs, depending on environmental conditions.

The following description of the main supergroups of eukaryotes is based on the most up-to-date revised phylogenetic tree of eukaryotes (Fig. 6.1; (Burki et al. 2020)). TSAR (Telonemids, Stramenopiles, Alveolates, and Rhizaria), one of the major supergroups of protists, has been estimated to comprise up to half of the diversity of eukaryotic species (Grattepanche et al. 2018). This supergroup includes several lineages of unicellular algae (i.e., diatoms, dinoflagellates, chrysophytes) and alveolates (i.e., ciliates, dinoflagellates, apicomplexans and perkinseans). As part of the alveolates, the ciliates are abundant and diverse, and are primarily

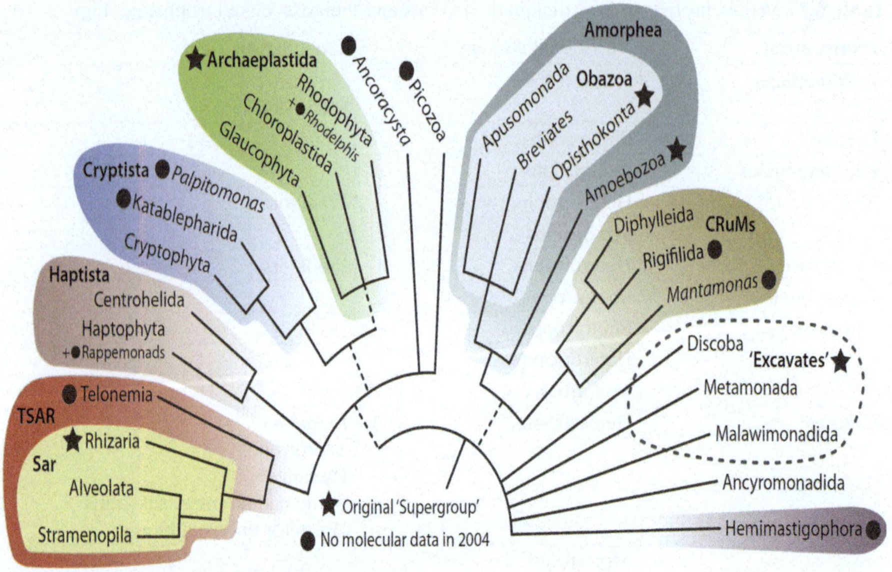

Fig. 6.1 Phylogenetic tree of eukaryotes from Burki et al. (2020) was constructed from a consensus of recent phylogenomic studies. The original 'supergroups' (i.e. group of organisms that share one common ancestor) are identified with the start symbol and correspond to the taxa that were previously considered as supergroups. The current supergroups are indicated by the colored grouping. Dashed lines indicate uncertainties about the monophyly of some groups and unresolved branching orders among lineages are displayed as multifurcations. Rappemonads (in parentheses) are placed on the basis of plastid rRNA data only. (Figure courtesy of Dr. Fabien Burki)

heterotrophs with the exceptions of some mixotrophic species. Heterotrophic ciliates include herbivores, omnivores and many bacterial predators. Dinoflagellates are either heterotrophs, phototrophs or mixotrophs common in marine and freshwater plankton, forming occasional blooms. Apicomplexans and perkinseans are relatively understudied lineages of the TSAR group, and include numerous parasites of plankton and other aquatic organisms. Among the rhizarians, cercozoans are mostly freshwater heterotrophic organisms while foraminifera are amoeboid organisms that are abundant in marine benthos.

Stramenopiles include a wide diversity of protists featuring metabolisms ranging from strict heterotrophs to phototrophs. Among them, oomycetes and bicosoecids are parasites, while diatoms, chrysophytes, synurophytes and xanthophytes are widely distributed phototrophs in marine and freshwater environments. More specifically, diatoms are diverse unicellular algae dominant in most aquatic habitats, and are important primary producers. Although a large number of Chrysophyceae are mixotrophs (i.e. chrysophytes), combining both phagotrophic and phototrophic modes of nutrition, some are also known to be strict phototrophs, such as within the genus of Synura (Beisser et al. 2017).

Archaeplastida consists of organisms from Chloroplastida, Rhodophyta and Glaucophyta. Chloroplastida include chlorophytes (all unicellular, mostly free living) and streptophytes (including unicellular organisms and land plants). In addition, glaucophytes are unicellular and mainly present in freshwater in low abundance while rhodophytes are single or multicellular algae essentially found in marine environments.

Amorphea comprises opisthokonts (animals, fungi and several protist lineages), amoebozoans (amoeboid protists) and two small lineages of heterotrophic flagellates. Haptista includes the haptophytes and centrohelids. Although the majority of haptophytes are photosynthetic, a few are mixotrophs. Haptophytes can form blooms in both marine and freshwater ecosystems. Cryptista contain the cryptomonads and two enigmatic groups of heterotrophic flagellates i.e., katablepharids and *Palpitomonas*. Other main eukaryotic lineages are CruMs (collodictyonids, *Rigifilida*, *Mantamonas*), Discoba (Euglenozoa, Heterolobosea, Jokobida, Tsukubamonas), Metamonada (various anaerobic protists) and Hemimastigophora.

Protists Diversity Across Biomes

As highlighted by recent meta-analyses of molecular data, the diversity of protists differs between environments from soils, biofilms, brackish sediment and freshwaters (Burki et al. 2021; Debroas et al. 2017; Grossmann et al. 2016; Singer et al. 2021; Xiong et al. 2021). A key driver of protist taxonomic diversity is the salinity gradient as soil and freshwater communities are more similar to each other than to marine communities (Singer et al. 2021). In terms of composition, amoebozoans, cercozoans and apicomplexans are more abundant in soil samples while dinoflagellates, diatoms and radiolarians tend to dominate protist communities in marine ecosystems. In comparison, haptophytes, synurophytes, and ciliates are predominant in lakes. Further, a meta-analysis of the surface waters from 366 Canadian freshwater lakes across a north temperate to subarctic range revealed lower protist diversity in hypereutrophic lakes compared to nutrient-poor systems (Garner et al. 2022). The first studies of lake protist diversity with 18S rRNA gene targeted sequencing revealed the seasonal dynamics of these communities (Mangot et al. 2013; Nolte et al. 2010) and new insights into how biogeographic distance and environmental selection shapes their communities (Filker et al. 2016; Kammerlander et al. 2015; Triadó-Margarit and Casamayor 2012).

Overall, lake protist diversity appears to be distinct from other environments, for instance eukaryotic phototrophs dominate freshwater ecosystems while consumers dominate soil and marine ecosystems (Singer et al. 2021). Local environmental conditions can shape protist diversity, thereby leading to variability in the composition of protists between lakes. Greater variability in the taxonomic diversity of heterotrophs and mixotrophs are observed among eutrophic lakes while phototroph assemblages are more similar under high-nutrient conditions (Garner et al. 2022), such discrepancy demonstrates that the impacts of environmental selection on protist diversity depend on the functional group.

Role of Protists in Aquatic Ecosystems

Protists are essential to aquatic ecosystems, ensuring several functions that include the transfer of energy between multiple trophic levels (e.g., phototrophs, mixotrophs, predators), the regulation of populations (e.g., parasites) and sustaining populations as hosts or symbionts.

Phototrophic Protists

Phototrophic protists, also called eukaryotic microalgae or phytoplankton, are able to use light to produce oxygen and organic matter that is subsequently utilized by heterotrophic organisms as a source of energy. Eukaryotic phototrophs are distributed across several lineages. Some taxonomic groups are strictly phototrophs such as the chlorophytes and diatoms; others are mixotrophs such as the dinoflagellates; and some comprise both mixotrophic and strictly phototrophic species such as the chrysophytes (Table 6.1). Due to their high taxonomic and genetic diversity, autotrophic protists have different requirements for light intensity and nutrient levels to thrive (Wetzel 2001). Most chlorophytes are competitive under high light conditions. In contrast, diatoms can thrive under low light and high nutrient conditions (Marzetz et al. 2020), but do not use light as efficiently as cyanobacteria (see Chap. 5 of this volume). Eukaryotic microalgae often compete for light and nutrients with cyanobacteria, and the interplay between these two phytoplanktonic groups is strongly regulated by thermal stratification, mixing regime and the biogeochemical balance of lakes (Posch et al. 2012).

Heterotrophic Protists

Heterotrophic protists, which include ciliates, amoebozoans, perkinseans, cercozoans and oomycetes, represent a diverse and abundant fraction of planktonic and benthic assemblages in both freshwater and marine environments (Debroas et al. 2017; Grossmann et al. 2016; Likens 2010). Heterotrophic protists comprise a wide range of trophic strategies that include predators, phagotrophs, and parasites (Lefranc et al. 2005; Mitra et al. 2016; Oikonomou et al. 2015; Singer et al. 2021).

Phagotrophic protists can act as the main predators of the microbial loop of freshwater food webs. They participate in the remobilization and recycling of carbon and nutrients (Pernthaler and Posch 2009). In freshwaters, phagotrophic protists can become the primary consumers of bacteria (Sanders 2009; Sherr and Sherr 2002) and shape the structure of microbial communities (Boenigk and Arndt 2002; Lischke et al. 2016). At times, their biomass can even exceed that of phototrophic eukaryotes (Pick and Caron 1987). As such they can exert a strong grazing pressure with potential cascading effects on both the planktonic and benthic food webs (Caron and Hutchins 2013).

Parasitic forms of protists are found in several lineages that include ciliates, perkinseans, and pirsonia. Parasites can regulate populations and influence the community structure and composition of their hosts (Bråte et al. 2010; Lepère et al. 2008). The role of parasitic protists in food webs and their response to environmental changes needs further investigations. Nonetheless, several species of ciliates, such as *Ichthyophthirius multifiliis*, *Cryptocaryon irritans*, *Balantidium ctenopharyngodoni*, parasites of fish (Li et al. 2023) or *Tetrahymena glochidiophila*, a parasite of mollusks (Lynn et al. 2018; Prosser et al. 2018) have been well studied. These parasites can be strong regulators of their host populations' dynamics and contribute to the organic matter cycling through an increased mortality of the host species.

Among the heterotrophic protists, some can display mixotrophic life strategies by either sequestering the chloroplasts from ingested algae, as known as kleptoplasty, or by harboring a photosynthetic endosymbiont capable of replication within the host, and transmission to succeeding protist generations (Jones 2000). The role, ecology, and dynamic of mixotrophic protists is largely hypothetical due to the lack of in situ experiments, as most of our knowledge of mixotrophic life strategies is currently based on cultured organisms in controlled laboratory settings (Millette et al. 2023). Nonetheless, it has been argued that a significant proportion of the protists in the photic zone are mixotrophic and that they can represent an important link in the flux of nutrients and energy through planktonic food webs (Jones 2000; Mitra et al. 2014). In marine environments, mixotrophic protists tend to dominate stable, more-mature ecosystems (Mitra et al. 2014). In freshwater environments, these trends remain to be investigated.

SedDNA Analysis of Protist Communities

Despite their remarkable diversity, protists are often underestimated and, as part of the microorganisms, are often referred to as the "Hidden diversity" – along with many bacterial and archaeal lineages. The application of molecular biological techniques to sedDNA to recover the genetic signal from protist communities has opened new avenues in the field of microbial ecology, providing a unique temporal perspective on the biodiversity, community assemblages and populations dynamic of lake protists (Barouillet et al. 2023; Domaizon et al. 2017). This temporal perspective offers the potential to answer some timely research questions (Weisse and Montagnes 2022). For example, it allows the evaluation of communities and population acclimation to long-term environmental changes. The generated genetic dataset also allows to investigate the biogeography and endemicity of protists communities across biomes including in remote understudied places, where routine water monitoring cannot be conducted.

In the following section, we provide a general overview of the methodology applied with sedDNA approaches with specificities related to the study of protists communities, the taphonomic processes influencing their past molecular signal in sediments, the potential bias due to the presence of living protists in sediments, and their responses to long-term environmental changes and potential use as bioindicators.

Workflow of the sedDNA Approach to Study Past Lake Protists

The workflow of the sedDNA approach to investigate the past dynamics of protists is similar to the overall workflow of the sedDNA approach applied to lake sediments, which has been extensively described in Chap. 3 of this volume. Methodological developments associated with the study of whole protist communities – based on the DNA metabarcoding approach – have been primarily made from marine sediments (Coolen et al. 2013) and later applied to lake sediments (Capo et al. 2016; Capo et al. 2017a). We summarize the main steps of the sedDNA workflow, highlighting some specificities related to the recovery and characterization of protist sedDNA in the following section.

Selection of Coring Sites

As for every paleolimnological investigation, the selection of the coring sites and the use of sediment core replicates depend on the physical characteristics of the lake of interest and the ecological questions and hypotheses under investigation. Nonetheless, it is noteworthy that, for lakes that are characterized by simple morphological characteristics, the small-scale spatial variability of microbial communities across the deep part of the sediment basin has been shown to be negligible (Weisbrod et al. 2020). As such, the selection of one single core at the deepest site appears to be sufficient when studying the temporal dynamics of pelagic protist communities of such lakes.

Sediment Quantity

Protists represent a large fraction of the microbial biomass in lakes (Pick and Caron 1987). While less than 1 g of lake (wet) sediment is generally sufficient to unravel protist diversity, the exact amount of sediment needed, in part, depends on the extraction protocol applied. The DNA signal of protist communities is sensitive to the extraction protocol and initial mass of the sediment sample extracted (Capo et al. 2021a). Comparison between different sediment masses (0.75 g vs 4 g) using various extraction protocols (total DNA extraction versus extracellular DNA extraction) demonstrated that, regardless of the extraction protocol, extracting 0.75 g of sediment yielded similar DNA concentrations for large and deep hard-water lakes, whereas for shallow, organic-rich high-altitude lakes, the 4 g protocol yielded much higher DNA concentrations. As such, one might want to test protocols before extractions, especially when applying such techniques on as yet understudied types of lakes (e.g., tropical lakes, shallow lakes). As kits and protocols are rapidly evolving, they will not be listed in this chapter; however, a non-exhaustive list is available in Capo et al. (Capo et al. 2021a).

DNA Extraction from the Sediment Matrix

DNA extraction methods have been optimized to specifically recover the DNA signal from eukaryotes (Armbrecht et al. 2020). Nonetheless, several groups of protists are known to form resting stages thus raising the question of the need to extract and sequence intracellular, extracellular or both DNA fractions to fully characterize the protist diversity from an ecosystem. A comparison of the DNA inventories obtained using two different protocols, one extracting specifically extracellular DNA and the other extracting both intracellular and extracellular DNA, has been recently conducted (Appendix, Case Study 3 from (Capo et al. 2021a)). Differences were detected in both total DNA concentrations and the composition of microbial eukaryotes across the DNA extraction protocols, thus demonstrating the necessity to further develop our understanding of the methodologies around DNA extraction protocols.

One assumption regarding post-burial DNA preservation in sediments is that extracellular DNA is less protected than intracellular DNA, which is physically protected within resistant cells. Resting stages probably influence the sinking of cells, and associated DNA, through the water column. However, aside from a study that suggested that DNA from siliceous diatoms is better preserved over Holocene time scales than DNA from soft bodied micro-algae (Kirkpatrick et al. 2016), no study has clearly demonstrated that the DNA of cyst-forming taxa is better preserved in sediments. Capo et al. (2017b), who studied the effects of early diagenesis on the preservation of protist DNA in lake sediments, found no correlation between the relative abundance of protist taxa and the putative level of cellular DNA protection (capacity of taxa to form resistant cells, e.g., spores, cysts). Additionally, even though intracellular DNA is protected against external factors inside protective resting stages such as cysts or spores, DNA molecules can nonetheless be degraded by intracellular nucleases upon cell death. In contrast, extracellular DNA released into the sediment upon cell lysis can be quickly absorbed by clay minerals, which significantly promotes its preservation. Once bound to minerals, DNA molecules are protected from physico-chemical degradation processes (Kanbar et al. 2020), also making the DNA molecules less accessible to bacterial grazing (Corinaldesi et al. 2005).

Molecular Techniques

The choice of molecular technique primarily depends on the research questions (e.g., community changes vs. population genomics, targeted analysis vs. overall community changes, semi-quantitative analysis). To date, qPCR and DNA

Table 6.2 List of genic regions targeted to study the past abundance and diversity of protists and fungi from lake ecosystems

Organisms	Genic regions	Length (bp)	References
All protists	18S V1-V3	560	Coolen et al. (2013)
All protists	18S V4	300–540	Kisand et al. (2018)
All protists	18S V4	400–600	Brasell et al. (2022)
All protists	18S V6-V8	na	Wurzbacher et al. (2017)
All protists	18S V7	260	Capo et al. (2016)
All protists	18S V9	130	More et al. (2018)
Phototrophic protists	23S	410	Li et al. (2016)
Phototrophic protists	18S	260	Huo et al. (2022a)
Chlorophytes	ITS3, ITS4	650	Fonseca et al. (2022)
Ciliates	18S	378	Barouillet et al. (2022)
Diatoms	*rbcL* gene	67–76	Stoof-Leichsenring et al. (2012)
Diatoms	*rbcL* gene	577	Stoof-Leichsenring et al. (2014)
Diatoms	18S V4	400	Pawlowski et al. (2012)
Diatoms	18S	556	Coolen et al. (2004)
Diatoms	18S	458	Randlett et al. (2014)
Dinoflagellates	18S	465	Boere et al. (2009)
Fungi	ITS2	300–350	Talas et al. (2021)
Fungi	ITS2	250–500	Tõnno et al. (2021)
Fungi	ITS1	500	Seeber et al. (2022)
Haptophytes	18S	498	Coolen et al. (2004)

Note: Although fungi are not considered as protists, they are often included in sedDNA studies of protists communities and as such their markers have been added to this table

metabarcoding have been the most applied methods to study the abundance and diversity of protist communities in lakes (Barouillet et al. 2023; Capo et al. 2017a; Huo et al. 2022a; Keck et al. 2020; Zhang et al. 2021). Both methods require the development of primers targeting overall protist communities or specific protist groups (Table 6.2). Primers selected for the detection of protists require amplifying relatively short DNA fragments (cf. size range in Table 6.2). Additionally, the metabarcoding approach requires careful selection of the targeted genic regions to be able to differentiate all protists present in a given environment. In paleolimnological studies of overall protist communities, the 18S rRNA gene is the most targeted gene up to now. For group-specific studies within the protist communities 18S rRNA, 23S rRNA, ITS, or *rbc*L genic regions are usually targeted (cf. Table 6.2). More specifically for the 18S rRNA gene, several variable genic sequences have been identified, with the V4 and V7 regions as the most targeted regions in freshwater sedDNA studies (e.g., (Brasell et al. 2022; Capo et al. 2016; Kisand et al. 2018)) while the V9 region tends to be more frequently used in marine sedDNA studies (More et al. 2018; Pawlowski et al. 2018; Siano et al. 2021). In eukaryotes, the V4 region is the largest variable region and is better represented in public reference databases, and, as such, a well-suited region marker for biodiversity assessment (Hadziavdic et al. 2014), while the V7 region offers a good trade-off between taxonomic resolution and marker fragment length (Capo et al. 2016). More recently, advances in metagenomic approaches (shotgun sequencing, hybridization capture) have been used to recover the DNA signal of protists from aquatic sedimentary archives (Armbrecht et al. 2021a, b, 2022; Garner et al. 2020; Moguel et al. 2021; Zimmermann et al. 2023). Such sedimentary metagenomic approaches successfully recovered the mitochondrial and chloroplast genomes of a lake alga living during the Ice Age (Lammers et al. 2021) and specific photosynthetic genes (*psb*O gene, (Armbrecht et al. 2022)). Additionally, one of the advantages of datasets generated by shotgun sequencing is that it allows for capturing ultra-short DNA fragments and characterizing DNA damage patterns, which are important factors in assessing the authenticity of the genetic signal although it requires a good genome database to compare ancient DNA fragments to the taxa of interest (cf. Chap. 3 of this volume).

Resurrection Ecology

Many protist groups can produce cysts, and some remain encysted most of their life waiting for the appropriate conditions for excystment (Foissner 1999). These cysts allow for the study of the ecology and variability in the protist communities' response to environmental change at the population level. For instance, some studies successfully combined germination techniques with specific molecular biology protocols to investigate population dynamics and biogeography of dinoflagellates (Klouch et al. 2016; Miyazono et al. 2012) and diatoms (Härnström et al. 2011; Sanyal et al. 2022). Briefly, germination techniques for dinoflagellate cysts require the isolation of the cysts from the sediment matrix using the combination of various methods (i.e., sonication and sieving). The viability of the isolated cysts are tested and viables cysts are then germinated (Klouch et al. 2016). For diatom cysts, the first step of germination consists in the germination of cysts by applying a subsample of sediment slurry on a medium. The successfully germinated colonies of diatoms are then transferred into a fresh medium to allow further propagation (Härnström et al. 2011). In both cases, to allow for the temporal reconstruction of the population dynamics, the protocol is applied on each sediment sample separately. Once mono-clonal cultures of the targeted strains are successfully obtained, the cultures can be filtered (optional step) and DNA extracted. Given that these techniques provide access to a fresh sample of DNA, such germination techniques allow to circumvent some of the common issues encountered in sedDNA studies associated with the degradation of ancient DNA molecules, thus providing important and novel information on the temporal dynamics of protists populations. Additionally, combined with Sanger sequencing, as the genetic sequences obtained, including those of past populations, contribute to the construction of reference databases.

Technical Replicates

In sedDNA studies targeting protists, although two DNA extraction replicates are generally used, three are recommended (Ibrahim et al. 2021). For metabarcoding approaches, PCR replicates can also be used. Technical replicates (i.e., PCR and DNA extraction replicates) often result in highly similar outputs. This is especially true for the dominant OTUs (Operational Taxonomic Units or ASVs (Amplicon Sequence Variants), while higher variability can be found for the rare OTUs. Nonetheless, technical replicates are important in all molecular ecology studies to assess the degree of reproducibility (Zinger et al. 2019).

Taphonomy of Protist DNA from Sources to Sediments

The taphonomy, the study of processes affecting the molecular signal of lake protists from their death to their decomposition, is of particular importance for protists' sedDNA studies because of the high diversity in morphologies and ecological behaviors. The DNA signal present in lake sediments is a mix of allochthonous (e.g., lake catchment, atmospheric deposition, runoff waters) and autochthonous (e.g., lake water, surface sediments, subsurface sediments) sources (Capo et al. 2022). DNA of protists living in the water column is transferred to the sediments by sinking particles (organic particles, cell remains, cysts) as intracellular DNA (within dead or viable cells) or extracellular DNA (free or bound to particles). The efficiency of the transfer of the molecular signal from planktonic protists to the sediment may differ between taxa, as discussed in Chap. 2 of this volume. We focus here on the factors that can affect the taphonomy of protist DNA at the water-sediment interface and during burial in sediments. The presence of oxygen at the water-sediment interface is a key factor affecting the preservation of DNA molecules in sediments. When the surface sediment is oxygenated, macrofauna feeding on sinking particles from the water column may result in organic matter remineralization and bioturbation, thereby affecting the vertical transport of oxygen and particles to the sediments (Krantzberg 1985). In the few centimeters below the water-sediment interface, the active microbial community can recycle the pool of extracellular DNA, as part of the organic matter, buried in the sediments (Dell'Anno and Danovaro 2005). At deeper depths, microorganisms enter a stationary growth state due to a shortage in electron acceptors required for microbial metabolism. In parallel, the microbial activity tends to decrease as sediment depth increases and the composition of the active microbes (bacteria or archaea) able to metabolize various types of organic matter changes through depth associated with the redox gradient. In deep sediment layers (usually below 30 cm in lakes), DNA molecules are degraded by chemical reactions into shorter DNA fragments with post-mortem damages and cross-linking to other molecules and are thus technically ancient DNA fragments (Orlando et al. 2021). This long-term decay of DNA in aquatic sediments lasts over geological times (Boere et al. 2011a; Kirkpatrick et al. 2016) until DNA fragments become too short to be analyzed. A recent study revealed that 2-million years old DNA fragments could be recovered from sedimentary archives (Kjær et al. 2022).

To evaluate the preservation of the sedDNA signal from protists in lake sediments, the molecular signal of two sediment cores collected at a 6-year interval were compared (Capo et al. 2017b). Briefly, protist DNA was sequenced from 12 sediment layers covering a period of ~33 years from each core and samples with corresponding dates between the two cores were analyzed. This work revealed that the molecular signal, as seen through the composition and the structure of the protist community, appears to be stable after 15 years of burial in sediments. In addition, this work provided evidence of early diagenesis, as molecular signals undergo some modifications over the first years of burying and then stabilize over time. Additionally, comparing DNA amplification using two different lengths of barcodes (~400 and ~ 260 bp) provided further evidence of DNA fragmentation in old sediment samples (Capo et al. 2016). More specifically, this study detected a loss signal in the sediment cores from ~900 years ago when using longer fragments, while the 260-bp-long DNA fragment was successfully amplified over the entire sediment cores (up to ~2100 years ago). On a longer time frame, the decay of chloroplast DNA was shown to occur mainly over the first 100–200 thousand years of deposition in Bering Sea sediments (Kirkpatrick et al. 2016). In this same study, diatom DNA was detected through the whole sediment core spanning ~1.4 million years. The authors suggested the detection of siliceous diatom microfossils as one explanation for the better preservation of diatom DNA over other soft-bodied micro-algae. Similarly, Boere et al. (Boere et al. 2011b) revealed a stronger persistence of the sedDNA signal from diatoms compared to dinoflagellates in a 2700-year-old sediment record from Ellis fjord (Antarctica).

Reliability of the Molecular Signal

Several studies have assessed whether the DNA signal recovered from the sedimentary archives is reliable to track ancient protist communities. For example, comparison of DNA inventories from the water column, sediment traps and sediment cores demonstrated that the DNA signal recovered from the sediments reflects a part of the signal observed in the water column (Capo et al. 2015; Gauthier et al. 2021). Some groups are particularly well represented in the DNA sedimentary archives such as the chrysophytes, chlorophytes, dinoflagellates, ciliates, chytrids and apicomplexans. Comparisons of the sedDNA datasets with other sedimentary proxies, such as lipids (Sinninghe et al., 2004) and morphological remains (Anslan et al. 2022; Dulias et al. 2017; Huang et al. 2020; Stoof-Leichsenring et al. 2014, 2020a), revealed the presence of concomitant changes in time series related to environmental perturbations. However, some discrepancies are observed between the two approaches especially in the taxonomic assignments (e.g., (Anslan et al. 2022)). These differences also highlight the complementarity of the two approaches. While the lack of references in public databases can limit the taxonomic resolution in genetic datasets, sedDNA dataset can help to uncover previously hidden genetic and phylogenetic information, as well as population changes in response to environmental changes, that were not detectable by microscopic based approaches (Barouillet et al. 2022).

One important factor to consider when using DNA approaches to reconstruct the structure and describe changes in the relative proportion of taxa in protist communities is the variability of the copy number of a gene. The presence of numerous copies of a gene in a single cell is common in several protist lineages and is considered as one of the most limiting factors and biggest challenge when using molecular approaches to survey protist communities, especially when the sequencing depth is low and the aim is to assess changes in the relative proportion of taxa. This technical bias impedes the quantitative estimates of protist communities. Giner et al. (Giner et al. 2016) suggest that sequencing data can provide reliable estimates for the relative abundance of certain protists but also warns about considering high-throughput sequencing data as a direct proxy of cell abundance. The development of a database that contains information about the number of copies of a gene that can be found in a single cell for various protist species might help to develop standardization methods for the correction of DNA sequences abundance (Martin et al. 2022).

Signal from Sediment-Living Protists

Numerous protists are able to thrive in organic-rich surface sediments and are an important part of benthic food webs (Coolen and Shtereva 2009; Forster et al. 2016). However, their potential ability to thrive in energy-limited subsurface deep sediments is still understudied and raises important questions for molecular paleoreconstructions of protists. Additionally, we are still lacking information about the taxonomic composition of the active community of protists in sediments, especially in lakes.

An indirect clue of the presence of sediment-living protists can be obtained through the recovery, PCR amplification and/or sequencing of RNA molecules from protists in subsurface sediments. In marine ecosystems, several studies suggest considerable activity of protists in deep subsurface sediments (Ciobanu et al. 2014; Edgcomb et al. 2011; Orsi et al. 2013; Pachiadaki et al. 2016), with fungi dominating the active eukaryotic assemblages. The adaptability of certain microbial eukaryotes to withstand energy-starved subsurface conditions appear to be limited to some yeasts (e.g., *Cryptococcus* sp., *Trichosporon* sp.) and filamentous fungi (*Fusarium* sp., *Trichoderma* sp.) (Pachiadaki et al. 2016), two groups that are not classified as protists. Edgcomb et al. (Edgcomb et al. 2011) revealed the presence of transcripts from unexpected eukaryotes, such as animals, green algae, and red algae in deep marine sediments. The detection of RNA sequences from photosynthetic organisms, higher plants and multicellular organisms was also revealed in the work of

Orsi et al. (2013) and suggested that, although this RNA signal had been preserved in sediments over time, it did not reflect an active subsurface eukaryotic community. Rather, it shows the presence of resting stages containing RNA. It is possible that some protists are active in sediments but still overlooked by current approaches. Chambouvet et al. (2014) revealed for instance the presence of active perkinseans (parasites) in marine sediments. Compared to marine environments, the presence of active protists in lake subsurface sediments has been scarcely studied.

Resurrection ecology research from aquatic sediments has revealed the presence of protists that are able to survive in subsurface sediments, including amoebae, ciliates, diatoms and dinoflagellates ((Ellegaard et al. 2020) for a synthesis; (Delebecq et al. 2020)). For instance, Sanyal et al. (Sanyal et al. 2022) revealed for the first time the revival, viability, and germination rate of resting spores of the diatom *Chaetoceros* deposited in sub-seafloor Baltic Sea sediments up to ~6600 years old. Although such studies provide interesting findings about the genomics of past organisms, such approaches do not provide evidence for the presence of active protists in aquatic subsurface sediments.

Ecology and responses of protists to environmental changes revealed by sedimentary DNA.

The temporal dynamics of phytoplanktonic protists and their response to environmental changes have been under investigation for decades using the paleolimnological approach and through limnological surveys (Smol et al. 2001; Suthers et al. 2019). Complementing traditional methods based on morphological observation, the sequencing of protist DNA from lake sediments brings additional information about the diversity, distribution and relative importance of protists, including some understudied heterotrophic groups of protists, and can also shed light on communities from remote locations (e.g., (Fonseca et al. 2022)). SedDNA also offers the promise of exploring how protist biodiversity and lake ecosystem functioning (e.g., lake productivity) – using a trait-based functional community approach (Barouillet et al. 2022) – are changing as a response to environmental changes. Importantly, most of the lake sedDNA studies to date that have focused on protist communities provided insights into the relative change in community composition rather than absolute quantitative estimates (Barouillet et al. 2023). Several 100 years-long time series have been obtained with the 18S metabarcoding approach, mainly in temperate and subtropical lake systems. Analysis of published 18S sedDNA records of eukaryotic communities across a large variety of lakes revealed that the response of phytoplanktonic groups to anthropogenic stressors over the last ~200 years is lake-dependent (Fig. 6.2). The scarcity of relevant data, that is 18S metabarcoding data from sediment layers and measurements or inference of past environmental conditions (air and water temperature, precipitations, phosphorus concentrations), in a number of lakes also hampers any firm conclusions thus far. Nonetheless, the recent application of sedDNA approaches has expanded both the time perspective and the diversity of protist lineages studied (Domaizon et al. 2017). We summarize below findings from the few sedDNA studies that investigated the responses of lake protists to environmental changes during the Anthropocene.

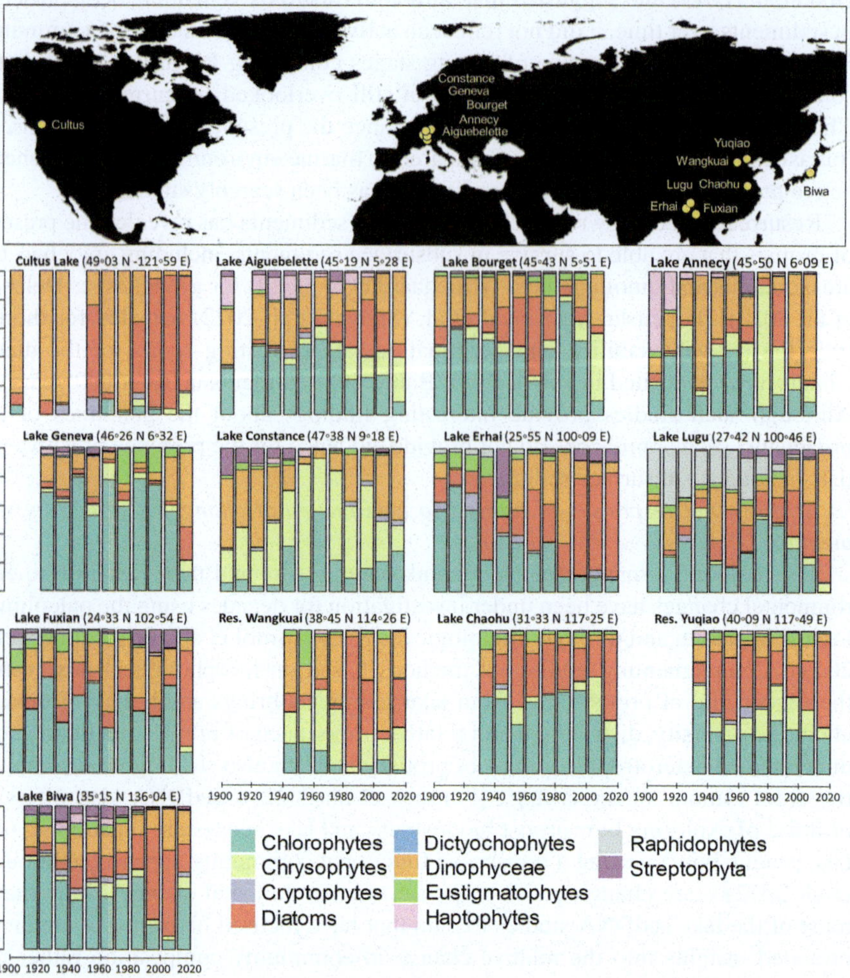

Fig. 6.2 Temporal changes over the past century in the eukaryotic phytoplankton in lakes determined by sedimentary DNA research. (Data were compiled from several research works ((Capo et al. 2017a; Gauthier et al. 2022; Huo et al. 2022a; Ibrahim et al. 2021); and unpublished works from authors))

Environmental Changes in the Anthropocene

Several sedDNA investigations have focused on the response of protist communities to prevalent environmental changes over the Anthropocene, such as cultural eutrophication, climate and the introduction of invasive species. Reviewed in detail in Barouillet et al. (2023), these studies revealed new information on the long-term response of diatoms, dinoflagellates, ciliates, and perkinseans to anthropogenic changes. These studies also brought new insights into the sensitivity of some understudied heterotrophic protists to environmental changes and into the cumulative impacts of multiple stressors on protist communities. Interestingly, while the response of some protists to environmental changes is lake dependent (as shown in Fig. 6.2), some common trends were also observed through the innovative comparison of the sedDNA of pre- and post-disturbance samples (Keck et al. 2020).

SedDNA records revealed changes in some functional groups of protists in responses to recent anthropogenic-induced changes. More specifically, higher proportions of phototrophs and mixotrophs, while lower proportions of some phagotrophs and saprotrophs (i.e., amoebozoans, cercozoans and some stramenopiles, such as Gyrista and Bigyra) were found in recent decades, thereby suggesting that enhanced human activities favor autotrophic pathways (Keck et al. 2020). A similar increase in the relative abundance of mixotrophic ciliates has been observed across a large variety of lakes over the last decade (Barouillet et al. 2022), further suggesting that recent environmental changes are favoring mixotrophic life-strategies in freshwaters. Although the exact mechanisms remain elusive, given that mixotrophs have a competitive advantage under low nutrient and light conditions, it has been suggested that re-oligotrophication of freshwater lakes is the most likely driver of these observed changes. The influence of environmental changes to mixotrophic groups of protists was also demonstrated by Capo et al. (2017a), which showed that chrysophytes were favored by low phosphorus concentrations and increased temperatures. These long-term trends confirm patterns that had been previously observed in limnological (e.g., (Sandgren et al. 1995)) and paleolimnological studies (Smol 1985). Interestingly, analysis of DNA extracted from water column and sediment trap samples across a 3-year monthly time series provided new insights into the sensitivity of chrysophytes to fluctuations in water chemistry (Gauthier et al. 2021). More specifically, the authors found two species of chrysophytes, in the DNA samples from both the water column and sediment traps, that were strongly correlated with high ammonia concentration (Gauthier et al. 2021). Additionally, an increase in the relative abundance of chlorophytes during periods of high phosphorus concentrations were observed in large peri-alpine lakes (Capo et al. 2017a).

SedDNA studies also provide insights into the response of protists to environmental changes at the population level. Thus far, accessing information at the population level through sedimentary archives has only been possible through the germination and genetic analysis of isolated resting stages. Such resurrection ecology studies have been mostly applied to marine sediment cores (Härnström et al. 2011; Klouch et al. 2016; Miyazono et al. 2012; Sanyal et al. 2022). To date, McCarthy et al. (2011) is the only sedDNA analysis of cysts that has been conducted on a freshwater lake. This study showed promising results, and demonstrated the contrasting responses of two dinoflagellate species to cultural eutrophication and land-use change through time (McCarthy et al. 2011).

Holocene Records

Applying the sedDNA approach on Holocene-long records brought a new perspective into the baseline conditions of protist communities (Capo et al. 2021b; Kisand et al. 2018). These studies often highlight the importance of long-term climate-driven changes in protist communities. The millennium-long record of Lake Yamzhog Yumco, in the Southern Tibetan Plateau, showed that nutrient inputs, mediated by climate through higher precipitation, glacier melting and permafrost degradation, have been the major drivers of changes in the phytoplankton biomass and in the community composition over the last ~2000 years (Huo et al. 2022b). Similarly, Holocene records of *Isochrysis* (haptophytes) population dynamics of Lake Kusai (China) highlighted the sensitivity of these assemblages to climate-driven changes in nutrients, temperature, salinity and light intensity (Hou et al. 2014). The analysis demonstrated the strong influence of the Asian summer monsoon season and precipitation on the dominance of haptophytes relative to cyanobacteria, two algal groups commonly found in Tibetan lakes.

Such temporal perspectives can also provide new insights into the population response to large, long-term ecosystem changes. For instance, analyses of lipids and DNA revealed changes in diversity and community composition of haptophyte and diatom populations over the Holocene following the post-glacial development of Ace Lake (Antarctica) from a freshwater basin to a marine inlet (Coolen et al. 2004).

Biogeography

SedDNA studies unraveled a wealth of information on the relative influence of geographic versus local environmental factors in explaining patterns of protists biogeography and genetic plasticity (Stoof-Leichsenring et al. 2014, 2015, 2020b). Analysis of the diatom lineages in Siberian lakes along environmental gradients (i.e., across treeline transects) and across large spatial scales and time scales (i.e., last ~7000 years) showed that the relatedness of the identified diatom lineages reflected changes in environmental conditions (i.e., vegetation coverage, climatic and limnetic variables) through space and time rather than geographic distance (Stoof-Leichsenring et al. 2015). More specifically, the authors found that closely-related *Staurosira* lineages occur in environments that shared similar local conditions, while phylogenetically-distant lineages were found in dissimilar environments. Similarly, the genetic structure of the bloom-forming diatom *Skeletonema marinoi* was investigated using microsatellite markers applied on resurrected strains preserved in sediment for over ~100 years (Härnström et al. 2011). This latter study showed that the dispersal potential and generation time do not have a large impact on the genetic structure of the populations, but rather changes in environmental conditions, such as extreme eutrophication, are more important drivers. The influence of landscape setting on the environmental selections of protist taxa, as well as the persistence of certain taxa over millennial timescales and the subsequent effects on shaping lake microbial communities in more recent times, were also revealed through sedDNA studies (Capo et al. 2021b). Importantly, the development of such sedDNA datasets on various aquatic ecosystems across the globe will provide new insights into the true diversity of aquatic protists and might help elucidate the still ongoing debate on protist biogeographic distribution and endemicity (Finlay and Fenchel 1999; Foissner 1999).

Long-Term Perspective on Protist Biodiversity

SedDNA analysis has the advantage of providing a broad perspective on community changes, including some key insights into historical aquatic biodiversity. For instance, a universal assay recovered the whole eukaryotes biodiversity from sedDNA archives, from microorganisms to vascular plants and vertebrates, from both aquatic and terrestrial origins (Kisand et al. 2018). In parallel, using a group-specific assay, Ibrahim et al. (Ibrahim et al. 2021) highlighted a decline in diatom diversity in Lake Constance (Germany) as a response to an increase in phosphorus concentrations. Using a metabarcoding approach, one general trend that has been observed across several sedDNA datasets is the overall homogenization of protist communities in recent time (Barouillet et al. 2022; Keck et al. 2020; Zhang et al. 2021). The measures of alpha- and beta-diversity have been shown to be a powerful tool to evaluate the impacts of anthropogenic stressors on lake ecosystems (Capo et al. 2017a). However, it has also been noted that, to better estimate the lakes sensitivity to anthropogenically-driven environmental changes (e.g., nutrient supply, climate change) through protist diversity and community changes, it is important to predetermine lake-specific interannual variability in environmental conditions (Capo et al. 2019). These biodiversity assessments considerably improve our understanding of the biodiversity response to multiple stressors, and participate in the global assessment of protist diversity.

Protists as Potential Paleo-Indicators

Environmental changes in lakes are multifaceted and impact several functional groups and trophic levels. There is a particularly high potential to find new paleo-indicators among protists, as they comprise organisms with various life strategies that are highly sensitive to environmental change. Furthermore, their position at the base of trophic networks is of key importance as any qualitative or quantitative change in their community is expected to affect upper trophic levels, including the macrofauna. Many protist groups have been widely used as bioindicators, such as diatoms, chrysophytes or ciliates (Sagova-Mareckova et al. 2021). Applying molecular biology approaches for the development of new bioindicators is still relatively young but showing promising perspectives for future application in environmental assessments (Vasselon et al. 2019). As such, the development of new paleo-indicators through protist sedDNA studies, while still requiring some methodological development, is a promising research field.

Through the combination of trait-based functional community approach with the sedDNA approach, protists could be used as proxies to reveal functional changes in lake ecosystems. Paleo-investigations that explored community changes through functional diversity rather than taxonomy-based diversity revealed some important insights into long-term changes in lake ecosystem functioning, including some shifts in the metabolic balance of lakes (Barouillet et al. 2022; Keck et al. 2020). While the exploration of the functional diversity of protists from natural archives is a still relatively recent approach, its development and expansion will likely provide new perspectives on lake response to environmental changes.

To unlock the potential of certain groups of protists as paleo-indicators, it would thus be important to increase the number of in-depth analysis of specific taxonomic and/or functional groups. Finally, protists have a high phenotypic and genotypic plasticity allowing them to quickly adapt to environmental changes. Hence, protists could be of particular interest in detecting and studying early warning signals in paleo-investigations with potential direct implications in the management and conservation of lakes.

Summary

The study of protist communities through sedDNA approaches has greatly improved our understanding of the long-term changes in their diversity, distribution, and relative importance in lake ecosystems. While technical and methodological advancements are constantly improving, they still require development. For instance, deepening our understanding of the active communities of protists in lake sediments is needed in order to discriminate them from archived communities. Moreover, the potential use of protists as paleo-indicators needs further investigations in order to expand the applicability of sedimentary records as a tool for environmental assessment. In particular, heterotrophic protists are pivotal in lake functioning; however, they are often overlooked when studying the response of aquatic organisms to environmental changes. Their paleo-reconstructions should offer a panel of unique information related to their role in lake ecosystems (i.e., remobilization and recycling of carbon and nutrients, grazers and consumers in aquatic food webs). Altogether, sedDNA studies provide a novel inventory of past communities and unlock new avenues for paleolimnological investigations to: (1) include the missing link represented by protists in paleo-reconstructions of lake food webs; (2) explore of the role of protists in the resistance and resilience of lakes facing disturbances; and (3) provide new information on lake ecosystem functioning. Ultimately, the information gained through sedDNA research of protist communities will help to gain key knowledge for lake management and preservation and conservation of biodiversity.

Acknowledgements We acknowledge financial support from the Pole Research & Development on Lacustrine Ecosystems (ECLA) of the French Biodiversity Agency (OFB) and the Severo Ochoa Excellence Program postdoctoral fellowship awarded in 2021 to Eric Capo (CEX2019-000928-S). We are very grateful to Kuldeep More, Thomas Weisse and Frances Pick for their critical reading of our chapter.

References

Adl SM, Bass D, Lane CE, Lukeš J, Schoch CL, Smirnov A, Agatha S, Berney C, Brown MW, Burki F, Cárdenas P, Čepička I, Chistyakova L, del Campo J, Dunthorn M, Edvardsen B, Eglit Y, Guillou L, Hampl V, Heiss AA, Hoppenrath M, James TY, Karnkowska A, Karpov S, Kim E, Kolisko M, Kudryavtsev A, Lahr DJG, Lara E, Le Gall L, Lynn DH, Mann DG, Massana R, Mitchell EAD, Morrow C, Park JS, Pawlowski JW, Powell MJ, Richter DJ, Rueckert S, Shadwick L, Shimano S, Spiegel FW, Torruella G, Youssef N, Zlatogursky V, Zhang Q (2019) Revisions to the classification, nomenclature, and diversity of eukaryotes. J Eukaryot Microbiol 66(1):4–119. https://doi.org/10.1111/jeu.12691

Anslan S, Kang W, Dulias K, Wünnemann B, Echeverría-Galindo P, Börner N, Schwarz A, Liu Y, Liu K, Künzel S, Kisand V, Rioual P, Peng P, Wang J, Zhu L, Vences M, Schwalb A (2022) Compatibility of diatom valve records with sedimentary ancient DNA amplicon data: a case study in a brackish, alkaline Tibetan lake. Front Earth Sci 10

Armbrecht L, Herrando-Pérez S, Eisenhofer R, Hallegraeff GM, Bolch CJS, Cooper A (2020) An optimized method for the extraction of ancient eukaryote DNA from marine sediments. Mol Ecol Resour 20(4):906–919. https://doi.org/10.1111/1755-0998.13162

Armbrecht L, Hallegraeff G, Bolch CJS, Woodward C, Cooper A (2021a) Hybridisation capture allows DNA damage analysis of ancient marine eukaryotes. Sci Rep 11:3220. https://doi.org/10.1038/s41598-021-82578-6

Armbrecht L, Eisenhofer R, Utge J, Sibert EC, Rocha F, Ward R, Pierella Karlusich JJ, Tirichine L, Norris R, Summers M, Bowler C (2021b) Paleo-diatom composition from Santa Barbara Basin deep-sea sediments: a comparison of 18S-V9 and diat-rbcL metabarcoding vs shotgun metagenomics. ISME Commun 1(1):1–10. https://doi.org/10.1038/s43705-021-00070-8

Armbrecht L, Weber ME, Raymo ME, Peck VL, Williams T, Warnock J, Kato Y, Hernández-Almeida I, Hoem F, Reilly B, Hemming S, Bailey I, Martos YM, Gutjahr M, Percuoco V, Allen C, Brachfeld S, Cardillo FG, Du Z, Fauth G, Fogwill C, Garcia M, Glüder A, Guitard M, Hwang J-H, Iizuka M, Kenlee B, O'Connell S, Pérez LF, Ronge TA, Seki O, Tauxe L, Tripathi S, Zheng X (2022) Ancient marine sediment DNA reveals diatom transition in Antarctica. Nat Commun 13(1):5787. https://doi.org/10.1038/s41467-022-33494-4

Barouillet C, Vasselon V, Keck F, Millet L, Etienne D, Galop D, Rius D, Domaizon I (2022) Paleoreconstructions of ciliate communities reveal long-term ecological changes in temperate lakes. Sci Rep 12(1):7899. https://doi.org/10.1038/s41598-022-12041-7

Barouillet C, Monchamp M-E, Bertilsson S, Brasell K, Domaizon I, Epp LS, Ibrahim A, Mejbel H, Nwosu EC, Pearman JK, Picard M, Thomson-Laing G, Tsugeki N, Von Eggers J, Gregory-Eaves I, Pick FR, Wood SA, Capo E (2023) Investigating the effects of anthropogenic stressors on lake biota using sedimentary DNA. Freshw Biol. https://doi.org/10.1111/fwb.14027

Beisser D, Graupner N, Bock C, Wodniok S, Grossmann L, Vos M, Sures B, Rahmann S, Boenigk J (2017) Comprehensive transcriptome analysis provides new insights into nutritional strategies and phylogenetic relationships of chrysophytes. PeerJ 5:e2832. https://doi.org/10.7717/peerj.2832

Bik HM, Porazinska DL, Creer S, Caporaso JG, Knight R, Thomas WK (2012) Sequencing our way towards understanding global eukaryotic biodiversity. Trends Ecol Evol 27(4):233–243. https://doi.org/10.1016/j.tree.2011.11.010

Bjorbækmo MFM, Evenstad A, Røsæg LL, Krabberød AK, Logares R (2020) The planktonic protist interactome: where do we stand after a century of research? ISME J 14(2):544–559. https://doi.org/10.1038/s41396-019-0542-5

Boenigk J, Arndt H (2002) Bacterivory by heterotrophic flagellates: community structure and feeding strategies. Antonie Van Leeuwenhoek 81(1):465–480. https://doi.org/10.1023/A:1020509305868

Boere AC, Abbas B, Rijpstra WIC, Versteegh GJM, Volkman JK, Sinninghe Damsté JS, Coolen MJL (2009) Late-Holocene succession of dinoflagellates in an Antarctic fjord using a multi-proxy approach: paleoenvironmental genomics, lipid biomarkers and palynomorphs. Geobiology 7(3):265–281. https://doi.org/10.1111/j.1472-4669.2009.00202.x

Boere AC, Rijpstra WIC, De Lange GJ, Sinninghe Damsté JS, Coolen MJL (2011a) Preservation potential of ancient plankton DNA in Pleistocene marine sediments. Geobiology 9(5):377–393. https://doi.org/10.1111/j.1472-4669.2011.00290.x

Boere AC, Sinninghe Damsté JS, Rijpstra WIC, Volkman JK, Coolen MJL (2011b) Source-specific variability in post-depositional DNA preservation with potential implications for DNA based paleoecological records. Org Geochem 42(10):1216–1225. https://doi.org/10.1016/j.orggeochem.2011.08.005

Brasell KA, Pochon X, Howarth J, Pearman JK, Zaiko A, Thompson L, Vandergoes MJ, Simon KS, Wood SA (2022) Shifts in DNA yield and biological community composition in stored sediment: implications for paleogenomic studies. Metabarcoding Metagenom 6:e78128. https://doi.org/10.3897/mbmg.6.78128

Bråte J, Logares R, Berney C, Ree DK, Klaveness D, Jakobsen KS, Shalchian-Tabrizi K (2010) Freshwater Perkinsea and marine-freshwater colonizations revealed by pyrosequencing and phylogeny of environmental rDNA. ISME J 4(9):1144–1153. https://doi.org/10.1038/ismej.2010.39

Burki F (2014) The eukaryotic tree of life from a global phylogenomic perspective. Cold Spring Harb Perspect Biol 6(5):a016147. https://doi.org/10.1101/cshperspect.a016147

Burki F, Roger AJ, Brown MW, Simpson AGB (2020) The new tree of eukaryotes. Trends Ecol Evol 35(1):43–55. https://doi.org/10.1016/j.tree.2019.08.008

Burki F, Sandin MM, Jamy M (2021) Diversity and ecology of protists revealed by metabarcoding. Curr Biol 31(19):R1267–R1280. https://doi.org/10.1016/j.cub.2021.07.066

Capo E, Debroas D, Arnaud F, Domaizon I (2015) Is planktonic diversity well recorded in sedimentary DNA? Toward the reconstruction of past protistan diversity. Microb Ecol 70(4):865–875. https://doi.org/10.1007/s00248-015-0627-2

Capo E, Debroas D, Arnaud F, Guillemot T, Bichet V, Millet L, Gauthier E, Massa C, Develle A-L, Pignol C, Lejzerowicz F, Domaizon I (2016) Long-term dynamics in microbial eukaryotes communities: a palaeolimnological view based on sedimentary DNA. Mol Ecol 25(23):5925–5943. https://doi.org/10.1111/mec.13893

Capo E, Debroas D, Arnaud F, Perga M-E, Chardon C, Domaizon I (2017a) Tracking a century of changes in microbial eukaryotic diversity in lakes driven by nutrient enrichment and climate warming. Environ Microbiol 19(7):2873–2892. https://doi.org/10.1111/1462-2920.13815

Capo E, Domaizon I, Maier D, Debroas D, Bigler C (2017b) To what extent is the DNA of microbial eukaryotes modified during burying into lake sediments? A repeat-coring approach on annually laminated sediments. J Paleolimnol 58(4):479–495. https://doi.org/10.1007/s10933-017-0005-9

Capo E, Rydberg J, Tolu J, Domaizon I, Debroas D, Bindler R, Bigler C (2019) How does environmental inter-annual variability shape aquatic microbial communities? A 40-year annual record of sedimentary DNA from a boreal Lake (Nylandssjön, Sweden). Front Ecol Evol 7

Capo E, Giguet-Covex C, Rouillard A, Nota K, Heintzman PD, Vuillemin A, Ariztegui D, Arnaud F, Belle S, Bertilsson S, Bigler C, Bindler R, Brown AG, Clarke CL, Crump SE, Debroas D, Englund G, Ficetola GF, Garner RE, Gauthier J, Gregory-Eaves I, Heinecke L, Herzschuh U, Ibrahim A, Kisand V, Kjær KH, Lammers Y, Littlefair J, Messager E, Monchamp M-E, Olajos F, Orsi W, Pedersen MW, Rijal DP, Rydberg J, Spanbauer T, Stoof-Leichsenring KR, Taberlet

P, Talas L, Thomas C, Walsh DA, Wang Y, Willerslev E, van Woerkom A, Zimmermann HH, Coolen MJL, Epp LS, Domaizon IG, Alsos I, Parducci L (2021a) Lake sedimentary DNA research on past terrestrial and aquatic biodiversity: overview and recommendations. Quaternary 4(1):6. https://doi.org/10.3390/quat4010006

Capo E, Ninnes S, Domaizon I, Bertilsson S, Bigler C, Wang X-R, Bindler R, Rydberg J (2021b) Landscape setting drives the microbial eukaryotic community structure in four Swedish Mountain lakes over the Holocene. Microorganisms 9(2):355. https://doi.org/10.3390/microorganisms9020355

Capo E, Monchamp M-E, Coolen MJL, Domaizon I, Armbrecht L, Bertilsson S (2022) Environmental paleomicrobiology: using DNA preserved in aquatic sediments to its full potential. Environ Microbiol 24(5):2201–2209. https://doi.org/10.1111/1462-2920.15913

Caron DA, Hu SK (2019) Are we overestimating protistan diversity in nature? Trends Microbiol 27(3):197–205. https://doi.org/10.1016/j.tim.2018.10.009

Caron DA, Hutchins DA (2013) The effects of changing climate on microzooplankton grazing and community structure: drivers, predictions and knowledge gaps. J Plankton Res 35(2):235–252. https://doi.org/10.1093/plankt/fbs091

Chambouvet A, Berney C, Romac S, Audic S, Maguire F, De Vargas C, Richards TA (2014) Diverse molecular signatures for ribosomally 'active' Perkinsea in marine sediments. BMC Microbiol 14(1):110. https://doi.org/10.1186/1471-2180-14-110

Ciobanu M-C, Burgaud G, Dufresne A, Breuker A, Rédou V, Ben Maamar S, Gaboyer F, Vandenabeele-Trambouze O, Lipp JS, Schippers A, Vandenkoornhuyse P, Barbier G, Jebbar M, Godfroy A, Alain K (2014) Microorganisms persist at record depths in the subseafloor of the Canterbury Basin. ISME J 8(7):1370–1380. https://doi.org/10.1038/ismej.2013.250

Coolen MJL, Shtereva G (2009) Vertical distribution of metabolically active eukaryotes in the water column and sediments of the Black Sea. FEMS Microbiol Ecol 70(3):525–539. https://doi.org/10.1111/j.1574-6941.2009.00756.x

Coolen MJL, Muyzer G, Rijpstra WIC, Schouten S, Volkman JK, Sinninghe Damsté JS (2004) Combined DNA and lipid analyses of sediments reveal changes in Holocene haptophyte and diatom populations in an Antarctic lake. Earth Planet Sci Lett 223:225–239. https://doi.org/10.1016/j.epsl.2004.04.014

Coolen MJL, Orsi WD, Balkema C, Quince C, Harris K, Sylva SP, Filipova-Marinova M, Giosan L (2013) Evolution of the plankton paleome in the Black Sea from the Deglacial to Anthropocene. Proc Natl Acad Sci 110(21):8609–8614. https://doi.org/10.1073/pnas.1219283110

Corinaldesi C, Danovaro R, Dell'Anno A (2005) Simultaneous recovery of extracellular and intracellular DNA suitable for molecular studies from marine sediments. Appl Environ Microbiol 71(1):46–50. https://doi.org/10.1128/AEM.71.1.46-50.2005

Debroas D, Domaizon I, Humbert J-F, Jardillier L, Lepère C, Oudart A, Taïb N (2017) Overview of freshwater microbial eukaryotes diversity: a first analysis of publicly available metabarcoding data. FEMS Microbiol Ecol 93(4):fix023. https://doi.org/10.1093/femsec/fix023

Delebecq G, Schmidt S, Ehrhold A, Latimier M, Siano R (2020) Revival of ancient marine dinoflagellates using molecular biostimulation. J Phycol 56(4):1077–1089. https://doi.org/10.1111/jpy.13010

Dell'Anno A, Danovaro R (2005) Extracellular DNA plays a key role in Deep-Sea ecosystem functioning. Science 309(5744):2179–2179. https://doi.org/10.1126/science.1117475

DeLong JP, Van Etten JL, Al-Ameeli Z, Agarkova IV, Dunigan DD (2023) The consumption of viruses returns energy to food chains. Proc Natl Acad Sci 120(1):e2215000120. https://doi.org/10.1073/pnas.2215000120

Domaizon I, Winegardner A, Capo E, Gauthier J, Gregory-Eaves I (2017) DNA-based methods in paleolimnology: new opportunities for investigating long-term dynamics of lacustrine biodiversity. J Paleolimnol 58(1):1–21. https://doi.org/10.1007/s10933-017-9958-y

Dulias K, Stoof-Leichsenring KR, Pestryakova LA, Herzschuh U (2017) Sedimentary DNA versus morphology in the analysis of diatom-environment relationships. J Paleolimnol 57(1):51–66. https://doi.org/10.1007/s10933-016-9926-y

Edgcomb VP, Beaudoin D, Gast R, Biddle JF, Teske A (2011) Marine subsurface eukaryotes: the fungal majority. Environ Microbiol 13(1):172–183. https://doi.org/10.1111/j.1462-2920.2010.02318.x
Ellegaard M, Clokie MRJ, Czypionka T, Frisch D, Godhe A, Kremp A, Letarov A, McGenity TJ, Ribeiro S, John Anderson N (2020) Dead or alive: sediment DNA archives as tools for tracking aquatic evolution and adaptation. Commun Biol 3(1):1–11. https://doi.org/10.1038/s42003-020-0899-z
Filker S, Sommaruga R, Vila I, Stoeck T (2016) Microbial eukaryote plankton communities of high-mountain lakes from three continents exhibit strong biogeographic patterns. Mol Ecol 25(10):2286–2301. https://doi.org/10.1111/mec.13633
Finlay BJ, Fenchel T (1999) Divergent perspectives on protist species richness. Protist 150(3):229–233. https://doi.org/10.1016/S1434-4610(99)70025-8
Finlay BJ, Fenchel T (2004) Cosmopolitan metapopulations of free-living microbial eukaryotes. Protist 155(2):237–244. https://doi.org/10.1078/143446104774199619
Foissner W (1999) Protist diversity: estimates of the near-imponderable. Protist 150(3):363–368. https://doi.org/10.1016/S1434-4610(99)70037-4
Fonseca BM, Câmara PEAS, Ogaki MB, Pinto OHB, Lirio JM, Coria SH, Vieira R, Carvalho-Silva M, Amorim ET, Convey P, Rosa LH (2022) Green algae (Viridiplantae) in sediments from three lakes on Vega Island, Antarctica, assessed using DNA metabarcoding. Mol Biol Rep 49(1):179–188. https://doi.org/10.1007/s11033-021-06857-1
Forster D, Dunthorn M, Mahé F, Dolan JR, Audic S, Bass D, Bittner L, Boutte C, Christen R, Claverie J-M, Decelle J, Edvardsen B, Egge E, Eikrem W, Gobet A, Kooistra WHCF, Logares R, Massana R, Montresor M, Not F, Ogata H, Pawlowski J, Pernice MC, Romac S, Shalchian-Tabrizi K, Simon N, Richards TA, Santini S, Sarno D, Siano R, Vaulot D, Wincker P, Zingone A, de Vargas C, Stoeck T (2016) Benthic protists: the under-charted majority. FEMS Microbiol Ecol 92(8):fiw120. https://doi.org/10.1093/femsec/fiw120
Garner RE, Gregory-Eaves I, Walsh DA (2020) Sediment metagenomes as time capsules of lake microbiomes. mSphere 5(6):e00512-20. https://doi.org/10.1128/mSphere.00512-20
Garner RE, Kraemer SA, Onana VE, Huot Y, Gregory-Eaves I, Walsh DA (2022) Protist diversity and metabolic strategy in freshwater lakes are shaped by trophic state and watershed land use on a continental scale. mSystems 7(4):e00316-22. https://doi.org/10.1128/msystems.00316-22
Gauthier J, Walsh D, Selbie DT, Bourgeois A, Griffiths K, Domaizon I, Gregory-Eaves I (2021) Evaluating the congruence between DNA-based and morphological taxonomic approaches in water and sediment trap samples: analyses of a 36-month time series from a temperate monomictic lake. Limnol Oceanogr 66(8):3020–3039. https://doi.org/10.1002/lno.11856
Gauthier J, Walsh D, Selbie DT, Domaizon I, Gregory-Eaves I (2022) Sedimentary DNA of a human-impacted lake in Western Canada (Cultus Lake) reveals changes in micro-eukaryotic diversity over the past 200 years. Environmental DNA 4(5):1106–1119. https://doi.org/10.1002/edn3.310
Giner CR, Forn I, Romac S, Logares R, de Vargas C, Massana R (2016) Environmental sequencing provides reasonable estimates of the relative abundance of specific picoeukaryotes. Appl Environ Microbiol 82(15):4757–4766. https://doi.org/10.1128/AEM.00560-16
Grattepanche JD, Walker LM, Ott BM, Paim Pinto DL, Delwiche, CF, Lane CE, Katz LA (2018) Microbial Diversity in the Eukaryotic SAR Clade: Illuminating the Darkness Between Morphology and Molecular Data. Biological Sciences: Faculty Publications, Smith College, Northampton, MA. https://scholarworks.smith.edu/bio_facpubs/91
Grossmann L, Jensen M, Heider D, Jost S, Glücksman E, Hartikainen H, Mahamdallie SS, Gardner M, Hoffmann D, Bass D, Boenigk J (2016) Protistan community analysis: key findings of a large-scale molecular sampling. ISME J 10(9):2269–2279. https://doi.org/10.1038/ismej.2016.10
Hadziavdic K, Lekang K, Lanzen A, Jonassen I, Thompson EM, Troedsson C (2014) Characterization of the 18S rRNA gene for designing universal eukaryote specific primers. PLoS One 9(2):e87624. https://doi.org/10.1371/journal.pone.0087624

Härnström K, Ellegaard M, Andersen TJ, Godhe A (2011) Hundred years of genetic structure in a sediment revived diatom population. Proc Natl Acad Sci 108(10):4252–4257. https://doi.org/10.1073/pnas.1013528108

Hou W, Dong H, Li G, Yang J, Coolen MJL, Liu X, Wang S, Jiang H, Wu X, Xiao H, Lian B, Wan Y (2014) Identification of photosynthetic plankton communities using sedimentary ancient DNA and their response to late-Holocene climate change on the Tibetan plateau. Sci Rep 4(1):6648. https://doi.org/10.1038/srep06648

Huang S, Herzschuh U, Pestryakova LA, Zimmermann HH, Davydova P, Biskaborn BK, Shevtsova I, Stoof-Leichsenring KR (2020) Genetic and morphologic determination of diatom community composition in surface sediments from glacial and thermokarst lakes in the Siberian Arctic. J Paleolimnol 64(3):225–242. https://doi.org/10.1007/s10933-020-00133-1

Huo S, Zhang H, Monchamp M-E, Wang R, Weng N, Zhang J, Zhang H, Wu F (2022a) Century-long homogenization of algal communities is accelerated by nutrient enrichment and climate warming in lakes and reservoirs of the north temperate zone. Environ Sci Technol 56(6):3780–3790. https://doi.org/10.1021/acs.est.1c06958

Huo S, Zhang H, Wang J, Chen J, Wu F (2022b) Temperature and precipitation dominates millennium changes of eukaryotic algal communities in Lake Yamzhog Yumco, southern Tibetan Plateau. Sci Total Environ 829:154636. https://doi.org/10.1016/j.scitotenv.2022.154636

Ibrahim A, Capo E, Wessels M, Martin I, Meyer A, Schleheck D, Epp LS (2021) Anthropogenic impact on the historical phytoplankton community of lake constance reconstructed by multimarker analysis of sediment-core environmental DNA. Mol Ecol 30(13):3040–3056. https://doi.org/10.1111/mec.15696

Jones RI (2000) Mixotrophy in planktonic protists: an overview. Freshw Biol 45(2):219–226. https://doi.org/10.1046/j.1365-2427.2000.00672.x

Kammerlander B, Breiner H-W, Filker S, Sommaruga R, Sonntag B, Stoeck T (2015) High diversity of protistan plankton communities in remote high mountain lakes in the European Alps and the Himalayan mountains. FEMS Microbiol Ecol 91(4):fiv010. https://doi.org/10.1093/femsec/fiv010

Kanbar HJ, Olajos F, Englund G, Holmboe M (2020) Geochemical identification of potential DNA-hotspots and DNA-infrared fingerprints in lake sediments. Appl Geochem 122:104728. https://doi.org/10.1016/j.apgeochem.2020.104728

Keck F, Millet L, Debroas D, Etienne D, Galop D, Rius D, Domaizon I (2020) Assessing the response of micro-eukaryotic diversity to the great acceleration using lake sedimentary DNA. Nat Commun 11(1):3831. https://doi.org/10.1038/s41467-020-17682-8

Kirkpatrick J, Walsh E, D'Hondt S (2016) Fossil DNA persistence and decay in marine sediment over hundred-thousand-year to million-year time scales. Graduate School of Oceanography Faculty Publications https://doi.org/10.1130/G37933.1

Kisand V, Talas L, Kisand A, Stivrins N, Reitalu T, Alliksaar T, Vassiljev J, Liiv M, Heinsalu A, Seppä H, Veski S (2018) From microbial eukaryotes to metazoan vertebrates: wide spectrum paleo-diversity in sedimentary ancient DNA over the last ~14,500 years. Geobiology 16(6):628–639. https://doi.org/10.1111/gbi.12307

Kjær KH, Winther Pedersen M, De Sanctis B, De Cahsan B, Korneliussen TS, Michelsen CS, Sand KK, Jelavić S, Ruter AH, Schmidt AMA, Kjeldsen KK, Tesakov AS, Snowball I, Gosse JC, Alsos IG, Wang Y, Dockter C, Rasmussen M, Jørgensen ME, Skadhauge B, Prohaska A, Kristensen JÅ, Bjerager M, Allentoft ME, Coissac E, Rouillard A, Simakova A, Fernandez-Guerra A, Bowler C, Macias-Fauria M, Vinner L, Welch JJ, Hidy AJ, Sikora M, Collins MJ, Durbin R, Larsen NK, Willerslev E (2022) A 2-million-year-old ecosystem in Greenland uncovered by environmental DNA. Nature 612(7939):283–291. https://doi.org/10.1038/s41586-022-05453-y

Klouch KZ, Schmidt S, Andrieux-Loyer F, Le Gac M, Hervio-Heath D, Qui-Minet ZN, Quéré J, Bigeard E, Guillou L, Siano R (2016) Historical records from dated sediment cores reveal the multidecadal dynamic of the toxic dinoflagellate Alexandrium minutum in the Bay of Brest (France). FEMS Microbiol Ecol 92(7):fiw101. https://doi.org/10.1093/femsec/fiw101

Krantzberg G (1985) The influence of bioturbation on physical, chemical and biological parameters in aquatic environments: a review. Environ Pollut Ser A, Ecol Biol 39(2):99–122. https://doi.org/10.1016/0143-1471(85)90009-1

Lammers Y, Heintzman PD, Alsos IG (2021) Environmental palaeogenomic reconstruction of an ice age algal population. Commun Biol 4(1):1–11. https://doi.org/10.1038/s42003-021-01710-4

Lefranc M, Thénot A, Lepère C, Debroas D (2005) Genetic diversity of small eukaryotes in lakes differing by their trophic status. Appl Environ Microbiol 71(10):5935–5942. https://doi.org/10.1128/AEM.71.10.5935-5942.2005

Lepère C, Domaizon I, Debroas D (2008) Unexpected importance of potential parasites in the composition of the freshwater small-eukaryote community. Appl Environ Microbiol 74(10):2940–2949. https://doi.org/10.1128/AEM.01156-07

Li G, Dong H, Hou W, Wang S, Jiang H, Yang J, Wu G (2016) Temporal succession of ancient phytoplankton community in Qinghai lake and implication for paleo-environmental change. Sci Rep 6(1):19769. https://doi.org/10.1038/srep19769

Li M, Bastos Gomes G, Zhao W, Hu G, Huang K, Yoshinaga T, Clark TG, Li W, Zou H, Wu S, Wang G (2023) Cultivation of fish ciliate parasites: progress and prospects. Rev Aquac 15(1):142–162. https://doi.org/10.1111/raq.12708

Likens GE (ed) (2010) Plankton of inland waters: a derivative of encyclopedia of inland waters. Academic Press/Elsevier, San Diego

Lischke B, Weithoff G, Wickham SA, Attermeyer K, Grossart HP, Scharnweber K, HIlt S, Gaedke U (2016) Large biomass of small feeders: ciliates may dominate herbivory in eutrophic lakes. J Plankton Res 38(1):2–15. https://doi.org/10.1093/plankt/fbv102

Lynn DH, Doerder FP, Gillis PL, Prosser RS (2018) Tetrahymena glochidiophila n. sp., a new species of Tetrahymena (Ciliophora) that causes mortality to glochidia larvae of freshwater mussels (Bivalvia). Dis Aquat Org 127(2):125–136. https://doi.org/10.3354/dao03188

Mangot J-F, Domaizon I, Taib N, Marouni N, Duffaud E, Bronner G, Debroas D (2013) Short-term dynamics of diversity patterns: evidence of continual reassembly within lacustrine small eukaryotes. Environ Microbiol 15(6):1745–1758. https://doi.org/10.1111/1462-2920.12065

Martin JL, Santi I, Pitta P, John U, Gypens N (2022) Towards quantitative metabarcoding of eukaryotic plankton: an approach to improve 18S rRNA gene copy number bias. Metabarcoding Metagenom 6:e85794. https://doi.org/10.3897/mbmg.6.85794

Marzetz V, Spijkerman E, Striebel M, Wacker A (2020) Phytoplankton community responses to interactions between light intensity, light variations, and phosphorus supply. Front Environ Sci 8

McCarthy FMG, Mertens KN, Ellegaard M, Sherman K, Pospelova V, Ribeiro S, Blasco S, Vercauteren D (2011) Resting cysts of freshwater dinoflagellates in southeastern Georgian Bay (Lake Huron) as proxies of cultural eutrophication. Rev Palaeobot Palynol 166(1):46–62. https://doi.org/10.1016/j.revpalbo.2011.04.008

Millette NC, Gast RJ, Luo JY, Moeller HV, Stamieszkin K, Andersen KH, Brownlee EF, Cohen NR, Duhamel S, Dutkiewicz S, Glibert PM, Johnson MD, Leles SG, Maloney AE, Mcmanus GB, Poulton N, Princiotta SD, Sanders RW, Wilken S (2023) Mixoplankton and mixotrophy: future research priorities. J Plankton Res:fbad020. https://doi.org/10.1093/plankt/fbad020

Mitra A, Flynn KJ, Burkholder JM, Berge T, Calbet A, Raven JA, Granéli E, Glibert PM, Hansen PJ, Stoecker DK, Thingstad F, Tillmann U, Våge S, Wilken S, Zubkov MV (2014) The role of mixotrophic protists in the biological carbon pump. Biogeosciences 11(4):995–1005. https://doi.org/10.5194/bg-11-995-2014

Mitra A, Flynn KJ, Tillmann U, Raven JA, Caron D, Stoecker DK, Not F, Hansen PJ, Hallegraeff G, Sanders R, Wilken S, McManus G, Johnson M, Pitta P, Våge S, Berge T, Calbet A, Thingstad F, Jeong HJ, Burkholder J, Glibert PM, Granéli E, Lundgren V (2016) Defining planktonic Protist functional groups on mechanisms for energy and nutrient acquisition: incorporation of diverse mixotrophic strategies. Protist 167(2):106–120. https://doi.org/10.1016/j.protis.2016.01.003

Miyazono A, Nagai S, Kudo I, Tanizawa K (2012) Viability of Alexandrium tamarense cysts in the sediment of Funka Bay, Hokkaido, Japan: over a hundred year survival times for cysts. Harmful Algae 16:81–88. https://doi.org/10.1016/j.hal.2012.02.001

Moguel B, Pérez L, Alcaraz LD, Blaz J, Caballero M, Muñoz-Velasco I, Becerra A, Laclette JP, Ortega-Guerrero B, Romero-Oliva CS, Herrera-Estrella L, Lozano-García S (2021) Holocene life and microbiome profiling in ancient tropical Lake Chalco. Mexico Sci Rep 11(1):13848. https://doi.org/10.1038/s41598-021-92981-8

More KD, Orsi WD, Galy V, Giosan L, He L, Grice K, Coolen MJL (2018) A 43 kyr record of protist communities and their response to oxygen minimum zone variability in the Northeastern Arabian Sea. Earth Planet Sci Lett 496:248–256. https://doi.org/10.1016/j.epsl.2018.05.045

Nanney DL, Park C, Preparata R, Simon EM (1998) Comparison of sequence differences in a variable 23S rRNA domain among sets of cryptic species of ciliated protozoa. J Eukaryot Microbiol 45(1):91–100. https://doi.org/10.1111/j.1550-7408.1998.tb05075.x

Nolte V, Pandey RV, Jost S, Medinger R, Ottenwälder B, Boenigk J, Schlötterer C (2010) Contrasting seasonal niche separation between rare and abundant taxa conceals the extent of protist diversity. Mol Ecol 19(14):2908–2915. https://doi.org/10.1111/j.1365-294X.2010.04669.x

Oikonomou A, Filker S, Breiner H-W, Stoeck T (2015) Protistan diversity in a permanently stratified meromictic Lake (lake Alatsee, SW Germany). Environ Microbiol 17(6):2144–2157. https://doi.org/10.1111/1462-2920.12666

Orlando L, Allaby R, Skoglund P, Der Sarkissian C, Stockhammer PW, Ávila-Arcos MC, Fu Q, Krause J, Willerslev E, Stone AC, Warinner C (2021) Ancient DNA analysis. Nat Rev Methods Primers 1(1):1–26. https://doi.org/10.1038/s43586-020-00011-0

Orsi W, Biddle JF, Edgcomb V (2013) Deep sequencing of subseafloor eukaryotic rRNA reveals active fungi across marine subsurface provinces. PLoS One 8(2):e56335. https://doi.org/10.1371/journal.pone.0056335

Pachiadaki MG, Rédou V, Beaudoin DJ, Burgaud G, Edgcomb VP (2016) Fungal and prokaryotic activities in the marine subsurface biosphere at Peru margin and Canterbury Basin inferred from RNA-based analyses and microscopy. Front Microbiol 7

Pawlowski J, Audic S, Adl S, Bass D, Belbahri L, Berney C, Bowser SS, Cepicka I, Decelle J, Dunthorn M, Fiore-Donno AM, Gile GH, Holzmann M, Jahn R, Jirků M, Keeling PJ, Kostka M, Kudryavtsev A, Lara E, Lukeš J, Mann DG, Mitchell EAD, Nitsche F, Romeralo M, Saunders GW, Simpson AGB, Smirnov AV, Spouge JL, Stern RF, Stoeck T, Zimmermann J, Schindel D, Vargas C de (2012) CBOL Protist Working Group: Barcoding Eukaryotic Richness beyond the Animal, Plant, and Fungal Kingdoms. PLOS Biology 10(11):e1001419. https://doi.org/10.1371/journal.pbio.1001419

Pawlowski J, Kelly-Quinn M, Altermatt F, Apothéloz-Perret-Gentil L, Beja P, Boggero A, Borja A, Bouchez A, Cordier T, Domaizon I, Feio MJ, Filipe AF, Fornaroli R, Graf W, Herder J, van der Hoorn B, Iwan Jones J, Sagova-Mareckova M, Moritz C, Barquín J, Piggott JJ, Pinna M, Rimet F, Rinkevich B, Sousa-Santos C, Specchia V, Trobajo R, Vasselon V, Vitecek S, Zimmerman J, Weigand A, Leese F, Kahlert M (2018) The future of biotic indices in the ecogenomic era: integrating (e)DNA metabarcoding in biological assessment of aquatic ecosystems. Sci Total Environ 637–638:1295–1310. https://doi.org/10.1016/j.scitotenv.2018.05.002

Pernthaler J, Posch T (2009) Microbial food webs. In: Likens GE (ed) Encyclopedia of inland waters. Academic, Oxford, pp 244–251

Pick FR, Caron DA (1987) Picoplankton and Nanoplankton biomass in Lake Ontario: relative contribution of phototrophic and heterotrophic communities. Can J Fish Aquat Sci 44:2164–2172. https://doi.org/10.1139/f87-265

Posch T, Köster O, Salcher MM, Pernthaler J (2012) Harmful filamentous cyanobacteria favoured by reduced water turnover with lake warming. Nature Clim Change 2(11):809–813. https://doi.org/10.1038/nclimate1581

Prosser RS, Lynn DH, Salerno J, Bennett J, Gillis PL (2018) The facultatively parasitic ciliated protozoan, Tetrahymena glochidiophila (Lynn, 2018), causes a reduction in viability of freshwater mussel glochidia. J Invertebr Pathol 157:25–31. https://doi.org/10.1016/j.jip.2018.07.012

Randlett M-È, Coolen MJL, Stockhecke M, Pickarski N, Litt T, Balkema C, Kwiecien O, Tomonaga Y, Wehrli B, Schubert CJ (2014) Alkenone distribution in Lake Van sediment over the last 270 ka: influence of temperature and haptophyte species composition. Quat Sci Rev 104:53–62. https://doi.org/10.1016/j.quascirev.2014.07.009

Sagova-Mareckova M, Boenigk J, Bouchez A, Cermakova K, Chonova T, Cordier T, Eisendle U, Elersek T, Fazi S, Fleituch T, Frühe L, Gajdosova M, Graupner N, Haegerbaeumer A, Kelly A-M, Kopecky J, Leese F, Nõges P, Orlic S, Panksep K, Pawlowski J, Petrusek A, Piggott JJ, Rusch JC, Salis R, Schenk J, Simek K, Stovicek A, Strand DA, Vasquez MI, Vrålstad T, Zlatkovic S, Zupancic M, Stoeck T (2021) Expanding ecological assessment by integrating microorganisms into routine freshwater biomonitoring. Water Res 191:116767. https://doi.org/10.1016/j.watres.2020.116767

Sanders RW (2009) Protists. In: Likens GE (ed) Encyclopedia of inland waters. Academic, Oxford, pp 252–260

Sandgren CD, Smol JP, Kristiansen J (eds) (1995) Chrysophyte algae: ecology, phylogeny and development. Cambridge University Press, Cambridge

Sanyal A, Larsson J, van Wirdum F, Andrén T, Moros M, Lönn M, Andrén E (2022) Not dead yet: diatom resting spores can survive in nature for several millennia. Am J Bot 109(1):67–82. https://doi.org/10.1002/ajb2.1780

Seeber PA, von Hippel B, Kauserud H, Löber U, Stoof-Leichsenring KR, Herzschuh U, Epp LS (2022) Evaluation of lake sedimentary ancient DNA metabarcoding to assess fungal biodiversity in Arctic paleoecosystems. Environmental DNA 4(5):1150–1163. https://doi.org/10.1002/edn3.315

Sherr EB, Sherr BF (2002) Significance of predation by protists in aquatic microbial food webs. Antonie Van Leeuwenhoek 81(1):293–308. https://doi.org/10.1023/A:1020591307260

Siano R, Lassudrie M, Cuzin P, Briant N, Loizeau V, Schmidt S, Ehrhold A, Mertens KN, Lambert C, Quintric L, Noël C, Latimier M, Quéré J, Durand P, Penaud A (2021) Sediment archives reveal irreversible shifts in plankton communities after world war II and agricultural pollution. Curr Biol 31(12):2682–2689.e7. https://doi.org/10.1016/j.cub.2021.03.079

Singer D, Seppey CVW, Lentendu G, Dunthorn M, Bass D, Belbahri L, Blandenier Q, Debroas D, de Groot GA, de Vargas C, Domaizon I, Duckert C, Izaguirre I, Koenig I, Mataloni G, Schiaffino MR, Mitchell EAD, Geisen S, Lara E (2021) Protist taxonomic and functional diversity in soil, freshwater and marine ecosystems. Environ Int 146:106262. https://doi.org/10.1016/j.envint.2020.106262

Smol JP (1985) The ratio of diatom frustules to chrysophycean statospores: a useful paleolimnological index. Hydrobiologia 123(3):199–208. https://doi.org/10.1007/BF00034378

Smol JP, Birks HJB, Last WM, Bradley RS, Alverson K (eds) (2001) Tracking environmental change using Lake sediments: terrestrial, algal, and siliceous indicators. Springer, Dordrecht

Stoof-Leichsenring KR, Epp LS, Trauth MH, Tiedemann R (2012) Hidden diversity in diatoms of Kenyan Lake Naivasha: a genetic approach detects temporal variation. Mol Ecol 21(8):1918–1930. https://doi.org/10.1111/j.1365-294X.2011.05412.x

Stoof-Leichsenring KR, Bernhardt N, Pestryakova LA, Epp LS, Herzschuh U, Tiedemann R (2014) A combined paleolimnological/genetic analysis of diatoms reveals divergent evolutionary lineages of Staurosira and Staurosirella (Bacillariophyta) in Siberian lake sediments along a latitudinal transect. J Paleolimnol 52(1):77–93. https://doi.org/10.1007/s10933-014-9779-1

Stoof-Leichsenring KR, Herzschuh U, Pestryakova LA, Klemm J, Epp LS, Tiedemann R (2015) Genetic data from algae sedimentary DNA reflect the influence of environment over geography. Sci Rep 5(1):12924. https://doi.org/10.1038/srep12924

Stoof-Leichsenring KR, Liu S, Jia W, Li K, Pestryakova LA, Mischke S, Cao X, Liu X, Ni J, Neuhaus S, Herzschuh U (2020a) Plant diversity in sedimentary DNA obtained from high-latitude (Siberia) and high-elevation lakes (China). Biodivers Data J 8:e57089. https://doi.org/10.3897/BDJ.8.e57089

Stoof-Leichsenring KR, Pestryakova LA, Epp LS, Herzschuh U (2020b) Phylogenetic diversity and environment form assembly rules for Arctic diatom genera – a study on recent and ancient sedimentary DNA. J Biogeogr 47(5):1166–1179. https://doi.org/10.1111/jbi.13786

Suthers I, Rissik D, Richardson A (eds) (2019) Plankton: a guide to their ecology and monitoring for water quality. CSIRO publishing

Talas L, Stivrins N, Veski S, Tedersoo L, Kisand V (2021) Sedimentary ancient DNA (sedaDNA) reveals fungal diversity and environmental drivers of community changes throughout the Holocene in the present boreal Lake Lielais Svētiņu (eastern Latvia). Microorganisms 9(4):719. https://doi.org/10.3390/microorganisms9040719

Tõnno I, Talas L, Freiberg R, Kisand A, Belle S, Stivrins N, Alliksaar T, Heinsalu A, Veski S, Kisand V (2021) Environmental drivers and abrupt changes of phytoplankton community in temperate lake Lielais Svētiņu, eastern Latvia, over the last post-glacial period from 14.5 kyr. Quat Sci Rev 263:107006. https://doi.org/10.1016/j.quascirev.2021.107006

Triadó-Margarit X, Casamayor EO (2012) Genetic diversity of planktonic eukaryotes in high mountain lakes (Central Pyrenees, Spain). Environ Microbiol 14(9):2445–2456. https://doi.org/10.1111/j.1462-2920.2012.02797.x

Triadó-Margarit X, Casamayor EO (2015) High protists diversity in the plankton of sulfurous lakes and lagoons examined by 18s rRNA gene sequence analyses. Environ Microbiol Rep 7(6):908–917. https://doi.org/10.1111/1758-2229.12324

Vasselon V, Rimet F, Domaizon I, Monnier O, Reyjol Y, Bouchez A (2019) Assessing pollution of aquatic environments with diatoms' DNA metabarcoding: experience and developments from France water framework directive networks. Metabarcoding Metagenom 3:e39646. https://doi.org/10.3897/mbmg.3.39646

Weisbrod B, Wood SA, Steiner K, Whyte-Wilding R, Puddick J, Laroche O, Dietrich DR (2020) Is a central sediment sample sufficient? Exploring spatial and temporal microbial diversity in a small lake. Toxins 12(9):580. https://doi.org/10.3390/toxins12090580

Weisse T, Montagnes DJS (2022) Ecology of planktonic ciliates in a changing world: concepts, methods, and challenges. J Eukaryot Microbiol 69(5):e12879. https://doi.org/10.1111/jeu.12879

Wetzel RG (2001) Limnology: Lake and river ecosystems. Elsevier

Wurzbacher C, Fuchs A, Attermeyer K, Frindte K, Grossart H-P, Hupfer M, Casper P, Monaghan MT (2017) Shifts among Eukaryota, bacteria, and archaea define the vertical organization of a lake sediment. Microbiome 5(1):41. https://doi.org/10.1186/s40168-017-0255-9

Xiong W, Jousset A, Li R, Delgado-Baquerizo M, Bahram M, Logares R, Wilden B, de Groot GA, Amacker N, Kowalchuk GA, Shen Q, Geisen S (2021) A global overview of the trophic structure within microbiomes across ecosystems. Environ Int 151:106438. https://doi.org/10.1016/j.envint.2021.106438

Zhang H, Huo S, Cao X, Ma C, Zhang J, Wu F (2021) Homogenization of reservoir eukaryotic algal and cyanobacterial communities is accelerated by dam construction and eutrophication. J Hydrol 603:126842. https://doi.org/10.1016/j.jhydrol.2021.126842

Zimmermann HH, Stoof-Leichsenring KR, Dinkel V, Harms L, Schulte L, Hütt M-T, Nürnberg D, Tiedemann R, Herzschuh U (2023) Marine ecosystem shifts with deglacial sea-ice loss inferred from ancient DNA shotgun sequencing. Nat Commun 14(1):1650. https://doi.org/10.1038/s41467-023-36845-x

Zinger L, Bonin A, Alsos IG, Bálint M, Bik H, Boyer F, Chariton AA, Creer S, Coissac E, Deagle BE, De Barba M, Dickie IA, Dumbrell AJ, Ficetola GF, Fierer N, Fumagalli L, Gilbert MTP, Jarman S, Jumpponen A, Kauserud H, Orlando L, Pansu J, Pawlowski J, Tedersoo L, Thomsen PF, Willerslev E, Taberlet P (2019) DNA metabarcoding – need for robust experimental designs to draw sound ecological conclusions. Mol Ecol 28(8):1857–1862. https://doi.org/10.1111/mec.15060

Chapter 7
Diatom DNA from Lake Sediments

Katharina Dulias, Laura S. Epp, and Kathleen R. Stoof-Leichsenring

Keywords Diatoms · Diatom-environment relationships · Lakes · Sediments · Sedimentary ancient DNA

Introduction

Diatoms (Bacillariophyta) are an important group of eukaryotic algae, contributing to approximately 25–45 % of the primary production in lakes and oceans, as well as to global carbon, nitrogen and silica cycles (Cavicchioli et al. 2019; Scibona et al. 2022). These single celled (unicellular or colonies) eukaryotic algae are characterised by their species-specific siliceous cell walls (frustule, composed of two overlapping thecae or valves). They are globally ubiquitous and important members of the benthic and planktonic algal communities, with habitats spanning from freshwater and marine to terrestrial ecosystems (Mann and Droop 1996). As they are abundant in almost all water bodies and are ecologically sensitive, diatoms serve as ideal biological indicators for the assessment of the water bodies' current ecological status (Battarbee et al. 2001a; Smol and Stoermer 2010). The preservation of their silica cell walls over geological time periods has also established them as classical indicators in paleoecology. However, the preservation of diatom valves in sediments is not always optimal (Flower 1993; Anslan et al. 2022), the analysis of

K. Dulias (✉)
Institute of Geosystems and Bioindication, Technische Universität Braunschweig, Braunschweig, Germany
e-mail: k.dulias@tu-braunschweig.de

L. S. Epp
Limnological Institute, Universität Konstanz, Konstanz, Germany
e-mail: laura.epp@uni-konstanz.de

K. R. Stoof-Leichsenring
Division Geosciences|Polar Terrestrial Environmental Systems, Alfred-Wegener-Institute, Potsdam, Germany
e-mail: kathleen.stoof-leichsenring@awi.de

E. Capo et al. (eds.), *Tracking Environmental Change Using Lake Sediments*, Developments in Paleoenvironmental Research 21,
https://doi.org/10.1007/978-3-031-43799-1_7

morphological remains is often time consuming and requires expert taxonomic knowledge. These shortcomings potentially limit sample size and biodiversity assessments, especially since low abundance taxa are often missed out. This calls for additional tools to track compositional changes, specifically when interested in analyses involving large sample numbers, multiple datasets or conducting detailed biodiversity monitoring. Furthermore, the taxonomy of diatoms is far from resolved, and they contain many undescribed species (Guiry 2012; Malviya et al. 2016), while the nomenclature of described species progress and can change over time (Adl et al. 2019). Additionally, cryptic species, i.e., closely related species that cannot be identified based on morphology alone (Leliaert et al. 2014; Mann et al. 2021), can be discriminated and can thus potentially refine the analyses of ecosystems.

A strategy to overcome these challenges is the use of DNA for the detection and identification of diatom genetic signatures preserved in sedimentary archives, including low abundance taxa. The development of standardized lab protocols, bioinformatic and data analysis tools have promoted the use of sedDNA for (paleo-) ecological analysis of sediments (e.g., Capo et al. 2021, Pawlowski et al. 2022, Chap. 3 of this volume). The present chapter highlights how diatom sedDNA approaches have contributed to paleolimnological research and the advantages and challenges related to the use of sedDNA for diatoms. It further introduces current standard methods and discusses the potential of sedDNA-based approaches such as metagenomics (shotgun sequencing), hybridization capture and the sedRNA-based metatranscriptomics. These methods enable better taxonomic identification, potentially facilitate a robust quantification of diatom signals across many taxa, allow the investigation of genomic variation within species and can integrate information on diatom-related taxa, functional processes and biotic networks, and their changes through time.

Diatoms as Ecological Indicators

Many diatom taxa are sensitive to ecological conditions (Smol and Stoermer 2010). The diatom community composition thus responds rapidly to environmental changes, both natural and anthropogenic. The composition of diatoms, both in modern ecosystems and extracted from the past, can inform on ecological conditions and the status of water bodies. Decades of intense work allow for the application of diatoms as indicators for environmental changes in flowing waters and lakes (e.g., Water Framework Directive in Europe (WFD)). The use of diatoms as ecological indicators is performed globally in many aquatic habitats, including extreme environments, such as the Arctic, Antarctic, and alpine lacustrine environments, as well as in marine and estuarine environments (for a detailed collection of diatom applications please refer to Smol and Stoermer (2010) and Table 7.1).

In lakes, diatoms live in a range of different habitats and their proliferation depends on a combination of biological, chemical, and physical conditions. The most important key controls on diatom communities used for ecological studies and

Table 7.1 Application of diatom assemblage data (based both on morphological or genetic data) for the analysis of different environments with associated references

Application of diatoms as bioindicators for the following environmental effects	Environment	References
Eutrophication	Lake, River	e.g., Hausmann et al. (2002), Stoof-Leichsenring et al. (2011), Cvetkoska et al. (2018), Liu et al. (2018), Rivera et al. (2018), Lu et al. (2020), Ibrahim et al. (2021), Khedairia et al. (2022), and Nunes et al. (2022)
Assessment of environmental conditions	Lake, River, Ocean	e.g., Pienitz and Smol (1993), Rühland et al. (2003), Härnström et al. (2011), Biskaborn et al. (2012, 2019), Pestryakova et al. (2012), Desrosiers et al. (2013), Stoof-Leichsenring et al. (2020a, b), Busseni et al. (2020), Apothéloz-Perret-Gentil et al. (2021), Costa and Schneck (2022) and Luo et al. (2022)
Long-term environmental changes	Lake	e.g., Korhola et al. (2000), Smol and Cumming (2000), Schmidt et al. (2002), Stoof-Leichsenring et al. (2011), Pestryakova et al. (2012), and Zalat et al. (2018)
Water acidity	Lake, River, Ocean	e.g., Charles (1985), Weckström et al. (1997), Laing et al. (1999), Smol and Cumming (2000), Puusepp and Kangur (2010), Biskaborn et al. (2012), and Saros and Anderson (2015)
Water-level changes	Lake	e.g., Yang and Duthie (1995), Yang et al. (2003), Puusepp and Kangur (2010), Kasper et al. (2013), Zalat et al. (2018) and Laug et al. (2021)
Changes in oceanographic conditions	Ocean	e.g., Abrantes and Moita (1999), Nees et al. (1999), Jiang et al. (2004), Okazaki et al. (2005), Witak et al. (2005), Gil et al. (2006), Lopes et al. (2010), and Pawłowska et al. (2020)
Paleoenvironmental reconstructions	Lake	e.g., Kienel et al. (1999)), Yang et al. (2003), Schmidt et al. (2004), Biskaborn et al. (2012), Pestryakova et al. (2012), Maslennikova (2020), Krahn et al. (2021), Laug et al. (2021), and Wetterich et al. (2021)

(paleo-)reconstructions are physico-chemical variables (e.g., light, temperature, turbulence, ice cover, pH, DOC, nutrients and salinity). The relationship between these environmental variables and individual diatom taxa can be described in an ecological response function (Birks et al. 2012). Usually, surface sediment calibration training sets combine modern diatom and water chemistry data from selected lakes in the area of interest and along an environmental gradient of interest (e.g., salinity, pH, DOC, total phosphorus). This approach assumes that ecological preferences of diatom taxa have not changed through time. Such diatom-environment relationships can be quantified (Weckström et al. 1997; Reed 1998; Paull et al. 2008; Hobbs et al. 2010) and are referred to as transfer functions (Fritz et al. 1991; Bennion et al. 1996;

Reed 1998; Yang et al. 2003; Laird et al. 2010; Maslennikova 2020, 2022). A key to transfer function development for a specific environmental factor is that this factor should explain a significant part of the overall diatom variation. This can be estimated using statistical methods, e.g., partial canonical correspondence analysis (CCA). The most commonly used technique for the numerical development of a diatom-based transfer function is the weighted-averaging regression (WA) (e.g., Hall and Smol 1992; Yang and Duthie 1995; Roberts and McMinn 1998; Adler and Hübener 2007). This approach assumes that taxa will be most abundant nearest to their optimum of a given environmental variable. In the reconstruction step it is then assumed that ecological preferences and optima for the variable of interest have been stable for the diatom taxa in the dataset. Thus, based on the past compositions of diatom communities, the temporal variation in the environmental variable can be reconstructed (Weckström et al. 1997; Reed 1998; Paull et al. 2008; Hobbs et al. 2010) and a range of environmental variables can now be confidently used as transfer functions (Fritz et al. 1991; Bennion et al. 1996; Reed 1998; Yang et al. 2003; Laird et al. 2010; Maslennikova 2020, 2022; Zou et al. 2021). This development allowed scientists to focus more on questions of ecological response to environmental changes rather than just the descriptive reconstruction of past environmental change.

Evolution, Taxonomy and Identification of Diatoms

The evolutionary history of diatoms (Bacillariophyta) reaches back 150 to 200 million years ago, to the Jurassic (Medlin 2016), with a major radiation into the three major clades Coscinodiscophyceae (radial centrics), Mediophyceae (polar centrics and radial Thalassiosirales) and Bacillariophyceae (pennates) occurring during the late Cretaceous. Their major defining feature, the silica frustule, evolved and diversified, presumably allowing them to efficiently exploit more nutrient-rich lower layers of the water column through an enhanced sinking ability (Smetacek 1985). The evolution of the silica cell wall also provided an enhanced protection, such as from grazers (Hamm et al. 2003), and it allowed the different diatom taxa to evolve a vast diversity of shapes - with respective advantages in different environments and a boost for the radiation of the group (Ryabov et al. 2021). In fact, diatoms have become the most species-rich group of autotrophic algae, with an estimated 100 000 species distributed amongst diverse habitats across the globe (Mann et al. 2017). Among these, many taxa remain undescribed, and the taxonomy of diatoms is far from resolved (Medlin 2018).

Identification of diatom species is classically performed microscopically according to morphological features, in particular of their frustule. Diatom valves can survive on geological timescales, providing a rich fossil record in sedimentary deposits. The preservation of the valves, however, is not always optimal (Flower 1993; Anslan et al. 2022), and can make secure identifications challenging. Furthermore, taxa that are in low abundance in the living communities can easily be lost or overlooked in these records. Additionally, and although the diatoms display such a wide variety of visible differences, it has become increasingly clear that many of the morphotaxa are in fact species complexes containing more than one cryptic species (e.g., Amato et al. 2007; Lundholm et al. 2012; Çiftçi et al. 2022). Such cryptic species are not always functionally equivalent (Mann and Vanormelingen 2013), and this can be relevant for the use of diatoms as indicators (Poulíčková et al. 2017).

Environmental and Sedimentary DNA of Diatoms

Modern environmental DNA (eDNA) and sedimentary (ancient) DNA (sedDNA), utilizing genetic information obtained directly from environmental samples, have emerged as reliable alternatives for the identification of diatoms. eDNA can be recovered from environmental samples collected in modern ecosystems, like water columns, surface soils or surface sediments, and from historical sedimentary archives, sediment core strata representing different ages of sedDNA deposition. Both DNA sample types allow a taxonomic assignment independent of microscopic diatom analyses. The earliest publication including "diatoms", "fossil DNA", "lake" and "sediments" as keywords on Scopus dates back to 2004 (Coolen et al. 2004). Since this first publication in 2004 a total of 13 research articles investigating lake diatom sedDNA have been published to date. Coolen et al. (2004) showed the unique potential of using sedimentary ancient DNA to reconstruct plankton changes in an Antarctic lake. Since then eDNA and sedDNA approaches have been shown to generate novel and interesting information on the taxonomic composition of diatom communities.

The main obstacle in assigning taxonomy to eDNA sequences is currently attributed to gaps and misidentifications in DNA sequence reference databases (Weigand et al. 2019; Monchamp et al. 2023). Since 2012, a collaborative effort has been made to provide a curated diatom sequence reference database, Diat.barcode (Rimet et al. 2019). Diat.barcode provides barcode sources, the associated metadata, describes data curation procedures, and gives information on data storage and accessibility (Rimet et al. 2019). This effort has consistently closed existing gaps and reduced misidentifications, yet database completion and the refinement of taxonomic information still need improvements. However, in published studies to date the comparison between diatom communities detected with sedDNA and microscopic data generally shows a large overlap (Dulias et al. 2017; Stoof-Leichsenring et al. 2020a; Kang et al. 2021; Anslan et al. 2022).

eDNA and sedDNA can improve the taxonomic resolution (Stoof-Leichsenring et al. 2012). Distinct genetic lineages among very similar morphotypes were detected through DNA metabarcoding, such as the (pseudo)-cryptic diversity within e.g., *Staurosira* (Dulias et al. 2017), *Chaetoceros* (De Luca et al. 2019), *Aulacoseira* (Stoof-Leichsenring et al. 2020b), and *Pseudo-nitzschia* (Ruggiero et al. 2022). While errors introduced into the DNA sequences through the experimental procedures inflate the observed variation (Brandt et al. 2021) and strict filtering criteria need to be adhered to, the analyses of cryptic diversity and/or population-level variation from environmental DNA is a new and highly promising avenue to obtain highly resolved biodiversity data (Sigsgaard et al. 2020). Because environmental filtering plays a key role in structuring diatom communities (Stoof-Leichsenring et al. 2020a; Jovanovska et al. 2022) the identification of cryptic diversity, that putatively represent functional and ecological adaptations (Godhe and Rynearson 2017) could considerably improve ecological inferences. For example, identifying cryptic diatom diversity from sedDNA made it possible to recover intra-lake diversity (Stoof-Leichsenring et al. 2020a) and the impact of lake types (e.g., glacial and thermokarst lakes) on diatom lineages (Huang et al. 2020).

The general congruence between eDNA/sedDNA and microscopic approaches in identifying diatom community changes and the identification of hidden diversity motivates the application of taxonomy free approaches applied on eDNA data (Apothéloz-Perret-Gentil et al. 2017; Porter and Hajibabaei 2018). By calculating a molecular index directly from the eDNA data and calibrating it based on the microscopic analysis of diatom assemblages, the genetic data can be used for biomonitoring and biodiversity assessments without taxonomic assignments (Apothéloz-Perret-Gentil et al. 2017). Such taxonomy-independent assignments would facilitate fast and cost-effective assessments of diatom composition, especially useful for monitoring (Pawlowski et al. 2018) and can overcome limitations posed by current gaps in sequence reference databases (Weigand et al. 2019). It also holds the potential to remove all biases introduced by taxonomic knowledge via reference databases that rely on morphological approaches. Further, it facilitates data reproducibility as it is not affected by errors introduced by differences in personal identification strategies, which is helpful for the computation of ecological indices in modern training datasets. However, this calls for a standardization of bioinformatic analyses of the DNA data to guarantee the comparability of biotic assemblages between different studies (Bailet et al. 2020).

Although eDNA and sedDNA can be very similar in terms of sample type, the quality of recovered DNA is different due to the effect of DNA degradation, which can lead to the fragmentation of DNA molecules into short fragments. Moreover, the chemical modifications of DNA molecules causing post-mortem damage at the extremity of DNA fragments can hamper the identification of diatom taxa from sedimentary archives. Such degradations of the DNA molecules in sedimentary archives can lead to technical challenges when applying the molecular biology approaches. For instance 1) when using a metabarcoding approach - a method based on the use of primers to amplify specific genic regions that have conserved sequences for the targeted biological groups - the targeted short and potentially damaged DNA sequences are likely less amplifiable by PCR than modern eDNA; and 2) when using shotgun sequencing approach - based on the sequencing of all DNA sequences from a given DNA extract - the DNA sequences can be misidentify when compared to a reference database because of the presence of post-mortem damages. Therefore, an important aspect when considering the analysis of sedDNA is the question of its persistence on geological timescales. Generally, DNA will be more damaged over time, which means that DNA fragments get shorter and modifications on the DNA molecule accumulate. Such modifications will reduce the detectability, amplification success and taxonomic identification of sedDNA. However, environmental conditions during and after DNA deposition affect its preservation. Cold and anoxic conditions are suspected to favour DNA preservation in lake sediments. In addition, the lake water pH and conductivity (Jia et al. 2022) are important factors influencing the absorption of DNA to organic and mineral particles, and thus its preservation (for details on DNA preservation in sediments please refer to Chap. 2 of this volume). Even under optimal conditions, there is a time limit to which DNA can persist in the environment. While the oldest fossil evidence for diatoms in the geological record dates to the early Jurassic (Rothpletz 1896, 1900) and early Cretaceous (Harwood and Gersonde 1990), specific molecular analysis focusing on diatom sedDNA has to date successfully been performed on samples spanning the Holocene in lake sediment (Coolen et al. 2004; Epp et al. 2015; Stoof-Leichsenring et al. 2015), and in marine sediments the Late Pleistocene (~ 30,000 years) (Zimmermann et al. 2020, 2021) and reach even further up to hundred thousand years ago (Kirkpatrick et al. 2016). Similarly, recent sedDNA works based on shotgun sequencing approach detected diatom DNA in ~ 540 000 years-old marine sediments (Armbrecht et al. 2022; Zimmermann et al. 2023) and ~ 2 million years-old lake sediments (Kjær et al. 2022). The relatively few sedDNA diatom studies on lake sediments to date are thus by far not representative of the potential this method has to offer and the field of lake diatom sedDNA research will surely expand in the future.

Established Metabarcoding Approaches for Diatoms

Environmental samples, such as lake sediments, are very complex and the extraction of DNA and the detection of genetic material from specific taxa can be very challenging (see Chap. 3 of this volume). The development of approaches based on taxon-specific DNA markers is one such methodological challenge, including primer biases, but also particularities of the genetic region regarding the gene copy number, database completion and the extraction, amplification and sequencing biases creating false positive or negative identifications.

Overall, diatom eDNA metabarcoding studies rely on the molecular analysis of DNA sequences found in five genes: the D2/D3 region of the LSU rRNA gene, cox1 gene, the *rbc*L gene, the V4 region of the 18S rRNA gene, and the ITS region (Visco et al. 2015) (Table 7.1). Additionally, targeting regions of the 16S rRNA gene based-analysis has recently also been tested for biomonitoring and offers the potential of re-analysing existing V4 16S rRNA gene data (initially generated for microbial community analysis) with a specific focus on diatoms using the PhytoREF database, allowing to use this standard marker for prokaryotes to inform diatom distribution, phylogeny and ecological assessments (Bonfantine et al. 2021). The ideal molecular methods to study diatom DNA may not exist, but the targeting of the V4

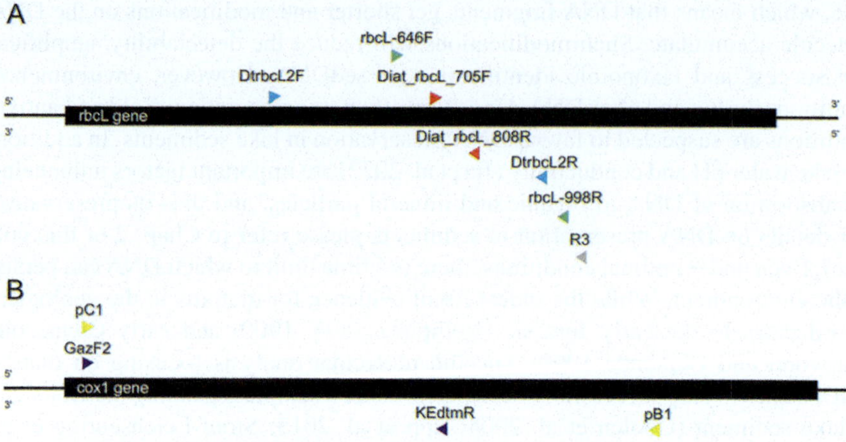

Fig. 7.1 Position of commonly used *rbc*L and cox1 primers for the analysis of sediment samples. See Table 7.1 for the sequences and references

16 rRNA gene has been proposed as more appropriate for applied studies on environmental samples than 18S rRNA gene and *rbc*L. In contrast, highly discriminating barcodes such as cox1 and ITS are more suitable for species-specific studies (Visco et al. 2015). For sedDNA, diatoms are eukaryotic algae, as such their compositional changes have been successfully reconstructed using 18S metabarcoding approach and general eukaryotic primers (e.g., Capo et al. 2017; Keck et al. 2020; Ibrahim et al. 2021). However, studies have shown that diatom monitoring using the 18S rRNA gene, compared to the *rbc*L gene, provides lower taxonomic resolution (Apothéloz-Perret-Gentil et al. 2021). As such, analyses of *rbc*L and *cox*1 are most commonly used in sedDNA studies (Fig. 7.1).

The design and evaluation of diatom specific primers for sedDNA fragments of different lengths was and still is essential to the success of this research field. A high taxonomic resolution is generally achieved using longer barcodes, which calls for a trade-off when dealing with the shorter sedDNA fragments. One such pioneer study for primers specific to diatom DNA fragments from lake sediments was done by Epp et al. (2011) using short fragments (~ 100 bp) of the small subunit ribosomal (SSU) gene. DNA molecules from environmental samples, such as lake sediments, are not only exposed to degradation, leading to small fragments (i.e. smaller than 100 bp), they also exist in small amounts. As such, these samples have a high contamination risk (Willerslev and Cooper 2005). Therefore, a follow-up study developed a genetic survey technique based on a short region (76-67 bp) of the ribulose-1,5-bisphosphate carboxylase/oxygenase large subunit (*rbc*L) gene as genetic barcode (Stoof-Leichsenring et al. 2012), as *rbc*L is a single locus plastid gene which occurs in multiple copies per diatom cell. This is particularly advantageous when working with degraded DNA. The designed barcode for a short region of the *rbc*L gene has proven very useful for the analysis of short and fragmented sedDNA. The 18S rRNA genes have slightly more copies per cell, however the *rbc*L genes have a higher discrimination capacity (Rimet et al. 2019). As an increasing number of diatom reference DNA sequences for the *rbc*L gene are becoming available in GenBank, the short fragment of the *rbc*L gene is a very good candidate to analyse diatom diversity from lake sediments. In the future, the available Diat.barcode database (Rimet et al. 2019) should also be implemented in sedDNA analysis to improve taxonomic assignments. Nonetheless, the dependency on complete reference databases and the time it takes to create them, might hamper the development of new barcodes.

Another advantage of the *rbc*L gene is that, as a chloroplast gene, the likelihood of amplifying DNA from non-photosynthetic organisms is greatly reduced (Evans et al. 2007; Stoof-Leichsenring et al. 2012), which is important as lake sediments contain a mixture of genetic material from various other organisms. Other DNA fragments of the *rbc*L gene (with primers rbcL-646F/rbcL-998R (Kelly et al. 2018) and Diat_rbcL_708F/R3 (Vasselon et al. 2017), initially developed for freshwater metabarcoding of diatoms, were tested on sedimentary DNA of lake surface sediments (Kang et al. 2021). Of these the primer pair rbcL-646F/rbcL-998R, which is amplifying a 331 bp long gene fragment, performed better than the slightly shorter primer pair developed by Vasselon et al. (2017). Similar results using this primer pair were demonstrated for biomonitoring in rivers (Pérez-Burillo et al. 2022). However, even though the longer barcodes provide the highest taxonomic resolution and allow the analysis of intraspecific genetic diversity and cryptic species on a detailed level, their application to fragmented and degraded sedDNA is limited.

A study of mock communities identified methodological biases in DNA extraction and PCR amplification, which created false-negatives, whereas errors in pyrosequencing generated false positives in closely related species (Kermarrec et al. 2013). The selection of the marker gene influences the observed diatom diversity. Indeed, different markers work better for certain taxa and the fragmented and damaged sedDNA is prone to primer bias, as well-preserved sedDNA molecules are preferentially amplified. Moreover, the taxonomic coverage of the DNA reference database is crucial for accurate identification of diatom taxa (Kermarrec et al. 2013). Hence, both the markers and available reference libraries influence which taxa are dominantly found in the investigated samples. This bias can be tackled by using different marker combinations or using other methods than metabarcoding, e.g., metagenomics (shotgun sequencing). A study by Armbrecht et al. (2021a) demonstrated that the diatom composition differed significantly between metabarcoding and shotgun sequencing data due to differing read lengths and the choice of reference database in metabarcoding data. The authors recommendation for future studies was to pursue metagenomic shotgun sequencing with a high sequencing depth to avoid amplification biases, as well as the combination of both metabarcoding and shotgun sequencing for a few samples to evaluate the bias between target gene region length and average shotgun sequencing DNA sequence lengths (Table 7.2).

Table 7.2 Commonly used diatom (meta)barcoding primers for sediment samples. The GazF2 primer has been developed for brown algae. The positions of these primers are given in Fig. 7.1

Gene or genomic region	Code forward Primer	Code reverse Primer	Forward primer sequence 5′-3′	Reverse primer sequence 5′-3′	References
18S	DIV4for	DIV4rev3	GCGGTAATTCCAGCTCCAATAG	CTCTGACAATGGAATACGAATA	Visco et al. (2015)
*rbc*L	DtrbcL2F	DtrbcL2R	TCAGAACGGACTCGAATTAAA	CCAATWGTACCACCACCYAAT	Forward: Lang and Kaczmarska (2011) Reverse: NDrbcL8; Daugbjerg and Andersen (1997)
*rbc*L	DtrbcL3F	DtrbcL3R	GATTTATTTGAGGAAGSTTCD	ACACCWGACATACGCATCCA	Lang and Kaczmarska (2011)
*rbc*L	Diat_rbcL_705F	Diat_rbcL_808R	AACAGGTGAAGTTAAAGGTTCATAYTT	TGTAACCCATAACTAAATCGATCAT	Stoof-Leichsenring et al. (2012)
*rbc*L	rbcL-646F	rbcL-998R	ATGCGTTGGAGAGARCGTTTC	GATCACCTTCTAATTTACCWACAACTG	Kelly et al. (2018)
*rbc*L	R3_1	R3_2	CCTTCTAATTTACCWACWACTG	CCTTCTAATTTACCWACAACAG	Vasselon et al. (2017)
*cox*1	GazF2	KEdtmR	CAACCAYAAAGATATWGGTAC	AAACTTCWGGRTGACCAAAAA	Forward: Hamsher et al. (2011) Reverse: Evans et al. (2007)
*cox*1	pB1	pC1	GCWACWACRTARTAWGTRTCRTG	TGGTTNTTYTCNACNAAYCAYAARGAYAT	Ehara et al. (2000)
LSU D2/D3	T16N	T24U	AMAAGTACCRYGAGGGAAAG	SCWCTAATCATTCGCTTTACC	Forward: Sherwood and Presting (2007) Reverse: Hamsher et al. (2011) modified from Sherwood and Presting (2007)
ITS	ITS1	ITS2; ITS4	TCCGTAGGTGAACCTGCGG	GCTGCGTTCTTCATCGATGC; TCCTCCGCTTATTGATATGC	White et al. (1990)

Comparison of Microfossil and sedDNA Diatom Data and Their Limitations and Biases

The compatibility of the genetic data with the classical morphological identification of the retrieved diatom species assemblages has been a central question of sedDNA studies. Coolen et al. (2004) were the first to combine DNA data (molecular inventories) and microscopic data (diatom valves counting) to study the long-term changes in a lake diatom assemblage. Since then, several studies have investigated the similarities between molecular and morphological diatom inventories from lake sediments and showed comparable findings in terms of relative frequencies and beta diversity (Stoof-Leichsenring et al. 2012, 2014, 2020a; Epp et al. 2015; Dulias et al. 2017; Kang et al. 2021; Anslan et al. 2022). Also, the implications of diatom-environment relationships have been shown to overlap largely in comparative studies (Dulias et al. 2017; Huang et al. 2020) (Fig. 7.2). However, all of these studies also reported mismatches between metabarcoding data and microscopy, which can mostly be attributed to the incompleteness and erroneousness (Zimmermann et al. 2015) of the reference sequence databases rather than the limitations in taxonomic resolution of the developed genetic barcodes. Additionally, large diatom taxa might also be underrepresented or missed during morphological identification, which would also cause discrepancies between the two approaches.

Ever since the first sedDNA diatom study (Coolen et al. 2004), improving the reference database and available target genes has still been a primary concern when dealing with diatom sedDNA. A good quality reference database, such as Diat.barcode (Rimet et al. 2019) or NCBI Genbank, is particularly important to ensure that

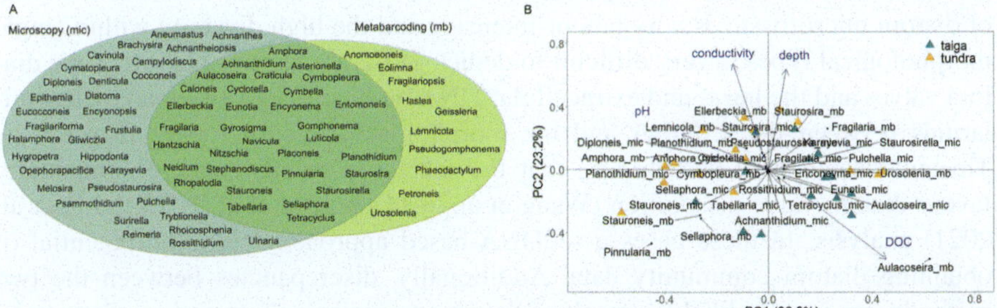

Fig. 7.2 (**a**) Venn diagram of 42 genetically identified diatom sequence types and 64 morphologically identified diatom genera. In both approaches 32 genera were identified, ~76 % of the genetically identified diatom sequences were concordant with the morphological data and 50 % of the morphologically identified data were concordant with the genetically identified data. 10 genera were uniquely identified by the genetic approach and 32 by the morphological identification. Reproduced from Dulias et al. (Dulias et al. 2017). (**b**) Principal Component Analysis of the morphologically (indicated by suffix "_mic") and genetically (indicated by suffix "_mb", for metabarcoding) retrieved genera from surface sediment samples of thermokarst lakes in tundra and taiga sites of the Taymyr Peninsula (Siberia, Russia) and projected explanatory variables including water depth, conductivity, DOC and pH. The figure is adopted from Dulias et al. (2017)

as many diatom sequence reads as possible can not only be identified, but separated from contaminant reads in the sequencing data (Pinder and Töpel 2022). Thus, strategies for the curation, management and completion of these reference libraries have been developed (for details please see Rimet et al. 2021).

Although the completion of the reference database is an important task and still a major bias for sedDNA analysis, other limitations and biases should be considered as well. Due to taxonomic changes, discrepancies exist between the reference database and morphological taxonomy (e.g., Ács et al. 2016). Barcode length and primer bias pose a further limitation to sedDNA diatom analysis, as the separation of closely related taxa due to barcode limitations might not be possible. Moreover, DNA extraction methods might extract certain taxa preferentially over others, depending on valve size and robustness of their frustulum. Furthermore, the bioinformatic processing might induce biases, such as the algorithms for taxonomic assignments or the differing approaches of operational taxonomic units (OTU) versus individual sequence units (ISU). Another bias is continuously evolving nomenclature which can result in different versions of nomenclature (e.g., old or not yet widely applied new versions) being used creating confusions in taxonomic classification (Adl et al. 2019), a problem that already exists with the current reference databases or calibration datasets for transfer functions. As such, the taxonomic nomenclature needs to be unified in order to maximize the power of taxonomic assignment. In order to overcome this taxonomic disparity, a harmonization method has been developed to utilize already existing sedDNA data and create repeatability for future analysis (Potapova et al. 2022).

One major potential of sedDNA is to overcome the effect of dissolution of diatom frustules that limits microscopic analyses (Kang et al. 2021), rendering diatom analysis possible even in cases, in which the microfossil remains cannot or can only poorly be identified, such as e.g., alkaline environments. Taxonomic identification of diatom microfossils is very labour-intensive and the high diversity within single morphological types is very difficult to depict. Differential preservation of the diatom valves and the intra- and extracellular DNA molecules can also lead to dissimilarities between the sedDNA and the morphological records, as studies on the Tibetan Plateau have shown e.g., that the increased salinity and pH potentially favour diatom valve dissolution (Wang et al. 2011; Kasper et al. 2013; Laug et al. 2021) analysis. In these cases a sedDNA-based approach offers the potential of obtaining diatom community data. Additionally, discrepancies between the two methods are, in part, likely, a result of detection efficiency between small-size and large-size taxa. Small taxa that tend to be preferentially counted under the microscope, while larger taxa have higher gene copy numbers and are thus, more easily detected by the DNA approach. During the morphological slide preparation silica valves of fragile taxa (e.g., *Fistulifera saprophila*) may be destroyed (Zgrundo et al. 2013) and the morphological approach is limited (usually) to 300–400 valve counts (due to time constraints), the genetic approach has no comparable limits. Thus, the analysis depth of the two approaches is different.

While the two approaches might provide different taxonomic resolution and richness estimates of diatom taxa, the compositional change is mostly comparable between both methods. The comparability of the approaches has also been confirmed in similar diatom-environment relationships detected in recent ecosystems, promoting the application of diatom environmental DNA for lake and river monitoring and water quality assessment (Bailet et al. 2020; Apothéloz-Perret-Gentil et al. 2021; Nistal-García et al. 2021; Kutty et al. 2022; Pérez-Burillo et al. 2022).

Diatom Resting Stages and Their Potential to Investigate Past Diatom Population Genetics

Another very exciting prospect is the information gained from the dormant resting stages that some diatom species form, and can be revived from the archives after extended time periods of up to millennia. As reported by Sanyal et al. (2022) who successfully revived and amplified DNA from resting spores of *Chaetoceros muelleri* that had been archived ~ 1300 and 7200 years ago. While traditional ancient DNA studies have the disadvantage of DNA degradation and contamination, the revival of dormant cells would allow access to phenotypic responses to various environmental conditions and allow for comparisons between ancient and recent strains (Pinder and Töpel 2022). Furthermore, the revival of resting stages and their germination and propagation, where possible, could also provide plentiful DNA to perform genotypic studies and thus further connect modern with ancient molecular research, as well as with (paleo-)ecological research. One such interdisciplinary field would be *resurrection ecology*, which studies either the genetic material of dormant organisms or dormant organisms themselves, in order to gain information about the past (Kerfoot et al. 1999; Pinder and Töpel 2022). Studying living organisms from the past allows us to assess the impact of past events and their genetic effects by comparing pre- and post-event populations (Härnström et al. 2011). This can help shed light on the specific mechanisms diatom strains have evolved to survive environmental changes in the past and inform about potential future population developments.

Advances for Paleoecological Research

The sedDNA approach offers significant qualitative advantages both for the workflow and the depth of analyses that can be undertaken. The molecular methodology allows a standardized and replicable workflow, independent of expert taxonomic knowledge and thus, overcoming misidentification problems. It can be streamlined for the analysis of a multitude of samples, which drastically improves the analysis time. Thus, larger temporal and spatial scales can be investigated with a higher sampling rate. Furthermore, this approach allows a higher taxonomic resolution, as inter- and intra-species diversity can be investigated in detail offering new insights into species composition and population diversity shifts. This improves the estimates of diversity, as this previously hidden diversity can now be detected and included. In the ecosystems where fossil remains are not well-preserved, sedDNA might also offer the potential to study diatom communities thus allowing access to a wider range of ecosystems.

Emerging Topics and Future Directions

Technical Advancements in Metabarcoding

Without doubt the number of sedDNA metabarcoding studies targeting diatoms (or protists in general, see Chap. 6 of this volume) are increasing, similar to plant metabarcoding studies in recent years (see dataset Von Eggers et al. 2022, Chap. 10 of this volume). However, the taxonomic complexity within diatoms, cryptic diversity, and differences in the resolution and length of established markers make decisions for a standardised approach more difficult, and the incompleteness of sequence reference databases spread over different markers present an additional challenge. The reference database situation is improving (Rimet et al. 2021), but the inherent difficulties of diatom taxonomy and systematics will remain for the foreseeable future. This has prompted the development of taxonomy-free approaches (e.g., Apothéloz-Perret-Gentil et al. 2017; Feio et al. 2020; Tapolczai et al. 2021). A standardized metabarcoding protocol would allow for coherent delimitation of genetic lineages, and an integration of these datasets as training data for such approaches. Defined genetic lineages and their relationship to the environment from a modern dataset could be used to implement transfer functions to quantify past environmental changes, which has been established on diatom microfossils for more than three decades.

Ecosystem Level Approaches Including Diatoms

In order to gain a more complete view of biodiversity changes than is possible with single metabarcoding markers, metagenomic shotgun sequencing of total sedDNA are being more commonly used (Armbrecht et al. 2021a, 2022; Wang et al. 2021; Courtin et al. 2022). These approaches can provide ecosystem-level information, and recover information from non-fossilized taxa, allowing a joint analysis of turnover and associations among taxa. Shotgun sequencing approaches of total sedimentary DNA provide a more complete picture of multiple organismic groups without the limitations due to primer choice, PCR biases, and the limited databases per barcode, thereby allowing to provide a broader overview of ecosystem changes and community diversity. These approaches come with their own limitations and biases, many of which are yet to be described. Nonetheless, as we continue to improve and develop analytical methods of shotgun sequencing data, including the development of efficient bioinformatic tools to process and interpret the data, this type of data promises to provide a novel view on ecosystem dynamics. Metagenomic sequencing libraries from sedDNA can also be enriched for certain taxa through hybridization capture approaches (Slon et al. 2017; Schulte et al. 2021). This method allows an enrichment of specific molecules in the pool of total DNA fragments by using specific baits hybridizing to the target DNA. Such an approach could be used to enrich metagenomic information of key taxa to maximize DNA recovery (Murchie et al. 2021), to investigate genetic variation among closely related lineages (Schulte et al. 2021, 2022) and to improve ancient DNA damage pattern detection (Armbrecht et al. 2021b). Another approach is multi-barcode hybridization capture approach, which enriches multiple diagnostic barcodes at once to increase taxonomic assignment in data-poor ecosystems like deep sea sediments (Günther et al. 2022) and rare taxa identification in eDNA samples (Giebner et al. 2020). In all of these experimental procedures, the resulting data remains more complex than in PCR-based approaches, permitting an ecosystem-level view through the analyses of DNA.

At the ecosystem-level, diatoms also have a functional importance. A powerful and not yet very often applied method to study their functional activity and the associated species diversity is the metatranscriptomic analysis of gene expression in the species community and in separate populations (Broman et al. 2017). Early studies have investigated the effects of oxygen concentration on the metabolic processes of marine diatoms from hypoxic sediments, with oxygenation not significantly resulting in major shifts or changes in the relative proportion of diatoms (Broman et al. 2017). The metatranscriptomic profiles of diatoms from coastal waters off Greenland showed that diatoms have a more plastic gene expression compared to dinoflagellates and that they fill their ecological niches mostly through variable resource use (Elferink et al. 2020). Importantly, a study of the sediment microbiome using metatranscriptomics showed that sample storage has a significantly altering effect on the results (Tennant et al. 2022), and should thus be considered for future studies employing this promising approach.

Quantitative Use of Environmental DNA

One aspect that is not yet solved is the question of how environmental DNA data can best be used quantitatively. It is a great challenge to remove the discrepancies between the number of DNA sequences, which are assigned to a given species in a metabarcoding dataset, and the observed number of specimens of the same species in the microscopic analysis (Vasselon et al. 2019). The same issue has been observed when sequencing mock communities (Visco et al. 2015) and explained through technical biases introduced at any stage from DNA extraction, PCR amplification and sequencing, as well as by biological factors like the variation of gene copies, which depend on differences in cell size, genome size and number of organelles.

Abundance estimates are, however, the classical basis for transfer functions, with which diatom-environment relationships can be inferred back in time (Hall and Smol 1992; Gasse et al. 1997; Battarbee et al. 2001a; Ramstack et al. 2003; Cumming et al. 2015; Zou et al. 2021). These rely on modern datasets (either from water column or surface sediment data), from which diatom-environment relationships are deduced in order to reconstruct the environment from sedimentary assemblages. Such an approach can also be envisioned using sedDNA data. The preservation of both diatom frustules and sedDNA are not equally dependent on the same environmental factors and, therefore, need individual strategies for abundance estimation. For reconstruction using sedDNA-based transfer functions from lake sediment cores will require the development of calibration sets based on sedDNA surface samples, in similar ways as it has been developed for microscopic-based transfer functions (e.g. European calibration datasets, EDDI; Battarbee et al. (2001b)). An alternative approach could be to rely solely on DNA from water column samples to create the calibration sets which is then applied to sedDNA-based paleoreconstructions, in order to avoid the known limitations of surface sediment DNA.

The metabarcoding approach has shown that the compositional signal of morphological and genetic analyses is overall very similar. However, numbers of sequencing reads cannot be transformed to abundance estimations based on microscopic count, for two main reasons: 1) the primers used in the metabarcoding reaction might introduce a taxon-specific bias, distorting the amount of amplified DNA for different taxa; and 2) different diatom species inherently contain different copy numbers of the markers used that are dependent on biovolumes of the cells (Godhe et al. 2008; Vasselon et al. 2018). While the first aspect has to be tackled during experimental design, which is highly significant (Pierella Karlusich et al. 2023), the second aspect can potentially be addressed through introducing a calibration, such as a correction factor for biovolume (Vasselon et al. 2018). Another option using the link between cell biovolume and *rbc*L gene copy number is the calibration of the genetic data to calculate the Trophic Diatom Index (TDI) for ecological assessments carried out for freshwater monitoring (Kelly et al. 2018). In this approach a variant of the TDI is derived specifically for the DNA data, which is calibrated to optimally fit the morphological data. The bypassing of the taxonomic assignment, which allows researchers to include unclassified sequences missing from the reference database, by producing new indices for ecological assessments is another promising avenue (Tapolczai et al. 2019). This method uses individual sequence units (ISUs) from metabarcoding, together with minimal filtering steps for ecological index estimation and the ISU-based indices, and outperformed methods based on operational taxonomic units (OTUs) in terms of accuracy and predictive power (Tapolczai et al. 2019). These new approaches were thus able to reveal intra-species and intra-population variability otherwise hidden in the clustered OTUs (Tapolczai et al. 2019). These approaches allow semi-quantitative analyses providing consistent information for the computation of ecological indices without the need for correction.

As a recent alternative relying on metagenomic sequencing, Pierella Karlusich et al. (2023) proposed the use of a new marker, the photosynthetic gene *psb*O. In the analysed dataset of the *Tara* Oceans expedition, with a very high sequencing depth, this approach was highly suitable as a proxy of relative cell abundance (Pierella Karlusich et al. 2023). The increasing power of sequencers, coupled with dropping costs, has the potential to robustly deliver quantitative information based on sequencing data.

Summary

Generally, the analysis of sedDNA of lake sediments has increased in the last decade and is now an established tool for addressing questions of ecological and environmental changes related to anthropogenic activities, biodiversity and climate change. The focus on diatom sedDNA is an important part of this research field. The molecular assessment of diatom abundance and diversity has the potential to improve efficiency, reduce analytical error through standardisation and automation, and thus, provides an alternative to the time-consuming and expert-dependent traditional species identification approach. The identification of cryptic diatom diversity and their adaptation to different environments is of great importance when interpreting diatom community compositions. The use of sedDNA provides the possibility to unlock new taxonomic information not previously accessible through morphologically-based methods.

The reconstruction of diatom population dynamic, diversity and structure through sedDNA approaches has a great potential to grow in the future. The expansion of metagenomics, shotgun sequencing would provide a more complete picture of multiple organismic groups without the limitations due to primer choice, PCR biases, and the limited databases per barcode. Likewise, ongoing, phylogenetic and evolutionary studies on diatoms will likely benefit from sedDNA data.

Acknowledgements We would like to thank Valentin Vasselon, Cécilia Barouillet, John Smol and Eric Capo for their thorough review of this manuscript.

References

Abrantes F, Moita M (1999) Water column and recent sediment data on diatoms and coccolithophorids, off Portugal, confirm sediment record of upwelling events. Oceanol Acta 22(1):67–84. https://doi.org/10.1016/S0399-1784(99)80034-6

Ács É, Ari E, Duleba M, Dreßler M, Genkal SI, Jakó É, Rimet F, Ector L, Kiss KT (2016) Pantocsekiella, a new centric diatom genus based on morphological and genetic studies. Fottea 16(1):56–78. https://doi.org/10.5507/fot.2015.028

Adl SM, Bass D, Lane CE, Lukeš J, Schoch CL, Smirnov A, Agatha S, Berney C, Brown MW, Burki F, Cárdenas P, Čepička I, Chistyakova L, del Campo J, Dunthorn M, Edvardsen B, Eglit Y, Guillou L, Hampl V, Heiss AA, Hoppenrath M, James TY, Karnkowska A, Karpov S, Kim E, Kolisko M, Kudryavtsev A, Lahr DJG, Lara E, Le Gall L, Lynn DH, Mann DG, Massana R, Mitchell EAD, Morrow C, Park JS, Pawlowski JW, Powell MJ, Richter DJ, Rueckert S, Shadwick L, Shimano S, Spiegel FW, Torruella G, Youssef N, Zlatogursky V, Zhang Q (2019) Revisions to the classification, nomenclature, and diversity of eukaryotes. J Eukaryot Microbiol 66(1):4–119. https://doi.org/10.1111/jeu.12691

Adler S, Hübener T (2007) Spatial variability of diatom assemblages in surface lake sediments and its implications for transfer functions. J Paleolimnol 37(4):573–590. https://doi.org/10.1007/s10933-006-9054-1

Amato A, Kooistra WHCF, Levialdi Ghiron JH, Mann DG, Pröschold T, Montresor M (2007) Reproductive isolation among sympatric cryptic species in marine diatoms. Protist 158(2):193–207. https://doi.org/10.1016/j.protis.2006.10.001

Anslan S, Kang W, Dulias K, Wünnemann B, Echeverría-Galindo P, Börner N, Schwarz A, Liu Y, Liu K, Künzel S, Kisand V, Rioual P, Peng P, Wang J, Zhu L, Vences M, Schwalb A (2022) Compatibility of diatom valve records with sedimentary ancient DNA amplicon data: a case study in a Brackish, Alkaline Tibetan Lake. Front Earth Sci:10

Apothéloz-Perret-Gentil L, Cordonier A, Straub F, Iseli J, Esling P, Pawlowski J (2017) Taxonomy-free molecular diatom index for high-throughput eDNA biomonitoring. Mol Ecol Resour 17(6):1231–1242. https://doi.org/10.1111/1755-0998.12668

Apothéloz-Perret-Gentil L, Bouchez A, Cordier T, Cordonier A, Guéguen J, Rimet F, Vasselon V, Pawlowski J (2021) Monitoring the ecological status of rivers with diatom eDNA metabarcoding: A comparison of taxonomic markers and analytical approaches for the inference of a molecular diatom index. Mol Ecol 30(13):2959–2968. https://doi.org/10.1111/mec.15646

Armbrecht L, Eisenhofer R, Utge J, Sibert EC, Rocha F, Ward R, Pierella Karlusich JJ, Tirichine L, Norris R, Summers M, Bowler C (2021a) Paleo-diatom composition from Santa Barbara Basin deep-sea sediments: a comparison of 18S-V9 and diat-rbcL metabarcoding vs shotgun metagenomics. ISME COMMUN 1(1):1–10. https://doi.org/10.1038/s43705-021-00070-8

Armbrecht L, Hallegraeff G, Bolch CJS, Woodward C, Cooper A (2021b) Hybridisation capture allows DNA damage analysis of ancient marine eukaryotes. Sci Rep 11:3220. https://doi.org/10.1038/s41598-021-82578-6

Armbrecht L, Weber ME, Raymo ME, Peck VL, Williams T, Warnock J, Kato Y, Hernández-Almeida I, Hoem F, Reilly B, Hemming S, Bailey I, Martos YM, Gutjahr M, Percuoco V, Allen C, Brachfeld S, Cardillo FG, Du Z, Fauth G, Fogwill C, Garcia M, Glüder A, Guitard M, Hwang J-H, Iizuka M, Kenlee B, O'Connell S, Pérez LF, Ronge TA, Seki O, Tauxe L, Tripathi S, Zheng X (2022) Ancient marine sediment DNA reveals diatom transition in Antarctica. Nat Commun 13(1):5787. https://doi.org/10.1038/s41467-022-33494-4

Bailet B, Apothéloz-Perret-Gentil L, Baričević A, Chonova T, Franc A, Frigerio J-M, Kelly M, Mora D, Pfannkuchen M, Proft S, Ramon M, Vasselon V, Zimmermann J, Kahlert M (2020) Diatom DNA metabarcoding for ecological assessment: Comparison among bioinformatics pipelines used in six European countries reveals the need for standardization. Sci Total Environ 745:140948. https://doi.org/10.1016/j.scitotenv.2020.140948

Battarbee RW, Jones VJ, Flower RJ et al (2001a) Diatoms. In: Smol JP, Birks HJB, Last WM (eds) Tracking environmental change using lake sediments. Terrestrial, algal and siliceous indicators, vol 3. Kluwer Academic Publisher, New York, pp 155–202

Battarbee RW, Juggins S, Gasse F, et al (2001b) European Diatom Database (EDDI). An information system for palaeoenvironmental reconstruction

Bennion H, Juggins S, Anderson NJ (1996) Predicting epilimnetic phosphorus concentrations using an improved diatom-based transfer function and its application to lake eutrophication management. Environ Sci Technol 30(6):2004–2007. https://doi.org/10.1021/es9508030

Birks HJB, Lotter AF, Juggins S, Smol JP (2012) Tracking Environmental Change Using Lake Sediments: Data Handling and Numerical Techniques. Springer Science & Business Media

Biskaborn BK, Herzschuh U, Bolshiyanov D, Savelieva L, Diekmann B (2012) Environmental variability in northeastern Siberia during the last ~13,300yr inferred from lake diatoms and sediment–geochemical parameters. Palaeogeogr Palaeoclimatol Palaeoecol 329–330:22–36. https://doi.org/10.1016/j.palaeo.2012.02.003

Biskaborn BK, Nazarova L, Pestryakova LA, Syrykh L, Funck K, Meyer H, Chapligin B, Vyse S, Gorodnichev R, Zakharov E, Wang R, Schwamborn G, Bailey HL, Diekmann B (2019) Spatial distribution of environmental indicators in surface sediments of Lake Bolshoe Toko, Yakutia, Russia. Biogeosciences 16(20):4023–4049. https://doi.org/10.5194/bg-16-4023-2019

Bonfantine KL, Trevathan-Tackett SM, Matthews TG, Neckovic A, Gan HM (2021) Dumpster diving for diatom plastid 16S rRNA genes. PeerJ 9:e11576. https://doi.org/10.7717/peerj.11576

Brandt MI, Trouche B, Quintric L, Günther B, Wincker P, Poulain J, Arnaud-Haond S (2021) Bioinformatic pipelines combining denoising and clustering tools allow for more comprehensive prokaryotic and eukaryotic metabarcoding. Mol Ecol Resour 21(6):1904–1921. https://doi.org/10.1111/1755-0998.13398

Broman E, Sachpazidou V, Dopson M, Hylander S (2017) Diatoms dominate the eukaryotic metatranscriptome during spring in coastal 'dead zone' sediments. Proc R Soc B Biol Sci 284(1864):20171617. https://doi.org/10.1098/rspb.2017.1617

Busseni G, Caputi L, Piredda R, Fremont P, Hay Mele B, Campese L, Scalco E, de Vargas C, Bowler C, d'Ovidio F, Zingone A, Ribera d'Alcalà M, Iudicone D (2020) Large scale patterns of marine diatom richness: Drivers and trends in a changing ocean. Glob Ecol Biogeogr 29(11):1915–1928. https://doi.org/10.1111/geb.13161

Capo E, Debroas D, Arnaud F, Perga M-E, Chardon C, Domaizon I (2017) Tracking a century of changes in microbial eukaryotic diversity in lakes driven by nutrient enrichment and climate warming. Environ Microbiol 19(7):2873–2892. https://doi.org/10.1111/1462-2920.13815

Capo E, Giguet-Covex C, Rouillard A, et al (2021) Lake sedimentary DNA research on past terrestrial and aquatic biodiversity: Overview and recommendations. Quaternary 4(1):6. https://doi.org/10.3390/quat4010006

Cavicchioli R, Ripple WJ, Timmis KN, Azam F, Bakken LR, Baylis M, Behrenfeld MJ, Boetius A, Boyd PW, Classen AT, Crowther TW, Danovaro R, Foreman CM, Huisman J, Hutchins DA, Jansson JK, Karl DM, Koskella B, Mark Welch DB, Martiny JBH, Moran MA, Orphan VJ, Reay DS, Remais JV, Rich VI, Singh BK, Stein LY, Stewart FJ, Sullivan MB, van Oppen MJH, Weaver SC, Webb EA, Webster NS (2019) Scientists' warning to humanity: microorganisms and climate change. Nat Rev Microbiol 17(9):569–586. https://doi.org/10.1038/s41579-019-0222-5

Charles DF (1985) Relationships between Surface Sediment Diatom Assemblages and Lakewater Characteristics in Adirondack Lakes. Ecology 66(3):994–1011. https://doi.org/10.2307/1940561

Çiftçi O, Alverson AJ, van Bodegom P, Roberts WR, Mertens A, Van de Vijver B, Trobajo R, Mann DG, Pirovano W, van Eijk I, Gravendeel B (2022) Phylotranscriptomics reveals the reticulate evolutionary history of a widespread diatom species complex. J Phycol 58(5):643–656. https://doi.org/10.1111/jpy.13281

Coolen MJL, Muyzer G, Rijpstra WIC, Schouten S, Volkman JK, Sinninghe Damsté JS (2004) Combined DNA and lipid analyses of sediments reveal changes in Holocene haptophyte and diatom populations in an Antarctic lake. Earth Planet Sci Lett 223:225–239. https://doi.org/10.1016/j.epsl.2004.04.014

Costa APT, Schneck F (2022) Diatoms as indicators in running waters: trends of studies on biological assessment and monitoring. Environ Monit Assess 194(10):695. https://doi.org/10.1007/s10661-022-10383-3

Courtin J, Perfumo A, Andreev AA, Opel T, Stoof-Leichsenring KR, Edwards ME, Murton JB, Herzschuh U (2022) Pleistocene glacial and interglacial ecosystems inferred from ancient DNA analyses of permafrost sediments from Batagay megaslump, East Siberia. Environ DNA 4(6):1265–1283. https://doi.org/10.1002/edn3.336

Cumming B, Laird K, Gregory-Eaves I, Simpson K, Sokal M, Nordin R, Walker I (2015) Tracking past changes in lake-water phosphorus with a 251-lake calibration dataset in British Columbia: tool development and application in a multiproxy assessment of eutrophication and recovery in Osoyoos Lake, a transboundary lake in Western North America. Front Ecol Evol 3

Cvetkoska A, Pavlov A, Jovanovska E, Tofilovska S, Blanco S, Ector L, Wagner-Cremer F, Levkov Z (2018) Spatial patterns of diatom diversity and community structure in ancient Lake Ohrid. Hydrobiologia 819(1):197–215. https://doi.org/10.1007/s10750-018-3637-5

Daugbjerg N, Andersen RA (1997) A molecular phylogeny of the heterokont algae based on analyses of chloroplast-encoded rbcL sequence data. J Phycol 33(6):1031–1041. https://doi.org/10.1111/j.0022-3646.1997.01031.x

De Luca DD, Kooistra WHCF, Sarno D, Gaonkar CC, Piredda R (2019) Global distribution and diversity of Chaetoceros (Bacillariophyta, Mediophyceae): integration of classical and novel strategies. PeerJ 7:e7410. https://doi.org/10.7717/peerj.7410

Desrosiers C, Leflaive J, Eulin A, Ten-Hage L (2013) Bioindicators in marine waters: Benthic diatoms as a tool to assess water quality from eutrophic to oligotrophic coastal ecosystems. Ecol Indic 32:25–34. https://doi.org/10.1016/j.ecolind.2013.02.021

Dulias K, Stoof-Leichsenring KR, Pestryakova LA, Herzschuh U (2017) Sedimentary DNA versus morphology in the analysis of diatom-environment relationships. J Paleolimnol 57(1):51–66. https://doi.org/10.1007/s10933-016-9926-y

Ehara M, Inagaki Y, Watanabe KI, Ohama T (2000) Phylogenetic analysis of diatom coxI genes and implications of a fluctuating GC content on mitochondrial genetic code evolution. Curr Genet 37(1):29–33. https://doi.org/10.1007/s002940050004

Elferink S, Wohlrab S, Neuhaus S, Cembella A, Harms L, John U (2020) Comparative metabarcoding and metatranscriptomic analysis of microeukaryotes within coastal surface waters of west greenland and northwest iceland. frontiers in marine. Science 7

Epp LS, Stoof-Leichsenring KR, Trauth MH, Tiedemann R (2011) Molecular profiling of diatom assemblages in tropical lake sediments using taxon-specific PCR and Denaturing High-Performance Liquid Chromatography (PCR-DHPLC). Mol Ecol Resour 11(5):842–853. https://doi.org/10.1111/j.1755-0998.2011.03022.x

Epp LS, Gussarova G, Boessenkool S, Olsen J, Haile J, Schrøder-Nielsen A, Ludikova A, Hassel K, Stenøien HK, Funder S, Willerslev E, Kjær K, Brochmann C (2015) Lake sediment multi-taxon DNA from North Greenland records early post-glacial appearance of vascular plants and accurately tracks environmental changes. Quat Sci Rev 117:152–163. https://doi.org/10.1016/j.quascirev.2015.03.027

Evans KM, Wortley AH, Mann DG (2007) An assessment of potential diatom "barcode" genes (cox1, rbcL, 18S and ITS rDNA) and their effectiveness in determining relationships in Sellaphora (Bacillariophyta). Protist 158(3):349–364. https://doi.org/10.1016/j.protis.2007.04.001

Feio MJ, Serra SRQ, Mortágua A, Bouchez A, Rimet F, Vasselon V, Almeida SFP (2020) A taxonomy-free approach based on machine learning to assess the quality of rivers with diatoms. Sci Total Environ 722:137900. https://doi.org/10.1016/j.scitotenv.2020.137900

Flower RJ (1993) Diatom preservation: experiments and observations on dissolution and breakage in modern and fossil material. In: van Dam H (ed) Twelfth international diatom symposium. Springer, Dordrecht, pp 473–484

Fritz SC, Juggins S, Battarbee RW, Engstrom DR (1991) Reconstruction of past changes in salinity and climate using a diatom-based transfer function. Nature 352(6337):706–708. https://doi.org/10.1038/352706a0

Gasse F, Barker P, Gell PA, Fritz SC, Chalie F (1997) Diatom-inferred salinity in palaeolakes: An indirect tracer of climate change. Quat Sci Rev 16(6):547–563. https://doi.org/10.1016/S0277-3791(96)00081-9

Giebner H, Langen K, Bourlat SJ, Kukowka S, Mayer C, Astrin JJ, Misof B, Fonseca VG (2020) Comparing diversity levels in environmental samples: DNA sequence capture and metabarcoding approaches using 18S and COI genes. Mol Ecol Resour 20(5):1333–1345. https://doi.org/10.1111/1755-0998.13201

Gil IM, Abrantes FF, Hebbeln D (2006) The North Atlantic Oscillation forcing through the last 2000 years: spatial variability as revealed by high-resolution marine diatom records from N and SW Europe. Mar Micropaleontol 60(2):113–129. https://doi.org/10.1016/j.marmicro.2006.03.003

Godhe A, Rynearson T (2017) The role of intraspecific variation in the ecological and evolutionary success of diatoms in changing environments. Philos Trans Royal Soc B Biol Sci 372(1728):20160399. https://doi.org/10.1098/rstb.2016.0399

Godhe A, Asplund ME, Härnström K, Saravanan V, Tyagi A, Karunasagar I (2008) Quantification of diatom and dinoflagellate biomasses in coastal marine seawater samples by real-time PCR. Appl Environ Microbiol 74(23):7174–7182. https://doi.org/10.1128/AEM.01298-08

Guiry MD (2012) How many species of algae are there? J Phycol 48(5):1057–1063. https://doi.org/10.1111/j.1529-8817.2012.01222.x

Günther B, Marre S, Defois C, Merzi T, Blanc P, Peyret P, Arnaud-Haond S (2022) Capture by hybridization for full-length barcode-based eukaryotic and prokaryotic biodiversity inventories of deep sea ecosystems. Mol Ecol Resour 22(2):623–637. https://doi.org/10.1111/1755-0998.13500

Hall RI, Smol JP (1992) A weighted—averaging regression and calibration model for inferring total phosphorus concentration from diatoms in British Columbia (Canada) lakes. Freshw Biol 27(3):417–434. https://doi.org/10.1111/j.1365-2427.1992.tb00551.x

Hamm CE, Merkel R, Springer O, Jurkojc P, Maier C, Prechtel K, Smetacek V (2003) Architecture and material properties of diatom shells provide effective mechanical protection. Nature 421(6925):841–843. https://doi.org/10.1038/nature01416

Hamsher SE, Evans KM, Mann DG, Poulíčková A, Saunders GW (2011) Barcoding diatoms: exploring alternatives to COI-5P. Protist 162(3):405–422. https://doi.org/10.1016/j.protis.2010.09.005

Härnström K, Ellegaard M, Andersen TJ, Godhe A (2011) Hundred years of genetic structure in a sediment revived diatom population. Proc Natl Acad Sci 108(10):4252–4257. https://doi.org/10.1073/pnas.1013528108

Harwood DM, Gersonde R (1990) Lower Cretaceous diatoms from ODP Leg 113 Site 693 (Weddell Sea). Part 2: resting spores, chrysophycean cysts, an endoskeletal dinoflagellate, and notes on the origin of diatoms. In: Barker PF, Kennett JP, Al E (eds) Proceedings of the ocean drilling program, scientific results. pp 403–425. https://doi.org/10.2973/odp.proc.sr.113.201.1990

Hausmann S, Lotter AF, van Leeuwen JFN, Ohlendorf C, Lemcke G, Grönlund E, Sturm M (2002) Interactions of climate and land use documented in the varved sediments of Seebergsee in the Swiss Alps. The Holocene 12(3):279–289. https://doi.org/10.1191/0959683602hl544rp

Hobbs WO, Telford RJ, Birks HJB, Saros JE, Hazewinkel RRO, Perren BB, Saulnier-Talbot É, Wolfe AP (2010) Quantifying recent ecological changes in remote lakes of North America and Greenland using sediment diatom assemblages. PLoS ONE 5(4):e10026. https://doi.org/10.1371/journal.pone.0010026

Huang S, Herzschuh U, Pestryakova LA, Zimmermann HH, Davydova P, Biskaborn BK, Shevtsova I, Stoof-Leichsenring KR (2020) Genetic and morphologic determination of diatom community composition in surface sediments from glacial and thermokarst lakes in the Siberian Arctic. J Paleolimnol 64(3):225–242. https://doi.org/10.1007/s10933-020-00133-1

Ibrahim A, Capo E, Wessels M, Martin I, Meyer A, Schleheck D, Epp LS (2021) Anthropogenic impact on the historical phytoplankton community of Lake Constance reconstructed by multimarker analysis of sediment-core environmental DNA. Mol Ecol 30(13):3040–3056. https://doi.org/10.1111/mec.15696

Jia W, Liu X, Stoof-Leichsenring KR, Liu S, Li K, Herzschuh U (2022) Preservation of sedimentary plant DNA is related to lake water chemistry. Environmental DNA 4(2):425–439. https://doi.org/10.1002/edn3.259

Jiang H, Zheng Y, Ran L, Seidenkrantz M-S (2004) Diatoms from the surface sediments of the South China Sea and their relationships to modern hydrography. Mar Micropaleontol 53(3):279–292. https://doi.org/10.1016/j.marmicro.2004.06.005

Jovanovska E, Hauffe T, Stelbrink B, Cvetkoska A, Levkov Z, Wagner B, Lacey JH, Ognjanova-Rumenova N, Hamilton PB, Brandenburg KM, Albrecht C, Wilke T (2022) Environmental filtering drives assembly of diatom communities over evolutionary time-scales. Glob Ecol Biogeogr 31(5):954–967. https://doi.org/10.1111/geb.13471

Kang W, Anslan S, Börner N, Schwarz A, Schmidt R, Künzel S, Rioual P, Echeverría-Galindo P, Vences M, Wang J, Schwalb A (2021) Diatom metabarcoding and microscopic analyses from sediment samples at Lake Nam Co, Tibet: the effect of sample-size and bioinformatics on the identified communities. Ecol Indic 121:107070. https://doi.org/10.1016/j.ecolind.2020.107070

Kasper T, Frenzel P, Haberzettl T, Schwarz A, Daut G, Meschner S, Wang J, Zhu L, Mäusbacher R (2013) Interplay between redox conditions and hydrological changes in sediments from Lake Nam Co (Tibetan Plateau) during the past 4000cal BP inferred from geochemical and micro-paleontological analyses. Palaeogeogr Palaeoclimatol Palaeoecol 392:261–271. https://doi.org/10.1016/j.palaeo.2013.09.027

Keck F, Millet L, Debroas D, Etienne D, Galop D, Rius D, Domaizon I (2020) Assessing the response of micro-eukaryotic diversity to the Great Acceleration using lake sedimentary DNA. Nat Commun 11(1):3831. https://doi.org/10.1038/s41467-020-17682-8

Kelly M, Boonham N, Juggins S, Kille P, Mann D, Pass D, Sapp M, Sato S, Glover R (2018) A DNA based diatom metabarcoding approach for Water Framework Directive classification of rivers

Kerfoot WC, Robbins JA, Weider LJ (1999) A new approach to historical reconstruction: Combining descriptive and experimental paleolimnology. Limnol Oceanogr 44(5):1232–1247. https://doi.org/10.4319/lo.1999.44.5.1232

Kermarrec L, Bouchez A, Rimet F, Humbert J-F (2013) First evidence of the existence of semi-cryptic species and of a phylogeographic structure in the Gomphonema parvulum (Kützing) Kützing complex (Bacillariophyta). Protist 164(5):686–705. https://doi.org/10.1016/j.protis.2013.07.005

Khedairia T, Okhapkin AG, Genkal SI, Gluschenko AM, Kulikovskiy MS (2022) Composition and ecological characteristics of diatoms of Algocenoses in the benthal of the mouth of a large eutrophic–hypertrophic river (Oka River, Russia). Inland Water Biol 15(4):415–423. https://doi.org/10.1134/S1995082922040320

Kienel U, Siegert C, Hahne J (1999) Late quaternary palaeoenvironmental reconstructions from a permafrost sequence (North Siberian Lowland, SE Taymyr Peninsula)—a multidisciplinary case study. Boreas 28(1):181–193. https://doi.org/10.1111/j.1502-3885.1999.tb00213.x

Kirkpatrick J, Walsh E, D'Hondt S (2016) Fossil DNA persistence and decay in marine sediment over hundred-thousand-year to million-year time scales. Graduate School of Oceanography Faculty Publications. https://doi.org/10.1130/G37933.1

Kjær KH, Winther Pedersen M, De Sanctis B, De Cahsan B, Korneliussen TS, Michelsen CS, Sand KK, Jelavić S, Ruter AH, Schmidt AMA, Kjeldsen KK, Tesakov AS, Snowball I, Gosse JC, Alsos IG, Wang Y, Dockter C, Rasmussen M, Jørgensen ME, Skadhauge B, Prohaska A, Kristensen JÅ, Bjerager M, Allentoft ME, Coissac E, Rouillard A, Simakova A, Fernandez-Guerra A, Bowler C, Macias-Fauria M, Vinner L, Welch JJ, Hidy AJ, Sikora M, Collins MJ, Durbin R, Larsen NK, Willerslev E (2022) A 2-million-year-old ecosystem in Greenland uncovered by environmental DNA. Nature 612(7939):283–291. https://doi.org/10.1038/s41586-022-05453-y

Korhola A, Weckström J, Holmström L, Erästö P (2000) A quantitative holocene climatic record from diatoms in Northern Fennoscandia. Quat Res 54(2):284–294. https://doi.org/10.1006/qres.2000.2153

Krahn KJ, Tucci M, Urban B, Pilgrim J, Frenzel P, Soulié-Märsche I, Schwalb A (2021) Aquatic and terrestrial proxy evidence for Middle Pleistocene palaeolake and lake-shore development at two Lower Palaeolithic sites of Schöningen, Germany. Boreas 50(3):723–745. https://doi.org/10.1111/bor.12523

Kutty SN, Loh RK, Bannister W, Taylor D (2022) Evaluation of a diatom eDNA-based technique for assessing water quality variations in tropical lakes and reservoirs. Ecol Indic 141:109108. https://doi.org/10.1016/j.ecolind.2022.109108

Laing TE, Rühland KM, Smol JP (1999) Past environmental and climatic changes related to tree-line shifts inferred from fossil diatoms from a lake near the Lena River Delta, Siberia. The Holocene 9(5):547–557. https://doi.org/10.1191/095968399675614733

Laird KR, Kingsbury MV, Cumming BF (2010) Diatom habitats, species diversity and water-depth inference models across surface-sediment transects in Worth Lake, northwest Ontario, Canada. J Paleolimnol 44(4):1009–1024. https://doi.org/10.1007/s10933-010-9470-0

Lang I, Kaczmarska I (2011) A protocol for a single-cell PCR of diatoms from fixed samples: method validation using Ditylum brightwellii (T. West) Grunow. Diatom Res 26(1):43–49. https://doi.org/10.1080/0269249X.2011.57370

Laug A, Haberzettl T, Pannes A, Schwarz A, Turner F, Wang J, Engels S, Rigterink S, Börner N, Ahlborn M, Ju J, Schwalb A (2021) Holocene paleoenvironmental change inferred from two sediment cores collected in the Tibetan lake Taro Co. J Paleolimnol 66(3):171–186. https://doi.org/10.1007/s10933-021-00198-6

Leliaert F, Verbruggen H, Vanormelingen P, Steen F, López-Bautista JM, Zuccarello GC, De Clerck O (2014) DNA-based species delimitation in algae. Eur J Phycol 49(2):179–196. https://doi.org/10.1080/09670262.2014.904524

Liu B, McLean CE, Long DT, Steinman AD, Stevenson RJ (2018) Eutrophication and recovery of a Lake inferred from sedimentary diatoms originating from different habitats. Sci Total Environ 628–629:1352–1361. https://doi.org/10.1016/j.scitotenv.2018.02.174

Lopes C, Mix AC, Abrantes F (2010) Environmental controls of diatom species in northeast Pacific sediments. Palaeogeogr Palaeoclimatol Palaeoecol 297(1):188–200. https://doi.org/10.1016/j.palaeo.2010.07.029

Lu X, Liu Y, Fan Y (2020) Diatom taxonomic composition as a biological indicator of the ecological health and status of a river basin under agricultural influence. Water 12(7):2067. https://doi.org/10.3390/w12072067

Lundholm N, Bates SS, Baugh KA, Bill BD, Connell LB, Léger C, Trainer VL (2012) Cryptic and pseudo-cryptic diversity in diatoms—with descriptions of Pseudo-Nitzschia Hasleana Sp. Nov. and P. Fryxelliana Sp. Nov.1. J Phycol 48(2):436–454. https://doi.org/10.1111/j.1529-8817.2012.01132.x

Luo H, Li J, Li P, Rioual P, Zou Y, Zhang J, Yan Y, Li J, Wang L (2022) Variation in the seasonal response to climate change during the past 1000 years as inferred from a Maar Lake sediment record, northeast China. J Paleolimnol 68(1):133–154. https://doi.org/10.1007/s10933-021-00228-3

Malviya S, Scalco E, Audic S, Vincent F, Veluchamy A, Poulain J, Wincker P, Iudicone D, de Vargas C, Bittner L, Zingone A, Bowler C (2016) Insights into global diatom distribution and diversity in the world's ocean. Proc Natl Acad Sci 113(11):E1516–E1525. https://doi.org/10.1073/pnas.1509523113

Mann DG, Droop SJM (1996) 3. Biodiversity, biogeography and conservation of diatoms. Hydrobiologia 336(1):19–32. https://doi.org/10.1007/BF00010816

Mann DG, Vanormelingen P (2013) An inordinate fondness? The number, distributions, and origins of diatom species. J Eukaryot Microbiol 60(4):414–420. https://doi.org/10.1111/jeu.12047

Mann DG, Crawford RM, Round FE (2017) Bacillariophyta. In: Archibald JM, Simpson AGB, Slamovits CH (eds) Handbook of the Protists. Springer International Publishing, Cham, pp 205–266

Mann DG, Trobajo R, Sato S, Li C, Witkowski A, Rimet F, Ashworth MP, Hollands RM, Theriot EC (2021) Ripe for reassessment: A synthesis of available molecular data for the speciose diatom family Bacillariaceae. Mol Phylogenet Evol 158:106985. https://doi.org/10.1016/j.ympev.2020.106985

Maslennikova AV (2020) Development and application of an electrical conductivity transfer function, using diatoms from lakes in the Urals, Russia. J Paleolimnol 63(2):129–146. https://doi.org/10.1007/s10933-019-00106-z

Maslennikova AV (2022) Dataset of diatom abundances and diatom-inferred total phosphorus and electrical conductivity for Lake Tavatui (Middle Urals, Russi) over the last 11.7 cal ka BP. Data Brief 42:108300. https://doi.org/10.1016/j.dib.2022.108300

Medlin LK (2016) Evolution of the diatoms: major steps in their evolution and a review of the supporting molecular and morphological evidence. Phycologia 55(1):79–103. https://doi.org/10.2216/15-105.1

Medlin LK (2018) Mini review: diatom species as seen through a molecular window. Braz J Bot 41(2):457–469. https://doi.org/10.1007/s40415-018-0444-1

Monchamp M-E, Taranu ZE, Garner RE, Rehill T, Morissette O, Iversen LL, Fugère V, Littlefair JE, da Costa NB, Desforges JE, Sánchez Schacht JR, Derry AM, Cooke SJ, Barrett RDH, Walsh DA, Ragoussis J, Albert M, Cristescu ME, Gregory-Eaves I (2023) Prioritizing taxa for genetic reference database development to advance inland water conservation. Biol Conserv 280:109963. https://doi.org/10.1016/j.biocon.2023.109963

Murchie TJ, Kuch M, Duggan AT, Ledger ML, Roche K, Klunk J, Karpinski E, Hackenberger D, Sadoway T, MacPhee R, Froese D, Poinar H (2021) Optimizing extraction and targeted capture of ancient environmental DNA for reconstructing past environments using the PalaeoChip Arctic-1.0 bait-set. Quat Res 99:305–328. https://doi.org/10.1017/qua.2020.59

Nees S, Armand L, De Deckker P, Labracherie M, Passlow V (1999) A diatom and benthic foraminiferal record from the South Tasman Rise (southeastern Indian Ocean): implications for palaeoceanographic changes for the last 200,000 years. Mar Micropaleontol 38(1):69–89. https://doi.org/10.1016/S0377-8398(99)00039-0

Nistal-García A, García-García P, García-Girón J, Borrego-Ramos M, Blanco S, Bécares E (2021) DNA metabarcoding and morphological methods show complementary patterns in the metacommunity organization of lentic epiphytic diatoms. Sci Total Environ 786:147410. https://doi.org/10.1016/j.scitotenv.2021.147410

Nunes M, Lemley DA, Adams JB (2022) Benthic diatom diversity and eutrophication in temporarily closed estuaries. Estuar Coasts. https://doi.org/10.1007/s12237-022-01126-1

Okazaki Y, Takahashi K, Katsuki K, Ono A, Hori J, Sakamoto T, Uchida M, Shibata Y, Ikehara M, Aoki K (2005) Late quaternary paleoceanographic changes in the southwestern Okhotsk Sea: evidence from geochemical, radiolarian, and diatom records. Deep-Sea Res Part II Top Stud Oceanogr 52(16–18):2332–2350. https://doi.org/10.1016/j.dsr2.2005.07.007

Paull TM, Hamilton PB, Gajewski K, LeBlanc M (2008) Numerical analysis of small Arctic diatoms (Bacillariophyceae) representing the Staurosira and Staurosirella species complexes. Phycologia 47(2):213–224. https://doi.org/10.2216/07-17.1

Pawłowska J, Łącka M, Kucharska M, Pawlowski J, Zajączkowski M (2020) Multiproxy evidence of the Neoglacial expansion of Atlantic Water to eastern Svalbard. Clim Past 16(2):487–501. https://doi.org/10.5194/cp-16-487-2020

Pawlowski J, Kelly-Quinn M, Altermatt F, Apothéloz-Perret-Gentil L, Beja P, Boggero A, Borja A, Bouchez A, Cordier T, Domaizon I, Feio MJ, Filipe AF, Fornaroli R, Graf W, Herder J, van der Hoorn B, Iwan Jones J, Sagova-Mareckova M, Moritz C, Barquín J, Piggott JJ, Pinna M, Rimet F, Rinkevich B, Sousa-Santos C, Specchia V, Trobajo R, Vasselon V, Vitecek S, Zimmerman J, Weigand A, Leese F, Kahlert M (2018) The future of biotic indices in the ecogenomic era: Integrating (e)DNA metabarcoding in biological assessment of aquatic ecosystems. Sci Total Environ 637–638:1295–1310. https://doi.org/10.1016/j.scitotenv.2018.05.002

Pawlowski J, Bruce K, Panksep K, Aguirre FI, Amalfitano S, Apothéloz-Perret-Gentil L, Baussant T, Bouchez A, Carugati L, Cermakova K, Cordier T, Corinaldesi C, Costa FO, Danovaro R, Dell'Anno A, Duarte S, Eisendle U, Ferrari BJD, Frontalini F, Frühe L, Haegerbaeumer A, Kisand V, Krolicka A, Lanzén A, Leese F, Lejzerowicz F, Lyautey E, Maček I, Sagova-Marečková M, Pearman JK, Pochon X, Stoeck T, Vivien R, Weigand A, Fazi S (2022) Environmental DNA metabarcoding for benthic monitoring: a review of sediment sampling and DNA extraction methods. Sci Total Environ 818:151783. https://doi.org/10.1016/j.scitotenv.2021.151783

Pérez-Burillo J, Mann DG, Trobajo R (2022) Evaluation of two short overlapping rbcL markers for diatom metabarcoding of environmental samples: effects on biomonitoring assessment and species resolution. Chemosphere 307:135933. https://doi.org/10.1016/j.chemosphere.2022.135933

Pestryakova LA, Herzschuh U, Wetterich S, Ulrich M (2012) Present-day variability and Holocene dynamics of permafrost-affected lakes in central Yakutia (Eastern Siberia) inferred from diatom records. Quat Sci Rev 51:56–70. https://doi.org/10.1016/j.quascirev.2012.06.020

Pienitz R, Smol JP (1993) Diatom assemblages and their relationship to environmental variables in lakes from the boreal forest-tundra ecotone near Yellowknife, Northwest Territories, Canada. Hydrobiologia 269(1):391–404. https://doi.org/10.1007/BF00028037

Pierella Karlusich JJ, Pelletier E, Zinger L, Lombard F, Zingone A, Colin S, Gasol JM, Dorrell RG, Henry N, Scalco E, Acinas SG, Wincker P, de Vargas C, Bowler C (2023) A robust approach to estimate relative phytoplankton cell abundances from metagenomes. Mol Ecol Resour 23(1):16–40. https://doi.org/10.1111/1755-0998.13592

Pinder MI, Töpel M (2022) Ancient Diatom DNA. In: The molecular life of diatoms. Springer, Cham, pp 87–108

Porter TM, Hajibabaei M (2018) Scaling up: A guide to high-throughput genomic approaches for biodiversity analysis. Mol Ecol 27(2):313–338. https://doi.org/10.1111/mec.14478

Potapova MG, Lee SS, Spaulding SA, Schulte NO (2022) A harmonized dataset of sediment diatoms from hundreds of lakes in the northeastern United States. Sci Data 9(1):540. https://doi.org/10.1038/s41597-022-01661-3

Poulíčková A, Letáková M, Hašler P, Cox E, Duchoslav M (2017) Species complexes within epiphytic diatoms and their relevance for the bioindication of trophic status. Sci Total Environ 599–600:820–833. https://doi.org/10.1016/j.scitotenv.2017.05.034

Puusepp L, Kangur M (2010) Linking diatom community dynamics to changes in terrestrial vegetation: a palaeolimnological case study of Lake Ķūži, Vidzeme Heights (Central Latvia). Est J Ecol 59(4):259–280

Ramstack JM, Fritz SC, Engstrom DR, Heiskary SA (2003) The application of a diatom-based transfer function to evaluate regional water-quality trends in minnessota since 1970. J Paleolimnol 29(1):79–94. https://doi.org/10.1023/A:1022869205291

Reed JM (1998) A diatom-conductivity transfer function for Spanish salt lakes. J Paleolimnol 19(4):399–416. https://doi.org/10.1023/A:1007934627134

Rimet F, Gusev E, Kahlert M, Kelly MG, Kulikovskiy M, Maltsev Y, Mann DG, Pfannkuchen M, Trobajo R, Vasselon V, Zimmermann J, Bouchez A (2019) Diat.barcode, an open-access curated barcode library for diatoms. Sci Rep 9(1):15116. https://doi.org/10.1038/s41598-019-51500-6

Rimet F, Aylagas E, Borja A, Bouchez A, Canino A, Chauvin C, Chonova T Jr, Costa FO, BJD F, Gastineau R, Goulon C, Gugger M, Holzmann M, Jahn R, Kahlert M, Kusber W-H, Laplace-Treyture C, Leese F, Leliaert F, Mann DG, Marchand F, Méléder V, Pawlowski J, Rasconi S, Rivera S, Rougerie R, Schweizer M, Trobajo R, Vasselon V, Vivien R, Weigand A, Witkowski A, Zimmermann J, Ekrem T (2021) Metadata standards and practical guidelines for specimen and DNA curation when building barcode reference libraries for aquatic life. Metabarcod Metagenom 5:e58056. https://doi.org/10.3897/mbmg.5.58056

Rivera SF, Vasselon V, Jacquet S, Bouchez A, Ariztegui D, Rimet F (2018) Metabarcoding of lake benthic diatoms: from structure assemblages to ecological assessment. Hydrobiologia 807(1):37. https://doi.org/10.1007/s10750-017-3381-2

Roberts D, McMinn A (1998) A weighted-averaging regression and calibration model for inferring lakewater salinity from fossil diatom assemblages in saline lakes of the Vestfold Hills: a new tool for interpreting Holocene lake histories in Antarctica. J Paleolimnol 19(2):99–113. https://doi.org/10.1023/A:1007947402927

Rothpletz A (1896) Ueber die Flysch-Fucoiden und einige andere fossile Algen, sowie über liasische, Diatomeen führende Hornschwämme. Z Dtsch Geol Ges 854–914

Rothpletz A (1900) Über einen neuen jurassischen Hornschwamm und die darin eingeschlossenen Diatomeen. Zeitschrift der Dtsch Geol Gesellschaft 52:154–160

Ruggiero MV, Kooistra WHCF, Piredda R, Sarno D, Zampicinini G, Zingone A, Montresor M (2022) Temporal changes of genetic structure and diversity in a marine diatom genus discovered via metabarcoding. Environmental DNA 4(4):763–775. https://doi.org/10.1002/edn3.288

Rühland K, Priesnitz A, Smol JP (2003) Paleolimnological evidence from diatoms for recent environmental changes in 50 lakes across Canadian Arctic Treeline. Arct Antarct Alp Res 35(1):110–123

Ryabov A, Kerimoglu O, Litchman E, Olenina I, Roselli L, Basset A, Stanca E, Blasius B (2021) Shape matters: the relationship between cell geometry and diversity in phytoplankton. Ecol Lett 24(4):847–861. https://doi.org/10.1111/ele.13680

Sanyal A, Larsson J, van Wirdum F, Andrén T, Moros M, Lönn M, Andrén E (2022) Not dead yet: diatom resting spores can survive in nature for several millennia. Am J Bot 109(1):67–82. https://doi.org/10.1002/ajb2.1780

Saros JE, Anderson NJ (2015) The ecology of the planktonic diatom Cyclotella and its implications for global environmental change studies. Biol Rev Camb Philos Soc 90(2):522–541. https://doi.org/10.1111/brv.12120

Schmidt R, Koinig KA, Thompson R, Kamenik C (2002) A multi proxy core study of the last 7000 years of climate and alpine land-use impacts on an Austrian mountain lake (Unterer Landschitzsee, Niedere Tauern). Palaeogeogr Palaeoclimatol Palaeoecol 187(1):101–120. https://doi.org/10.1016/S0031-0182(02)00511-4

Schmidt R, Kamenik C, Lange-Bertalot H, Klee R (2004) Fragilaria and Staurosira (Bacillariophyceae) from sediment surfaces of 40 lakes in the Austrian Alps in relation to environmental variables, and their potential for palaeoclimatology. J Limnol 63(2):171–189. https://doi.org/10.4081/jlimnol.2004.171

Schulte L, Bernhardt N, Stoof-Leichsenring K, Zimmermann HH, Pestryakova LA, Epp LS, Herzschuh U (2021) Hybridization capture of larch (Larix Mill.) chloroplast genomes from sedimentary ancient DNA reveals past changes of Siberian forest. Mol Ecol Resour 21(3):801–815. https://doi.org/10.1111/1755-0998.13311

Schulte L, Meucci S, Stoof-Leichsenring KR, Heitkam T, Schmidt N, von Hippel B, Andreev AA, Diekmann B, Biskaborn BK, Wagner B, Melles M, Pestryakova LA, Alsos IG, Clarke C, Krutovsky KV, Herzschuh U (2022) Larix species range dynamics in Siberia since the Last Glacial captured from sedimentary ancient DNA. Commun Biol 5(1):1–11. https://doi.org/10.1038/s42003-022-03455-0

Scibona A, Nizzoli D, Hupfer M, Valerio G, Pilotti M, Viaroli P (2022) Decoupling of silica, nitrogen and phosphorus cycling in a meromictic subalpine lake (Lake Iseo, Italy). Biogeochemistry 159(3):371–392. https://doi.org/10.1007/s10533-022-00933-9

Sherwood AR, Presting GG (2007) Universal primers amplify a 23S rDNA plastid marker in eukaryotic algae and cyanobacteria. J Phycol 43(3):605–608. https://doi.org/10.1111/j.1529-8817.2007.00341.x

Sigsgaard EE, Jensen MR, Winkelmann IE, Møller PR, Hansen MM, Thomsen PF (2020) Population-level inferences from environmental DNA—current status and future perspectives. Evol Appl 13(2):245–262. https://doi.org/10.1111/eva.12882

Slon V, Hopfe C, Weiß CL, Mafessoni F, de la Rasilla M, Lalueza-Fox C, Rosas A, Soressi M, Knul MV, Miller R, Stewart JR, Derevianko AP, Jacobs Z, Li B, Roberts RG, Shunkov MV, de Lumley H, Perrenoud C, Gušić I, Kućan Ž, Rudan P, Aximu-Petri A, Essel E, Nagel S, Nickel B, Schmidt A, Prüfer K, Kelso J, Burbano HA, Pääbo S, Meyer M (2017) Neandertal and Denisovan DNA from Pleistocene sediments. Science 356(6338):605–608. https://doi.org/10.1126/science.aam9695

Smetacek VS (1985) Role of sinking in diatom life-history cycles: ecological, evolutionary and geological significance. Mar Biol 84(3):239–251. https://doi.org/10.1007/BF00392493

Smol JP, Cumming BF (2000) Tracking long-term changes in climate using algal indicators in lake sediments. J Phycol 36(6):986–1011. https://doi.org/10.1046/j.1529-8817.2000.00049.x

Smol JP, Stoermer EF (2010) Applications and uses of diatoms: prologue. In: Stoermer EF, Smol JP (eds) The Diatoms: Applications for the Environmental and Earth Sciences, 2nd edn. Cambridge University Press, Cambridge, pp 3–7

Stoof-Leichsenring KR, Junginger A, Olaka LA, Tiedemann R, Trauth MH (2011) Environmental variability in Lake Naivasha, Kenya, over the last two centuries. J Paleolimnol 45(3):353

Stoof-Leichsenring KR, Epp LS, Trauth MH, Tiedemann R (2012) Hidden diversity in diatoms of Kenyan Lake Naivasha: a genetic approach detects temporal variation. Mol Ecol 21(8):1918–1930. https://doi.org/10.1111/j.1365-294X.2011.05412.x

Stoof-Leichsenring KR, Bernhardt N, Pestryakova LA, Epp LS, Herzschuh U, Tiedemann R (2014) A combined paleolimnological/genetic analysis of diatoms reveals divergent evolutionary lineages of Staurosira and Staurosirella (Bacillariophyta) in Siberian lake sediments along a latitudinal transect. J Paleolimnol 52(1):77–93. https://doi.org/10.1007/s10933-014-9779-1

Stoof-Leichsenring KR, Herzschuh U, Pestryakova LA, Klemm J, Epp LS, Tiedemann R (2015) Genetic data from algae sedimentary DNA reflect the influence of environment over geography. Sci Rep 5(1):12924. https://doi.org/10.1038/srep12924

Stoof-Leichsenring KR, Dulias K, Biskaborn BK, Pestryakova LA, Herzschuh U (2020a) Lake-depth related pattern of genetic and morphological diatom diversity in boreal Lake Bolshoe Toko, Eastern Siberia. PLoS ONE 15(4):e0230284. https://doi.org/10.1371/journal.pone.0230284

Stoof-Leichsenring KR, Pestryakova LA, Epp LS, Herzschuh U (2020b) Phylogenetic diversity and environment form assembly rules for Arctic diatom genera—a study on recent and ancient sedimentary DNA. J Biogeogr 47(5):1166–1179. https://doi.org/10.1111/jbi.13786

Tapolczai K, Keck F, Bouchez A, Rimet F, Kahlert M, Vasselon V (2019) Diatom DNA metabarcoding for biomonitoring: strategies to avoid major taxonomical and bioinformatical biases limiting molecular indices capacities. Front Ecol Evol 7

Tapolczai K, Selmeczy GB, Szabó B, Béres V, Keck F, Bouchez A, Rimet F, Padisák J (2021) The potential of exact sequence variants (ESVs) to interpret and assess the impact of agricultural pressure on stream diatom assemblages revealed by DNA metabarcoding. Ecol Indic 122:107322. https://doi.org/10.1016/j.ecolind.2020.107322

Tennant RK, Power AL, Burton SK, Sinclair N, Parker DA, Jones RT, Lee R, Love J (2022) In-situ sequencing reveals the effect of storage on lacustrine sediment microbiome demographics and functionality. Environm Microbiome 17(1):5. https://doi.org/10.1186/s40793-022-00400-w

Vasselon V, Domaizon I, Rimet F, Kahlert M, Bouchez A (2017) Application of high-throughput sequencing (HTS) metabarcoding to diatom biomonitoring: do DNA extraction methods matter? Freshw Sci 36(1):162–177. https://doi.org/10.1086/690649

Vasselon V, Bouchez A, Rimet F, Jacquet S, Trobajo R, Corniquel M, Tapolczai K, Domaizon I (2018) Avoiding quantification bias in metabarcoding: application of a cell biovolume correction factor in diatom molecular biomonitoring. Methods Ecol Evol 9(4):1060–1069. https://doi.org/10.1111/2041-210X.12960

Vasselon V, Rimet F, Domaizon I, Monnier O, Reyjol Y, Bouchez A (2019) Assessing pollution of aquatic environments with diatoms' DNA metabarcoding: experience and developments from France Water Framework Directive networks. Metabarcod Metagenom 3:e39646. https://doi.org/10.3897/mbmg.3.39646

Visco JA, Apothéloz-Perret-Gentil L, Cordonier A, Esling P, Pillet L, Pawlowski J (2015) Environmental Monitoring: Inferring the Diatom Index from Next-Generation Sequencing Data. Environ Sci Technol 49(13):7597–7605. https://doi.org/10.1021/es506158m

Von Eggers J, Monchamp M-E, Capo E, Giguet-Covex C, Spanbauer T, Heintzman PD (2022) Inventory of ancient environmental DNA from sedimentary archives: locations, methods, and target taxa. Zenodo. https://doi.org/10.5281/ZENODO.6847522

Wang R, Yang X, Langdon P, Zhang E (2011) Limnological responses to warming on the Xizang Plateau, Tibet, over the past 200 years. J Paleolimnol 45(2):257–271. https://doi.org/10.1007/s10933-011-9496-y

Wang Y, Pedersen MW, Alsos IG, De Sanctis B, Racimo F, Prohaska A, Coissac E, Owens HL, Merkel MKF, Fernandez-Guerra A, Rouillard A, Lammers Y, Alberti A, Denoeud F, Money D, Ruter AH, McColl H, Larsen NK, Cherezova AA, Edwards ME, Fedorov GB, Haile J, Orlando L, Vinner L, Korneliussen TS, Beilman DW, Bjørk AA, Cao J, Dockter C, Esdale J, Gusarova G, Kjeldsen KK, Mangerud J, Rasic JT, Skadhauge B, Svendsen JI, Tikhonov A, Wincker P, Xing Y, Zhang Y, Froese DG, Rahbek C, Bravo DN, Holden PB, Edwards NR, Durbin R, Meltzer DJ, Kjær KH, Möller P, Willerslev E (2021) Late quaternary dynamics of Arctic biota from ancient environmental genomics. Nature 600(7887):86–92. https://doi.org/10.1038/s41586-021-04016-x

Weckström J, Korhola A, Blom T (1997) Diatoms as quantitative indicators of pH and water temperature in subarctic Fennoscandian lakes. Hydrobiologia 347(1):171–184. https://doi.org/10.1023/A:1003091923476

Weigand H, Beermann AJ, Čiampor F, Costa FO, Csabai Z, Duarte S, Geiger MF, Grabowski M, Rimet F, Rulik B, Strand M, Szucsich N, Weigand AM, Willassen E, Wyler SA, Bouchez A, Borja A, Čiamporová-Zaťovičová Z, Ferreira S, Dijkstra K-DB, Eisendle U, Freyhof J, Gadawski P, Graf W, Haegerbaeumer A, van der Hoorn BB, Japoshvili B, Keresztes L, Keskin E, Leese F, Macher JN, Mamos T, Paz G, Pešić V, Pfannkuchen DM, Pfannkuchen MA, Price BW, Rinkevich B, Teixeira MAL, Várbíró G, Ekrem T (2019) DNA barcode reference libraries for the monitoring of aquatic biota in Europe: gap-analysis and recommendations for future work. Sci Total Environ 678:499–524. https://doi.org/10.1016/j.scitotenv.2019.04.247

Wetterich S, Rudaya N, Nazarova L, Syrykh L, Pavlova M, Palagushkina O, Kizyakov A, Wolter J, Kuznetsova T, Aksenov A, Stoof-Leichsenring KR, Schirrmeister L, Fritz M (2021) Paleoecology of the yedoma ice complex on sobo-sise Island (EasternLena Delta, Siberian Arctic). Front Earth Sci 9

White T, Bruns T, Lee S, Taylor J, Innis M, Gelfand D, Sninsky J (1990) Amplification and direct sequencing of fungal ribosomal RNA genes for phylogenetics. In: Pcr Protocols: a guide to methods and applications, pp 315–322

Willerslev E, Cooper A (2005) Review Paper. Ancient DNA. Proc Royal Soc B Biol Sci 272(1558):3–16. https://doi.org/10.1098/rspb.2004.2813

Witak M, Wachnicka A, Kuijpers A, Troelstra S, Prins MA, Witkowski A (2005) Holocene North Atlantic surface circulation and climatic variability: evidence from diatom records. The Holocene 15(1):85–96. https://doi.org/10.1191/0959683605hl768rp

Yang J-R, Duthie HC (1995) Regression and weighted averaging models relating surficial sedimentary diatom assemblages to water depth in lake Ontario. J Great Lakes Res 21(1):84–94. https://doi.org/10.1016/S0380-1330(95)71023-1

Yang X, Kamenik C, Schmidt R, Wang S (2003) Diatom-based conductivity and water-level inference models from eastern Tibetan (Qinghai-Xizang) Plateau lakes. J Paleolimnol 30(1):1–19. https://doi.org/10.1023/A:1024703012475

Zalat A, Welc F, Nitychoruk J, Marks L, Chodyka M, Zbucki Ł (2018) Last two millennia water level changes of the Młynek Lake (northern Poland) inferred from diatoms and chrysophyte cysts record. Studia Quaternaria 35, Nr 2. https://doi.org/10.2478/squa-2018-0006

Zgrundo A, Lemke P, Pniewski F, Cox EJ, Latała A (2013) Morphological and molecular phylogenetic studies on *Fistulifera saprophila*. Diatom Res 28(4):431–443. https://doi.org/10.1080/0269249X.2013.833136

Zimmermann J, Glöckner G, Jahn R, Enke N, Gemeinholzer B (2015) Metabarcoding vs. morphological identification to assess diatom diversity in environmental studies. Mol Ecol Resour 15(3):526–542. https://doi.org/10.1111/1755-0998.12336

Zimmermann HH, Stoof-Leichsenring KR, Kruse S, Müller J, Stein R, Tiedemann R, Herzschuh U (2020) Changes in the composition of marine and sea-ice diatoms derived from sedimentary ancient DNA of the eastern Fram Strait over the past 30,000 years. Ocean Sci 16(5):1017–1032. https://doi.org/10.5194/os-16-1017-2020

Zimmermann HH, Stoof-Leichsenring KR, Kruse S, Nürnberg D, Tiedemann R, Herzschuh U (2021) Sedimentary ancient DNA from the subarctic north pacific: how sea ice, salinity, and insolation dynamics have shaped diatom composition and richness over the past 20,000 years. Paleoceanogr Paleoclimatol 36(4):e2020PA004091. https://doi.org/10.1029/2020PA004091

Zimmermann HH, Stoof-Leichsenring KR, Dinkel V, Harms L, Schulte L, Hütt M-T, Nürnberg D, Tiedemann R, Herzschuh U (2023) Marine ecosystem shifts with deglacial sea-ice loss inferred from ancient DNA shotgun sequencing. Nat Commun 14(1):1650. https://doi.org/10.1038/s41467-023-36845-x

Zou Y, Wang L, He H, Liu G, Zhang J, Yan Y, Gu Z, Zheng H (2021) Application of a diatom transfer function to quantitative paleoclimatic reconstruction – a case study of Yunlong lake, Southwest China. Front Earth Sci 9. https://doi.org/10.3389/feart.2021.700194

Chapter 8
Aquatic Vegetation DNA from Lake Sediments

Aloïs Revéret, Inger G. Alsos, and Peter D. Heintzman

Keywords Sedimentary DNA · Lakes · Macrophytes · Metabarcoding · Trophic status

Introduction

Importance to Reconstruct Aquatic Vegetation

Aquatic vegetation plays a key role in lake ecosystems, as it consists of major primary producers, fosters high biodiversity and trophic complexity, and has a central place in freshwater biogeochemical cycles. Aquatic vegetation can be broadly divided into three taxonomic categories: macroalgae, bryophytes, and macrophytes. Macroalgae – multicellular algae larger than periphyton – are a polyphyletic group mainly represented in freshwater by Chlorophyta and Charophyta, that can produce large amounts of biomass when the system is in a eutrophic state (Messyasz et al. 2018). Aquatic bryophytes (i.e. mosses) serve as bioindicators for monitoring water quality within the European Water Framework Directive (Directive 2000/60/EC;

A. Revéret (✉) · I. G. Alsos
The Arctic University Museum of Norway, UiT—The Arctic University of Norway, Tromsø, Norway
e-mail: alois.n.reveret@uit.no

P. D. Heintzman
Department of Geological Sciences, Stockholm University, Stockholm, Sweden

Centre for Palaeogenetics, Stockholm, Sweden

E. Capo et al. (eds.), *Tracking Environmental Change Using Lake Sediments*, Developments in Paleoenvironmental Research 21,
https://doi.org/10.1007/978-3-031-43799-1_8

Ceschin et al. 2012) and bioaccumulators for assessing heavy metals pollution (Vásquez et al. 2019). Macrophytes – aquatic vascular plants – are globally distributed but show relatively high endemicity (Murphy et al. 2019) and exhibit a large variety of growth forms (e.g., Tyrrell et al. 2022). Macrophyte communities face worldwide decrease due to human activities, eutrophication and invasive species; and little is known about the status and trend of macrophyte populations in the mountainous and boreal regions (Botrel and Maranger 2023). Overall, aquatic plants are major contributors to carbon (Reitsema et al. 2018) and nutrient cycling in waterbodies (Dan et al. 2021; Zhang et al. 2021; Wang et al. 2022). Thus, by studying aquatic macrophytes of the present and past, we can build knowledge on both local and large-scale changes in the environment and associated drivers of biological change.

Proxies for Aquatic Vegetation Reconstruction

The traditional approach to studying macrophytes is by direct observation, using either a rake to collect benthic and pelagic aquatics or a bathyscope to survey submerged plants from a boat or the lake shore, while remote sensing observations are used for large-scale surveys. However, these approaches are time consuming and can be challenging due to poor visibility in turbulent or deeper water. Past aquatic macrophytes have been studied using macrofossils or pollen preserved in lake sediments; however, macrophytes may be poorly represented in pollen records, and taxonomic resolution can be low for some key taxa as e.g. Potamogetonaceae (Moore et al. 1991; Zhao et al. 2006). Macrofossils may give a higher taxonomic resolution, but their preservation and abundance are variable among sites, and the majority of macrofossils are deposited near where the plant grows, which can cause heterogeneous distributions (Birks 2000; Zhao et al. 2006). Sedimentary DNA (sedDNA) and sedimentary ancient DNA (*sed*aDNA; Crump 2021) have the potential to show increased taxonomic resolution compared to pollen analyses, and offer a more persistent detection than macrofossils (Alsos et al. 2018; Clarke et al. 2020; Capo et al. 2021). Therefore, they are a valuable tool for monitoring biodiversity and reconstructing ecosystems and environments.

Aquatic Vegetation and Its Ecology

Macrophytes and macroalgae are primary producers that provide food for a large range of grazing organisms, including fish, waterbirds, immature amphibians (tadpoles), gastropods, bivalves, arthropods, nematodes, and annelids. Aquatic vegetation also contributes greatly to water oxygenation through photosynthesis and

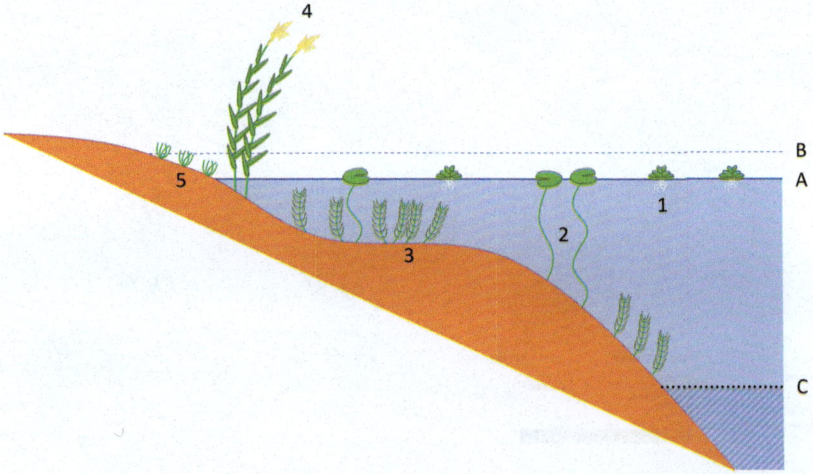

Fig. 8.1 Traditional growth form classification of lacustrine aquatic macrophytes and their habitat. Aquatic macrophytes *sensu stricto* can be free-floating (1), floating-leaved (2), or submerged (3); here we also include in the definition the emergent (4) and marginal (5) taxa. Some species are restricted to a depth category, while others may tolerate a broader depth range, or some emersion time (daily / yearly). (A) is the mean water level, (B) the highest water level, where marginal plants are temporarily inundated; (C) is the depth limit for aquatic macrophyte growth, which depends on water clarity and temperature. Species exemplified in the diagram are (1) *Pistia stratiotes* L., (2) *Nymphaea alba* (L.) Sm., (3) *Myriophyllum alterniflorum* DC., (4) *Phragmites australis* (Cav.) Trin. ex Steud., and (5) *Carex rostrata* Stokes

provides key habitat for a large variety of lake organisms. Macrophytes can occupy a wide variety of niches within the biotope thanks to their many growth forms (Fig. 8.1), which can in turn shape community dynamics and composition. For instance, the free-floating duckweed *Lemna*, the fern *Salvinia*, and the floating-leafed giant water lily *Victoria*, despite having very different sizes and distributions, can monopolise light resources by spreading across the entire surface of a waterbody (Netten et al. 2010). This can have important consequences for the trophic system in the water column, especially phytoplankton, as macrophytes coexist and compete with other aquatic photosynthetic organisms (e.g. macro- and microalgae, cyanobacteria). Such interspecific competition for crucial resources, e.g. light and dissolved nutrients, is shaped by various factors, including water depth, flow, physico-chemistry, or clarity (Hughes et al. 2018). Reproductive and dispersal strategies of aquatic vegetation heavily rely on water transport (i.e. hydrogamy and hydrochory), although spermatophytes (e.g. Nymphaeaceae) are more reliant on insect pollination.

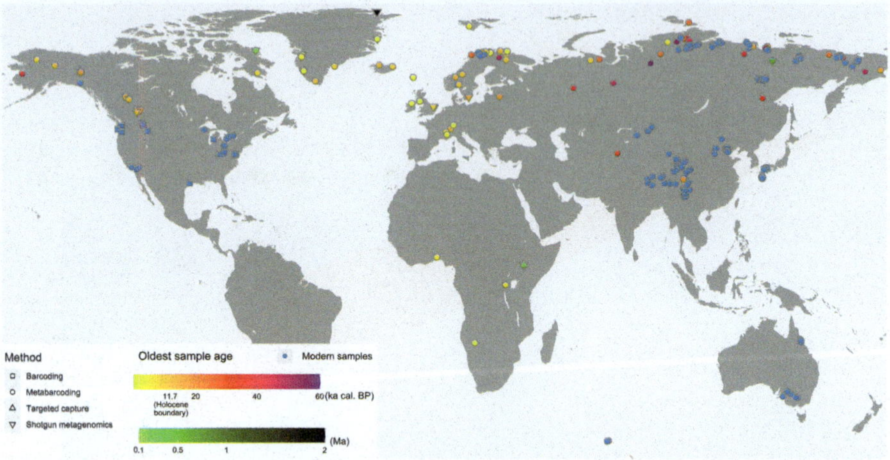

Fig. 8.2 Summary map of sites (n = 450) where environmental DNA of aquatic macrophytes has been detected in 62 studies. Modern samples are shown in blue, whereas ancient samples are shown with a colour gradient (yellow to purple, for Holocene to Late Pleistocene samples; green to black, for samples 100,000 BP and older). Samples consist of water, lake and river sediment, soil including permafrost, and peat. (Modified from Revéret et al. 2023)

Grounds for Further Research

Aquatic vegetation can be detected in ancient samples from various substrates and across many biomes (Fig. 8.2). Despite this, macrophytes remain the freshwater taxon least investigated with environmental DNA (eDNA), representing just 2% of the concerned literature (Belle et al. 2019; Schenekar 2023). A recent review compiled all studies reporting the detection of aquatic macrophytes through eDNA (mixed samples derived from multiple taxa) and/or *sed*aDNA. From the 62 identified studies, 34 used lake sediment (Revéret et al. 2023). These studies include metagenomic, barcoding and metabarcoding datasets from 239 lakes and 211 other sites. From lakes, these datasets include 183 modern sediment samples and 56 sediment records spanning from 0 to 130 thousand years before present (BP). In addition, several other papers have re-used these *sed*aDNA datasets for proxy comparison, total floristic reconstruction, or other purposes (e.g. Alsos et al. 2022; Bjune et al. 2022; Elliott et al. 2023). Hence, there is growing evidence that applying eDNA methods to sedimentary lacustrine archives can unravel the patterns and drivers of aquatic plant diversity through time.

Detection and Identification of Aquatic Macrophytes with Molecular Methods

Methodological Approaches (Metabarcoding, Shotgun Sequencing, Targeted Capture)

Environmental DNA (eDNA) from macrophytes can be detected from a wide range of substrates: water, lake and river sediment, permafrost and other soils, and peat. Most studies of macrophytes in ancient lake sediments are from records spanning the Holocene (ca. 11,700 BP to present), with the oldest reported from the Last Interglacial (ca. 129,000–116,000 BP; Crump et al. 2021). However, identifying *sed*aDNA is restricted by its much lower concentration and preservation, which explains why most palaeobotanical studies have so far focused on colder environments from the northern boreal and Arctic, as well as high-elevation areas (Fig. 8.2). Yet, the detection of macrophyte *sed*aDNA can be achieved by several methodological approaches, namely metagenomics (i.e. shotgun sequencing), barcoding assays and metabarcoding, and targeted capture (also called "hybridisation capture" or "target enrichment"; see Chap. 3 of this volume for definitions). These methods have variable success rates and their recommended utilisation largely depends on the research question and type of material under study.

Metabarcoding markers should ideally be chosen for their taxonomic resolution: the high discriminatory power of standard plant barcodes *rbc*L and *mat*K relies on their length (respectively 654 and 862–910 base pairs (bp); Hollingsworth et al. 2011), so that the primers can amplify across vascular plant diversity and the amplicon is long enough to effectively delineate species. Shotgun sequencing is an untargeted approach where volumes of sequenced *sed*aDNA molecules are identified against global sequence repositories. At present, this approach often has lower taxonomic resolution, as compared to metabarcoding (Fig. 8.3), because most DNA sequences are very short and uninformative, and genome-wide databases are incomplete. Barcoding approaches are often used in conservation science to detect a single species with taxon-specific assays (e.g. Fujiwara et al. 2016; Anglès d'Auriac et al. 2019; Kuehne et al. 2020). Targeted capture has insofar seldom been used in palaeoecology, detecting aquatic macrophytes only in one African palaeolake and three Alaskan permafrost sites (Figs. 8.2 and 8.3; Krueger et al. 2021; Murchie et al. 2021). Environmental DNA as a tool for the conservation and monitoring of present-day freshwater ecosystems has become a popular approach, but its implementation remains heterogeneous and would benefit from standardisation (Thomsen and Willerslev 2015; Beng and Corlett 2020; Monchamp et al. 2023). Metabarcoding provides a robust tool for surveying macrophyte community composition (Kuzmina et al. 2018; Coghlan et al. 2021) whereas single-species barcoding can be used to track invasive or endangered species (Anglès d'Auriac et al. 2019; Tsukamoto et al. 2021).

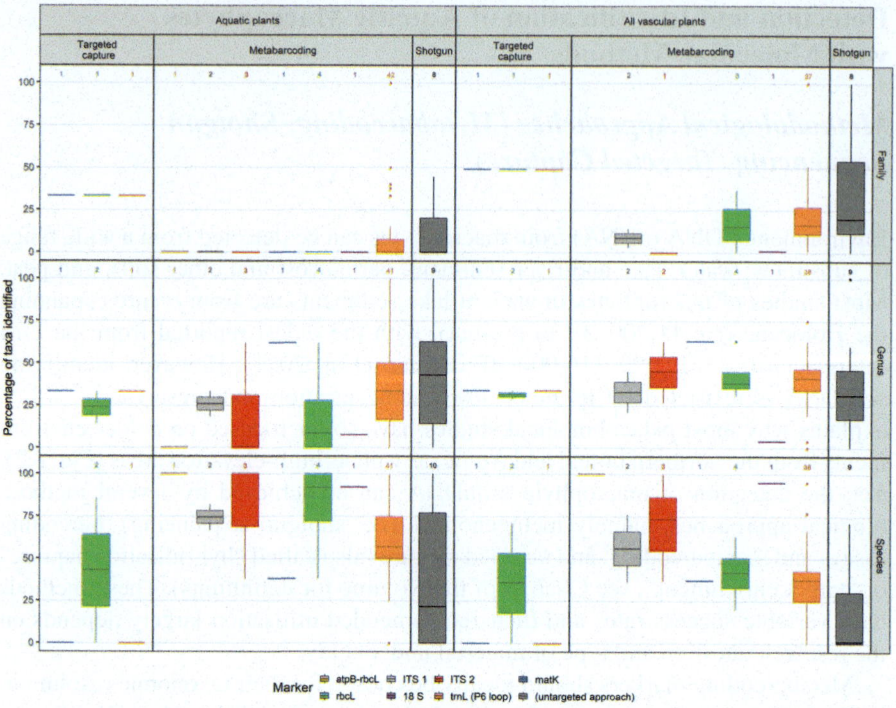

Fig. 8.3 Taxonomic resolution, i.e. percentage of taxa assigned to a given taxonomic level, obtained with targeted capture, metabarcoding, and metagenomics. Single-species targeted capture studies were excluded because their assumed resolution is by definition 100%. Taxonomic resolution is shown for aquatic vascular plants only (left), and for all vascular plants detected including all aquatics (right). Numbers above each boxplot indicate the number of studies involved. Each of the three taxonomic levels shown include taxa identified to lower subdivisions: "family" (upper) includes subfamilies, tribes, and subtribes; "genus" (centre) includes sections; and "species" includes varieties or subspecies (lower). (Originally published in Revéret et al. 2023)

The scarcer and more fragmented nature of *sed*aDNA challenges the above methodologies. Standard DNA barcodes are too long for fragmented ancient DNA and therefore do not yield data of palaeoecological use (Fig. 8.3). Hence the current standard is to use a much shorter marker, such as the P6 loop of the chloroplast *trn*L intron (ca. 86 bp; Taberlet et al. 2007; Hollingsworth et al. 2011), which has repeatedly been shown to be effective and benefits from excellent coverage in reference libraries for some regions (Fig. 8.3). Shotgun sequencing is currently less successful than metabarcoding for confidently identifying plant taxa from *sed*aDNA (Murchie et al. 2021; Courtin et al. 2022). Targeted capture results are too scarce to this day to be compared with the other methods.

Species identification from eDNA remains especially challenging for aquatic taxa because of this group's larger genetic diversity. Even in families that include both terrestrial and aquatic representatives (e.g. Ranunculaceae), taxa are genetically close and therefore cannot always be delineated from one another, resulting in identifications at lower taxonomic resolution (i.e. genus level) and hampering further inferences about habitat origin. Additionally, introgressive hybridisation between species is particularly common within aquatic families such as Potamogetonaceae, which increases both sequence sharing and intraspecific variability. For instance, a reference library for circum-Arctic vascular plants lists 6 haplotypes of the *trn*L P6 loop for the 9 considered species in this family, with 2 different haplotypes unique to *Stuckenia filiformis* (Sønstebø et al. 2010). Some of these haplotypes are homopolymeric sequences, which are especially prone to amplification and sequencing errors (Schirmer et al. 2016), especially with such short markers. Therefore, it is of paramount importance that reference libraries cover intraspecific as well as interspecific variability for the currently established markers (Hollingsworth et al. 2011). Alternatively, the use of other metabarcoding markers has been explored; and early attempts to design species-specific markers targeting monophyletic aquatic taxa have yielded promising results, hinting towards ITS and *mat*K regions (Du et al. 2011; Scriver et al. 2015).

The identification of aquatic bryophytes in *seda*DNA remains problematic because their detection is usually achieved using the *trn*L P6 loop primers that target vascular plants. Bryophytes are therefore considered as bycatch in such experiments (Taberlet et al. 2007). This is confounded by the incompleteness of bryophyte taxa in reference libraries. We are only aware of one regional P6 loop reference library for bryophytes, which contains 455 of the ca. 900 known Arctic and boreal bryophyte species (Soininen et al. 2015; Lewis et al. 2017). Detections may also be improved by designing bryophyte-specific primers, or by deeper sequencing of P6 loop datasets.

Representation of Aquatic Plants in sedDNA Data

The most straightforward approach to assess how representative sedDNA data are for aquatic macrophytes that grow in a lake or its catchment is to compare the sedDNA data with traditional vegetation surveys. A study of 11 lakes in northern Norway showed that many aquatic macrophyte families (Haloragaceae, Lentibulariaceae, Menyanthaceae, Nymphaeaceae, Plantaginaceae, Potamogetonaceae, Sparganiaceae) were well detected by P6 loop sedDNA data, including some that were not recorded in the vegetation surveys. Indeed, it was possible to detect 88% of both dominant and common taxa, 33% of scattered, and even 62% of rare taxa using the P6 loop; as compared to e.g. detection of 60% of dominant taxa in the terrestrial vegetation surveys (Alsos et al. 2018). The actual proportion of detections may have been even higher, but these potential additions were removed due to bioinformatic filtering criteria such as low counts of DNA reads and/or PCR replicates (Alsos et al. 2018). Such reliable detection of aquatic taxa may be due to the hypothesis that, when released, DNA will diffuse around, but remain within the water body and sediments. Therefore, one sample from the centre of a relatively small lake would be sufficient to detect species and diversity of aquatic plants, while spatial heterogeneity for larger or more complex lakes may require the sampling of multiple sedimentation basins (Ibrahim et al. 2022). A similar study used two markers (P6 loop, 18S), but with a less complete DNA reference library, to show good detection of aquatic plant taxa (11–100%) in six floodplains of the Murray River, Australia (Shackleton et al. 2019). Thus, with good reference libraries, it may be possible to detect species overlooked by vegetation surveys and may therefore be a valuable complement to, or even replacement for, traditional surveys.

Factors Affecting the Detection and Preservation of Aquatic Macrophyte sedDNA

Aquatic vegetation grows directly over the sediment, then sheds DNA, together with possible macroscopic organic remains (i.e. macrofossils) that undergo short to no transport. Even though the distribution of aquatics within the waterbody is restricted mostly by bathymetry (Fig. 8.1; also turbidity and temperature), they remain well-detected in sedDNA and can even be detected in a single core taken at the deepest point of the waterbody (Alsos et al. 2018), which is the traditional palaeolimnological approach. Aquatic plant eDNA is therefore likely to be present in high initial concentrations in lake sediment. Indeed, when detected by metabarcoding, aquatic plants are often detected in multiple PCR replicates, i.e. >6 out of 8 (Fig. 8.4), which is suggestive of a high abundance of DNA templates in the sediment. Similarly, detection of Potamogetonaceae in P6 loop experiments hints towards good DNA preservation overall, as the Potamogetonaceae template is more sensitive to fragmentation. This is because the P6 loop marker in this family is longer (86 bp, excluding primer binding sites) than average (~50 bp). In general, eukaryotic freshwater taxa seem to benefit from especially good representation within sedDNA and *sed*aDNA: for example, it is possible to study the genomic palaeodynamics of algal populations as old as the Last Glacial Maximum (Lammers et al. 2021). This study reconstructed entire chloroplast and mitochondrial genomes of a 19,000-year-old unicellular microalgae, which showcased the considerable potential of lake *sed*aDNA for palaeoecology and palaeogenomics of aquatic photosynthetic taxa.

Applications of *sed*aDNA to Macrophyte Ecological Research

Detecting Plant Dispersal, Establishment, and Survival

Aquatic macrophytes are known for their rapid responses to changing environmental conditions. They benefit from efficient dispersal mechanisms by birds (especially waterfowl such as ducks and geese) and by waterways (rivers and streams; Figuerola and Green 2002). In northern Fennoscandia, aquatic macrophytes are not

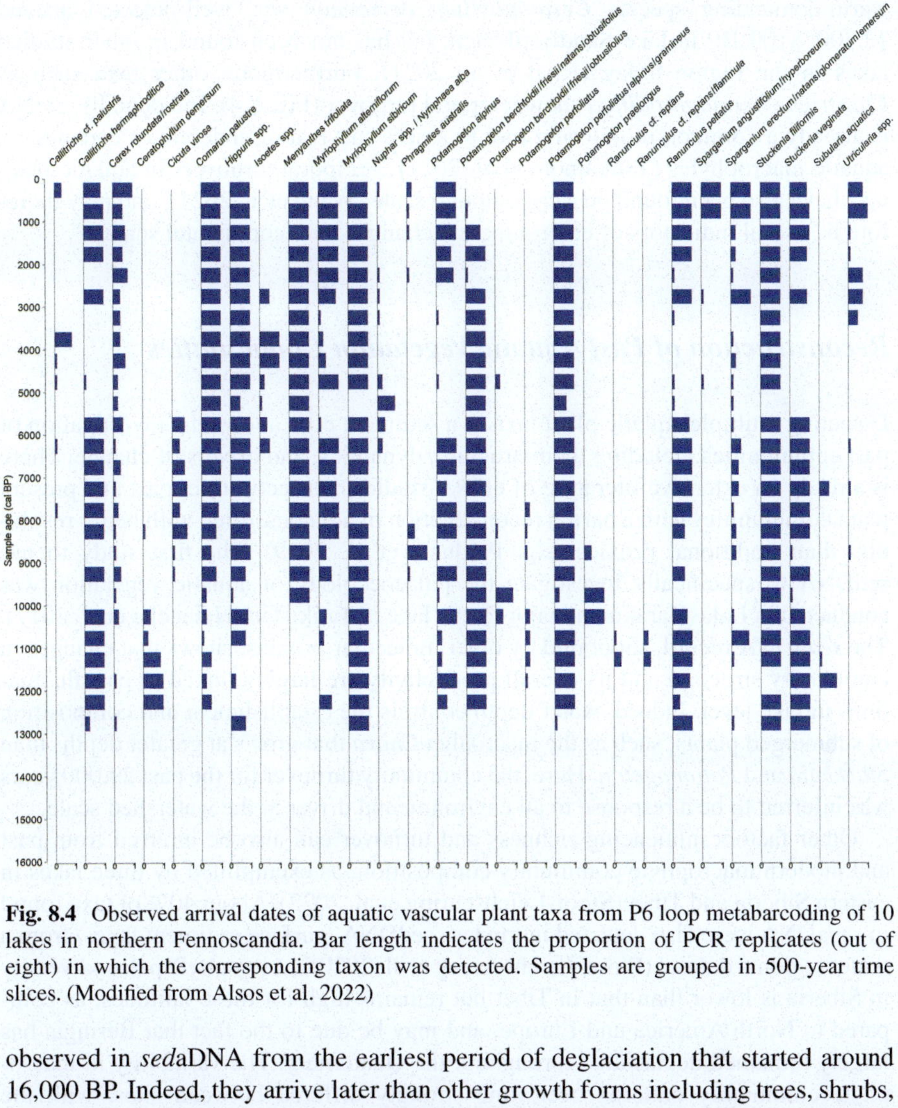

Fig. 8.4 Observed arrival dates of aquatic vascular plant taxa from P6 loop metabarcoding of 10 lakes in northern Fennoscandia. Bar length indicates the proportion of PCR replicates (out of eight) in which the corresponding taxon was detected. Samples are grouped in 500-year time slices. (Modified from Alsos et al. 2022)

observed in *sed*aDNA from the earliest period of deglaciation that started around 16,000 BP. Indeed, they arrive later than other growth forms including trees, shrubs, dwarf shrubs, and ferns (Alsos et al. 2022). However, there is a sudden appearance of aquatic macrophytes from the onset of the Younger Dryas at 12,900 BP (Fig. 8.4), which then rapidly accumulate and make up around 30% of the relative proportion of PCR replicates by 12,000–10,000 BP, before reducing to 20% thereafter (Alsos et al. 2022). In fact, 17 out of 25 taxa arrived between 12,900–12,350 BP, demonstrating the high dispersal ability of aquatic macrophytes even to this region that was far from the Last Glacial Maximum refugia areas in southern-most Scandinavia and north-western-most Siberia.

All detected aquatic plant taxa are also found in the region with modern investigations, but their persistence varies among lakes. Notably, the relatively warm-demanding species *Ceratophyllum demersum* was well-detected around 12,300–8100 BP in lake Sandfjorddalen, but has not been found in other studied lakes in the region today (Rijal et al. 2021). Furthermore, other taxa such as *Callitriche hermaphroditica* show temporal turnover (Fig. 8.4). Incidentally, eDNA is already a common monitoring tool to assess survival of endemic or endangered aquatic macrophytes (Tsukamoto et al. 2021). Temporal turnovers in aquatic plant taxa have also been found on a circumpolar scale (Wang et al. 2021), and may therefore be useful indicators of environmental changes over continental scales.

Reconstruction of Past Aquatic Vegetation Communities

Detecting multiple aquatic plant taxa in a sediment core allows the investigation of past communities including their structure, dynamics, and drivers of change. There is already an extensive literature of eDNA studies that reconstruct past and present plant communities with a partial reconstruction of aquatics, often with better resolution than traditional proxies (e.g. Parducci et al. 2019). The first study to use sedDNA to specifically investigate the palaeoecology of aquatic vegetation was conducted in Lake Karakul, a major Pamir lake in Tajikistan (Heinecke et al. 2017). The *sed*aDNA record, supported by other molecular proxies, shows that changes in community structure and its overall productivity are largely linked to past fluctuations in lake level. Indeed, water depth controls the establishment and composition of submerged plants, such as the macroalga *Chara* that grows at greater depths than *Stuckenia* and *Potamogeton*. Here, the community turnover for the past 29,000 years was inferred to be a response to an environmental driver at the watershed scale.

Other factors influencing richness and turnover can also be inferred from past and modern macrophyte community composition, as exemplified by three lakes in eastern Siberia and Tibet (Stoof-Leichsenring et al. 2022). About 40% of taxa found in *sed*aDNA were also detected in modern sedDNA (surface sediment layer) from a wider regional dataset (Stoof-Leichsenring et al. 2020). Macrophyte palaeodiversity in Siberia is lower than that in Tibet but remains high for these latitudes, as compared to North America and Europe, and may be due to the fact that Beringia has largely remained deglaciated throughout the Quaternary. The variability in macrophyte community patterns over time strongly differed between sites, with the Tibetan glacial lake (Naleng) having the highest richness and turnover, and the glacial lake in Siberia (Ilirney) yielding more diversity than the thermokarst lake (Omoloy). Lake Ilirney record spans the Last Glacial Maximum, a period at which a lower diversity of aquatic macrophytes is observed, whereas all sites show fluctuations in composition during the Middle (8300–4250 BP) and Late Holocene (4250 BP–present). From the regional modern dataset, summer air temperature, water conductivity, and lake water depth have been identified as major drivers underlying macrophyte community composition (Stoof-Leichsenring et al. 2022). These environmental variables can be linked to the macrophyte *sed*aDNA record in order to infer their history in their respective catchments. For instance, the limited presence of submerged taxa like *Potamogeton* in Siberian lakes is likely due to harsher winter temperatures limiting growth, whereas the Omoloy and Ilirney records are dominated by emergent macrophytes (e.g. *Hippuris*) that are tolerant to water level changes. In contrast, the authors hypothesise that the major compositional shifts observed around 14,000 BP and since 5000 BP in lake Naleng may be explained by stadial/interstadial thermal variations, in combination with the influence of glacial activity which would influence nutrient availability through erosion dynamics in the catchment. This is supported by findings that identify local climatic conditions and glacial presence in the catchment as drivers of vegetation change (Elliott et al. 2023).

Reconstructing Past Environments

Beyond community reconstruction, the taxonomic composition of the macrophyte *sed*aDNA signal can also allow for the reconstruction of abiotic factors, i.e. environmental conditions, through time by leveraging known traits of aquatic plant taxa. Macrophytes are commonly used as environmentally indicative taxa (Poikane et al. 2018; Tyrrell et al. 2022), and their occurrence at a site is used to infer local environmental parameters. Indeed, abiotic factors are known to be the most influential drivers of macrophyte assemblage richness and composition (Alahuhta, Lindholm, et al. 2018, Alahuta, Rosbakh, et al. 2020). Macrophytes already support palaeoenvironmental inferences of past temperature and water level (Isarin & Bohncke 1999; Väliranta, Kultti, et al. 2005, Väliranta, Salonen, et al. 2015), but these analyses were based on macrofossils and/or pollen. Utilising the *sed*aDNA signal for this purpose therefore stands as a new proxy to be used for ecological and environmental inferences.

Indeed, this has been shown through the reconstruction of postglacial environmental conditions across northern Fennoscandia over the last 13,000 years using P6 loop metabarcoding data of aquatic vascular plants (Revéret et al. 2023). The ten studied lakes supported a regional scale reconstruction, showing a long-term trend of acidification and the early establishment of a dominant environmental niche characterised by oligotrophy, half-shade, and alpine temperatures. Climate features can also be reconstructed, as exemplified by an inferred cooling between ca. 5000–3000 BP, which is in accordance with the other climatic proxies for this region (Sjögren 2021). Focusing on aquatic plants also allowed for environmental reconstruction at a local scale, thus characterising a specific lake environment through time. Some lakes display a very high compositional stability, suggesting niche persistence and stable environmental conditions, whereas abrupt changes occur in other lakes. Such shifts in freshwater parameters can also be linked to local events happening within the lake catchment. For instance, the acidification of lake Horntjernet coincides with the establishment of a pine forest around 8000 years ago (Revéret et al. 2023).

However, the possibility of such reconstructions from aquatic plant *sed*aDNA is constrained by two factors. First, a database of plant traits must be available to use these taxa as environmental indicators. The respective niches of aquatic plants can be informative for inferring their drivers, i.e. abiotic factors that are ecologically meaningful (e.g. water temperature, depth, pH, turbidity, dissolved nutrient availability, salinity). In this regard, the aquatic flora may offer more possibilities than land plants in the number of traits constructible: the submerged vegetation is more tightly constrained by the above-mentioned factors than terrestrial vegetation, because its presence is largely controlled by water physico-chemistry. Second, the reference library used must be nearly complete and specific to the study area, in order to maximise taxonomic resolution to support a reliable niche-based reconstruction. Indeed, using a global sequence repository, such as NCBI GenBank database, increases the risk of erroneous assignments to a closely related taxon from an exotic biogeographic area, or simply overlooking endemic species from lack of an available reference sequence. Additionally, the superior taxonomic resolution afforded by *sed*aDNA, as compared to other proxies, contributes to the accuracy of the reconstruction.

Detection and Monitoring of Invasive Plants

Targeted capture is a common methodology in modern aquatic molecular ecology and could be applied to palaeoecological studies in order to retrace the history of past invasive events and their ensuing consequences. Past aquatic plant communities can also be reconstructed by targeted capture using a set of baits covering a wide taxonomic spectrum (Murchie et al. 2021), but the low taxonomic resolution, as compared to metabarcoding, makes the single-species approach more relevant for *sed*aDNA. Indeed, it is usual with present-day soil or water samples to focus on single-species tracking for conservation purposes, but the recent sediment (upper ~1–2 cm) can also be a powerful tool for monitoring modern aquatic plant diversity (Ji et al. 2021). In addition, the sediment might temporally accumulate and concentrate the DNA signal from invasive species better than water samples, which are the current standard.

Investigating the ecology of past species invasions requires more than the mere detection of the species. Biomass quantification methods are needed to supersede the presence/absence data in this context. Current palaeoecological interpretations of *sed*aDNA records use the proportion of PCR replicates and DNA reads in which a taxon is detected as a semi-quantitative measure of its relative abundance in the community (Alsos et al. 2022; Heinecke et al. 2017; Parducci et al. 2017). For instance, the allochthonous water caltrop *Trapa* sp. was detected in *sed*aDNA metabarcoding data from sediments of Lake Aiguebelette that were older than the first appearance in historical vegetation observations, thus indicating its presence even before the first confirmed field record (Giguet-Covex et al. 2022). Long-term changes in environmental conditions also influence the dispersion range of invasive macrophytes (Hoveka et al. 2016), thus palaeoclimatic models may help to reconstruct changes in their putative spatial distribution.

Reconstruct Past Trophic Networks

Lake eutrophication is a major problem in freshwater systems, which is often induced by anthropogenic activities and strongly affects the development and dynamics of aquatic vegetation (Zhang et al. 2017). Shifts in the trophic status of a waterbody and the ensuing community changes can be reconstructed from *sed*aDNA at fine spatiotemporal scales (Ibrahim et al. 2022) and be correlated, through geochemical proxies, with changes occurring in agricultural practices, soil erosion, and urbanisation within the catchment (Giguet-Covex, Pansu, et al. 2014, Giguet-Covex, Ficetola, et al. 2019).

Furthermore, the reconstruction of long-term changes of trophic networks can be enhanced by sedDNA datasets (Barouillet et al. 2023), as emphasised by metabarcoding data from Lake Aiguebelette, France (Giguet-Covex et al. 2022). Over the last 1800 years, richness and composition of the macrophyte community have been influenced by human activities that increase nutrient input and thus cause eutrophication. In turn, the establishment of new macrophyte species increased habitat complexity and fostered further complexification of the lake ecosystem. As an example, the presence of the roach *Rutilus* sp. coincides with a higher richness and abundance of submerged macrophytes, as pondweeds provide a breeding and feeding ground for fish. On another trophic level, macrophyte abundance also controls ecological interactions such as resource competition with microalgae, macroinvertebrate abundance and diversity (by providing food and habitat), or phytoplanktonic productivity. Indeed, the prominent role of aquatic vegetation in lake trophic networks may support much wider ecological inferences based on community composition than is currently realised.

Summary

Aquatic vegetation is unevenly identified in *sed*aDNA datasets: macrophytes are especially well detected and identified with the P6 loop short metabarcode, but macroalgae and bryophytes are underrepresented using the same experimental approach. High-throughput sequencing of *sed*aDNA effectively complements and largely enhances ecological inferences derived from traditional palaeolimnological proxies. Taxon detection could be improved by further populating reference databases to enable better taxonomic resolution of present and future metagenomics data, and by designing new metabarcoding primers specific to these taxa (e.g. Chlorophyta, Bryophyta). Identifying these taxa with optimal resolution can only be accomplished with a comprehensive reference library. Such libraries are, to date, unfortunately missing for most non-vascular aquatic plants and this shortage is especially dramatic in low latitude regions, where freshwater ecosystems are the most diverse but also the most critically endangered.

By reconstructing the trait palaeodiversity of aquatic plants, one can also make larger inferences at the scale of the lake ecosystem to understand its palaeodynamics. Aquatic vegetation encompasses a large variety of crucial ecological functions that support the establishment, sustainability, and resilience of a much more complex network of trophic interactions. Additionally, knowledge of the respective niches of aquatic plant species can be harnessed for the temporal reconstruction of environmental factors. This exploration of past variation in lake ecosystem composition may also unlock the understanding of drivers of change, as the *sed*aDNA signal can be interpreted as a response to environmental changes. Thus, it could allow us to identify the biotic and abiotic factor(s) that cause regime shifts in aquatic vegetation ecosystems.

Acknowledgements We thank our reviewers Sandra Garcés-Pastor and Qi Lin for providing useful comments and suggestions on our manuscript.

References

Alahuhta J, Lindholm M, Bove CP, Chappuis E, Clayton J, de Winton M, Feldmann T, Ecke F, Gacia E, Grillas P, Hoyer MV, Johnson LB, Kolada A, Kosten S, Lauridsen T, Lukács BA, Mjelde M, Mormul RP, Rhazi L, Rhazi M, Sass L, Søndergaard M, Xu J, Heino J (2018) Global patterns in the metacommunity structuring of lake macrophytes: regional variations and driving factors. Oecologia 188(4):1167–1182. https://doi.org/10.1007/s00442-018-4294-0

Alahuhta J, Rosbakh S, Chepinoga V, Heino J (2020) Environmental determinants of lake macrophyte communities in Baikal Siberia. Aquat Sci 82(2):39. https://doi.org/10.1007/s00027-020-0710-8

Alsos IG, Lammers Y, Yoccoz NG, Jørgensen T, Sjögren P, Gielly L, Edwards ME (2018) Plant DNA metabarcoding of lake sediments: how does it represent the contemporary vegetation. PLoS One 13(4):e0195403. https://doi.org/10.1371/journal.pone.0195403

Alsos IG, Rijal DP, Ehrich D, Karger DN, Yoccoz NG, Heintzman PD, Brown AG, Lammers Y, Pellissier L, Alm T, Bråthen KA, Coissac E, Merkel MKF, Alberti A, Denoeud F, Bakke J, PHYLONORWAY CONSORTIUM (2022) Postglacial species arrival and diversity buildup of northern ecosystems took millennia. Science. Advances 8(39):eabo7434. https://doi.org/10.1126/sciadv.abo7434

Anglès d'Auriac MB, Strand DA, Mjelde M, Demars BOL, Thaulow J (2019) Detection of an invasive aquatic plant in natural water bodies using environmental DNA. PLoS One 14(7):e0219700. https://doi.org/10.1371/journal.pone.0219700

Barouillet C, Monchamp M, Bertilsson S, Brasell K, Domaizon I, Epp LS, Ibrahim A, Mejbel H, Nwosu EC, Pearman JK, Picard M, Thomson-Laing G, Tsugeki N, Von Eggers J, Gregory-Eaves I, Pick FR, Wood SA, Capo E (2023) Investigating the effects of anthropogenic stressors on lake biota using sedimentary DNA. Freshw Biol fwb.14027. https://doi.org/10.1111/fwb.14027

Belle CC, Stoeckle BC, Geist J (2019) Taxonomic and geographical representation of freshwater environmental DNA research in aquatic conservation. Aquat Conserv Mar Freshwat Ecosyst 29(11):1996–2009. https://doi.org/10.1002/aqc.3208

Beng KC, Corlett RT (2020) Applications of environmental DNA (eDNA) in ecology and conservation: opportunities, challenges and prospects. Biodivers Conserv 29(7):2089–2121. https://doi.org/10.1007/s10531-020-01980-0

Birks HH (2000) Aquatic macrophyte vegetation development in Kråkenes Lake, western Norway, during the late-glacial and early-Holocene. J Paleolimnol 23(1):7–19. https://doi.org/10.1023/A:1008025116936

Bjune AE, Greve Alsos I, Brendryen J, Edwards ME, Haflidason H, Johansen MS, Mangerud J, Paus A, Regnéll C, Svendsen J-I, Clarke CL (2022) Rapid climate changes during the Lateglacial and the early Holocene as seen from plant community dynamics in the Polar Urals, Russia. J Quat Sci 37(5):805–817. https://doi.org/10.1002/jqs.3352

Botrel M, Maranger R (2023) Global historical trends and drivers of submerged aquatic vegetation quantities in lakes. Glob Chang Biol 29(9):2493–2509. https://doi.org/10.1111/gcb.16619

Capo E, Giguet-Covex C, Rouillard A, Nota K, Heintzman PD, Vuillemin A, Ariztegui D, Arnaud F, Belle S, Bertilsson S, Bigler C, Bindler R, Brown AG, Clarke CL, Crump SE, Debroas D, Englund G, Ficetola GF, Garner RE, Gauthier J, Gregory-Eaves I, Heinecke L, Herzschuh U, Ibrahim A, Kisand V, Kjær KH, Lammers Y, Littlefair J, Messager E, Monchamp M-E, Olajos F, Orsi W, Pedersen MW, Rijal DP, Rydberg J, Spanbauer T, Stoof-Leichsenring KR, Taberlet P, Talas L, Thomas C, Walsh DA, Wang Y, Willerslev E, van Woerkom A, Zimmermann HH, Coolen MJL, Epp LS, Domaizon IG, Alsos I, Parducci L (2021) Lake sedimentary DNA research on past terrestrial and aquatic biodiversity: overview and recommendations. Quaternary 4(1):6. https://doi.org/10.3390/quat4010006

Ceschin S, Aleffi M, Bisceglie S, Savo V, Zuccarello V (2012) Aquatic bryophytes as ecological indicators of the water quality status in the Tiber River basin (Italy). Ecol Indic 14(1):74–81. https://doi.org/10.1016/j.ecolind.2011.08.020

Clarke CL, Alsos IG, Edwards ME, Paus A, Gielly L, Haflidason H, Mangerud J, Regnéll C, Hughes PDM, Svendsen JI, Bjune AE (2020) A 24,000-year ancient DNA and pollen record from the Polar Urals reveals temporal dynamics of arctic and boreal plant communities. Quat Sci Rev 247:106564. https://doi.org/10.1016/j.quascirev.2020.106564

Coghlan SA, Shafer ABA, Freeland JR (2021) Development of an environmental DNA metabarcoding assay for aquatic vascular plant communities. Environ DNA 3(2):372–387. https://doi.org/10.1002/edn3.120

Courtin J, Perfumo A, Andreev AA, Opel T, Stoof-Leichsenring KR, Edwards ME, Murton JB, Herzschuh U (2022) Pleistocene glacial and interglacial ecosystems inferred from ancient DNA analyses of permafrost sediments from Batagay megaslump, East Siberia. Environ DNA 4(6):1265–1283. https://doi.org/10.1002/edn3.336

Crump SE (2021) Sedimentary ancient DNA as a tool in paleoecology. Nat Rev Earth Environ 2(4):Article 4. https://doi.org/10.1038/s43017-021-00158-8

Crump SE, Fréchette B, Power M, Cutler S, de Wet G, Raynolds MK, Raberg JH, Briner JP, Thomas EK, Sepúlveda J, Shapiro B, Bunce M, Miller GH (2021) Ancient plant DNA reveals High Arctic greening during the last interglacial. Proc Natl Acad Sci 118(13):e2019069118. https://doi.org/10.1073/pnas.2019069118

Dan Z, Chuan W, Qiaohong Z, Xingzhong Y (2021) Sediments nitrogen cycling influenced by submerged macrophytes growing in winter. Water Sci Technol 83(7):1728–1738. https://doi.org/10.2166/wst.2021.081

Directive 2000/60/EC of the European Parliament and of the Council of 23 October 2000 establishing a framework for Community action in the field of water policy, L327 2000/60/EC 1 (2000). https://eur-lex.europa.eu/eli/dir/2000/60/oj

Du Z-Y, Qimike A, Yang C-F, Chen J-M, Wang Q-F (2011) Testing four barcoding markers for species identification of Potamogetonaceae. J Syst Evol 49(3):246–251. https://doi.org/10.1111/j.1759-6831.2011.00131.x

Elliott LD, Rijal DP, Brown AG, Bakke J, Topstad L, Heintzman PD, Alsos IG (2023) Sedimentary ancient DNA reveals local vegetation changes driven by glacial activity and climate. Quaternary 6(1):7. https://doi.org/10.3390/quat6010007

Figuerola J, Green AJ (2002) Dispersal of aquatic organisms by waterbirds: a review of past research and priorities for future studies. Freshw Biol 47(3):483–494. https://doi.org/10.1046/j.1365-2427.2002.00829.x

Fujiwara A, Matsuhashi S, Doi H, Yamamoto S, Minamoto T (2016) Use of environmental DNA to survey the distribution of an invasive submerged plant in ponds. Freshw Sci 35(2):748–754. https://doi.org/10.1086/685882

Giguet-Covex C, Pansu J, Arnaud F, Rey P-J, Griggo C, Gielly L, Domaizon I, Coissac E, David F, Choler P, Poulenard J, Taberlet P (2014) Long livestock farming history and human landscape shaping revealed by lake sediment DNA. Nat Commun 5(1):3211. https://doi.org/10.1038/ncomms4211

Giguet-Covex C, Ficetola GF, Walsh K, Poulenard J, Bajard M, Fouinat L, Sabatier P, Gielly L, Messager E, Develle AL, David F, Taberlet P, Brisset E, Guiter F, Sinet R, Arnaud F (2019) New insights on lake sediment DNA from the catchment: importance of taphonomic and analytical issues on the record quality. Sci Rep 9:14676. https://doi.org/10.1038/s41598-019-50339-1

Giguet-Covex C, Messager E, Arthaud F, Gielly L, Jenny J-P (2022) Aquatic plant dynamic and fishes over the last 1800 years in the Lake Aiguebelette (Northern French Alps). ScienceOpen Posters. https://doi.org/10.14293/S2199-1006.1.SOR-.PPHTSYA.v1

Heinecke L, Epp LS, Reschke M, Stoof-Leichsenring KR, Mischke S, Plessen B, Herzschuh U (2017) Aquatic macrophyte dynamics in Lake Karakul (Eastern Pamir) over the last 29 cal ka revealed by sedimentary ancient DNA and geochemical analyses of macrofossil remains. J Paleolimnol 58(3):403–417. https://doi.org/10.1007/s10933-017-9986-7

Hollingsworth PM, Graham SW, Little DP (2011) Choosing and using a plant DNA barcode. PLoS One 6(5):e19254. https://doi.org/10.1371/journal.pone.0019254

Hoveka LN, Bezeng BS, Yessoufou K, Boatwright JS, Van der Bank M (2016) Effects of climate change on the future distributions of the top five freshwater invasive plants in South Africa. South African journal of botany

Hughes JMR, Clarkson BR, Castro-Castellon AT, Hess LL (2018) Wetland plants and aquatic macrophytes. In: Hughes J (ed) Freshwater ecology and conservation: approaches and techniques. Oxford University Press

Ibrahim A, Höckendorff S, Schleheck D, Epp L, van Kleunen M, Meyer A (2022) Vegetation changes over the last centuries in the Lower Lake Constance region reconstructed from sediment-core environmental DNA. Environ DNA 4(4):830–845. https://doi.org/10.1002/edn3.292

Isarin RFB, Bohncke SJP (1999) Mean July temperatures during the Younger Dryas in Northwestern and Central Europe as Inferred from climate indicator plant species. Quat Res 51(2):158–173. https://doi.org/10.1006/qres.1998.2023

Ji F, Yan L, Yan S, Qin T, Shen J, Zha J (2021) Estimating aquatic plant diversity and distribution in rivers from Jingjinji region, China, using environmental DNA metabarcoding and a traditional survey method. Environ Res 199:111348. https://doi.org/10.1016/j.envres.2021.111348

Krueger J, Foerster V, Trauth MH, Hofreiter M, Tiedemann R (2021) Exploring the past biosphere of chew Bahir/Southern Ethiopia: cross-species hybridization capture of ancient sedimentary DNA from a deep drill core. Front Earth Sci 9:id.775

Kuehne LM, Ostberg CO, Chase DM, Duda JJ, Olden JD (2020) Use of environmental DNA to detect the invasive aquatic plants Myriophyllum spicatum and Egeria densa in lakes. Freshw Sci 39(3):521–533. https://doi.org/10.1086/710106

Kuzmina ML, Braukmann TWA, Zakharov EV (2018) Finding the pond through the weeds: eDNA reveals underestimated diversity of pondweeds. Appl Plant Sci 6(5):e01155. https://doi.org/10.1002/aps3.1155

Lammers Y, Heintzman PD, Alsos IG (2021) Environmental palaeogenomic reconstruction of an Ice Age algal population. Commun Biol 4(1):1–11. https://doi.org/10.1038/s42003-021-01710-4

Lewis LR, Ickert-Bond SM, Biersma EM, Convey P, Goffinet B, Hassel K, Kruijer H(JD), Farge CL, Metzgar J, Stech M, Villarreal JC, McDaniel SF (2017) Future directions and priorities for Arctic bryophyte research. Arct Sci 3(3):475–497. https://doi.org/10.1139/as-2016-0043

Messyasz B, Pikosz M, Treska E (2018) Biology of freshwater macroalgae and their distribution. In: Chojnacka K, Wieczorek PP, Schroeder G, Michalak I (eds) Algae biomass: characteristics and applications: towards algae-based products. Springer, Cham, pp 17–31

Monchamp M-E, Taranu ZE, Garner RE, Rehill T, Morissette O, Iversen LL, Fugère V, Littlefair JE, da Costa NB, Desforges JE, Sánchez Schacht JR, Derry AM, Cooke SJ, Barrett RDH, Walsh DA, Ragoussis J, Albert M, Cristescu ME, Gregory-Eaves I (2023) Prioritizing taxa for genetic reference database development to advance inland water conservation. Biol Conserv 280:109963. https://doi.org/10.1016/j.biocon.2023.109963

Moore PD, Webb JA, Collinson ME (1991) Pollen analysis. Blackwell Scientific Publications

Murchie TJ, Kuch M, Duggan AT, Ledger ML, Roche K, Klunk J, Karpinski E, Hackenberger D, Sadoway T, MacPhee R, Froese D, Poinar H (2021) Optimizing extraction and targeted capture of ancient environmental DNA for reconstructing past environments using the PalaeoChip Arctic-1.0 bait-set. Quat Res 99:305–328. https://doi.org/10.1017/qua.2020.59

Murphy K, Efremov A, Davidson TA, Molina-Navarro E, Fidanza K, Crivelari Betiol TC, Chambers P, Tapia Grimaldo J, Varandas Martins S, Springuel I, Kennedy M, Mormul RP, Dibble E, Hofstra D, Lukács BA, Gebler D, Baastrup-Spohr L, Urrutia-Estrada J (2019) World distribution, diversity and endemism of aquatic macrophytes. Aquat Bot 158:103127. https://doi.org/10.1016/j.aquabot.2019.06.006

Netten JJCA, Nes EH, Scheffer M, Roijackers RMM (2010) Effect of temperature and nutrients on the competition between free-floating Salvinia natans and submerged Elodea nuttallii in mesocosms. Fundam Appl Limnol 177:125–132. https://doi.org/10.1127/1863-9135/2010/0177-0125

Parducci L, Bennett KD, Ficetola GF, Alsos IG, Suyama Y, Wood JR, Pedersen MW (2017) Ancient plant DNA in lake sediments. New Phytol 214(3):924–942. https://doi.org/10.1111/nph.14470

Parducci L, Alsos IG, Unneberg P, Pedersen MW, Han L, Lammers Y, Salonen JS, Väliranta MM, Slotte T, Wohlfarth B (2019) Shotgun environmental DNA, pollen, and macrofossil analysis of Lateglacial Lake sediments from southern Sweden. Front Ecol Evol 7:1–15.

Poikane S, Portielje R, Denys L, Elferts D, Kelly M, Kolada A, Mäemets H, Phillips G, Søndergaard M, Willby N, van den Berg MS (2018) Macrophyte assessment in European lakes: diverse approaches but convergent views of 'good' ecological status. Ecol Indic 94:185–197. https://doi.org/10.1016/j.ecolind.2018.06.056

Reitsema RE, Meire P, Schoelynck J (2018) The future of freshwater macrophytes in a changing world: dissolved organic carbon quantity and quality and its interactions with macrophytes. Front Plant Sci 9:1–15.

Revéret A, Rijal DP, Heintzman PD, Brown AG, Stoof-Leichsenring KR, Alsos IG (2023) Environmental DNA of aquatic macrophytes: The potential for reconstructing past and present vegetation and environments. Freshw Biol 2023;00:1–22. https://doi.org/10.1111/fwb.14158

Rijal DP, Heintzman PD, Lammers Y, Yoccoz NG, Lorberau KE, Pitelkova I, Goslar T, Murguzur FJA, Salonen JS, Helmens KF, Bakke J, Edwards ME, Alm T, Bråthen KA, Brown AG, Alsos IG (2021) Sedimentary ancient DNA shows terrestrial plant richness continuously increased over the Holocene in northern Fennoscandia. Science. Advances 7(31):eabf9557. https://doi.org/10.1126/sciadv.abf9557

Schenekar T (2023) The current state of eDNA research in freshwater ecosystems: are we shifting from the developmental phase to standard application in biomonitoring? Hydrobiologia 850(6):1263–1282. https://doi.org/10.1007/s10750-022-04891-z

Schirmer M, D'Amore R, Ijaz UZ, Hall N, Quince C (2016) Illumina error profiles: resolving fine-scale variation in metagenomic sequencing data. BMC Bioinform 17(1):125. https://doi.org/10.1186/s12859-016-0976-y

Scriver M, Marinich A, Wilson C, Freeland J (2015) Development of species-specific environmental DNA (eDNA) markers for invasive aquatic plants. Aquat Bot 122:27–31. https://doi.org/10.1016/j.aquabot.2015.01.003

Shackleton ME, Rees GN, Watson G, Campbell C, Nielsen D (2019) Environmental DNA reveals landscape mosaic of wetland plant communities. Glob Ecol Conserv 19:e00689. https://doi.org/10.1016/j.gecco.2019.e00689

Sjögren PJE (2021) An overview of Holocene climate reconstructions in northernmost Fennoscandia: a contribution to the Stone Age Demographics project. Septentrio Reports. https://doi.org/10.7557/7.5747

Soininen EM, Gauthier G, Bilodeau F, Berteaux D, Gielly L, Taberlet P, Gussarova G, Bellemain E, Hassel K, Stenøien HK, Epp L, Schrøder-Nielsen A, Brochmann C, Yoccoz NG (2015) Highly overlapping diet in two sympatric lemming species during winter revealed by DNA metabarcoding. PLoS One. https://doi.org/10.1371/journal.pone.0115335

Sønstebø JH, Gielly L, Brysting AK, Elven R, Edwards M, Haile J, Willerslev E, Coissac E, Rioux D, Sannier J, Taberlet P, Brochmann C (2010) Using next-generation sequencing for molecular reconstruction of past Arctic vegetation and climate. Mol Ecol Resour 10(6):1009–1018. https://doi.org/10.1111/j.1755-0998.2010.02855.x

Stoof-Leichsenring KR, Liu S, Jia W, Li K, Pestryakova LA, Mischke S, Cao X, Liu X, Ni J, Neuhaus S, Herzschuh U (2020) Plant diversity in sedimentary DNA obtained from high-latitude (Siberia) and high-elevation lakes (China). Biodivers Data J 8:e57089. https://doi.org/10.3897/BDJ.8.e57089

Stoof-Leichsenring KR, Huang S, Liu S, Jia W, Li K, Liu X, Pestryakova LA, Herzschuh U (2022) Sedimentary DNA identifies modern and past macrophyte diversity and its environmental drivers in high-latitude and high-elevation lakes in Siberia and China. Limnol Oceanogr 67(5):1126–1141. https://doi.org/10.1002/lno.12061

Taberlet P, Coissac E, Pompanon F, Gielly L, Miquel C, Valentini A, Vermat T, Corthier G, Brochmann C, Willerslev E (2007) Power and limitations of the chloroplast trnL (UAA) intron for plant DNA barcoding. Nucleic Acids Res 35(3):e14. https://doi.org/10.1093/nar/gkl938

Thomsen PF, Willerslev E (2015) Environmental DNA – an emerging tool in conservation for monitoring past and present biodiversity. Biol Conserv 183:4–18. https://doi.org/10.1016/j.biocon.2014.11.019

Tsukamoto Y, Yonezawa S, Katayama N, Isagi Y (2021) Detection of endangered aquatic plants in rapid streams using environmental DNA. Front Ecol Evol 8. https://doi.org/10.3389/fevo.2020.622291

Tyrrell CD, Chambers PA, Culp JM (2022) Harnessing aquatic plant growth forms to apply European nutrient-enrichment bioindicators to Canadian waters. Appl Plant Sci 10(4):e11487. https://doi.org/10.1002/aps3.11487

Väliranta M, Kultti S, Nyman M, Sarmaja-Korjonen K (2005) Holocene development of aquatic vegetation in shallow Lake Njargajavri, Finnish Lapland, with evidence of water-level fluctuations and drying. J Paleolimnol 34(2):203–215. https://doi.org/10.1007/s10933-005-1840-7

Väliranta M, Salonen JS, Heikkilä M, Amon L, Helmens K, Klimaschewski A, Kuhry P, Kultti S, Poska A, Shala S, Veski S, Birks HH (2015) Plant macrofossil evidence for an early onset of the Holocene summer thermal maximum in northernmost Europe. Nat Commun 6(1):6809. https://doi.org/10.1038/ncomms7809

Vásquez C, Calva J, Morocho R, Donoso DA, Benítez Á (2019) Bryophyte communities along a tropical urban river respond to heavy metal and arsenic pollution. Water 11(4):813. https://doi.org/10.3390/w11040813

Wang Y, Pedersen MW, Alsos IG, De Sanctis B, Racimo F, Prohaska A, Coissac E, Owens HL, Merkel MKF, Fernandez-Guerra A, Rouillard A, Lammers Y, Alberti A, Denoeud F, Money D, Ruter AH, McColl H, Larsen NK, Cherezova AA, Edwards ME, Fedorov GB, Haile J, Orlando L, Vinner L, Korneliussen TS, Beilman DW, Bjørk AA, Cao J, Dockter C, Esdale J, Gusarova G, Kjeldsen KK, Mangerud J, Rasic JT, Skadhauge B, Svendsen JI, Tikhonov A, Wincker P, Xing Y, Zhang Y, Froese DG, Rahbek C, Bravo DN, Holden PB, Edwards NR, Durbin R, Meltzer DJ, Kjær KH, Möller P, Willerslev E (2021) Late quaternary dynamics of Arctic biota from ancient environmental genomics. Nature 600(7887):86–92. https://doi.org/10.1038/s41586-021-04016-x

Wang L, Yang T, Hei P, Zhang J, Yang J, Luo T, Zhou G, Liu C, Wang R, Chen F (2022) Internal phosphorus cycling in macrophyte-dominated eutrophic lakes and its implications. J Environ Manag 306:114424. https://doi.org/10.1016/j.jenvman.2021.114424

Zhang Y, Jeppesen E, Liu X, Qin B, Shi K, Zhou Y, Thomaz SM, Deng J (2017) Global loss of aquatic vegetation in lakes. Earth Sci Rev 173:259–265. https://doi.org/10.1016/j.earscirev.2017.08.013

Zhang J, Hei P, Shang Y, Yang J, Wang L, Yang T, Zhou G, Chen F (2021) Internal nitrogen cycle in macrophyte-dominated eutrophic lakes: mechanisms and implications for ecological restoration. ACS EST Water 1(11):2359–2369. https://doi.org/10.1021/acsestwater.1c00203

Zhao Y, Sayer CD, Birks HH, Hughes M, Peglar SM (2006) Spatial representation of aquatic vegetation by macrofossils and pollen in a small and shallow lake. J Paleolimnol 35(2):335–350. https://doi.org/10.1007/s10933-005-1336-5

Chapter 9
Aquatic Animal DNA from Lake Sediments

Irene Gregory-Eaves, Marie-Eve Monchamp, and Zofia E. Taranu

Keywords Sedimentary DNA · Metazoa · Paleolimnology · Metabarcoding · Genomics · Zooplankton · Fish

Introduction

Animals play key roles in aquatic ecosystems, both as connectors between primary producers and consumers, as well as mobile vectors, moving nutrients and other elements between habitats. Furthermore, many aquatic animals are associated with highly-valued ecosystem services including commercial and recreational fisheries, bird watching and other naturalist activities. Over the past few decades, numerous reports have appeared in the media of different wildlife populations declining (e.g., freshwater migratory fish in Canada[1]) or recovering from an earlier decline (e.g., cormorants in Great Lakes-St. Lawrence region[2]). These assessments are possible in

[1] https://www.theglobeandmail.com/canada/article-canada-is-failing-its-freshwater-fish-populations/

[2] https://www.cbc.ca/radio/ideas/championing-the-cormorant-why-experts-argue-this-water-bird-is-unfairly-vilified-1.6200538

I. Gregory-Eaves (✉) · M.-E. Monchamp
Department of Biology, McGill University, Montreal, Quebec, Canada

Groupe de Recherche Interuniversitaire en Limnologie, Montreal, Quebec, Canada
e-mail: irene.gregory-eaves@mcgill.ca

Z. E. Taranu
Department of Biology, McGill University, Montreal, Quebec, Canada

Groupe de Recherche Interuniversitaire en Limnologie, Montreal, Quebec, Canada

Environment and Climate Change Canada, Aquatic Contaminants Research Division, Montreal, Quebec, Canada

E. Capo et al. (eds.), *Tracking Environmental Change Using Lake Sediments*, Developments in Paleoenvironmental Research 21,
https://doi.org/10.1007/978-3-031-43799-1_9

some cases where long-term records of population dynamics exist, but often such data are restricted to only a small proportion of aquatic animals and many observations are discontinuous over time. Long-term data can, however, be developed from lake sediment archives, providing key insights into unknown or unmeasured population- and community-level changes. Furthermore, as sediment archives can date back over centuries, it is possible to examine baseline conditions in aquatic ecosystems that predate major anthropogenic change.

Traditional paleolimnological methods have largely focused on organisms or parts of organisms that are microscopic in size and are preserved in the sediments (e.g., pollen, diatoms, or algal pigments; Gregory-Eaves and Smol 2024). There has also been a long-standing effort to study zooplankton subfossils represented both as pieces of the carapace or as resting stages in sediment cores. In addition, classical paleolimnological methods include the study of chironomid subfossil head capsules, which may be quite abundant in profundal sediments and have been examined over a large geographic area (Engels et al. 2020). In contrast, subfossils of fish and other animals (e.g., molluscs and waterbirds) are generally very sparse in sediment cores and thus have not been the subject of extensive study (but see Davidson et al. 2003 and Walton et al. 2021).

Analyses of sedimentary DNA (sedDNA) have the potential to track how many populations have fluctuated over time, which was not possible with traditional paleolimnological approaches when no or too few morphological remains were found. Earlier genetic analyses in paleolimnology began with the identification and genotyping of propagules (e.g., gametes, spores, resting eggs) left by organisms in their environment as a way to reconstruct past population dynamics and infer relationships between aquatic organisms (Pollard et al. 2003; Orsini et al. 2013). Now, sedDNA approaches have largely been developed through modifications of environmental DNA (eDNA; Taberlet et al. 2018) and ancient DNA approaches (Orlando et al. 2021).

Through both metabarcoding and shotgun sequencing (i.e., metagenomics) of environmental samples it is possible to detect a broader range of organisms, which has the advantage of considering multiple trophic levels (Ficetola and Taberlet 2023). The primer sets used in metabarcoding are typically designed to detect particular groups, with cytochrome *c* oxidase subunit I (COI) gene primers often used for invertebrate assemblages and 12S rRNA gene primers used for fish assemblages (although 16S rRNA and 18S rRNA gene universal primers can also detect invertebrates and animals in general). Shotgun sequencing allows the study of all DNA in an environmental sample, and is advantageous as it removes the potential for polymerase chain reaction (PCR) and primer biases that are associated with metabarcoding. Furthermore, to study a broad range of organisms there is often the need of conducting metabarcoding with multiple specific markers, but this is often not ideal because of the limited amounts of DNA template available from sediment extraction. Finally, a high level of replication is recommended to detect rare species and to reduce the number of false positives, but this means that costs can quickly become rather unwieldy (Ficetola et al. 2015). Shotgun sequencing can allow for substantial sequencing depth, is becoming less expensive and requires less template DNA compared to a multi-marker approach, while increasing the breath of taxa than can be discovered (see Ficetola and Taberlet (2023) for a review of the pros and cons of these different approaches).

In this chapter, we have approached the topic of aquatic animals by developing two main themes that are based on analyses of DNA preserved in lake sediments: (a) zooplankton population and community dynamics; and (b) aquatic vertebrate dynamics. We end the chapter with a summary of the key findings and provide a perspective on topics that might be addressed in the near future.

Zooplankton Population and Community Dynamics

Zooplankton occupy a central position in the aquatic food web as many are herbivores of phytoplankton (primary producers) and act as prey to planktivorous fish (note: some zooplankton also take advantage of the microbial loop or are predators of other zooplankton). Consequently, zooplankton communities are impacted by both top-down and bottom-up processes, and have been found to change across gradients in trophic status, temperature, and other lake variables (Paquette et al. 2022).

Preservation of Zooplankton Resting Stages and DNA in Sediments

Zooplankton offer a unique opportunity for understanding past ecological and evolutionary dynamics because several species can produce dormant forms (resting eggs; Cuenca-Cambronero et al. 2022), which can survive harsh conditions and remain preserved for long periods of time in aquatic sediments (Radzikowski 2013). A substantial diversity of zooplankton resting stages can be found in lake sediments, from well-preserved cladocerans (Vandekerkhove et al. 2004), to less-well preserved rotifers and copepods. These resting eggs, however, offer limited possibilities for identification via morphological characters (Briski et al. 2011) and restricted success in reconstructing past zooplankton population dynamics. To date, the resting eggs of cladoceran *Daphnia* have been the most targeted in zooplankton population dynamics studies. *Daphnia* eggs are encapsulated in hard protective cases called ephippia which makes them more resistant to degradation in the environment. These heavy cases often settle at the bottom of lakes where they accumulate in sediments, acting as a reservoir for zooplankton populations once the environment becomes more favorable. Genetic analyses of resting eggs are typically done on individuals hatched in the laboratory from eggs retrieved from surface sediments or cores. However, this requires some time investment as resting eggs must first be isolated. More importantly, this approach may introduce biases in population estimates due to variable hatching success among genotypes and because conditions for successful hatching are unknown for most resting eggs (Schwartz and Hebert 1987; Briski et al. 2011). To overcome the often challenging limitation of ephippia egg hatching, investigators may extract DNA from the eggs directly, without hatching them first (Montero-Pau et al. 2008; Lack et al. 2018; Cordellier et al. 2021). Methods have also been developed for extracting DNA directly from the ephippial cases in which the eggs are enclosed (Ishida et al. 2012), meaning that *Daphnia* species can be determined even in the absence of eggs or from a cracked or incomplete ephippia, but success rates tend to be low. Finally, like other groups of organisms, zooplankton also leave pieces of their DNA behind in the environment which can be analyzed by tapping directly into bulk sediments (sedDNA) even in the absence of visible fossilized remains or propagules (see section "Moving beyond ephippial DNA" below).

Examining External Stressors as Drivers of Zooplankton Population or Community Dynamics

DNA-based paleolimnological evaluations of resting eggs have been broadly used to evaluate the temporal changes in the genetic structure of *Daphnia* populations in response to various external stressors such as eutrophication (e.g., Weider et al. 1997; Brede et al. 2009; Frisch et al. 2014; Möst et al. 2015). A frequent outcome of eutrophication and climate warming is the proliferation and accumulation of cyanobacteria in lakes. Because cyanobacteria are generally a poor food source for zooplankton compared to green algae, their proliferation can have a strong selection pressure on *Daphnia* populations (Rohrlack et al. 1999). Experimental studies and the hatching of resting eggs in the laboratory have shown that small-bodied *Daphnia* and *Bosmina* are better adapted to the presence of cyanobacteria (even toxic taxa; Hairston et al. 1999), and may initially outcompete larger-bodied *Daphnia* at the onset of toxic algal blooms (Jiang et al. 2017). In general, there is growing evidence that *Daphnia* may evolve resistance to toxic cyanobacteria (Ger et al. 2014).

Apart from eutrophication and climate warming, other forms of anthropogenic stressors in aquatic systems have been shown to impact *Daphnia* population dynamics using sedDNA, such as increases in aquatic contaminants. A study by Turko et al. (2016) suggested rapid adaptation to lead pollution in the invasive *Daphnia galeata* population of the temperate peri-Alpine Lake Greifensee (Switzerland) based on genotyping of hatched resting eggs. *D. galeata* acquired resistance to lead pollution and became dominant, but populations showed a loss of resistance to lead after concentrations decreased. Likewise, DNA sequencing of the 12S rRNA gene from resting eggs revealed that a past invasion of the Eurasian species *D. curvirostris* in Lake Onondaga (New York), a temperate dimictic lake, was at least partially facilitated by industrial contamination arising from the local chemical industry (Duffy et al. 2000). The invader largely disappeared once the lake water quality was restored.

Zooplankton-Parasite Studies

Another topic of interest which can be explored in zooplankton is host-parasite interactions, which is an important structuring force in natural populations (Decaestecker et al. 2007). Using genotyping of hatched resting eggs from sediment cores collected in a natural pond (Heverlee, Belgium), Decaestecker et al. (2007) experimentally reconstructed the coevolutionary dynamics of *Daphnia* and its bacterial endoparasite *Pasteuria ramosa* providing evidence of *Daphnia* – *Pasteuria* coevolution. More recently, Rajarajan et al. (2023) investigated the role of natural genetic variability of the host *Daphnia galeata* in determining the phylogenetic structure of the associated microbial community with a 5-year long common garden experiment (i.e., an experiment that allows one to disentangle the effects of genetic and environmental variation on phenotypes by comparing genetic differences among populations subjected to identical environmental conditions). Specifically, the team reared *D. galeata* hatched from eggs recovered at two time periods of contrasting environmental conditions archived in a sediment core of Lake Greifensee (Switzerland), and after the common garden experiment, conducted a 16S rRNA gene metabarcoding analysis of *D. galeata* microbiomes. Their results showed that bacterial communities in the gut and body of *D. galeata* differed significantly among *Daphnia* genotypes, but that bacterial community beta-diversity was not correlated with the genetic distance of the hosts and these microbiome communities were not clearly distinguished by the sediment layer of origin. These results led the authors to suggest that factors other than environmental selection and host genetics (such as stochasticity) were at play in shaping *Daphnia* microbiome over time. In addition, genetic analyses of sediment records have been used to unlock information on the composition, prevalence and effect of viral infections on plankton (Coolen 2011).

Shotgun sequencing of DNA from *Daphnia* resting eggs can also provide insights into the differentiation of populations among time points, or identify the presence of functional genes and viral pathogens (Hewson et al. 2013; Cordellier et al. 2021; O'Grady et al. 2022). For example, Hewson et al. (2013) used a shotgun metagenomic approach on DNA from resting eggs and bulk sedDNA from two temperate lakes to identify viruses suspected to be involved in the ecology of *Daphnia* spp. They were able to detect viruses in resting eggs retrieved from ~30 year-old sediments of Oneida Lake (New York, USA); here they found a high prevalence of potentially fatal viruses (DMClaHV+) prior to important population declines in *D. retrocurva*.

As mentioned above, genetic reconstructions based on the genotyping or sequencing of resting eggs are limited for taxa other than *Daphnia*. However, a few studies have successfully conducted genetic analyses on non-Daphnid resting eggs to investigate past population dynamics. For instance, Derry et al. (2010) examined the survival of *Leptodiaptomus minutus* calanoid copepods by hatching resting eggs from sediment intervals representing different levels of acidification over a period of up to 100 years. Likewise, Epp et al. (2010) reconstructed the history of the rotifer genus *Brachionus* from its resting stages archived in a sediment record from an alkaline-saline crater lake (Lake Sonachi, Kenya) to elucidate the persistence and temporal dynamics of *Brachionus* haplotypes over a period of past environmental change. Analyses of the mitochondrial COI gene showed that the dominant haplotype in Lake Sonachi changed concomitantly to major environmental perturbations (volcanic eruption and major lakewater level decrease).

Moving Beyond Ephippial DNA

To date, there has been a limited amount of metabarcoding work on bulk sedDNA to track changes in zooplankton communities (Table 9.1). Much of the initial work has compared sedDNA to time series generated through the analysis of zooplankton net tows of the water column or the study of subfossils preserved in sediments. For example, Tsugeki et al. (2022) demonstrated a strong congruence between quantitative (q)PCR analysis of sedDNA and morphological time-series data of resting eggs from *D. galeata* and *D. pulicaria* populations over the last 100 year in Lake Biwa (Japan). More recently, Nakane et al. (2023) applied qPCR to sedDNA extracts from Lake Biwa sediment cores to generate paleogenetically-derived time series of *Eodiaptomus japonicus*, a copepod. They found that their concentrations of *E. japonicus* gene copies per gram dry sediment were significantly correlated to the biomass estimates of *E. japonicus* and counts of their subitaneous eggs (i.e., eggs that hatch within days and do not require a period of dormancy) from net tow samples.

Based on a three-year time series of paired sediment trap samples and zooplankton net hauls from a monomictic, temperate lake in British Columbia (Canada), Gauthier et al. (2021) demonstrated that a substantial number of 18S rRNA gene DNA reads generated were associated with copepods, but that the zooplankton composition based on reads was weakly correlated with the composition of net haul counts or biomass. In this lake, *Daphnia* were abundant in the water column, but the 18S rRNA gene in some *Daphnia* (among other zooplankton species) can have extensive intragenomic length variation and thus this zooplankton group could be biasedly excluded (Brown et al. 2015).

We are aware of only one sedDNA study where a full temporal sequence was developed from a sediment core and metabarcoded using a COI primer set, which is best adapted for invertebrate taxa. Specifically, Garcés-Pastor et al. (2019) applied a dual primer set approach, using both 18S rRNA gene and COI gene primers to reconstruct the history of eukaryotic communities in a Pyrenean peat bog ecosystem. After removal of DNA reads associated with taxa that might be living *in situ*, the authors found that most of the reads associated with COI from their peat bog core were associated with Arthropod taxa, including *Daphnia*. We believe that further analyses of sediment cores applying primers targeting the COI region could thus be a fruitful area to expand the pool of invertebrate taxa that can be analyzed from more traditional approaches.

Table 9.1 Summary of select metabarcoding studies addressing invertebrate fauna distributions and dynamics in sediments

Authors	Year	Target taxa	Brief description of lab approach	Main findings
Coolen et al.	2013	Copepods	18S rRNA (317 bp fragment) amplified with general eukaryote and copepod specific reverse primer (Cop317r)	Presence of particular copepod taxa suggest establishment of saline conditions in Black Sea
Garcès-Pastor et al.	2019	Invertebrates and more broadly eukaryotes	313 bp fragment of the mitochondrial COI, using the Leray-XT primer set and V7 region of nuclear-encoded ribosomal 18S rRNA gene (< 117 bp)	Relatively more arthropods detected with the COI vs 18S rRNA primer set. Some of the most prevalent arthropods were mites, collembola, coleoptera, maxillopoda and ostracoda
Echeverria-Galindo et al.	2021	Ostracods	16S rRNA, 18S rRNA and COI primer pairs tested *in silico* and on surface sediment samples	Greatest percent of ostracod sequences obtained with two different 18S rRNA primers, but sediment volume extracted had little effect
Gauthier et al.	2021	Eukaryotes	v7 region of 18S rRNA (~260 bp) results compared with zooplankton morphological counts in sediment traps and water	Three maxillopoda ASVs dominated water and sediment samples, whereas morphological taxa were dominated by brachiopoda. Greatest congruence between zooplankton sedDNA and morphology was apparent by looking at extracellular sedDNA fraction
Pearman et al.	2022	Eukaryotes	~313 bp fragment of the mitochondrial COI, using the Leray primer set in both water and surface sediment samples	Among large planktonic eukaryotes, copepods were among the dominant taxa in water, but in the surface sediments the team detected a higher relative abundance of algae (Ceratiaceae) and worms (from the families Naididae and Chaetonotidae)

Note 1: additional 18S rRNA metabarcoding studies have been published but we only highlight ones where animals were the focus. Note 2: additional 18S rRNA metabarcoding studies focusing on protists are featured in Chap. 6 of this volume. Note 3: *bp* base pairs

From a bulk sedDNA extract, shotgun sequencing can provide an ecosystem-wide perspective. However, without any manipulation, the metazoan signal contained within the output of reads from an environmental sample is rather small (i.e., less than 1% of taxonomically assigned reads in lake water metagenomes; Monchamp et al. 2022; or lake sediment metagenomes; Monchamp personal communication). Nonetheless, comparisons of classical zooplankton taxonomy analyses with metagenomic data based on eDNA extracted from water samples at the same sites yielded results that were moderately congruent (Monchamp et al. 2022).

Aquatic Vertebrate Population and Community Dynamics

There is substantial interest in understanding the long-term dynamics of numerous aquatic vertebrates, in part because these taxa tend to be more iconic, represent important commercial species and, in some cases, may be taxa at risk of extinction. In addition to "desirable" taxa, there are also taxa that are not native to a particular site and thus are monitored as part of programs designed for their early detection or to limit their growth and spread. While studying the distribution and dynamics of both native and non-native taxa is necessary, it is also expensive and time consuming. As such, substantial investments have been made in developing efficiencies in the study of eDNA and sedDNA for issues of conservation and management (Taberlet et al. 2018; Ficetola et al. 2019; Pawlowski et al. 2022). Analyses of sedDNA targeting aquatic vertebrates are still in their early days, but this is an area that we anticipate will grow in the future (Capo et al. 2021, Chap. 13 of this volume). New methods specifically designed to detect sedDNA of aquatic vertebrates have been recently published (Thomson-Laing et al. 2022; Sakata et al. 2022).

PCR-Based Approaches

Fishes

Typically, when one thinks of aquatic vertebrates, fish is the group of organisms that comes to mind and indeed they are a critical part of many aquatic food webs. Relative to all other vertebrates, fish is the most diverse group (Pereira et al. 2010; Burkhead 2012), with approximately 12,000 freshwater species (Facey et al. 2022). However, fish have also seen more extinctions in the twentieth century than any other vertebrates (Burkhead 2012). As a result, there is a growing need to understand the distribution and dynamics of a wide diversity of fishes.

To date, there has been a substantial number of papers that have applied eDNA approaches to study fish abundances, distributions and dynamics in the water column (e.g., Yates et al. 2021; Littlefair et al. 2021; Miya 2022; and reviewed in Rourke et al. 2022). This foundational work has not only advanced the field of eDNA research but has also served the paleolimnological community well in terms of developing primers, improving laboratory workflows with information on inhibitors, and streamlining bioinformatic pipelines. Experimental work has also been conducted in fishponds or experimental vessels to quantify the persistence of fish

DNA particles in water and sediments over time (e.g., Turner et al. 2015; Sakata et al. 2020). For example, Sakata et al. (2020) collected samples of water and sediments from a fishpond and found that the concentration of DNA particles was generally greater and more persistent in sediments than in water, but that species composition based on metabarcoding was not different between sample matrices. In contrast, a couple of field surveys, where quantitative PCR analyses of water and sediments were collected and compared to field estimates of fish or amphibian abundances, authors found higher detection probabilities in water samples relative to sediments, and this may be due to the greater presence of inhibitors or the nature of the sediments analyzed (Baldigo et al. 2017; Buxton et al. 2018). Overall, early results are encouraging for sedDNA studies as they show that sediments have the potential to archive a longer term and representative portrait of fish assemblages, but the field is still in its infancy when it comes to reconstructing past composition and dynamics of fish and other large aquatic animals from sedDNA archives.

Only a handful of studies have been published thus far with the intent of developing historical trajectories of fish dynamics. In all cases, these studies have taken a targeted approach where a genetic marker specific to a particular fish species was evaluated (see Table 9.2). The earliest study conducted by Matisoo-Smith et al. (2008) confirmed the presence of a native fish in numerous samples throughout a core collected from a lake in New Zealand. In later studies by Stager et al. (2015) and Nelson-Chorney et al. (2019), the authors used sedDNA to show that a focal fish was present in the lake ecosystem historically, rather than a relatively new addition to the food web. In two more recent studies, the authors have had access to complementary historical data on fish landings and, in some cases, fish scale records, which they used to evaluate the coherency among records (Kuwae et al. 2020; Sakata et al. 2022). For many of their focal fish time series, sedDNA concentrations were positively correlated with fish landings. Sakata et al. (2022) further corrected their sedDNA fish time series for diagenesis, using a model developed for chlorophyll *a* pigments and its derivatives (based on a first order degradation equation). The authors found that the correlation between the sedDNA concentrations and the paired fish landing time series improved once the diagenesis correction was applied.

The limited number of published fish sedDNA studies may partly be due to challenges associated with studying fish DNA from sediment archives. Some studies have pointed to limitations related to the diffusion of fish DNA from sediments to water (Dunker et al. 2016) and the decomposition of sedDNA (Wei et al. 2018), as well as differential efficiencies in sedDNA extraction methods (Thomson-Laing et al. 2022), especially in sediments with high organic content. Alternative or additional explanations for the low to lack of detection of fish DNA could be due to low densities of fish in the pelagic zone and/or the concentration of some fish communities in littoral zones. Monitoring fish is essential for many reasons (economical as well as ecological) and paleo-reconstructions based on sedDNA have the potential to inform on fish dynamics that can contribute to better management. The sedDNA community is currently working towards optimizing protocols for fish DNA detection and extraction from sediment archives (e.g., Thomson-Laing et al. 2022 and other efforts underway).

Table 9.2 Compilation of paleolimnological studies using a targeted sedDNA approach to track past fish population dynamics

Authors	Year	Target taxa	Brief description of lab approach	Main finding
Matisoo-Smith et al.	2008	Common bully (*Gobiomorphus cotidianus*)	150 bp sequence of 12S	Detected the presence of a native fish species
Stager et al.	2015	Yellow perch (YP; *Perca flavescens*)	Species-specific barcode marker for YP; 124 bp COI gene	Despite concern that YP might not be native, all samples from the 2200-year core tested positive for YP (negative controls were clean)
Olajos et al.	2018	Whitefish (*Coregonus lavaretus* L.)	Whitefish specific primers designed from mitochondrial sequences	Based on analyses of ~10,000 year sediment records from Swedish lakes, results show colonization of whitefish immediately after the region's deglaciation in one site, versus a much delayed colonization in the second site
Nelson-Chorney et al.	2019	Yellowstone cutthroat trout (YCTT; *Oncorhynchus clarkii bouvieri*) and Westslope cutthroat trout (WCTT; *Oncorhynchus clarkii lewisi*)	Two diagnostic loci with SNPs unique to focal taxa used	DNA results challenged previous notion of recent introduction of WCTT as it was present in sediment record for ~100 years
Kuwae et al.	2020	Japanese anchovy (*Engraulis japonicus*), Japanese sardine (*S. Melanostictus*), and jack mackerel (*Tranchurus japonicus*)	qPCR	Detected significant correlations between DNA concentrations and fish landings for three taxa, and modest correlation with sardine fish scale records
Sakata et al.	Preprint	*Plecoglossus altivelis* and *Gymnogobius isaza*	mitochondrial cytochrome b (cytb) gene (<150 bp)	Positive relationships between preservation-adjusted DNA concentrations and historical CPUE estimates for *Plecoglossus altivelis*

This does not include studies of only surface sediments. *CPUE* catch per unit effort; *SNP* single nucleotide polymorphism

Other Vertebrates and Their Parasites

Targeted work on other aquatic animals that complete part of their life cycle in water (e.g., aquatic birds and amphibians) is also emerging and could have important conservation implications. For example, sedDNA work on birds in Canada could be quite useful, as this nation is home to 144 species of birds requiring fresh water to complete at least part of their life cycle and 29 of these taxa are endangered or threatened (Desforges et al. 2022). Already, there has been a reasonable amount of paleolimnological work tracking bird dynamics indirectly by looking at tracers of guano (reviewed by Duda et al. 2021). Large colonies of birds produce a substantial amount of guano that is enriched in nutrients as well as numerous isotopes, heavy metals and pesticides, and these materials are transported to neighboring lakes. The DNA from birds could also be preserved in sediment archives and has been detected in a few cases. For example, bird DNA was extracted from materials found in cave deposits (Willerslev et al. 2003; Haile et al. 2007), which is not too dissimilar from lake sediments. Willerslev et al. (2003) successfully sequenced the DNA of birds from New Zealand in a deposit that was at least 10,000 years old. Most recently, a metabarcoding study of a sediment core from a Norwegian lake detected the presence of a few bird species over the course of the Holocene (i.e., last 10,000 years; Brown et al. 2022). Although the 16S rRNA gene primer set was designed to detect mammals, the authors found that non-target animals were also detected, including birds, fish and invertebrates. Most recently, Duda et al. (2023) conducted a stratigraphic study of an Andean Condor (*Vultur gryphus*) deposit from Patagonia (Argentina) dating back 2200 years and employed a wide variety of paleoecological analyses. Their droplet digital PCR (ddPCR) analysis of numerous intervals from the deposit identified a diet shift in condors from reliance on native *Lama* carrion (and marine carrion, inferred from other proxies) to one focused primarily on native sheep and cattle carrion). Finally, birds have also been detected in several metagenomic studies, although generally with low read counts (see details below).

A few recent papers have also tracked long-term changes in parasites commonly associated with birds and amphibians. For example, based on a recent review by Coombe et al. (2021), the avian influenza virus (AIV) has been detected across numerous aquatic sediment samples (n = 7 studies) and other soil-based substrates including soil, mud, and sand. AIV detection rates were found to vary between 0–56% in sediment- and soil-based substrates, and one study reported that the probability of AIV detection increased in sites with anthropogenic disturbance (Himsworth et al. 2020). Lang et al. (2008) reported on the potential to detect AIV in full sediment cores, but to our knowledge actual detection has not been published. Similarly, several papers have focused on extracting amphibian pathogens from sediments, especially the *Batrachochytrium* taxa (fungus) which have been associated with global population declines of amphibians. Earlier work showed that *Batrachochytrium* DNA was persistent in soil for months (Stegen et al. 2017). In a study comparing DNA from frog skin swabs to eDNA from water and sediment samples (collected in the ponds the frogs inhabited), Brannelly et al. (2020) found that sediment samples generally had lower *Batrachochytrium* DNA concentrations than water samples, and limited seasonality relative to the other sample matrices. A logical next step would be to extract a sediment core from a site where long-term data exist on *Batrachochytrium* infections to evaluate how congruent the sediment archive is with the epidemiological data.

Vertebrate Reads Detected via Shotgun Sequencing

Several paleo-metagenomic studies have been conducted and have recorded the presence of numerous aquatic animal taxa based on a modest number of DNA reads per sample (Table 9.3). For example, Pedersen et al. (2016) detected the presence of pike fish (*Esox* sp.) in a lake sediment record from western Canada (Charlie Lake) dating back 11,700 calendar years by adopting a metagenomic analysis. However, the greatest number of reads detected for pike was only 68, whereas the number of reads for some plant taxa were in the thousands. Other fish taxa detected in the record included stickleback (*Gasterosteus* sp.) and perch (*Perca* sp.). Likewise, in a

Table 9.3 Summary of key metabarcoding or metagenomic studies tracking vertebrate fauna dynamics in cores

Authors	Year	Target taxa	Brief description of lab approach	Main findings
Pedersen et al.	2016	All taxa	Metagenomics analyses of lake sediments from past Beringia sites of western Canada	Diversity of aquatic fauna detected as far back as 12,000 cal year. BP
Dommain et al.	2020	All taxa	Metagenomic analyses of swamp core from Uganda dating back 2200 cal years BP	Detected substantial diversity in metazoans, including fish, insects, nematodes, birds and mammals
Capo et al.	2021	Eukaryotes	Metabarcoding study comparing extraction approaches and sediment volumes using 18S rRNA v7 region	Lake-specific differences were much greater than extraction differences. When zooming in on metazoan reads, no clear patterns were associated with sample volume. Slight tendency for chordate reads to be more abundant in extracellular DNA fraction.
Courtin et al.	2022	All taxa	Metagenomic analyses of permafrost sediments from Siberia dating back to mid Pleistocene	Culicidae of the mosquito family detected in all samples, among other organisms from viruses to mammals
Brown et al.	2022	Vertebrates	Metabarcoding of 16S primer set (MamP007) designed to target 60–84 bp fragments	Birds, fish and invertebrates as well as a diversity of mammals were detected over the ~12,000 cal year. BP
Hebda et al	2022	All taxa	Metagenomic analysis of little Woss Lake from British Columbia	Chinook salmon (*Oncorhynchus tshawytscha*) and other vertebrates detected in sediments dating to ~14,000 cal year. BP
Murchie et al.	2022	Vertebrates	Capture enrichment metagenomics of permafrost core samples from the Yukon dating back ~40,000 cal years	Water birds as well as ground birds and numerous mammals (including the now extinct woolly mammoth, *Mammuthus*) detected.

metagenomic study of a 2200 calendar year sediment record from Uganda, Dommain et al. (2020) detected the presence of fish, birds and amphibians (as well as numerous mammals).

To enhance the signal of taxa that typically have low read counts, one might adopt a hybridization capture approach, whereby one exposes the metagenomic library to RNA probes that hybridize to target sedDNA molecules and then these can be isolated from other DNA through the selection of probes by bacteria (streptavidin) that can be enriched using a magnet (Armbrecht et al. 2021). Applying a comparative approach, Armbrecht et al. (2021) demonstrated an enhanced eukaryotic signal when hybridization was applied relative to extracts from marine sediments that did not receive this treatment. A recent application of such an approach was conducted by Murchie et al. (2022) to extracts obtained from Yukon (Canada) permafrost cores that spanned the last 40,000 years; their analyses focused on four major vertebrate clades including birds and allowed them to detect the Anatidae family (which includes ducks and geese). We anticipate that this approach will be applied more in the future.

Summary

For most aquatic ecosystems, there are no long-term monitoring data on population or community dynamics, and thus lake sediment archives can fill key gaps in our understanding. The study of aquatic animals based on genetic material archived in lake sediments is largely rooted in population-based analyses of zooplankton resting stages. The field is now greatly expanding with the appearance of numerous exciting developments (e.g., shotgun sequencing + RNA probes), which is enabling the analysis of many other taxa and the simultaneous study of multiple trophic levels. With such analyses, time series data that capture a wider trophic web can be developed and explored with sophisticated network analyses and joint-species distribution models (see Chap. 12 of this volume: An overview of biodiversity and network modeling approaches: applications to sedimentary DNA records).

Currently, there are several large initiatives underway that are developing more complete reference libraries for barcodes and full genomes (e.g., BIOSCAN: https://ibol.org/programs/bioscan/ and the Earth Biogenome project: https://www.earthbiogenome.org/). These datasets will greatly improve taxonomic and functional gene assignments, and allow investigators to obtain substantially more information from their sedDNA analyses. Likewise, there are efforts underway to prioritize the development of reference barcodes and whole genomes, and to focus our efforts on taxa at risk or of undetermined risk status due to a lack of data (Monchamp et al. 2023). Collaborative efforts, facilitated through organisations like the SedaDNA Scientific Society (https://sedadna.github.io/) will also be instrumental in advancing the field and enhancing communication of results.

Acknowledgements The authors acknowledge the following funding sources for their salary support: Canada Research Chairs program (awarded to IGE), NSERC Post-Doctoral Fellowship (awarded to MEM) and Environment and Climate Change Canada (supporting ZET). The authors acknowledge feedback from Yuanyu Cheng and support in formatting references from Michelle Gros as well as Susie Woods and Suzanne McGowan and Luke E. Holman for their useful feedback on the manuscript.

References

Armbrecht L, Hallegraeff G, Bolch CJS, Woodward C, Cooper A (2021) Hybridisation capture allows DNA damage analysis of ancient marine eukaryotes. Sci Rep 11:3220. https://doi.org/10.1038/s41598-021-82578-6

Baldigo BP, Sporn LA, George SD, Ball JA (2017) Efficacy of environmental DNA to detect and quantify brook trout populations in headwater streams of the Adirondack Mountains, New York. Trans Am Fish Soc 146(1):99–111. https://doi.org/10.1080/00028487.2016.1243578

Brannelly LA, Wetzel DP, Ohmer MEB, Zimmerman L, Saenz V, Richards-Zawacki CL (2020) Evaluating environmental DNA as a tool for detecting an amphibian pathogen using an optimized extraction method. Oecologia 194(1):267–281. https://doi.org/10.1007/s00442-020-04743-4

Brede N, Sandrock C, Straile D, Spaak P, Jankowski T, Streit B, Schwenk K (2009) The impact of human-made ecological changes on the genetic architecture of Daphnia species. Proc Natl Acad Sci 106(12):4758–4763. https://doi.org/10.1073/pnas.0807187106

Briski E, Cristescu ME, Bailey SA, MacIsaac HJ (2011) Use of DNA barcoding to detect invertebrate invasive species from diapausing eggs. Biol Invasions 13(6):1325–1340. https://doi.org/10.1007/s10530-010-9892-7

Brown EA, Chain FJJ, Crease TJ, MacIsaac HJ, Cristescu ME (2015) Divergence thresholds and divergent biodiversity estimates: can metabarcoding reliably describe zooplankton communities? Ecol Evol 5(11):2234–2251. https://doi.org/10.1002/ece3.1485

Brown T, Rijal DP, Heintzman PD, Clarke CL, Blankholm H-P, Høeg HI, Lammers Y, Bråthen KA, Edwards M, Alsos IG (2022) Paleoeconomy more than demography determined prehistoric human impact in Arctic Norway. PNAS Nexus 1(5):pgac209. https://doi.org/10.1093/pnasnexus/pgac209

Burkhead NM (2012) Extinction rates in North American freshwater fishes, 1900–2010. Bioscience 62(9):798–808. https://doi.org/10.1525/bio.2012.62.9.5

Buxton AS, Groombridge JJ, Griffiths RA (2018) Seasonal variation in environmental DNA detection in sediment and water samples. PLoS One 13(1):e0191737. https://doi.org/10.1371/journal.pone.0191737

Capo E, Giguet-Covex C, Rouillard A, Nota K, Heintzman PD, Vuillemin A, Ariztegui D, Arnaud F, Belle S, Bertilsson S, Bigler C, Bindler R, Brown AG, Clarke CL, Crump SE, Debroas D, Englund G, Ficetola GF, Garner RE, Gauthier J, Gregory-Eaves I, Heinecke L, Herzschuh U, Ibrahim A, Kisand V, Kjær KH, Lammers Y, Littlefair J, Messager E, Monchamp M-E, Olajos F, Orsi W, Pedersen MW, Rijal DP, Rydberg J, Spanbauer T, Stoof-Leichsenring KR, Taberlet P, Talas L, Thomas C, Walsh DA, Wang Y, Willerslev E, van Woerkom A, Zimmermann HH, Coolen MJL, Epp LS, Domaizon IG, Alsos I, Parducci L (2021) Lake sedimentary DNA research on past terrestrial and aquatic biodiversity: overview and recommendations. Quaternary 4(1):6. https://doi.org/10.3390/quat4010006

Coolen MJL (2011) 7000 years of Emiliania huxleyi viruses in the Black Sea. Science. https://doi.org/10.1126/science.1200072

Coolen MJL, Orsi WD, Balkema C, Quince C, Harris K, Sylva SP, Filipova-Marinova M, Giosan L (2013) Evolution of the plankton paleome in the Black Sea from the Deglacial to Anthropocene. Proc Natl Acad Sci 110(21):8609–8614. https://doi.org/10.1073/pnas.1219283110

Coombe M, Iwasawa S, Byers KA, Prystajecky N, Hsiao W, Patrick DM, Himsworth CG (2021) A systematic review and narrative synthesis of the use of environmentl samples for the surveillance of avian influenza viruses in wild waterbirds. J Wildl Dis 57(1):1–18. https://doi.org/10.7589/JWD-D-20-00082

Cordellier M, Wojewodzic MW, Wessels M, Kuster C, von Elert E (2021) Next-generation sequencing of DNA from resting eggs: signatures of eutrophication in a lake's sediment. Zoology (Jena) 145:125895. https://doi.org/10.1016/j.zool.2021.125895

Courtin J, Perfumo A, Andreev AA, Opel T, Stoof-Leichsenring KR, Edwards ME, Murton JB, Herzschuh U (2022) Pleistocene glacial and interglacial ecosystems inferred from ancient DNA analyses of permafrost sediments from Batagay megaslump, East Siberia. Environ DNA 4(6):1265–1283. https://doi.org/10.1002/edn3.336

Cuenca-Cambronero M, Courtney-Mustaphi CJ, Greenway R, Heiri O, Hudson CM, King L, Lemmen KD, Moosmann M, Muschick M, Ngoepe N, Seehausen O, Matthews B (2022) An integrative paleolimnological approach for studying evolutionary processes. Trends Ecol Evol 37(6):488–496. https://doi.org/10.1016/j.tree.2022.01.007

Davidson TA, Sayer CD, Perrow MR, Tomlinson ML (2003) Representation of fish communities by scale sub-fossils in shallow lakes: implications for inferring percid–cyprinid shifts. J Paleolimnol 30(4):441–449. https://doi.org/10.1023/B:JOPL.0000007247.97059.e2

Decaestecker E, Gaba S, Raeymaekers JAM, Stoks R, Van Kerckhoven L, Ebert D, De Meester L (2007) Host-parasite "Red Queen" dynamics archived in pond sediment. Nature 450(7171):870–873. https://doi.org/10.1038/nature06291

Derry AM, Arnott SE, Boag PT (2010) Evolutionary shifts in copepod acid tolerance in an acid-recovering lake indicated by resurrected resting eggs. Evol Ecol 24:133–145. https://doi.org/10.1007/s10682-009-9295-3

Desforges JE, Clarke J, Harmsen EJ, Jardine AM, Robichaud JA, Serré S, Chakrabarty P, Bennett JR, Hanna DEL, Smol JP, Rytwinski T, Taylor JJ, Martel AL, Winegardner AK, Marty J, Taylor MK, O'Connor CM, Robinson SA, Reid AJ, Creed IF, Gregory-Eaves I, Lapointe NWR, Cooke SJ (2022) The alarming state of freshwater biodiversity in Canada. Can J Fish Aquat Sci 79(2):352–365. https://doi.org/10.1139/cjfas-2021-0073

Dommain R, Andama M, McDonough MM, Prado NA, Goldhammer T, Potts R, Maldonado JE, Nkurunungi JB, Campana MG (2020) The challenges of reconstructing tropical biodiversity with sedimentary ancient DNA: a 2200-year-long metagenomic record from Bwindi impenetrable forest, Uganda. Front Ecol Evol 8. https://doi.org/10.3389/fevo.2020.00218

Duda MP, Hargan KE, Michelutti N, Blais JM, Grooms C, Gilchrist HG, Mallory ML, Robertson GJ, Smol JP (2021) Reconstructing long-term changes in avian populations using lake sediments: opening a window onto the past. Front Ecol Evol 9. https://doi.org/10.3389/fevo.2021.698175

Duda MP, Grooms C, Sympson L, Blais JM, Dagodzo D, Feng W, Hayward KM, Julius ML, Kimpe LE, Lambertucci SA, Layton-Matthews D, Lougheed SC, Massaferro J, Michelutti N, Pufahl PK, Vuletich A, Smol JP (2023) A 2200-year record of Andean Condor diet and nest site usage reflects natural and anthropogenic stressors. Proc R Soc B 290:20230106. https://doi.org/10.1098/rspb.2023.0106

Duffy MA, Perry LJ, Kearns CM, Weider LJ, Hairston NG Jr (2000) Paleogenetic evidence for a past invasion of Onondaga Lake, New York, by exotic Daphnia curvirostris using mtDNA from dormant eggs. Limnol Oceanogr 45(6):1409–1414. https://doi.org/10.4319/lo.2000.45.6.1409

Dunker KJ, Sepulveda AJ, Massengill RL, Olsen JB, Russ OL, Wenburg JK, Antonovich A (2016) Potential of environmental DNA to evaluate Northern Pike (Esox lucius) eradication efforts: an experimental test and case study. PLoS One 11(9):e0162277. https://doi.org/10.1371/journal.pone.0162277

Echeverría-Galindo P, Anslan S, Frenzel P, Künzel S, Vences M, Pérez L, Börner N, Kang W, Schwarz A, Wang J, Peng P, Zhu L, Schwalb A (2021) High-throughput identification of non-marine Ostracoda from the Tibetan Plateau: evaluating the success of various primers on sedimentary DNA samples. Environ DNA 3(5):982–996. https://doi.org/10.1002/edn3.222

Engels S, Medeiros AS, Axford Y, Brooks SJ, Heiri O, Luoto TP, Nazarova L, Porinchu DF, Quinlan R, Self AE (2020) Temperature change as a driver of spatial patterns and long-term trends in chironomid (Insecta: Diptera) diversity. Glob Chang Biol 26(3):1155–1169. https://doi.org/10.1111/gcb.14862

Epp LS, Stoof KR, Trauth MH, Tiedemann R (2010) Historical genetics on a sediment core from a Kenyan lake: intraspecific genotype turnover in a tropical rotifer is related to past environmental changes. J Paleolimnol 43(4):939–954. https://doi.org/10.1007/s10933-009-9379-7

Facey DE, Bowen BW, Collette BB, Helfman GS (2022) The diversity of fishes: biology, evolution, and ecology, 3rd edn

Ficetola GF, Taberlet P (2023) Towards exhaustive community ecology via DNA metabarcoding. Mol Ecol. https://doi.org/10.1111/mec.16881

Ficetola GF, Pansu J, Bonin A, Coissac E, Giguet-Covex C, De Barba M, Gielly L, Lopes CM, Boyer F, Pompanon F, Rayé G, Taberlet P (2015) Replication levels, false presences and the estimation of the presence/absence from eDNA metabarcoding data. Mol Ecol Resour 15(3):543–556. https://doi.org/10.1111/1755-0998.12338

Ficetola GF, Manenti R, Taberlet P (2019) Environmental DNA and metabarcoding for the study of amphibians and reptiles: species distribution, the microbiome, and much more. Amphibia-Reptilia 40(2):129–148. https://doi.org/10.1163/15685381-20191194

Frisch D, Morton PK, Chowdhury PR, Culver BW, Colbourne JK, Weider LJ, Jeyasingh PD (2014) A millennial-scale chronicle of evolutionary responses to cultural eutrophication in Daphnia. Ecol Lett 17(3):360–368. https://doi.org/10.1111/ele.12237

Garcés-Pastor S, Wangensteen OS, Pérez-Haase A, Pèlachs A, Pérez-Obiol R, Cañellas-Boltà N, Mariani S, Vegas-Vilarrúbia T (2019) DNA metabarcoding reveals modern and past eukaryotic communities in a high-mountain peat bog system. J Paleolimnol 62(4):425–441. https://doi.org/10.1007/s10933-019-00097-x

Gauthier J, Walsh D, Selbie DT, Bourgeois A, Griffiths K, Domaizon I, Gregory-Eaves I (2021) Evaluating the congruence between DNA-based and morphological taxonomic approaches in water and sediment trap samples: analyses of a 36-month time series from a temperate monomictic lake. Limnol Oceanogr 66(8):3020–3039. https://doi.org/10.1002/lno.11856

Ger KA, Hansson L-A, Lürling M (2014) Understanding cyanobacteria-zooplankton interactions in a more eutrophic world. Freshw Biol 59(9):1783–1798. https://doi.org/10.1111/fwb.12393

Gregory-Eaves I, Smol JP (2024) Chapter 30 - Paleolimnology: Approaches and Applications, Eds ID Jones, JP Smol, Wetzel's Limnology (Fourth Edition), Academic Press, 2024, 1015–1043. ISBN 9780128227015.https://doi.org/10.1016/B978-0-12-822701-5.00030-6 (https://www.sciencedirect.com/science/article/pii/B9780128227015000306)

Haile J, Holdaway R, Oliver K, Bunce M, Gilbert MTP, Nielsen R, Munch K, Ho SYW, Shapiro B, Willerslev E (2007) Ancient DNA chronology within sediment deposits: are paleobiological reconstructions possible and is DNA leaching a factor? Mol Biol Evol 24(4):982–989. https://doi.org/10.1093/molbev/msm016

Hairston NG, Lampert W, Cáceres CE, Holtmeier CL, Weider LJ, Gaedke U, Fischer JM, Fox JA, Post DM (1999) Rapid evolution revealed by dormant eggs. Nature 401(6752):446–446. https://doi.org/10.1038/46731

Hebda CFG, McLaren D, Mackie Q, Fedje D, Pedersen MW, Willerslev E, Brown KJ, Hebda RJ (2022) Late Pleistocene palaeoenvironments and a possible glacial refugium on northern Vancouver Island, Canada: evidence for the viability of early human settlement on the northwest coast of North America. Quat Sci Rev 279:107388. https://doi.org/10.1016/j.quascirev.2022.107388

Hewson I, Ng G, Li W, LaBarre BA, Aguirre I, Barbosa JG, Breitbart M, Greco AW, Kearns CM, Looi A, Schaffner LR, Thompson PD, Hairston NG Jr (2013) Metagenomic identification, seasonal dynamics, and potential transmission mechanisms of a Daphnia-associated single-stranded DNA virus in two temperate lakes. Limnol Oceanogr 58(5):1605–1620. https://doi.org/10.4319/lo.2013.58.5.1605

Himsworth CG, Duan J, Prystajecky N, Coombe M, Baticados W, Jassem AN, Tang P, Sanders E, Hsiao W (2020) Targeted resequencing of wetland sediment as a tool for avian influenza virus surveillance. J Wildl Dis 56(2):397–408

Ishida S, Ohtsuki H, Awano T, Tsugeki NK, Makino W, Suyama Y, Urabe J (2012) DNA extraction and amplification methods for ephippial cases of daphnia resting eggs in lake sediments: a novel approach for reconstructing zooplankton population structure from the past. Limnology 13(3):261–267. https://doi.org/10.1007/s10201-012-0380-x

Jiang X, Xie J, Xu Y, Zhong W, Zhu X, Zhu C (2017) Increasing dominance of small zooplankton with toxic cyanobacteria. Freshw Biol 62(2):429–443. https://doi.org/10.1111/fwb.12877

Kuwae M, Tamai H, Doi H, Sakata MK, Minamoto T, Suzuki Y (2020) Sedimentary DNA tracks decadal-centennial changes in fish abundance. Commun Biol 3(1):1–12. https://doi.org/10.1038/s42003-020-01282-9

Lack JB, Weider LJ, Jeyasingh PD (2018) Whole genome amplification and sequencing of a Daphnia resting egg. Mol Ecol Resour 18(1):118–127. https://doi.org/10.1111/1755-0998.12720

Lang AS, Kelly A, Runstadler JA (2008) Prevalence and diversity of avian influenza viruses in environmental reservoirs. J Gen Virol 89(Pt 2):509–519. https://doi.org/10.1099/vir.0.83369-0

Littlefair JE, Hrenchuk LE, Blanchfield PJ, Rennie MD, Cristescu ME (2021) Thermal stratification and fish thermal preference explain vertical eDNA distributions in lakes. Mol Ecol 30(13):3083–3096. https://doi.org/10.1111/mec.15623

Matisoo-Smith E, Roberts K, Welikala N, Tannock G, Chester P, Feek D, Flenley J (2008) Recovery of DNA and pollen from New Zealand lake sediments. Quat Int 184(1):139–149. https://doi.org/10.1016/j.quaint.2007.09.013

Miya M (2022) Environmental DNA Metabarcoding: a novel method for biodiversity monitoring of marine fish communities. Annu Rev Mar Sci 14(1):161–185. https://doi.org/10.1146/annurev-marine-041421-082251

Monchamp M-È, Walsh DA, Garner RE, Kraemer SA, Beisner BE, Cristescu ME, Gregory-Eaves I (2022) Comparative analysis of zooplankton diversity in freshwaters: what can we gain from metagenomic analysis? Environ DNA 4(6):1250–1264. https://doi.org/10.1002/edn3.335

Monchamp M-E, Taranu ZE, Garner RE, Rehill T, Morissette O, Iversen LL, Fugère V, Littlefair JE, da Costa NB, Desforges JE, Sánchez Schacht JR, Derry AM, Cooke SJ, Barrett RDH, Walsh DA, Ragoussis J, Albert M, Cristescu ME, Gregory-Eaves I (2023) Prioritizing taxa for genetic reference database development to advance inland water conservation. Biol Conserv 280:109963. https://doi.org/10.1016/j.biocon.2023.109963

Montero-Pau J, Gómez A, Muñoz J (2008) Application of an inexpensive and high-throughput genomic DNA extraction method for the molecular ecology of zooplanktonic diapausing eggs. Limnol Oceanogr Methods 6(6):218–222. https://doi.org/10.4319/lom.2008.6.218

Möst M, Oexle S, Marková S, Aidukaite D, Baumgartner L, Stich H-B, Wessels M, Martin-Creuzburg D, Spaak P (2015) Population genetic dynamics of an invasion reconstructed from the sediment egg bank. Mol Ecol 24(16):4074–4093. https://doi.org/10.1111/mec.13298

Murchie TJ, Karpinski E, Eaton K, Duggan AT, Baleka S, Zazula G, MacPhee RDE, Froese D, Poinar HN (2022) Pleistocene mitogenomes reconstructed from the environmental DNA of permafrost sediments. Curr Biol 32(4):851–860.e7. https://doi.org/10.1016/j.cub.2021.12.023

Nakane K, Liu X, Doi H, Dur G, Kuwae M, Ban S, Tsugeki N (2023) Sedimentary DNA can reveal the past population dynamics of a pelagic copepod. Freshw Biol n/a(n/a) https://doi.org/10.1111/fwb.14096

Nelson-Chorney HT, Davis CS, Poesch MS, Vinebrooke RD, Carli CM, Taylor MK (2019) Environmental DNA in lake sediment reveals biogeography of native genetic diversity. Front Ecol Environ 17(6):313–318. https://doi.org/10.1002/fee.2073

O'Grady CJ, Dhandapani V, Colbourne JK, Frisch D (2022) Refining the evolutionary time machine: an assessment of whole genome amplification using single historical Daphnia eggs. Mol Ecol Resour 22(3):946–961. https://doi.org/10.1111/1755-0998.13524

Olajos F, Bokma F, Bartels P, Myrstener E, Rydberg J, Öhlund G, Bindler R, Wang X-R, Zale R, Englund G (2018) Estimating species colonization dates using DNA in lake sediment. Methods Ecol Evol 9(3):535–543. https://doi.org/10.1111/2041-210X.12890

Orlando L, Allaby R, Skoglund P, Der Sarkissian C, Stockhammer PW, Ávila-Arcos MC, Fu Q, Krause J, Willerslev E, Stone AC, Warinner C (2021) Ancient DNA analysis. Nat Rev Methods Primers 1(1):1–26. https://doi.org/10.1038/s43586-020-00011-0

Orsini L, Schwenk K, De Meester L, Colbourne JK, Pfrender ME, Weider LJ (2013) The evolutionary time machine: using dormant propagules to forecast how populations can adapt to changing environments. Trends Ecol Evol 28(5):274–282. https://doi.org/10.1016/j.tree.2013.01.009

Paquette C, Gregory-Eaves I, Beisner BE (2022) Environmental drivers of taxonomic and functional variation in zooplankton diversity and composition in freshwater lakes across Canadian continental watersheds. Limnol Oceanogr 67(5):1081–1097. https://doi.org/10.1002/lno.12058

Pawlowski J, Bruce K, Panksep K, Aguirre FI, Amalfitano S, Apothéloz-Perret-Gentil L, Baussant T, Bouchez A, Carugati L, Cermakova K, Cordier T, Corinaldesi C, Costa FO, Danovaro R, Dell'Anno A, Duarte S, Eisendle U, Ferrari BJD, Frontalini F, Frühe L, Haegerbaeumer A, Kisand V, Krolicka A, Lanzén A, Leese F, Lejzerowicz F, Lyautey E, Maček I, Sagova-Marečková M, Pearman JK, Pochon X, Stoeck T, Vivien R, Weigand A, Fazi S (2022) Environmental DNA metabarcoding for benthic monitoring: a review of sediment sampling and DNA extraction methods. Sci Total Environ 818:151783. https://doi.org/10.1016/j.scitotenv.2021.151783

Pearman JK, Thomson-Laing G, Thompson L, Waters S, Vandergoes MJ, Howarth JD, Duggan IC, Hogg ID, Wood SA (2022) Human access and deterministic processes play a major role in structuring planktonic and sedimentary bacterial and eukaryotic communities in lakes. PeerJ 10:e14378. https://doi.org/10.7717/peerj.14378

Pedersen MW, Ruter A, Schweger C, Friebe H, Staff RA, Kjeldsen KK, Mendoza MLZ, Beaudoin AB, Zutter C, Larsen NK, Potter BA, Nielsen R, Rainville RA, Orlando L, Meltzer DJ, Kjær KH, Willerslev E (2016) Postglacial viability and colonization in North America's ice-free corridor. Nature 537(7618):45–49. https://doi.org/10.1038/nature19085

Pereira HM, Leadley PW, Proença V, Alkemade R, Scharlemann JPW, Fernandez-Manjarrés JF, Araújo MB, Balvanera P, Biggs R, Cheung WWL, Chini L, Cooper HD, Gilman EL, Guénette S, Hurtt GC, Huntington HP, Mace GM, Oberdorff T, Revenga C, Rodrigues P, Scholes RJ, Sumaila UR, Walpole M (2010) Scenarios for global biodiversity in the 21st century. Science 330(6010):1496–1501. https://doi.org/10.1126/science.1196624

Pollard HG, Colbourne JK, Keller W(B) (2003) Reconstruction of centuries-old daphnia communities in a lake recovering from acidification and metal contamination. ambi 32(3):214–218. https://doi.org/10.1579/0044-7447(2003)032[0214:ROCDCI]2.0.CO;2

Radzikowski J (2013) Resistance of dormant stages of planktonic invertebrates to adverse environmental conditions. J Plankton Res 35(4):707–723. https://doi.org/10.1093/plankt/fbt032

Rajarajan A, Wolinska J, Walser J-C, Dennis SR, Spaak P (2023) Host-associated bacterial communities vary between Daphnia galeata genotypes but not by host genetic distance. Microb Ecol 85(4):1578–1589. https://doi.org/10.1007/s00248-022-02011-x

Rohrlack T, Dittmann E, Henning M, Börner T, Kohl J-G (1999) Role of microcystins in poisoning and food ingestion inhibition of Daphnia galeata caused by the cyanobacterium Microcystis aeruginosa. Appl Environ Microbiol 65(2):737–739. https://doi.org/10.1128/AEM.65.2.737-739.1999

Rourke ML, Fowler AM, Hughes JM, Broadhurst MK, DiBattista JD, Fielder S, Wilkes Walburn J, Furlan EM (2022) Environmental DNA (eDNA) as a tool for assessing fish biomass: a review of approaches and future considerations for resource surveys. Environ DNA 4(1):9–33. https://doi.org/10.1002/edn3.185

Sakata MK, Yamamoto S, Gotoh RO, Miya M, Yamanaka H, Minamoto T (2020) Sedimentary eDNA provides different information on timescale and fish species composition compared with aqueous eDNA. Environ DNA 2(4):505–518. https://doi.org/10.1002/edn3.75

Sakata MK, Tsugeki N, Kuwae M, Ochi N, Hayami K, Osawa R, Morimoto T, Yasashimoto T, Takeshita D, Doi H, Minamoto T (2022) Fish environmental DNA in lake sediment overcomes the gap of reconstructing past fauna in lake ecosystems 2022.06.16.496507

Schwartz SS, Hebert PDN (1987) Methods for the activation of the resting eggs of Daphnia. Freshw Biol 17(2):373–379. https://doi.org/10.1111/j.1365-2427.1987.tb01057.x

Stager JC, Sporn LA, Johnson M, Regalado S (2015) Of paleo-genes and perch: what if an "alien" is actually a native? PLoS One 10(3):e0119071. https://doi.org/10.1371/journal.pone.0119071

Stegen G, Pasmans F, Schmidt BR, Rouffaer LO, Van Praet S, Schaub M, Canessa S, Laudelout A, Kinet T, Adriaensen C, Haesebrouck F, Bert W, Bossuyt F, Martel A (2017) Drivers of salamander extirpation mediated by Batrachochytrium salamandrivorans. Nature 544(7650):353–356. https://doi.org/10.1038/nature22059

Taberlet P, Bonin A, Zinger L, Coissac E, Taberlet P, Bonin A, Zinger L, Coissac E (2018) Environmental DNA: for biodiversity research and monitoring. Oxford University Press, Oxford/New York

Thomson-Laing G, Howarth JD, Vandergoes MJ, Wood SA (2022) Optimised protocol for the extraction of fish DNA from freshwater sediments. Freshw Biol 67(9):1584–1603. https://doi.org/10.1111/fwb.13962

Tsugeki N, Nakane K, Doi H, Ochi N, Kuwae M (2022) Reconstruction of 100-year dynamics in Daphnia spawning activity revealed by sedimentary DNA. Sci Rep 12(1):1741. https://doi.org/10.1038/s41598-021-03899-0

Turko P, Sigg L, Hollender J, Spaak P (2016) Rapid evolutionary loss of metal resistance revealed by hatching decades-old eggs. Evolution 70(2):398–407. https://doi.org/10.1111/evo.12859

Turner CR, Uy KL, Everhart RC (2015) Fish environmental DNA is more concentrated in aquatic sediments than surface water. Biol Conserv 183:93–102. https://doi.org/10.1016/j.biocon.2014.11.017

Vandekerkhove J, Declerck S, Vanhove M, Brendonck L, Jeppesen E, Porcuna JMC, Meester LD (2004) Use of ephippial morphology to assess richness of anomopods: potentials and pitfalls. J Limnol 63(s1):75–84. https://doi.org/10.4081/jlimnol.2004.s1.75

Walton RE, Sayer CD, Bennion H, Axmacher JC (2021) Once a pond in time: employing palaeoecology to inform farmland pond restoration. Restor Ecol 29(1):e13301. https://doi.org/10.1111/rec.13301

Wei N, Nakajima F, Tobino T (2018) A microcosm study of surface sediment environmental DNA: decay observation, abundance estimation, and fragment length comparison. Environ Sci Technol 52(21):12428–12435. https://doi.org/10.1021/acs.est.8b04956

Weider LJ, Lampert W, Wessels M, Colbourne JK, Limburg P (1997) Long-term genetic shifts in a microcrustacean egg bank associated with anthropogenic changes in the Lake Constance ecosystem. Proc Biol Sci 264(1388):1613–1618. https://doi.org/10.1098/rspb.1997.0225

Willerslev E, Hansen AJ, Binladen J, Brand TB, Gilbert MTP, Shapiro B, Bunce M, Wiuf C, Gilichinsky DA, Cooper A (2003) Diverse plant and animal genetic records from Holocene and Pleistocene sediments. Science 300(5620):791–795. https://doi.org/10.1126/science.1084114

Yates MC, Glaser DM, Post JR, Cristescu ME, Fraser DJ, Derry AM (2021) The relationship between eDNA particle concentration and organism abundance in nature is strengthened by allometric scaling. Mol Ecol 30(13):3068–3082. https://doi.org/10.1111/mec.15543

Chapter 10
Terrestrial Plant DNA from Lake Sediments

Sandra Garcés-Pastor, Kevin Nota, Dilli P. Rijal, Sisi Liu, Weihan Jia, Maria Leunda, Christoph Schwörer, Sarah E. Crump, Laura Parducci, and Inger G. Alsos

Keywords sedDNA · Vegetation reconstruction · Tracing single species · Human impact

Introduction

Advantages of Palaeoecological Studies Using sedDNA Compared with Pollen and Macrofossils

Predicting and anticipating the effects of current and future climate warming on plant communities requires a long-term perspective of past ecosystem dynamics (Fordham et al. 2020). Lake sediments can preserve plant remains for millennia and have been used in palaeoecology to reconstruct past vegetation dynamics for more

---

Dedicated to the memory of Dr. Sarah E. Crump.

The original version of the chapter "Terrestrial Plant DNA from Lake Sediments" has been revised. A correction to this chapter can be found at https://doi.org/10.1007/978-3-031-43799-1_14

S. Garcés-Pastor (✉)
Department of Evolutionary Biology, Ecology and Environmental Sciences, University of Barcelona, Barcelona, Spain

Marine Sciences Institute, ICM-CSIC, Barcelona, Spain
e-mail: sandragarces@ub.edu

K. Nota
Department of Evolutionary Genetics, Max Planck Institute for Evolutionary Anthropology, Leipzig, Germany
e-mail: kevin_nota@eva.mpg.de

E. Capo et al. (eds.), *Tracking Environmental Change Using Lake Sediments*, Developments in Paleoenvironmental Research 21,
https://doi.org/10.1007/978-3-031-43799-1_10

than 100 years (Bennett and Willis 2001; Birks 2019). Pollen analysis is considered the gold standard in Quaternary vegetation history and has been applied worldwide since the start of the field. It is used to reconstruct past vegetation, both quantitatively and qualitatively, on a local to continental scale. Furthermore, pollen can potentially be used for quantitative climate reconstructions (Chevalier et al. 2020). Other taxa produce spores such as ferns, fungi and algae. These and other non-pollen palynomorphs (NPP) can also be used to provide information about past abundances (Van Geel 2001). Another important proxy in palaeoecology is the analysis of macroscopic plant remains, i.e. macrofossils, which often provides a higher taxonomic resolution than pollen. While pollen can be dispersed by wind over long distances, plant macrofossils remain close to their source and provide insight into the local vegetation surrounding the study site (Birks 2001). However, unlike pollen grains, macrofossils are generally rare. Besides these classical palaeoecological methods, over the last decades advances in molecular techniques have allowed the analysis of DNA preserved in the sediment matrix (sedDNA) – including a portion that is ancient DNA (aDNA) to be used for the reconstruction of past floras (de Boer et al. 2022).

D. P. Rijal · I. G. Alsos
The Arctic University Museum of Norway, UiT the Arctic University of Norway, Tromsø, Norway
e-mail: dilli.p.rijal@uit.no; inger.g.alsos@uit.no

S. Liu
Polar Terrestrial Environmental Systems Research Group, Alfred Wegener Institute Helmholtz Centre for Polar and Marine Research, Potsdam, Germany
e-mail: sisi.liu@awi.de

W. Jia
Polar Terrestrial Environmental Systems Research Group, Alfred Wegener Institute Helmholtz Centre for Polar and Marine Research, Potsdam, Germany

Institute of Environmental Science and Geography, University of Potsdam, Potsdam, Germany
e-mail: weihan.jia@awi.de

M. Leunda
Department of Plant Biology and Ecology, University of the Basque Country (UPV/EHU), Leioa, Spain

Swiss Federal Institute for Forest, Snow and Landscape Research, Birmensdorf, Switzerland

Institute of Plant Sciences, Oeschger Centre for Climate Change Research, University of Bern, Bern, Switzerland
e-mail: maria.leunda@unibe.ch

C. Schwörer
Institute of Plant Sciences, Oeschger Centre for Climate Change Research, University of Bern, Bern, Switzerland
e-mail: christoph.schwoerer@unibe.ch

S. E. Crump
Department of Geology & Geophysics, University of Utah, Salt Lake City, UT, USA
e-mail: sarah.crump@utah.eu

L. Parducci
Department of Environmental Biology, Sapienza University of Rome, Rome, Italy

Department of Ecology and Genetics, Uppsala University, Uppsala, Sweden
e-mail: laura.parducci@uniroma1.it

Many studies have successfully used sedDNA to reconstruct past changes in floristic composition. The major advantages of sedDNA include often attaining higher taxonomic resolution compared to pollen, and more diversity compared to macrofossils (e.g. Clarke et al. 2020; Parducci et al. 2019; Bjune et al. 2022; van Vugt et al. 2022). Pollen and sedDNA studies typically show similar vegetation trends but often with different taxonomic richness patterns (Clarke et al. 2019; Garcés-Pastor et al. 2022). To illustrate, sedDNA offers a higher representation of forbs (herbaceous taxa with the exclusion of grasses, sedges, and rushes) and aquatic plants than traditional methods, bridging a knowledge gap that cannot be resolved by conventional approaches. The difficulty in detecting certain taxa using traditional methods can arise from various factors, such as taxonomic limitations (e.g. subfossil remains that are only identifiable to family or genus level), low pollen productivity (e.g. insect-pollinated species and aquatics), and poor preservation of macrofossils. Similar problems can also hamper DNA-based vegetation reconstructions. SedDNA studies have mainly focussed on temperate, boreal, alpine, and Arctic ecosystems and less on low- and mid-latitude regions and tropical ecosystems, where ancient DNA preservation is presumed to be limited (Hofreiter et al. 2015; Kistler et al. 2017). In contrast to cold high-latitude regions, some of those environments contain much higher vascular plant diversity but without the availability of comprehensive databases that can be used for taxon assignments (e.g. Boessenkool et al. 2014; Dommain et al. 2020). However, the method has also been successfully used to reconstruct past vegetation in tropical biomes, albeit on shorter timescales of a few millennia (Bremond et al. 2017; Dommain et al. 2020). Advances in wet laboratory procedures and new molecular techniques might make it possible to go beyond reconstructing past vegetation by assessing presence, studying changes in genetic diversity on a population level, and investigating aspects such as adaptation to environmental stressors over time (Schmid et al. 2017; Napier et al. 2020; Lammers et al. 2021; Schwörer et al. 2022).

In this chapter, we summarize the current state of vegetation reconstructions from lake sediments, discuss the applications of sedDNA, and explore future directions. DNA developments have opened new avenues in Quaternary palaeoecology, such as allowing to establish the presence of taxa which are under-represented in pollen assemblages and the investigation of demographic and evolutionary processes over time (Parducci et al. 2017; Schmid et al. 2017; Clarke et al. 2019; Schulte et al. 2022; Schwörer et al. 2022).

Representation of Terrestrial Taxa in sedDNA and Taphonomy of DNA

Ancient plant DNA in lake sediments represents a complex mixture of extracellular DNA molecules bound to inorganic or organic surface-reactive particles (e.g. clay minerals and humic substances) and intracellular DNA, which is present in intact

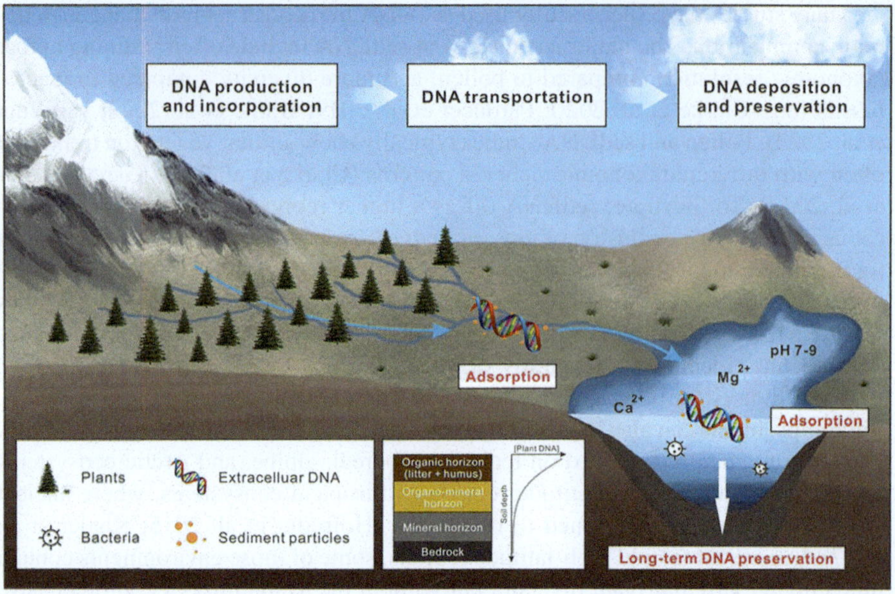

Fig. 10.1 Conceptual model of the taphonomic processes affecting extracellular terrestrial plant DNA in lake sediments. After having been released into the environment, plant DNA can be incorporated into soil particles by adsorbing to, for example, clay minerals and humus. The content of plant DNA decreases with catchment soil depth (Giguet-Covex et al. 2019). More plant DNA is expected to be transported to lake catchment with high erosion rates and well-developed hydrographic networks. During the deposition stage, plant DNA is better preserved in cold, deep lakes with neutral to slightly alkaline water pH and intermediate conductivity. After deposition and burial of plant DNA in the lake sediment, cold and anoxic environmental conditions favour long-term DNA preservation

cells or tissues (Parducci et al. 2018; Taberlet et al. 2018). An increasing number of studies have made efforts to understand the taphonomic processes that terrestrial plant DNA undergoes from its source to deposition in lake sediments. Understanding these processes is essential to ensure reliable interpretation of DNA assemblies (Sjögren et al. 2017; Alsos et al. 2018; Giguet-Covex et al. 2019; Dommain et al. 2020; Jia et al. 2022a, b) (Fig. 10.1). Investigations comparing historical vegetation maps with fossil pollen and plant DNA analyses from lake sediments suggest that pollen is not a significant source of DNA (Sjögren et al. 2017). Further, comparison of DNA inventories from lake surface sediments with contemporary vegetation surveys shows that the majority of detected taxa grow within the catchment (Alsos et al. 2018). Thus, sedDNA is mainly of local origin with a similar source area as macrofossils (Niemeyer et al. 2017; Parducci et al. 2017). Furthermore, comparisons with vegetation surveys show that the detection of DNA depends on the abundance of the taxa in the vegetation (Niemeyer et al. 2017; Alsos et al. 2018).

Although many aspects of the transport processes from source to sediment remain unclear, it is expected that lake catchments with high erosion rates and well-developed hydrographic networks will transport more plant DNA to the sediments

(Giguet-Covex et al. 2019) (see Chap. 2 of this volume). Furthermore, the distance between plant growth sites and lakes (transport distance) is important, as sedDNA detected a higher proportion of plants growing within 2 m from the lake compared to those growing further away (Alsos et al. 2018). This method can detect a similar percentage of the local flora as is possible with macrofossil or pollen analyses, which has been studied extensively (e.g. Prentice 1985; Sugita 1994). After DNA molecules reach the lake water, hydrochemical characteristics are considered to be the main factor influencing long-term DNA preservation. Lakes with intermediate water conductivities (100–500 $\mu S\ cm^{-1}$) and neutral to alkaline water (pH 7–9) may better preserve DNA molecules in sediments by adsorbing them onto mineral particles, which can protect them from bacterial nucleases (Jia et al. 2022b). Additionally, specific conditions in deep lakes, such as low temperatures, low oxygen and UV radiation levels, limited levels of bioturbation, sediment reworking and microbial activity, which may facilitate long-term DNA preservation (Jia et al. 2022a). Once sedDNA is absorbed onto sediment particles, water-saturation, in the absence of groundwater flow, seems to prevent downward leaching of DNA (Anderson-Carpenter et al. 2011; Giguet-Covex et al. 2014; Sjögren et al. 2017). In general, the main factors that impact DNA preservation in lake sediments are water temperature and burial age (Kistler et al. 2017; Dommain et al. 2020). This might prevent, so far, the full use of sedDNA analysis in tropical environments and over long time-scales.

Molecular Ecology Approaches Used to Detect Floral Diversity (Metabarcoding, Shotgun and Capture Probe)

Today, high-throughput sequencing (HTS) methods are considered the most promising for future development in the field of aDNA. In particular, there are two main HTS approaches for analyzing aDNA from sediments: metabarcoding, which is based on the amplification of a single locus or "barcode region" (chloroplast in plants); and metagenomics, which uses a shotgun sequencing approach, i.e. the sequencing of a non-discriminated pool of DNA extracted from sediments. These methods are well established and have been used to reconstruct palaeoenvironments in various environments, sediment types and time periods (Capo et al. 2021; Revéret et al. 2023). Metabarcoding is based on the DNA barcoding principle described by Hebert et al. (2003), combined with sequencing amplicon mixtures using HTS-methods (Taberlet et al. 2018). Barcoding loci can be both protein-coding or non-coding gene fragments (barcodes) and are amplified using PCR-based approaches with broad-range "universal" primers. These primers are designed to amplify the whole or part of a barcoding gene from the targeted taxa. Metabarcoding studies have the potential to use a combination of multiple primers to resolve taxa. Universal primers for vascular plants, such as the commonly used *trn*L g/h primers (Taberlet et al. 2007), can be used to obtain a general overview of plant diversity. This region has the highest overall resolution for the plant families (Revéret et al. 2023). Additional primers (e.g. specific primers for Cyperaceae, Poaceae and Asteraceae) may be added to increase resolution within families (Ait Baamrane et al. 2012; Willerslev et al. 2014). Following the PCR amplification, each amplicon variant is sequenced. The reads are then identified to taxonomic groups by comparing them to a reference database, such as the National Center for Biotechnology Information database (NCBI) (Sayers et al. 2022), the Barcode of Life Data system (BOLD) (Ratnasingham and Hebert 2007) or custom databases (Alsos et al. 2022; Garcés-Pastor et al. 2022) such as the ArctBorBryo reference databases (Sønstebø et al. 2010; Willerslev et al. 2014; Soininen et al. 2015). Currently, a major issue is the limited taxonomic coverage in reference databases covering regions outside Europe/Polar regions and limited taxonomic resolution of short barcoding genes. The development of high-quality customized DNA databases for ecosystems with high levels of plant diversity will be essential in the future, for unlocking the full potential of sedDNA studies in the tropical and mid-latitude regions.

Metagenomics has recently been suggested as an important tool to circumvent the amplification biases in metabarcoding (Pedersen et al. 2016; Wang et al. 2021; Kjær et al. 2022). This method uses the shotgun sequencing approach, i.e., sequencing all fragments present in the samples, regardless of their taxonomic origin (without prior target enrichment PCR). Sequences are identified using bioinformatic pipelines that assign taxonomy to the lowest common ancestor based on large databases containing available plant mitochondrial and chloroplast genomes as well as non-redundant nuclear databases. It is further possible to detect any genomic region that a plant can leave behind, which, in theory, should increase chances of detection. However, it has been shown that relatively few plant reads in sedDNA shotgun libraries can be assigned to mitochondrial and chloroplast genomes using this approach (Pedersen et al. 2016; Parducci et al. 2019). Genome skimming is a sequencing method that involves low-pass, shallow sequencing of a genome (up to 5%) to produce fragments of DNA, known as genome skims. Alsos et al. (2020) utilized this technique to generate nuclear genome skims from a significant number of arctic and boreal plants (PhyloNorway). Using PhyloNorway gave a 23-fold increase in detection, resulting in higher numbers of detected taxa and reads matching the database (Wang et al. 2021). Still, due to unequal representation of species in the genome skim data, identification beyond genus level was uncertain. Further, the high number of DNA reads generated by shotgun sequencing (often more than a billion), and the randomness of fragments originating from different genomes, complicates bioinformatic analysis. On the other hand, with the help of specific software (*mapDamage* – (Jónsson et al. 2013), *metaDMG* – (Michelsen et al. 2022), *MetaDamage* – (Everett and Cribdon 2023)), shotgun sequencing can detect damage patterns characteristic of ancient DNA molecules, validating the authenticity of the ancient reads (Briggs et al. 2007). In recent years, to increase the efficiency of the shotgun sequencing approach in detecting taxa of interest, hybridization capture methods have been introduced. This method uses magnetic baits (nucleotide strings complementary to the target DNA regions) to capture fragments of interest after library preparation and prior to sequencing (Slon et al. 2016; Schmid et al. 2017;

Armbrecht et al. 2021; Murchie et al. 2021; Schulte et al. 2022). When targeting specific species, the taxonomic resolution is at the species level or even within species. However, when targeting all vascular plants, the taxonomic resolution is low (Revéret et al. 2023).

Bias in sedDNA Methods on Past Floral Reconstructions

There are still challenges in the use of molecular methods to determine the diversity, composition, and abundance of biological communities from modern and sedimentary environmental samples. The choice of DNA primers in metabarcoding approaches targeting large groups of organisms have the inherent problem of possible nucleotide mutations in the primer binding regions of certain sub-groups, which can reduce amplification efficiency or prevent amplification of some taxa altogether. Using simulated data, Ficetola et al. (2015) showed that at least eight PCR replicates are needed per sample to minimize false absence in metabarcoding data. However, this might depend on the sample and level of DNA degradation. Nevertheless, rare taxa may still be underrepresented in the DNA metabarcoding record using the standard eight PCR repeats (Shirazi et al. 2021; Alsos et al. 2022), while some abundant species (e.g. *Salix* and certain aquatics) can be markedly overrepresented. Notably, abundant species can also be overrepresented with the shotgun approach (Wang et al. 2021), probably because they grow in the lake and streams or along their shore (Alsos et al. 2018).

Shotgun sequencing on sedDNA has some limitations, such as potential misidentifications due to lack of relevant species or unrecognized errors in the reference database. For most plant taxa, only small gene fragments are sequenced. Moreover, a large amount of microorganisms on Earth have not been sequenced yet, and even from the ones that have been sequenced, only a fraction have publicly available sequences for standard DNA barcodes (20–25%). As a consequence, reference libraries remain incomplete, and a high number of reads (>80%) cannot be assigned to a specific taxonomic group (Pedersen et al. 2016; Slon et al. 2017; Parducci 2019; Bell et al. 2022). The potential applications of aDNA metabarcoding and metagenomic studies can be increased by having relatively complete reference databases, in terms of both species and gene regions.

It has also been suggested that nuclear DNA is more abundant in sediment samples compared with the multicopy organellar DNA (Wang et al. 2021), implying that a large pool of plant DNA is not assessed when using only organelle DNA databases. Another limitation of shotgun sequencing is the overwhelming dominance of microbial DNA present in the DNA extracts (50–90%), which originate probably from living communities that are coeval with the rest of the DNA in a sample (Capo et al. 2022), and which reduces the chance that other taxonomic groups of interest are detected. Regardless of the method used, it is clear that there is a need for a customised, accurate and complete reference database containing all the potential information that can be obtained from sedDNA data.

The exact biases for the target capture approach are poorly studied to date. Fragments with low and high GC-content are difficult to capture, and might need different annealing temperatures during the amplification step. Overall, improved reference libraries and methods increase the information gained from sedDNA. Regarding metabarcoding, a multi-marker approach, using several primer-sets rather than using a single marker, should theoretically result in a more reliable estimation of species richness. Yet, a predominant portion of sedDNA studies targeting plants still use a single marker (*trn*L g/h, Revéret et al. 2023) and for the few studies that use more than one marker (Willerslev et al. 2014; Crump et al. 2019), the overall detection and taxonomic resolution does not improve much by the additional markers. Whether a better recovery of species richness (i.e. using multiple markers or the shotgun/capture approach) outweighs the costs will depend on the number of samples and the research question and aim of each study.

Applications of sedDNA Approaches

Reconstructing Past Vegetation

Originally, sedDNA was used as a way of reconstructing past vegetation composition at the local scale, complementing and expanding classical methods such as pollen or macrofossil analysis. As one of the first studies, Willerslev et al. (2003) was able to identify plant aDNA in Siberian permafrost samples up to 400,000 years old. With improved reference libraries, the field moved from identifying the presence of single taxa in sediment samples (Parducci et al. 2012), to reconstructing local terrestrial vegetation communities (e.g. Parducci et al. 2013; Pedersen et al. 2013; Alsos et al. 2016; Clarke et al. 2020; Crump et al. 2021; Garcés-Pastor et al. 2022; van Vugt et al. 2022).

In areas with relatively complete local reference libraries available, sedDNA can record more taxa at a higher taxonomic resolution than any other palaeoecological proxy (Clarke et al. 2020; Alsos et al. 2022; Bjune et al. 2022; Garcés-Pastor et al. 2022). In the European Alps, combined sedDNA and pollen/macrofossil studies show similar vegetation trajectories over time. At Lago di Sangiatto in the Italian Alps, the timing of arrival and expansion of characteristic tree species of montane forests, such as *Larix decidua,* matches well amongst all three proxies (van Vugt et al. 2022). However, due to the higher diversity of herbaceous species in the sedDNA record, the expansion of species-rich meadows linked to human land use since the Bronze Age is captured in greater detail (van Vugt et al. 2022). Additionally, while pollen and macrofossil records are largely dominated by tree species during periods of closed forests (e.g. European Early Holocene before human impacts), sedDNA offers a more complete picture by better capturing forbs and grasses (Willerslev et al. 2014; Garcés-Pastor et al. 2022; van Vugt et al. 2022). By capturing a broader sample of the local species pool, sedDNA studies are particularly suited to investigate changes in past plant species diversity. For example, recent studies based on sedDNA from Scandinavia and the Arctic indicate a continuous increase of plant diversity throughout the Holocene (Clarke et al. 2020; Voldstad et al. 2020; Rijal et al. 2021). In the European Alps, high species-diversity has been linked to anthropogenic impact over millennia, leading to the establishment of a mosaic of different plant communities (Garcés-Pastor et al. 2022). In the Tibetan Plateau, the warm-induced forest invasion of alpine habitat, regardless of land-use practices such as grazing, has directly resulted in a significant decrease in plant richness over the past 10,000 years (Fig. 10.2a). This suggests that ongoing alpine habitat loss could lead to further declines in Tibetan plant richness over the next 250 years (Fig. 10.2b, Liu et al. 2021).

Recent studies that compared pollen, macrofossil and sedDNA data indicate that the relative abundance of sedDNA reads and the number of positive PCR replicates can be used as a semiquantitative measure of local species abundance (Chen and Ficetola 2020; Garcés-Pastor et al. 2022; van Vugt et al. 2022; Alsos et al. 2022). However, it has been observed that sedDNA is sensitive to variations in biomass input into the lake (van Vugt et al. 2022) and to increased erosion in the catchment (Giguet-Covex et al. 2019). To enhance the accuracy of quantitative reconstructions of past vegetation dynamics in various environments, further studies are necessary to investigate the taphonomy of sedDNA (Chap. 2 of this volume) and to upscale from local to regional vegetation patterns with sedDNA. Ideally, different independent proxies, such as sedDNA, pollen and macrofossils, should be combined to arrive at a robust assessment of past vegetation dynamics.

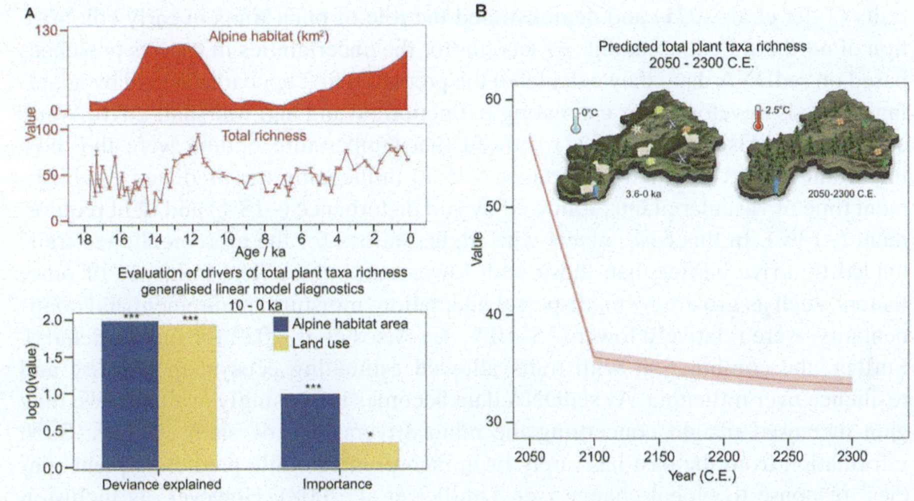

Fig. 10.2 Drivers of total plant taxon richness changes and potential pattern of future plant taxa richness. (**a**) The extent of the alpine habitat, rather than land use, directly drives the long-term total plant taxon richness recorded by sedDNA from sediments of Lake Naleng, located in the southeastern Tibetan Plateau, a global biodiversity hotspot. (**b**) A predicted decrease in total plant taxon richness between 2050 and 2300 C.E. under a 2.5 °C climate warming scenario, using a general linear model (GLM) based on the inferred past relationship between total plant taxon richness and alpine habitat area. (Modified from Liu et al. 2021)

Plant Dispersal and Establishment

Seed dispersal is one of the key driving processes that determines plant community assemblages and distributions (Kyrkjeeide et al. 2014; Beckman et al. 2020). Plants with specialized seeds for long-distance dispersal have been reported to establish more successfully than those without any specializations (Nathan 2006; van der Pijl 2012; Vargas et al. 2015; Arjona et al. 2018), although this may not apply to long-distance dispersal (Alsos et al. 2007, 2015). Species traits reflect habitat characteristics and have been used in ecology for interpreting Quaternary botanical data. However, there are concerns about the progress of this field due to taxonomic limitations of such data (see Birks 2020). As traits are measured at the species level, such concerns are reasonable when the data is generated using traditional approaches such as pollen analyses, where data are often dominated mainly by trees and taxonomic resolution of the ecologically important non-tree taxa are either low, with identifications at family or genus levels, or even indistinguishable (Birks and Birks 2000). However, sedDNA allows the identification of taxa to the species level in nearly 40–60% of the cases (Rijal et al. 2021; Alsos et al. 2022; Garcés-Pastor et al. 2022), making it possible to apply trait-based approaches to understand past and present biota.

By correlating the first appearance date of a species in a chronological sequence with its dispersal-related trait, it is possible to explore how species colonized a newly available habitat following a major disturbance, such as deglaciation (Fig. 10.3). Alsos et al. (2022) integrated sedDNA based species with vascular plant traits (Tyler et al. 2021) and demonstrated the role of plant traits in early colonization of northern Fennoscandia. To account for the uncertainties in species presences based on sedDNA data, they calculated the predicted first appearance date by adapting methods developed for estimating extinction (Wang and Marshall 2016; Wang et al. 2016). Alsos et al. (2022) showed that temperature optima were the most important factors (relative importance >45%) influencing the predicted establishment time of vascular plants, followed by soil disturbance (~18%) and light requirement (~13%). In this case, plants with higher values for the aforementioned traits tended to arrive earlier than those with lower trait values. The influence of other factors, such as growth form, dispersal adaptation, moisture requirement and continentality, were relatively lower (~5–10%, see Alsos et al. 2022 for further details). Further, the combination with traits allowed estimating ecosystem stability and resilience over millennia. As sedDNA data becomes increasingly available, we may gain increased insight concerning the main drivers of ecosystem changes. Such information from the past has rarely been incorporated while predicting plant species' response to global change (see Thuiller et al. 2008). However, its inclusion should be an essential element in the prediction of the future dispersal behaviour of plants.

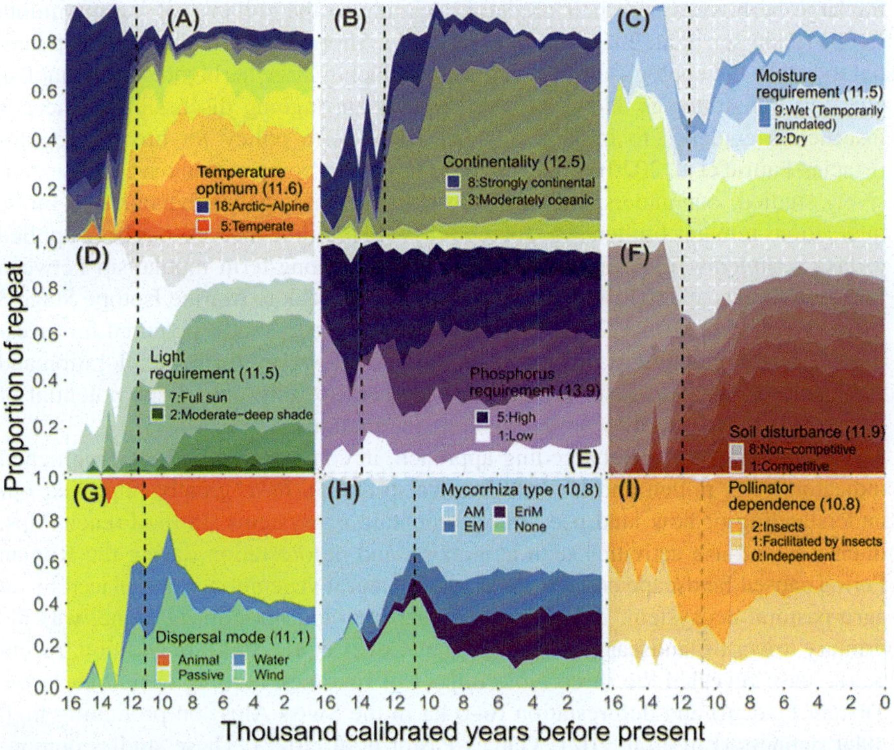

Fig. 10.3 Northern ecosystem reconstruction through 16 millennia based on abundance of plant taxa with different trait values: temperature optimum (**a**), continentality (**b**), moisture requirement (**c**), light requirement (**d**), phosphorous requirement (**e**), soil disturbance (**f**), dispersal mode (**g**), mycorrhiza type (**h**) and pollinator dependence (**i**). AM, arbuscular mycorrhiza; EM, ectomycorrhiza; ErM, ericoid mycorrhiza. Dashed vertical lines indicate the break point in diversity of each trait. The break points in cal ka (thousands years) BP are given in brackets. Taxa are quantified as detection in proportion of PCR repeats. Gray areas represent undefined traits for some taxa. For full description of traits and categories, see the Supplementary Materials from Alsos et al. (2022). (Figure from Alsos et al. 2022)

Reconstructing Arctic and Alpine Ecosystems

In Arctic regions, the causes and mechanisms of the extinction of the Pleistocene steppe-tundra biome, also known as the mammoth steppe, have challenged palaeoecologists. Pollen, dung fungus *Sporormiella*, and fossil remains have led to the notion that this biome is a top-down regulated ecosystem, where the hunting of large herbivores by humans was the primary cause of its disappearance during the late Pleistocene (15–11.7 ka) (Zimov et al. 1995; Johnson 2009; Malhi et al. 2016). Nevertheless, shotgun sequencing has emphasized an alternative viewpoint which suggests that climate had a direct impact on the shifts of vegetation composition from herbs to shrubs dominance. This in turn affected the availability of resources for large herbivores and their predators, leading to the collapse of the mammoth steppe (Pedersen et al. 2016; Wang et al. 2021). Shotgun sequencing has the potential to assess the ecological impacts of reintroducing large herbivores and establish conservation strategies. However, this remains an ongoing discussion as there is insufficient evidence to implement it as a long-term policy for broader regions (Macias-Fauria et al. 2020). While changes in species composition have been extensively studied, our understanding of the temporal and spatial dynamics of biotic interactions remains limited. Metabarcoding with multiple markers has been applied to lake sediments from Arctic Siberia, revealing a long-term mutualism between Pinaceae forests and their mycorrhizal fungi dating back to Marine Isotope Stage 3 (MIS3) (von Hippel et al. 2022). This underscores the method's potential for investigating biotic interactions in various ecosystems. However, further exploration and extension to other organism groups are necessary to fully unlock the potential of lake sediments as a tool for studying biotic interactions.

Studies using the metabarcoding approach, in combination with environmental indicators (e.g. pollen and bio/geochemical proxies), have greatly improved our understanding of how land use affects mountain ecosystems. In the French Alps, intensive land-use activities such as grazing and deforestation during the Roman Period caused landscape erosion. As a result, natural vegetation was replaced by an agro-pastoral ecosystem, indicating that land use, and not climate alone, was the primary driver of landscape change (Giguet-Covex et al. 2014). In addition, recent works have revealed the discernible impact of livestock herding (3.6–0 ka in the Tibetan Plateau) and deforestation (6–0 ka in the Swiss Alps) on promoting high plant richness (Liu et al. 2021; Garcés-Pastor et al. 2022). These studies demonstrate the potential of sedDNA in providing important baseline data for developing effective biodiversity conservation and management policies.

Moreover, sedDNA has expanded the detectable communities beyond the limited scope of traditional paleolimnological methods, which only measure microscopic organisms such as diatom, algae, chironomid taxa, and ostracods (Gregory-Eaves and Smol 2024). More than 60 studies report aquatic plant taxa (Revéret et al. 2023). In addition, aquatic biota recovered with sedDNA has focused on a wide variety of organisms, including bacteria, archaea, microeukaryotes (algae, fungi, diatoms, ciliates, and parasitic protists), zooplankton, macrophytes, and vertebrates (Capo et al. 2021; Brown et al. 2022a; Barouillet et al. 2023; Gregory-Eaves and Smol 2024). Consequently, sedDNA offers a more comprehensive view of long-term changes and key drivers in diversity, composition, and species co-occurrence at the community and ecosystem levels (Capo et al. 2021; Stoof-Leichsenring et al. 2022; Zimmermann et al. 2023). In particular, increasing sedDNA studies underlines the need for lake ecosystem protection from the anthropogenic perturbations that have shaped the turnover of aquatic biota such as microbial communities in Arctic lakes over the past ~200 years (Keck et al. 2020; Barouillet et al. 2023), as well as in alpine regions such as the Tibetan Plateau over the past millennium (Huo et al. 2022).

Tracing Single Species and Intraspecific Variation

Using sedDNA to target and trace single species is possible, but rarely conducted for vascular plants. There are three main wet-lab-methods for targeting single plant species in environmental samples: (i) a species-specific PCR approach with species-specific primers targeting one or more loci; and (ii) shotgun sequencing coupled with hybridization capture targeting genomic or plastid DNA from a single taxon (Chap. 3 of this volume; Capo et al. 2021). In some cases, it is possible to employ a third method by (iii) bioinformatically targeting a specific taxon in shotgun metagenomic data, similar to the approach taken by Lammers et al. (2021), in order to reconstruct ancient plastid genomes from *Nannochloropsis*. This approach requires a high abundance of DNA from the target taxon within the sample, accompanied by a high sequencing depth. Additionally, the presence of the taxon of interest in a sample remains unknown until the sequencing process is performed.

The species-specific PCR approaches can be very sensitive when well-designed. It has been estimated that, with highly optimised wet lab methods, a lower detection limit of fewer than 10 copies of the PCR target can be achieved (Klymus et al. 2020). In contrast, it has been reported that metabarcoding is less sensitive in detecting rare taxa due to the competitive nature of universal primers. According to Nichols et al. (2018), these primers have a tendency to amplify abundant sequences in a sample more readily than rare ones. Therefore, targeting a single taxon with species-specific primers is likely to be more effective in detecting the presence of rare taxa. Besides sensitivity, species-specific PCR has the additional benefits over metabarcoding when it comes to investigating intraspecific variation. In particular target capture but also shotgun approaches have been shown to utilize intraspecific variation to make population-level inferences from plant taxa directly from Holocene sediments (Meucci et al. 2021; Schulte et al. 2022). Studying the genetic diversity of single species over time allows for the reconstruction of their migration patterns and a better understanding of their distribution. By studying the genetic diversity of single species over time, it becomes possible to reconstruct their migration patterns and gain a better understanding of their distribution. Several studies have focused on the Siberian tundra, examining past genomic changes and the recruitment of species such as *Alnus alnobetula* and *Betula nana* over the Holocene (Meucci et al. 2021), as well as various species of *Larix* covering the period since the Last Glacial Maximum (Schulte et al. 2021, 2022).

The above results highlight the potential of studying plant species at the population level using sedDNA. It allows researchers to address research questions beyond detecting taxon presence. Recently, Kjær et al. (Kjær et al. 2022) used the high number of obtained chloroplast and mitochondrial DNA to phylogenetically place *Salix, Betula,* and *Populus* from 2 million year old permafrost samples, revealing their evolutionary relationships from basal to modern individuals. Moreover, they used molecular clock analysis using *Betula* chloroplast DNA for molecular dating of the sediments, establishing another line of evidence for the age of the sediments. This approach has also been applied to non-plant taxa, where mitochondrial and nuclear DNA from Neanderthals and Denisovans extracted from cave sediments showed similar phylogenetic placing as aDNA extracted from skeletal remains (Slon et al. 2017; Vernot et al. 2021). For most plant taxa, there are no well-preserved hard tissues comparable to bones from vertebrate taxa that can be used to study ancient individuals, except for woody structures. However, it was demonstrated in 2006 that authentic ancient DNA can be obtained from sub-fossilized wood remains (Liepelt et al. 2006). Despite this, several methodological obstacles still need to be overcome, including the challenge of extracting a very low amount of endogenous DNA (Lendvay et al. 2018; Wagner et al. 2018). Recently, Wagner et al. (2023) have made significant advancements in this field. They successfully obtained high amounts of nuclear DNA from approximately 500 macro remains and 3500-year-old waterlogged *Quercus* DNA. These breakthroughs allowed them to investigate population structure and phenology.

The advancement in sedDNA approaches may facilitate the linkage between present-day plant populations and ancient ones. This connection holds significant relevance for addressing questions such as: When did a taxon recolonize an area after disappearance due to glacial advances? Which population was responsible for the original recruitment? Were there multiple waves of recognizing individuals?

Detection of Human Impact on Vegetation and Domesticated Taxa

One of the main challenges in differentiating between modern and ancient DNA using metabarcoding is the authentication of domestic plants and mammals in sediments. Laboratory contaminants, such as cereals and edible vegetables like bananas, can be prevalent and may result in higher read counts in negative extraction or PCR controls compared to those found in lake sediments. Therefore, it is crucial to implement a rigorous extraction protocol and PCR experiment to prevent, detect, and control modern and cross-contamination (Parducci et al. 2017; Rohland et al. 2018) (see Chap. 3 of this volume for more details). Additionally, the Bayesian site occupancy detection modelling (SODM) approach can be employed to estimate species occupancy and evaluate the frequency of false detections, effectively accounting for the risk of false positives (Brown et al. 2022b).

Many palaeoecological studies have used domesticated plants and disturbance indicators to infer human activities such as forestry and crop cultivation. These inferences have predominantly relied on pollen records (e.g. Behre 1986). However, in some cases domesticated species belong to plant families or subfamilies that share similar pollen morphologies, which can complicate the clear differentiation from species that are not associated with human impacts. This limited taxonomic resolution of pollen often leads to the use of family-level classifications, such as Cichorioideae, Asteroideae, Brassicaceae, or Apiaceae, for ecological inferences.

In some studies, researchers have employed single genera, like *Plantago, Alchemilla,* and *Potentilla*, or even individual species like *Sanguisorba officinalis*, to infer human activities (Pansu et al. 2015; Liu et al. 2021). However, even a single genus can include several species that are not favoured by human activities and occur in undisturbed sites. A safer sedDNA approach would be the use of a group of well-characterised taxa or sequences to ensure robust ecological interpretation (Garcés-Pastor et al. 2022).

Tracing past animal-vegetation interactions can be challenging, as the presence of grazing animals can only be detected through animal remains or indirectly through coprophilous fungal spores, and to some extent, by using flora associated with grazing (van Geel and Aptroot 2006; Lee et al. 2022). However, a significant issue arises when it comes to distinguishing between wild and domestic herbivores using these methods. The use of sedDNA from mammals offers a solution by enabling us to differentiate between domesticated and wild animals, thereby providing a deeper understanding of their impact on plant species composition and the environment. By combining sedDNA from plants and mammals with pollen analysis and other traditional paleolimnological methods, it has become possible to reconstruct direct evidence of human occupation and their impact in various archaeological sites, including Celtic settlements in Ireland, Stonehenge, and Arctic Norway (Brown et al. 2022a, b; Hudson et al. 2022).

Future Directions

Until recently, the availability of DNA reference libraries has been a major limitation on identification at lower taxonomic levels, especially in tropical regions (Dommain et al. 2020). However, genomic reference libraries are currently being developed for various regions, including Norway and polar regions (PhyloNorway, n = 1899 taxa), the Alps and Carpathians (PhyloAlps, n = 4437 taxa, (Wang et al. 2021; Alsos et al. 2020, 2022; Garcés-Pastor et al. 2022), China (n = 1659 specimens, Li et al. 2019), and Australia (n = 530 taxa, (Nevill et al. 2020)). While these genomic references only partially cover the genome, limiting the taxonomic resolution of shotgun data (Wang et al. 2021), the nearly complete coverage of the P6 loop in the Norwegian and Alpine flora allows for 56% species-level identification in metabarcoding data (Alsos et al. 2022; Garcés-Pastor et al. 2022). This allows the combination of metabarcoding data with species trait databases, facilitating the reconstruction of various aspects of the ecosystem, such as abiotic conditions, pastoral and altitudinal indicators, pollinators, dispersal vectors, and mycorrhiza (Alsos et al. 2022; Garcés-Pastor et al. 2022). As reference libraries become more complete for other regions, the same approach can be applied. Moreover, the use of more complete genome skims, coupled with improved bioinformatic tools, will further enhance the taxonomic resolution achievable with shotgun sequence data.

Summary

Despite certain limitations arising from sequencing method biases, taphonomy, and proportions, the recovery and study of plant DNA from lake sediments offers immense potential for palaeoecology studies. This chapter highlights how the sedimentary DNA can provide a high level of taxonomic resolution, allowing for a detailed reconstruction of past terrestrial vegetation. This information can be used to establish dispersal patterns and other ecological inferences by utilizing plant traits, thereby filling a gap in our knowledge of Quaternary vegetation history. Additionally, we have evaluated the potential of sedDNA to infer human influence, study intraspecific variation at the species level, and discussed future prospects.

Acknowledgements We would like to express our gratitude to Mary Edwards, Rene Dommain, and Eric Capo for their valuable comments, which significantly enhanced the earlier versions of this chapter. Weihan Jia was supported by the China Scholarship Council (Grant No. 202008110205). Maria Leunda was supported by a 'María Zambrano' fellowship (Ministerio de Universidades/University of the Basque Country, financed by the European Union-Next Generation EU), Sandra Garcés Pastor, Dilli P Rijal and Inger G. Alsos were supported by Research Council of Norway grant 250963/F20 for the ECOGEN project. Inger G. Alsos was also supported by The European Research Council (ERC) under the European Union's Horizon 2020 research and innovation programme grant agreement No 819192 for the IceAGenT project. Sandra Garcés-Pastor was also supported by the Beatriu de Pinós Programme (BP-2021-00131).

References

Ait Baamrane MA, Shehzad W, Ouhammou A, Abbad A, Naimi M, Coissac E, Taberlet P, Znari M (2012) Assessment of the food habits of the Moroccan Dorcas Gazelle in M'Sabih Talaa, West Central Morocco, using the trnL approach. PLoS One 7(4):e35643. https://doi.org/10.1371/journal.pone.0035643

Alsos IG, Eidesen PB, Ehrich D, Skrede I, Westergaard K, Jacobsen GH, Landvik JY, Taberlet P, Brochmann C (2007) Frequent long-distance plant colonization in the changing Arctic. Science 316(5831):1606–1609. https://doi.org/10.1126/science.1139178

Alsos IG, Ehrich D, Eidesen PB, Solstad H, Westergaard KB, Schönswetter P, Tribsch A, Birkeland S, Elven R, Brochmann C (2015) Long-distance plant dispersal to North Atlantic islands: colonization routes and founder effect. AoB Plants 7:plv036. https://doi.org/10.1093/aobpla/plv036

Alsos IG, Sjögren P, Edwards ME, Landvik JY, Gielly L, Forwick M, Coissac E, Brown AG, Jakobsen LV, Føreid MK, Pedersen MW (2016) Sedimentary ancient DNA from Lake Skartjørna, Svalbard: assessing the resilience of arctic flora to Holocene climate change. The Holocene 26(4):627–642. https://doi.org/10.1177/0959683615612563

Alsos IG, Lammers Y, Yoccoz NG, Jørgensen T, Sjögren P, Gielly L, Edwards ME (2018) Plant DNA metabarcoding of lake sediments: how does it represent the contemporary vegetation. PLoS One 13(4):e0195403. https://doi.org/10.1371/journal.pone.0195403

Alsos IG, Lavergne S, Merkel MKF, Boleda M, Lammers Y, Alberti A, Pouchon C, Denoeud F, Pitelkova I, Puşcaş M, Roquet C, Hurdu B-I, Thuiller W, Zimmermann NE, Hollingsworth PM, Coissac E (2020) The treasure vault can be opened: large-scale genome skimming works well using herbarium and silica gel dried material. Plan Theory 9(4):432. https://doi.org/10.3390/plants9040432

Alsos IG, Rijal DP, Ehrich D, Karger DN, Yoccoz NG, Heintzman PD, Brown AG, Lammers Y, Pellissier L, Alm T, Bråthen KA, Coissac E, Merkel MKF, Alberti A, Denoeud F, Bakke J, PHYLONORWAY CONSORTIUM (2022) Postglacial species arrival and diversity buildup of northern ecosystems took millennia. Sci Adv 8(39):eabo7434. https://doi.org/10.1126/sciadv.abo7434

Anderson-Carpenter LL, McLachlan JS, Jackson ST, Kuch M, Lumibao CY, Poinar HN (2011) Ancient DNA from lake sediments: bridging the gap between paleoecology and genetics. BMC Evol Biol 11(1):30. https://doi.org/10.1186/1471-2148-11-30

Arjona Y, Nogales M, Heleno R, Vargas P (2018) Long-distance dispersal syndromes matter: diaspore–trait effect on shaping plant distribution across the Canary Islands. Ecography 41(5):805–814. https://doi.org/10.1111/ecog.02624

Armbrecht L, Hallegraeff G, Bolch CJS, Woodward C, Cooper A (2021) Hybridisation capture allows DNA damage analysis of ancient marine eukaryotes. Sci Rep 11:3220. https://doi.org/10.1038/s41598-021-82578-6

Barouillet C, Monchamp M-E, Bertilsson S, Brasell K, Domaizon I, Epp LS, Ibrahim A, Mejbel H, Nwosu EC, Pearman JK, Picard M, Thomson-Laing G, Tsugeki N, Von Eggers J, Gregory-Eaves I, Pick FR, Wood SA, Capo E (2023) Investigating the effects of anthropogenic stressors on lake biota using sedimentary DNA. Freshw Biol n/a(n/a). https://doi.org/10.1111/fwb.14027

Beckman NG, Aslan CE, Rogers HS (2020) Introduction to the special issue: the role of seed dispersal in plant populations: perspectives and advances in a changing world. AoB Plants 12(2):plaa010. https://doi.org/10.1093/aobpla/plaa010

Behre KE (1986) Anthropogenic indicators in pollen diagrams. AA Balkema, Rotterdam, p 232

Bell KL, Turo KJ, Lowe A, Nota K, Keller A, Encinas-Viso F, Parducci L, Richardson RT, Leggett RM, Brosi BJ, Burgess KS, Suyama Y, de Vere N (2022) Plants, pollinators and their interactions under global ecological change: the role of pollen DNA metabarcoding. Mol Ecol n/a(n/a). https://doi.org/10.1111/mec.16689

Bennett KD, Willis KJ (2001) Pollen. In: Smol JP, Birks HJB, Last WM, Bradley RS, Alverson K (eds) Tracking environmental change using lake sediments: terrestrial, algal, and siliceous indicators. Springer, Dordrecht, pp 5–32

Birks HH (2001) Plant macrofossils. In: Smol JP, Birks HJB, Last WM, Bradley RS, Alverson K (eds) Tracking environmental change using Lake sediments: terrestrial, algal, and siliceous indicators. Springer, Dordrecht, pp 49–74

Birks HJB (2019) Contributions of quaternary botany to modern ecology and biogeography. Plant Ecol Divers 12(3–4):189–385. https://doi.org/10.1080/17550874.2019.1646831

Birks HJB (2020) Reflections on the use of ecological attributes and traits in quaternary botany. Front Ecol Evol 8:166

Birks HH, Birks HJB (2000) Future uses of pollen analysis must include plant macrofossils. J Biogeogr 27(1):31–35. https://doi.org/10.1046/j.1365-2699.2000.00375.x

Bjune AE, Greve Alsos I, Brendryen J, Edwards ME, Haflidason H, Johansen MS, Mangerud J, Paus A, Regnéll C, Svendsen J-I, Clarke CL (2022) Rapid climate changes during the Lateglacial and the early Holocene as seen from plant community dynamics in the Polar Urals, Russia. J Quat Sci 37(5):805–817. https://doi.org/10.1002/jqs.3352

Boessenkool S, Mcglynn G, Epp LS, Taylor D, Pimentel M, Gizaw A, Nemomissa S, Brochmann C, Popp M (2014) Use of ancient sedimentary DNA as a novel conservation tool for high-altitude tropical biodiversity. Conserv Biol 28(2):446–455. https://doi.org/10.1111/cobi.12195

Bremond L, Favier C, Ficetola GF, Tossou MG, Akouégninou A, Gielly L, Giguet-Covex C, Oslisly R, Salzmann U (2017) Five thousand years of tropical lake sediment DNA records from Benin. Quat Sci Rev 170:203–211. https://doi.org/10.1016/j.quascirev.2017.06.025

Briggs AW, Stenzel U, Johnson PLF, Green RE, Kelso J, Prüfer K, Meyer M, Krause J, Ronan MT, Lachmann M, Pääbo S (2007) Patterns of damage in genomic DNA sequences from a Neandertal. Proc Natl Acad Sci U S A 104(37):14616–14621. https://doi.org/10.1073/pnas.0704665104

Brown T, Rijal DP, Heintzman PD, Clarke CL, Blankholm H-P, Høeg HI, Lammers Y, Bråthen KA, Edwards M, Alsos IG (2022a) Paleoeconomy more than demography determined prehistoric human impact in Arctic Norway. PNAS Nexus 1(5):pgac209. https://doi.org/10.1093/pnasnexus/pgac209

Brown AG, Van Hardenbroek M, Fonville T, Davies K, Mackay H, Murray E, Head K, Barratt P, McCormick F, Ficetola GF, Gielly L, Henderson ACG, Crone A, Cavers G, Langdon PG, Whitehouse NJ, Pirrie D, Alsos IG (2022b) Ancient DNA, lipid biomarkers and palaeoecological evidence reveals construction and life on early medieval lake settlements. Sci Rep 11(1):11807. https://doi.org/10.1038/s41598-021-91057-x

Capo E, Giguet-Covex C, Rouillard A, Nota K, Heintzman PD, Vuillemin A, Ariztegui D, Arnaud F, Belle S, Bertilsson S, Bigler C, Bindler R, Brown AG, Clarke CL, Crump SE, Debroas D, Englund G, Ficetola GF, Garner RE, Gauthier J, Gregory-Eaves I, Heinecke L, Herzschuh U, Ibrahim A, Kisand V, Kjær KH, Lammers Y, Littlefair J, Messager E, Monchamp M-E, Olajos F, Orsi W, Pedersen MW, Rijal DP, Rydberg J, Spanbauer T, Stoof-Leichsenring KR, Taberlet P, Talas L, Thomas C, Walsh DA, Wang Y, Willerslev E, van Woerkom A, Zimmermann HH, Coolen MJL, Epp LS, Domaizon IG, Alsos I, Parducci L (2021) Lake sedimentary DNA research on past terrestrial and aquatic biodiversity: overview and recommendations. Quaternary 4(1):6. https://doi.org/10.3390/quat4010006

Capo E, Monchamp ME, Coolen MJ, Domaizon I, Armbrecht L, Bertilsson S (2022) Environmental paleomicrobiology: using DNA preserved in aquatic sediments to its full potential. Environ Microbiol 24(5):2201–2209

Chen W, Ficetola GF (2020) Numerical methods for sedimentary-ancient-DNA-based study on past biodiversity and ecosystem functioning. Environ DNA 2(2):115–129. https://doi.org/10.1002/edn3.79

Chevalier M, Davis BAS, Heiri O, Seppä H, Chase BM, Gajewski K, Lacourse T, Telford RJ, Finsinger W, Guiot J, Kühl N, Maezumi SY, Tipton JR, Carter VA, Brussel T, Phelps LN, Dawson A, Zanon M, Vallé F, Nolan C, Mauri A, de Vernal A, Izumi K, Holmström L, Marsicek J, Goring S, Sommer PS, Chaput M, Kupriyanov D (2020) Pollen-based climate reconstruction techniques for late quaternary studies. Earth Sci Rev 210:103384. https://doi.org/10.1016/j.earscirev.2020.103384

Clarke CL, Edwards ME, Brown AG, Gielly L, Lammers Y, Heintzman PD, Ancin-Murguzur FJ, Bråthen K-A, Goslar T, Alsos IG (2019) Holocene floristic diversity and richness in Northeast Norway revealed by sedimentary ancient DNA (sedaDNA) and pollen. Boreas 48(2):299–316. https://doi.org/10.1111/bor.12357

Clarke CL, Alsos IG, Edwards ME, Paus A, Gielly L, Haflidason H, Mangerud J, Regnéll C, Hughes PDM, Svendsen JI, Bjune AE (2020) A 24,000-year ancient DNA and pollen record from the Polar Urals reveals temporal dynamics of arctic and boreal plant communities. Quat Sci Rev 247:106564. https://doi.org/10.1016/j.quascirev.2020.106564

Crump SE, Miller GH, Power M, Sepúlveda J, Dildar N, Coghlan M, Bunce M (2019) Arctic shrub colonization lagged peak postglacial warmth: molecular evidence in lake sediment from Arctic Canada. Glob Chang Biol 25(12):4244–4256. https://doi.org/10.1111/gcb.14836

Crump SE, Fréchette B, Power M, Cutler S, de Wet G, Raynolds MK, Raberg JH, Briner JP, Thomas EK, Sepúlveda J, Shapiro B, Bunce M, Miller GH (2021) Ancient plant DNA reveals high arctic greening during the last interglacial. Proc Natl Acad Sci 118(13):e2019069118. https://doi.org/10.1073/pnas.2019069118

de Boer H, Rydmark MO, Verstraete B, Gravendeel B (2022) Molecular identification of plants: from sequence to species. Adv Books 1:e98875. https://doi.org/10.3897/ab.e98875

Dommain R, Andama M, McDonough MM, Prado NA, Goldhammer T, Potts R, Maldonado JE, Nkurunungi JB, Campana MG (2020) The challenges of reconstructing tropical biodiversity with sedimentary ancient DNA: a 2200-year-long metagenomic record from Bwindi impenetrable Forest, Uganda. Front Ecol Evol 8:218

Everett R, Cribdon B (2023) MetaDamage tool: examining post-mortem damage in sedaDNA on a metagenomic scale. Front Ecol Evol 10:888421

Ficetola GF, Pansu J, Bonin A, Coissac E, Giguet-Covex C, De Barba M, Gielly L, Lopes CM, Boyer F, Pompanon F, Rayé G, Taberlet P (2015) Replication levels, false presences and the estimation of the presence/absence from eDNA metabarcoding data. Mol Ecol Resour 15(3):543–556. https://doi.org/10.1111/1755-0998.12338

Fordham DA, Jackson ST, Brown SC, Huntley B, Brook BW, Dahl-Jensen D, Gilbert MTP, Otto-Bliesner BL, Svensson A, Theodoridis S, Wilmshurst JM, Buettel JC, Canteri E, McDowell M, Orlando L, Pilowsky J, Rahbek C, Nogues-Bravo D (2020) Using paleo-archives to safeguard biodiversity under climate change. Science 369(6507):eabc5654. https://doi.org/10.1126/science.abc5654

Garcés-Pastor S, Coissac E, Lavergne S, Schwörer C, Theurillat J-P, Heintzman PD, Wangensteen OS, Tinner W, Rey F, Heer M, Rutzer A, Walsh K, Lammers Y, Brown AG, Goslar T, Rijal DP, Karger DN, Pellissier L, Heiri O, Alsos IG (2022) High resolution ancient sedimentary DNA shows that alpine plant diversity is associated with human land use and climate change. Nat Commun 13(1):6559. https://doi.org/10.1038/s41467-022-34010-4

Giguet-Covex C, Pansu J, Arnaud F, Rey P-J, Griggo C, Gielly L, Domaizon I, Coissac E, David F, Choler P, Poulenard J, Taberlet P (2014) Long livestock farming history and human landscape shaping revealed by lake sediment DNA. Nat Commun 5(1):3211. https://doi.org/10.1038/ncomms4211

Giguet-Covex C, Ficetola GF, Walsh K, Poulenard J, Bajard M, Fouinat L, Sabatier P, Gielly L, Messager E, Develle AL, David F, Taberlet P, Brisset E, Guiter F, Sinet R, Arnaud F (2019) New insights on lake sediment DNA from the catchment: importance of taphonomic and analytical issues on the record quality. Sci Rep 9:14676. https://doi.org/10.1038/s41598-019-50339-1

Gregory-Eaves I, Smol JP (2024, in press) Paleolimnology: methods and applications. In: Jones I, Smol JP (eds), Smol JP (tran) Wetzel's limnology: lake and river ecosystems (4th edn)

Hebert PDN, Cywinska A, Ball SL, deWaard JR (2003) Biological identifications through DNA barcodes. Proc Biol Sci 270(1512):313–321. https://doi.org/10.1098/rspb.2002.2218

Hofreiter M, Paijmans JLA, Goodchild H, Speller CF, Barlow A, Fortes GG, Thomas JA, Ludwig A, Collins MJ (2015) The future of ancient DNA: technical advances and conceptual shifts. BioEssays 37(3):284–293. https://doi.org/10.1002/bies.201400160

Hudson SM, Pears B, Jacques D, Fonville T, Hughes P, Alsos I, Snape L, Lang A, Brown A (2022) Life before Stonehenge: the hunter-gatherer occupation and environment of Blick Mead revealed by sedaDNA, pollen and spores. PLoS One 17(4):e0266789. https://doi.org/10.1371/journal.pone.0266789

Huo S, Zhang H, Wang J, Chen J, Wu F (2022) Temperature and precipitation dominates millennium changes of eukaryotic algal communities in Lake Yamzhog Yumco, Southern Tibetan Plateau. Sci Total Environ 829:154636. https://doi.org/10.1016/j.scitotenv.2022.154636

Jia W, Anslan S, Chen F, Cao X, Dong H, Dulias K, Gu Z, Heinecke L, Jiang H, Kruse S, Kang W, Li K, Liu S, Liu X, Liu Y, Ni J, Schwalb A, Stoof-Leichsenring KR, Shen W, Tian F, Wang J, Wang Y, Wang Y, Xu H, Yang X, Zhang D, Herzschuh U (2022a) Sedimentary ancient DNA reveals past ecosystem and biodiversity changes on the Tibetan Plateau: overview and prospects. Quat Sci Rev 293:107703. https://doi.org/10.1016/j.quascirev.2022.107703

Jia W, Liu X, Stoof-Leichsenring KR, Liu S, Li K, Herzschuh U (2022b) Preservation of sedimentary plant DNA is related to lake water chemistry. Environ DNA 4(2):425–439. https://doi.org/10.1002/edn3.259

Johnson C n (2009) Ecological consequences of late quaternary extinctions of megafauna. Proc R Soc B Biol Sci 276(1667):2509–2519. https://doi.org/10.1098/rspb.2008.1921

Jónsson H, Ginolhac A, Schubert M, Johnson PLF, Orlando L (2013) mapDamage2.0: fast approximate Bayesian estimates of ancient DNA damage parameters. Bioinformatics 29(13):1682–1684. https://doi.org/10.1093/bioinformatics/btt193

Keck F, Millet L, Debroas D, Etienne D, Galop D, Rius D, Domaizon I (2020) Assessing the response of micro-eukaryotic diversity to the great acceleration using lake sedimentary DNA. Nat Commun 11(1):3831. https://doi.org/10.1038/s41467-020-17682-8

Kistler L, Ware R, Smith O, Collins M, Allaby RG (2017) A new model for ancient DNA decay based on paleogenomic meta-analysis. Nucleic Acids Res 45(11):6310–6320. https://doi.org/10.1093/nar/gkx361

Kjær KH, Winther Pedersen M, De Sanctis B, De Cahsan B, Korneliussen TS, Michelsen CS, Sand KK, Jelavić S, Ruter AH, Schmidt AMA, Kjeldsen KK, Tesakov AS, Snowball I, Gosse JC, Alsos IG, Wang Y, Dockter C, Rasmussen M, Jørgensen ME, Skadhauge B, Prohaska A, Kristensen JÅ, Bjerager M, Allentoft ME, Coissac E, Rouillard A, Simakova A, Fernandez-Guerra A, Bowler C, Macias-Fauria M, Vinner L, Welch JJ, Hidy AJ, Sikora M, Collins MJ, Durbin R, Larsen NK, Willerslev E (2022) A 2-million-year-old ecosystem in Greenland uncovered by environmental DNA. Nature 612(7939):283–291. https://doi.org/10.1038/s41586-022-05453-y

Klymus KE, Ruiz Ramos DV, Thompson NL, Richter CA (2020) Development and testing of species-specific quantitative PCR assays for environmental DNA applications. J Vis Exp. https://doi.org/10.3791/61825

Kyrkjeeide MO, Stenøien HK, Flatberg KI, Hassel K (2014) Glacial refugia and post-glacial colonization patterns in European bryophytes. lnbg 37(2):47–59. https://doi.org/10.25227/linbg.01046

Lammers Y, Heintzman PD, Alsos IG (2021) Environmental palaeogenomic reconstruction of an Ice Age algal population. Commun Biol 4(1):1–11. https://doi.org/10.1038/s42003-021-01710-4

Lee CM, van Geel B, Gosling WD (2022) On the use of spores of coprophilous fungi preserved in sediments to indicate past herbivore presence. Quaternary 5(3):30. https://doi.org/10.3390/quat5030030

Lendvay B, Hartmann M, Brodbeck S, Nievergelt D, Reinig F, Zoller S, Parducci L, Gugerli F, Büntgen U, Sperisen C (2018) Improved recovery of ancient DNA from subfossil wood – application to the world's oldest Late Glacial pine forest. New Phytol 217(4):1737–1748. https://doi.org/10.1111/nph.14935

Li HT, Yi TS, Gao LM, Ma PF, Zhang T, Yang JB, Gitzendanner MA, Fritsch PW, Cai J, Luo Y, Wang H, van der Bank M, Zhang SD, Wang QF, Wang J, Zhang ZR, Fu CH, Yang J, Hollingsworth PM, Chase MW, Soltis DE, Soltis PS, Li DZ (2019) Origin of angiosperms and the puzzle of the Jurassic gap. Nat plant 5(5):461–470.

Liepelt S, Sperisen C, Deguilloux MF, Petit RJ, Kissling R, Spencer M, Beaulieu JL, Taberlet P, Gielly L, Ziegenhagen B (2006) Authenticated DNA from ancient wood remains. Ann Bot 98(5):1107. https://doi.org/10.1093/aob/mcl188

Liu S, Kruse S, Scherler D, Ree RH, Zimmermann HH, Stoof-Leichsenring KR, Epp LS, Mischke S, Herzschuh U (2021) Sedimentary ancient DNA reveals a threat of warming-induced alpine habitat loss to Tibetan Plateau plant diversity. Nat Commun 12(1). https://doi.org/10.1038/s41467-021-22986-4

Macias-Fauria M, Jepson P, Zimov N, Malhi Y (2020) Pleistocene Arctic megafaunal ecological engineering as a natural climate solution? Philos Trans R Soc B Biol Sci 375(1794):20190122. https://doi.org/10.1098/rstb.2019.0122

Malhi Y, Doughty CE, Galetti M, Smith FA, Svenning J-C, Terborgh JW (2016) Megafauna and ecosystem function from the Pleistocene to the Anthropocene. Proc Natl Acad Sci 113(4):838–846. https://doi.org/10.1073/pnas.1502540113

Meucci S, Schulte L, Zimmermann HH, Stoof-Leichsenring KR, Epp L, Bronken Eidesen P, Herzschuh U (2021) Holocene chloroplast genetic variation of shrubs (Alnus alnobetula, Betula nana, Salix sp.) at the siberian tundra-taiga ecotone inferred from modern chloroplast genome assembly and sedimentary ancient DNA analyses. Ecol Evol 11(5):2173–2193. https://doi.org/10.1002/ece3.7183

Michelsen C, Pedersen MW, Fernandez-Guerra A, Zhao L, Petersen TC, Korneliussen TS (2022) metaDMG – a fast and accurate ancient DNA damage toolkit for metagenomic data. 2022.12.06.519264

Murchie TJ, Monteath AJ, Mahony ME, Long GS, Cocker S, Sadoway T, Karpinski E, Zazula G, MacPhee RDE, Froese D, Poinar HN (2021) Collapse of the mammoth-steppe in Central Yukon as revealed by ancient environmental DNA. Nat Commun 12(1):7120. https://doi.org/10.1038/s41467-021-27439-6

Napier JD, de Lafontaine G, Chipman ML (2020) The evolution of paleoecology. Trends Ecol Evol 35(4):293–295. https://doi.org/10.1016/j.tree.2019.12.006

Nathan R (2006) Long-distance dispersal of plants. Science 313(5788):786–788. https://doi.org/10.1126/science.1124975

Nevill PG, Zhong X, Tonti-Filippini J, Byrne M, Hislop M, Thiele K, van Leeuwen S, Boykin LM, Small I (2020) Large scale genome skimming from herbarium material for accurate plant identification and phylogenomics. Plant Methods 16(1):1. https://doi.org/10.1186/s13007-019-0534-5

Nichols RV, Vollmers C, Newsom LA, Wang Y, Heintzman PD, Leighton M, Green RE, Shapiro B (2018) Minimizing polymerase biases in metabarcoding. Mol Ecol Resour 18(5):927–939. https://doi.org/10.1111/1755-0998.12895

Niemeyer B, Epp LS, Stoof-Leichsenring KR, Pestryakova LA, Herzschuh U (2017) A comparison of sedimentary DNA and pollen from lake sediments in recording vegetation composition at the Siberian treeline. Mol Ecol Resour 17(6):e46–e62. https://doi.org/10.1111/1755-0998.12689

Pansu J, Giguet-Covex C, Ficetola GF, Gielly L, Boyer F, Zinger L, Arnaud F, Poulenard J, Taberlet P, Choler P (2015) Reconstructing long-term human impacts on plant communities: an ecological approach based on lake sediment DNA. Mol Ecol 24(7):1485–1498. https://doi.org/10.1111/mec.13136

Parducci L, Jørgensen T, Tollefsrud MM, Elverland E, Alm T, Fontana SL, Bennett KD, Haile J, Matetovici I, Suyama Y, Edwards ME, Andersen K, Rasmussen M, Boessenkool S, Coissac E, Brochmann C, Taberlet P, Houmark-Nielsen M, Larsen NK, Orlando L, Gilbert MTP, Kjær KH, Alsos IG, Willerslev E (2012) Glacial survival of boreal trees in Northern Scandinavia. Science 335(6072):1083–1086. https://doi.org/10.1126/science.1216043

Parducci L, Matetovici I, Fontana SL, Bennett KD, Suyama Y, Haile J, Kjaer KH, Larsen NK, Drouzas AD, Willerslev E (2013) Molecular- and pollen-based vegetation analysis in lake sediments from Central Scandinavia. Mol Ecol 22(13):3511–3524. https://doi.org/10.1111/mec.12298

Parducci L, Bennett KD, Ficetola GF, Alsos IG, Suyama Y, Wood JR, Pedersen MW (2017) Ancient plant DNA in lake sediments. New Phytol 214(3):924–942. https://doi.org/10.1111/nph.14470

Parducci L, Nota K, Wood J (2018) Reconstructing past vegetation communities using ancient DNA from lake sediments. In: Lindqvist C, Rajora OP (eds) Paleogenomics: genome-scale analysis of ancient DNA. Springer, Cham, pp 163–187

Parducci L, Alsos IG, Unneberg P, Pedersen MW, Han LU, Lammers Y, Salonen JS, Väliranta MM, Slotte T, Wohlfarth B (2019) Shotgun environmental DNA, pollen, and macrofossil analysis of lateglacial lake sediments from southern Sweden. Front Ecol Evol 7:189. https://doi.org/10.3389/fevo.2019.00189

Pedersen MW, Ginolhac A, Orlando L, Olsen J, Andersen K, Holm J, Funder S, Willerslev E, Kjær KH (2013) A comparative study of ancient environmental DNA to pollen and macrofossils from lake sediments reveals taxonomic overlap and additional plant taxa. Quat Sci Rev 75:161–168. https://doi.org/10.1016/j.quascirev.2013.06.006

Pedersen MW, Ruter A, Schweger C, Friebe H, Staff RA, Kjeldsen KK, Mendoza MLZ, Beaudoin AB, Zutter C, Larsen NK, Potter BA, Nielsen R, Rainville RA, Orlando L, Meltzer DJ, Kjær KH, Willerslev E (2016) Postglacial viability and colonization in North America's ice-free corridor. Nature 537(7618):45–49. https://doi.org/10.1038/nature19085

Prentice IC (1985) Pollen representation, source area, and basin size: toward a unified theory of pollen analysis. Quat Res 23(1):76–86. https://doi.org/10.1016/0033-5894(85)90073-0

Ratnasingham S, Hebert PDN (2007) Bold: the barcode of life data system (http://www.barcodinglife.org). Mol Ecol Notes 7(3):355–364. https://doi.org/10.1111/j.1471-8286.2007.01678.x

Revéret A, Rijal DP, Heintzman PD, Brown AG, Stoof-Leichsenring KR, Alsos IG (2023) Environmental DNA of aquatic macrophytes: the potential for reconstructing past and present vegetation and environments:2023.03.27.533457

Rijal DP, Heintzman PD, Lammers Y, Yoccoz NG, Lorberau KE, Pitelkova I, Goslar T, Murguzur FJA, Salonen JS, Helmens KF, Bakke J, Edwards ME, Alm T, Bråthen KA, Brown AG, Alsos IG (2021) Sedimentary ancient DNA shows terrestrial plant richness continuously increased over the Holocene in northern Fennoscandia. Sci Adv 7(31):eabf9557. https://doi.org/10.1126/sciadv.abf9557

Rohland N, Glocke I, Aximu-Petri A, Meyer M (2018) Extraction of highly degraded DNA from ancient bones, teeth and sediments for high-throughput sequencing. Nat Protoc 13(11):2447–2461. https://doi.org/10.1038/s41596-018-0050-5

Sayers EW, Cavanaugh M, Clark K, Pruitt KD, Schoch CL, Sherry ST, Karsch-Mizrachi I (2022) GenBank. Nucleic Acids Res 50(D1):D161–D164. https://doi.org/10.1093/nar/gkab1135

Schmid S, Genevest R, Gobet E, Suchan T, Sperisen C, Tinner W, Alvarez N (2017) HyRAD-X, a versatile method combining exome capture and RAD sequencing to extract genomic information from ancient DNA. Methods Ecol Evol 8(10):1374–1388. https://doi.org/10.1111/2041-210X.12785

Schulte L, Bernhardt N, Stoof-Leichsenring K, Zimmermann HH, Pestryakova LA, Epp LS, Herzschuh U (2021) Hybridization capture of larch (Larix Mill.) chloroplast genomes from sedimentary ancient DNA reveals past changes of Siberian forest. Mol Ecol Resour 21(3):801–815. https://doi.org/10.1111/1755-0998.13311

Schulte L, Meucci S, Stoof-Leichsenring KR, Heitkam T, Schmidt N, von Hippel B, Andreev AA, Diekmann B, Biskaborn BK, Wagner B, Melles M, Pestryakova LA, Alsos IG, Clarke C, Krutovsky KV, Herzschuh U (2022) Larix species range dynamics in Siberia since the Last Glacial captured from sedimentary ancient DNA. Commun Biol 5(1):1–11. https://doi.org/10.1038/s42003-022-03455-0

Schwörer C, Leunda M, Alvarez N, Gugerli F, Sperisen C (2022) The untapped potential of macrofossils in ancient plant DNA research. New Phytol 235(2):391–401. https://doi.org/10.1111/nph.18108

Shirazi S, Meyer RS, Shapiro B (2021) Revisiting the effect of PCR replication and sequencing depth on biodiversity metrics in environmental DNA metabarcoding. Ecol Evol 11(22):15766–15779. https://doi.org/10.1002/ece3.8239

Sjögren P, Edwards ME, Gielly L, Langdon CT, Croudace IW, Merkel MKF, Fonville T, Alsos IG (2017) Lake sedimentary DNA accurately records 20th century introductions of exotic conifers in Scotland. New Phytol 213(2):929–941. https://doi.org/10.1111/nph.14199

Slon V, Glocke I, Barkai R, Gopher A, Hershkovitz I, Meyer M (2016) Mammalian mitochondrial capture, a tool for rapid screening of DNA preservation in faunal and undiagnostic remains, and its application to Middle Pleistocene specimens from Qesem Cave (Israel). Quat Int 398:210–218. https://doi.org/10.1016/j.quaint.2015.03.039

Slon V, Hopfe C, Weiß CL, Mafessoni F, de la Rasilla M, Lalueza-Fox C, Rosas A, Soressi M, Knul MV, Miller R, Stewart JR, Derevianko AP, Jacobs Z, Li B, Roberts RG, Shunkov MV, de Lumley H, Perrenoud C, Gušić I, Kućan Ž, Rudan P, Aximu-Petri A, Essel E, Nagel S, Nickel B, Schmidt A, Prüfer K, Kelso J, Burbano HA, Pääbo S, Meyer M (2017) Neandertal and Denisovan DNA from Pleistocene sediments. Science 356(6338):605–608. https://doi.org/10.1126/science.aam9695

Soininen EM, Gauthier G, Bilodeau F, Berteaux D, Gielly L, Taberlet P, Gussarova G, Bellemain E, Hassel K, Stenøien HK, Epp L, Schrøder-Nielsen A, Brochmann C, Yoccoz NG (2015) Highly overlapping diet in two sympatric lemming species during winter revealed by DNA metabarcoding. PLoS One. https://doi.org/10.1371/journal.pone.0115335

Sønstebø JH, Gielly L, Brysting AK, Elven R, Edwards M, Haile J, Willerslev E, Coissac E, Rioux D, Sannier J, Taberlet P, Brochmann C (2010) Using next-generation sequencing for molecular reconstruction of past Arctic vegetation and climate. Mol Ecol Resour 10(6):1009–1018. https://doi.org/10.1111/j.1755-0998.2010.02855.x

Stoof-Leichsenring KR, Huang S, Liu S, Jia W, Li K, Liu X, Pestryakova LA, Herzschuh U (2022) Sedimentary DNA identifies modern and past macrophyte diversity and its environmental drivers in high-latitude and high-elevation lakes in Siberia and China. Limnol Oceanogr 67(5):1126–1141. https://doi.org/10.1002/lno.12061

Sugita S (1994) Pollen representation of vegetation in quaternary sediments: theory and method in patchy vegetation. J Ecol 82(4):881–897. https://doi.org/10.2307/2261452

Taberlet P, Coissac E, Pompanon F, Gielly L, Miquel C, Valentini A, Vermat T, Corthier G, Brochmann C, Willerslev E (2007) Power and limitations of the chloroplast trnL (UAA) intron for plant DNA barcoding. Nucleic Acids Res 35(3):e14. https://doi.org/10.1093/nar/gkl938

Taberlet P, Bonin A, Zinger L, Coissac E, Taberlet P, Bonin A, Zinger L, Coissac E (2018) Environmental DNA: for biodiversity research and monitoring. Oxford University Press, Oxford/New York

Thuiller W, Albert CA, Araújo MB, Berry PM, Cabeza M, Guisan A, Hickler T, Midgley GF, Paterson J, Schurr FM, Sykes MT, Zimmermann NE (2008) Predicting global change impacts on plant species' distributions: future challenges. Perspect Plant Ecol Evol Syst 9(3–4):137–152. https://doi.org/10.1016/j.ppees.2007.09.004

Tyler T, Herbertsson L, Olofsson J, Olsson PA (2021) Ecological indicator and traits values for Swedish vascular plants. Ecol Indic 120:106923

van der Pijl L (2012) Principles of dispersal in higher plants. Springer

Van Geel B (2001) Non-pollen Palynomorphs. In: Smol JP, Birks HJB, Last WM, Bradley RS, Alverson K (eds) Tracking environmental change using lake sediments: terrestrial, algal, and siliceous indicators. Springer, Dordrecht, pp 99–119

van Geel B, Aptroot A (2006) Fossil ascomycetes in Quaternary deposits. Nova Hedwigia:313–329. https://doi.org/10.1127/0029-5035/2006/0082-0313

van Vugt L, Garcés-Pastor S, Gobet E, Brechbühl S, Knetge A, Lammers Y, Stengele K, Alsos IG, Tinner W, Schwörer C (2022) Pollen, macrofossils and sedaDNA reveal climate and land use impacts on Holocene mountain vegetation of the Lepontine Alps, Italy. Quat Sci Rev 296:107749. https://doi.org/10.1016/j.quascirev.2022.107749

Vargas P, Arjona Y, Nogales M, Heleno RH (2015) Long-distance dispersal to oceanic islands: success of plants with multiple diaspore specializations. AoB Plants 7:plv073. https://doi.org/10.1093/aobpla/plv073

Vernot B, Zavala EI, Gómez-Olivencia A, Jacobs Z, Slon V, Mafessoni F, Romagné F, Pearson A, Petr M, Sala N, Pablos A, Aranbur A, De Castro JMB, Carbonell E, Li B, Krajcarz MT, Krivoshapkin AI, Kolobova KA, Kozlikin MB, Shunkov MV, Derevianko AP, Viola B, Grote S, Essel E, Herraéz DL, Nagel S, Nickel B, Richter J, Schmidt A, Peter B, Kelso J, Roberts RG, Arsuaga JL, Meyer M (2021) Unearthing Neanderthal population history using nuclear and mitochondrial DNA from cave sediments. Science 372(6542). https://doi.org/10.1126/science.abf1667

Voldstad LH, Alsos IG, Farnsworth WR, Heintzman PD, Håkansson L, Kjellman SE, Rouillard A, Schomacker A, Eidesen PB (2020) A complete Holocene lake sediment ancient DNA record reveals long-standing high Arctic plant diversity hotspot in northern Svalbard. Quat Sci Rev 234:106207. https://doi.org/10.1016/j.quascirev.2020.106207

von Hippel B, Stoof-Leichsenring KR, Schulte L, Seeber P, Epp LS, Biskaborn BK, Diekmann B, Melles M, Pestryakova L, Herzschuh U (2022) Long-term fungus–plant covariation from multi-site sedimentary ancient DNA metabarcoding. Quat Sci Rev 295:107758. https://doi.org/10.1016/j.quascirev.2022.107758

Wagner S, Lagane F, Seguin-Orlando A, Schubert M, Leroy T, Guichoux E, Chancerel E, Bech-Hebelstrup I, Bernard V, Billard C, Billaud Y, Bolliger M, Croutsch C, Čufar K, Eynaud F, Heussner KU, Köninger J, Langenegger F, Leroy F, Lima C, Martinelli N, Momber G, Billamboz A, Nelle O, Palomo A, Piqué R, Ramstein M, Schweichel R, Stäuble H, Tegel W, Terradas X, Verdin F, Plomion C, Kremer A, Orlando L (2018) High-throughput DNA sequencing of ancient wood. Mol Ecol 27(5):1138–1154. https://doi.org/10.1111/mec.14514

Wagner S, Seguin-Orlando A, Leplé J-C, Leroy T, Lalanne C, Labadie K, Aury J-M, Poirier S, Wincker P, Plomion C, Kremer A, Orlando L (2023) Tracking population structure and phenology through time using ancient genomes from waterlogged white oak wood. Mol Ecol. https://doi.org/10.1111/mec.16859

Wang SC, Marshall CR (2016) Estimating times of extinction in the fossil record. Biol Lett 12(4):20150989. https://doi.org/10.1098/rsbl.2015.0989

Wang SC, Everson PJ, Zhou HJ, Park D, Chudzicki DJ (2016) Adaptive credible intervals on stratigraphic ranges when recovery potential is unknown. Paleobiology 42(2):240–256. https://doi.org/10.1017/pab.2015.37

Wang Y, Pedersen MW, Alsos IG, De Sanctis B, Racimo F, Prohaska A, Coissac E, Owens HL, Merkel MKF, Fernandez-Guerra A, Rouillard A, Lammers Y, Alberti A, Denoeud F, Money D, Ruter AH, McColl H, Larsen NK, Cherezova AA, Edwards ME, Fedorov GB, Haile J, Orlando L, Vinner L, Korneliussen TS, Beilman DW, Bjørk AA, Cao J, Dockter C, Esdale J, Gusarova G, Kjeldsen KK, Mangerud J, Rasic JT, Skadhauge B, Svendsen JI, Tikhonov A, Wincker P, Xing Y, Zhang Y, Froese DG, Rahbek C, Bravo DN, Holden PB, Edwards NR, Durbin R, Meltzer DJ, Kjær KH, Möller P, Willerslev E (2021) Late Quaternary dynamics of Arctic biota from ancient environmental genomics. Nature 600(7887):86–92. https://doi.org/10.1038/s41586-021-04016-x

Willerslev E, Hansen AJ, Binladen J, Brand TB, Gilbert MTP, Shapiro B, Bunce M, Wiuf C, Gilichinsky DA, Cooper A (2003) Diverse plant and animal genetic records from Holocene and Pleistocene sediments. Science 300(5620):791–795. https://doi.org/10.1126/science.1084114

Willerslev E, Davison J, Moora M, Zobel M, Coissac E, Edwards ME, Lorenzen ED, Vestergård M, Gussarova G, Haile J, Craine J, Gielly L, Boessenkool S, Epp LS, Pearman PB, Cheddadi R, Murray D, Bråthen KA, Yoccoz N, Binney H, Cruaud C, Wincker P, Goslar T, Alsos IG, Bellemain E, Brysting AK, Elven R, Sønstebø JH, Murton J, Sher A, Rasmussen M, Rønn R, Mourier T, Cooper A, Austin J, Möller P, Froese D, Zazula G, Pompanon F, Rioux D, Niderkorn V, Tikhonov A, Savvinov G, Roberts RG, MacPhee RDE, Gilbert MTP, Kjær KH, Orlando L, Brochmann C, Taberlet P (2014) Fifty thousand years of Arctic vegetation and megafaunal diet. Nature 506(7486):47–51. https://doi.org/10.1038/nature12921

Zimmermann HH, Stoof-Leichsenring KR, Dinkel V, Harms L, Schulte L, Hütt M-T, Nürnberg D, Tiedemann R, Herzschuh U (2023) Marine ecosystem shifts with deglacial sea-ice loss inferred from ancient DNA shotgun sequencing. Nat Commun 14(1):1650. https://doi.org/10.1038/s41467-023-36845-x

Zimov SA, Chuprynin VI, Oreshko AP, Chapin FS, Reynolds JF, Chapin MC (1995) Steppe-tundra transition: a herbivore-driven biome shift at the end of the Pleistocene. Am Nat 146(5):765–794. https://doi.org/10.1086/285824

Chapter 11
Terrestrial Fauna and Hominin DNA from Sedimentary Archives

Tyler J. Murchie, Charline Giguet-Covex, Peter D. Heintzman, Viviane Slon, and Yucheng Wang

Keywords Animal environmental DNA · Hominins · eDNA shedding · Caves · Lakes · Permafrost · Archaeological eDNA · Shotgun sequencing · Capture enrichment · Metabarcoding

The study of animals, including humans, has been at the forefront of ancient DNA research since the first Egyptian mummies and quagga were genetically studied in 1984 (Pääbo 1984; Higuchi et al. 1984). While multiple palaeoecological proxies

T. J. Murchie (✉)
McMaster Ancient DNA Centre, Department of Anthropology, McMaster University, Hamilton, ON, Canada

Hakai Institute, Heriot Bay, BC, Canada
e-mail: murchiet@mcmaster.ca

C. Giguet-Covex
EDYTEM, CNRS, Université Savoie Mont Blanc, Le Bourget-du-Lac, France
e-mail: charline.giguet-covex@univ-smb.fr

P. D. Heintzman
Centre for Palaeogenetics, Department of Geological Sciences, Stockholm University, Stockholm, Sweden
e-mail: peter.d.heintzman@geo.su.se

V. Slon
Department of Anatomy and Anthropology, Department of Human Molecular Genetics and Biochemistry, Faculty of Medicine, The Dan David Center for Human Evolution and Biohistory Research, Tel Aviv University, Tel Aviv, Israel
e-mail: viviane@tauex.tau.ac.il

Y. Wang
Department of Zoology, University of Cambridge, Cambridge, UK

State Key Laboratory of Tibetan Plateau Earth System, Resources and Environment (TPESRE), Institute of Tibetan Plateau Research, Chinese Academy of Sciences, Beijing, China
e-mail: yw502@cam.ac.uk

E. Capo et al. (eds.), *Tracking Environmental Change Using Lake Sediments*, Developments in Paleoenvironmental Research 21,
https://doi.org/10.1007/978-3-031-43799-1_11

exist for assessing the presence of plants, fungi, and microorganisms through time, studying Quaternary macro-fauna and humans has been based predominantly on visible remains like bones, teeth, and human-made artifacts. Although these provide the most direct measures of past animal presence, numerous taphonomic processes will transform and recycle these relatively rare materials over time, resulting in only a tiny fraction of past organisms being represented in macro-fossil records today. The key benefit of environmental DNA (eDNA) in this regard is that it is a continuously disseminated and high abundance micro-indicator that can be used to study the shifting presence of animal species through time, animal co-occurrences with other ecologically associated species, their evolutionary histories through genome reassembly and phylogenetics, and even estimate how long ago these animals lived based solely on the accumulation of genetic mutations relative to other individuals of known ages. All these applications (and more) are already possible by extracting eDNA from what are otherwise ordinary sediments—with or without any contextual association with visible animal remains.

The Environmental DNA of Ancient Animals

Woolly mammoth, horse, Shasta ground sloth, and moa were among the first ancient animals whose DNA was sequenced from permafrost sediments in Siberia and cave deposits from New Zealand and Arizona (Hofreiter et al. 2003; Willerslev et al. 2003). With the advent of next generation sequencing technologies (Goodwin et al. 2016) and their first applications to degraded ancient DNA (Green et al. 2006; Poinar et al. 2006), there was a growing recognition that not only can macro-fossils, such as bones and teeth, preserve genome-scale DNA in deep time, but that sedimentary strata themselves (e.g. soils, sediment, palaeosols, ice, tephras) can likewise serve as genetic archives of ancient life (Shokralla et al. 2012; Pedersen et al. 2015; Thomsen and Willerslev 2015). The field of ancient eDNA has grown rapidly since then, to the point now where diverse nuclear and organelle genomes can be reconstructed simultaneously from sediments, and whole communities of macro- and micro-organisms can be reconstructed from deposits irrespective of their macro-fossil preservation (Giguet-Covex et al. 2014; Willerslev et al. 2014; Graham et al. 2016; Pedersen et al. 2016, 2021; Slon et al. 2017; Zavala et al. 2021; Lammers et al. 2021; Murchie et al. 2021a; Wang et al. 2021; Armbrecht et al. 2022; Kjær et al. 2022).

Quaternary researchers have increasingly demonstrated that sediments and soils can preserve an immense amount of information about past life at the micro-scale (Swift et al. 2019). For decades, archaeologists routinely discarded faunal remains, ceramic sherds, and lithic debitage recovered during excavations until the rise of processual (scientific) archaeology, which began to reveal the potential of those overlooked materials for understanding the past (Shepard 1956; Aitken 1962; Binford and Binford 1968; Clarke 1968; Deetz 1977; Landon 2005). In much the same way, the depositional matrix of archaeological and palaeontological sites is often sifted through mesh screens during excavation to recover small remains (such as small bones, stone flakes and other artifacts, along with plant fossils), with the bulk of sediment then being discarded in disassociated back dirt piles off-site (although in many cases in the past and today, sediment samples are collected during fieldwork and stored for future research). It is likely that as ancient sedimentary DNA (sedDNA) methods increase in power and accessibility in the coming years—along with a growth in interdisciplinary collaborations between the Quaternary sciences—previously overlooked sedimentary matrixes will be reconceptualized as critical data sources in and of themselves. But as these techniques improve and become more cost-effective (van Dijk et al. 2014), so must the caution and broader awareness by which sedDNA is recovered, processed, analyzed, and interpreted.

In this chapter we discuss the magnitude of environmental DNA shedding by animals, the types of sites that researchers have found most informative for finding animal sedDNA, challenges associated with sedDNA research, differing approaches and considerations to be made in the research workflow, strategies for downstream computational analyses, and the types of end products one can reasonably expect. We conclude by highlighting recent applications and our thoughts on near future directions for the field, and what sorts of developments will help to further explore the wealth of human and faunal DNA preserved in environmental samples.

The Taphonomy and Magnitude of Animal eDNA Shedding

By the time ancient strands of animal eDNA find themselves floating in the lysing solution of a plastic microtube, those genetic fragments have already undergone a long history of extracellular chemical interactions forming organo-mineral complexes with minerals, organic compounds, and colloidal particles in the soil matrix (Romanowski et al. 1991; Lorenz and Wackernagel 1994; Blum et al. 1997; Cai et al. 2006a, b; Levy-Booth et al. 2007; Kanbar et al. 2020; Jelavic et al. 2023). These chemical bonds are thought to be the mechanism that enables the long-term persistence of DNA fragments outside of protective macro-tissues. A suite of chemical, physical, and biological processes will metabolise, degrade, or transform almost all DNA released into the environment. Of the infinitesimally tiny fraction of DNA that survives the dynamic processes of eDNA cycling in soil (see Levy-Booth et al. 2007), most of the remaining genetic material will have been fragmented into <100 base pair (bp) oligonucleotides due to various processes, such as hydrolytic depurination where DNA strands break (Briggs et al. 2007; Dabney et al. 2013; Kistler et al. 2017). Many of these now chemically exposed end-strand nucleotides will also become miscoded as a result of deaminated cytosines being misread by DNA polymerases as uracil (U), ultimately resulting in cytosine (C) to thymine (T) and guanine (G) to adenine (A) substitutions (Dabney et al. 2013). Additionally, many of the remaining fragments will have also been crosslinked and/or complexed with other environmental molecules, causing blocking lesions that inhibit polymerases from synthesizing complementary DNA in processes like the polymerase chain reaction (PCR) (Mullis et al. 1986; Bartlett and Stirling 2003; Hansen et al. 2006; Haile 2012; Dabney et al. 2013). These and other taphonomic processes began when dying cells were sloughed off as skin or hair, from waste products (faeces and urine) with gastrointestinal or mucosal tissues, with disseminated sex cells, or as part of blood or the overall decomposition of a carcass (Willerslev et al. 2004; Haile et al. 2007; Pedersen et al. 2015).

To put the scale of this sort of eDNA shedding into perspective, Roberts and Marks (1980) found that the average rate of passive desquamation (i.e., shedding of the outermost membrane of epidermal tissue) in humans is ~1000 cells per cm^2 per hour, with some body areas shedding at faster rates than others. If we take the average epidermal surface area of a human as 2 m^2 (which, as an aside, has been shown to actually be closer to 25 m^2 as skin is not a flat surface) (Gallo 2017), then the average person sheds ~20 million skin cells per hour, or about 0.48 billion per day. With a 6.4 billion bp nuclear and mitochondrial genome in each of those shed skin cells divided into 50 bp sized fragments (the typical strand length that ancient DNA will fragment down to over time) (Marciniak et al. 2015), this means that the average person sheds ~2.56×10^{15} ancient DNA sized fragments every day from skin alone. If intestinal epithelial cells are included, where a person sheds ~10 billion cells per day (Bullen et al. 2006), our potential eDNA output climbs to at least 1.28 x 10^{18}, or more than a quintillion ancient DNA sized fragments every day from a single human (Fig. 11.1). Nonetheless, the number of these DNA fragments that may remain sufficiently intact to be sequenced hundreds or thousands of years after release is surely very small.

Predicting DNA survival is a complex and relatively understudied area of research (Allentoft et al. 2012; Kistler et al. 2017). DNA decay rates appear to vary drastically based on environmental conditions, from a 521 year average half-life in bone (Allentoft et al. 2012) to a half-life of only a few hours in freshwater streams (Andruszkiewicz Allan et al. 2021). Factors such as the organic context (being protected in bone or micro-tissues, versus complexed with minerals and other environmental molecules, versus being unbound and extracellular), and environmental burial/release conditions (i.e. water content, bacterial composition/density, pH, salinity, oxygen content, UV exposure, and temperature) have been found to significantly impact eDNA survival (Kirkpatrick et al. 2016; Wei et al. 2018; Harrison et al. 2019; Andruszkiewicz Allan et al. 2021; Massilani et al. 2022; Mauvisseau et al. 2022). However, considering the hypothetically high eDNA output throughout life as calculated previously, even if only a billionth of one percent of a person's shed eDNA were to preserve, there would still be tens of millions of ancient DNA fragments strewn across the landscape throughout that person's day, everyday, along with a proportional output from every other organism in that ecosystem. In the end, only tens of DNA fragments are needed to identify the presence of an organism and only hundreds are needed to begin reassembling parts of its genome.

By contrast, an animal can only leave behind one body's worth of macro-fossils when it dies (e.g. two femora, one skull). Most of those remains will decompose and be recycled through local ecosystems, leaving almost no visible trace of their life histories. Traditional palaeontology and archaeology, focused on identifiable remains such as bones and teeth, is limited by the inherent biases of fossil records reconstructed from the exceptionally few remains that, against all odds, have

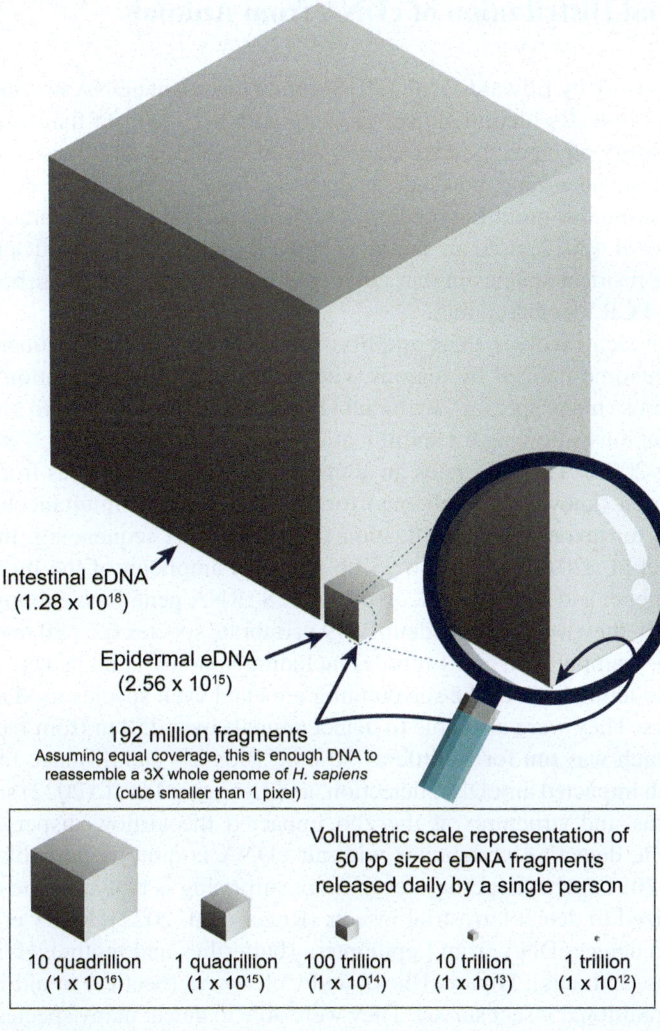

Fig. 11.1 Daily eDNA shedding by a single average human as represented by cubic volume. See references in the text

avoided the biosphere's nutrient cycle to date. In comparison, the continuous dissemination of genetic information into the local environment enables a far greater potential to find traces of ancient organisms by instead searching for evidence of DNA (in environments conducive to sedDNA survival) compared with their macrofossil potential from a finite number of hard tissues that rarely preserve or are found. In the case of eDNA, one needs to find the right sorts of deposits to sample where we might expect to find DNA from those animals of interest, where eDNA has the best chances of preserving, and where the stratigraphic context is trustworthy and informative.

The Spatial Distribution of eDNA from Animals

Modern soil work by Edwards et al. (2018) found that plant eDNA was highly local, with most genetically identified taxa growing within 1 meter of their soil samples. Animal mobility changes the expected spatial distribution and densities of eDNA dispersal on the landscape, however, as does varying rates of passive desquamation and other continuous moulting processes such as seasonal fur and feather shedding. Lynggaard et al. (2022) used air samplers at the Copenhagen Zoo to determine how many of the resident species in that zoo could be detected through airborne eDNA alone using PCR metabarcoding.

This approach involves PCR amplifying small, taxonomically diagnostic portions of a genome flanked by regions with strong genetic conservation (the same sequence across many species). Artificial DNA primers are designed to bind to these conserved regions, allowing for amplification using PCR (Mullis et al. 1986; Bartlett and Stirling 2003). This generates an abundance of DNA sequences from the PCR targeted region (known as amplicons) for multiple species simultaneously, which can be used for taxonomic identification following DNA sequencing. In this case, Lynggaard et al. (2022) targeted a 95 bp mammal amplicon of the mitochondrial 16S rRNA gene, and 97 bp amplicon of the 12S rRNA gene for vertebrates. Using this approach, they were able to identify 49 vertebrate species (i.e., 30 mammals, 13 birds, 4 fish, 1 amphibian, and 1 reptile) including not only animals kept at the zoo, but also wild animals from the surrounding area and even species used as food for the predators. They were also able to detect animals over 250 m from the sampling vacuum, which was run for as little as 30 min. Biomass and distance from the air sampler both impacted air eDNA detection, and Lynggaard et al. (2022) suggest that wind patterns and structures at the zoo impacted the airflow dispersal of their eDNA. While detecting vertebrates with air eDNA is quite remarkable (see also Clare et al. 2021), what is perhaps even more surprising is that airborne eDNA can also be utilized to detect terrestrial insects (Roger et al. 2022). Roger et al. (2022) were able to detect eDNA from Lepidoptera (butterflies and moths), Hymenoptera (bees, wasps, and ants), Diptera (flies), and Coleoptera (beetles) to aid in complementing a traditional insect survey. They were able to detect many species that were missed in their traditional surveys in a much less invasive manner. What these studies help demonstrate is that animal eDNA is ubiquitous in the environment.

But, how much of this airborne eDNA survives long enough in the environment to be complexed with minerals and other organic molecules in sediments and soils (and thereby preserved and become part of palaeo-records) is currently unclear. It is generally thought that most eDNA is derived from other forms of tissue shedding. Massilani et al. (2022) studied the microstratigraphic distribution of mammalian sedDNA in thin sections of resin impregnated sediments from archaeological caves in Eurasia. They demonstrated that animal derived micro-tissues—in this case, bone fragments and coprolites (ancient faeces)—were preserved in cave sediments, which yielded millimetre scale differences in the genetic faunal spectrum identified from a variety of species including hyenas, bears, horses, bovids, canids, and hominins. Massilani et al.'s (2022) research highlights that micro-tissues are a substantial source of animal sedDNA, but also that eDNA is not exactly homogeneously distributed. Rather, it seems to exist in discrete micro-pockets. This leads to the next question: how sensitive are these micro-pockets of eDNA over time?

Andersen et al. (2012) explored the sensitivity of 'dirt' eDNA for identifying vertebrates over the last 100 years at safari parks, farms, and zoological gardens in Denmark with constituents of non-native fauna. Soil and sediment samples were taken from the surface of these enclosures at either 10 or 2 cm intervals below the surface. Andersen and colleagues were trying to understand eDNA taxonomic richness relative to known faunal land use, the spatial variation in animal eDNA dispersal, and the impact of soil chemistry, texture, and structure on eDNA recovery, along with the propensity for downward leaching. They found that the haplotype diversity of eDNA accurately reflected above ground and historic taxonomic richness, with few exceptions—notably a genetically undetected camel that had been in the enclosure for two months. They also observed that eDNA was not uniformly dispersed in the enclosure areas, highlighting the need for multiple samples across a study site in order to accurately reconstruct biodiversity. Overall, animal biomass was found to be a better proxy indicator of eDNA presence than census counts of individual animals, and edaphic properties such as soil particle surface area, sediment structure, and pH all impacted eDNA recovery. Larger soil particles enabled more eDNA absorption, a finding that has been observed in other research (Frostegård et al. 1999; Lloyd-Jones and Hunter 2001; Cai et al. 2006a, b), and pH values above 6 seem to have allowed for increased eDNA adsorption to soil particles through cations (Nguyen and Elimelech 2007). As may be expected given the quantity of eDNA released by animals, Andersen and colleagues also found that their eDNA did not always remain in sedimentary strata that were contemporaneous with the habitation of animals in those enclosures.

Challenges in Contextual Authentication

Authenticity has been a fundamental concern in ancient DNA (aDNA) since its 'childhood' (Brown and Barnes 2015), particularly for ancient human DNA. Despite significant and rapid advances in sequencing technologies, determining whether a set of sequenced DNA reads are accurate genetic proxies of past organisms, and not a result of contamination, remains a substantial hurdle. This has maintained a hypercritical perspective within aDNA research since the characteristics of degraded DNA, and thus their vulnerability to contamination, were first identified (Hedges and Schweitzer 1995; Zischler et al. 1995). Both the DNA molecules originating from a target sample (i.e. ancient plant or animal DNA), and those from other contaminating sources (modern and ancient) are indiscriminately ligated (joined) with synthetic DNA adapters during the library preparation stage, allowing for downstream high-throughput sequencing (HTS) applications (Rogers and Venter 2005; Meyer and Kircher 2010; Kircher et al. 2012; Gansauge and Meyer 2013; Reuter et al. 2015; Wales et al. 2015; Gansauge et al. 2017; Kapp et al. 2021). This omnipresent threat has necessitated an abundance of caution when conducting aDNA field sampling and wet-lab work to minimize the potential for contaminant DNA to be unintentionally sequenced with the rest of the sample. The move towards high-throughput methods with massively parallel sequencing (millions of simultaneously read DNA strands) has allowed palaeo-geneticists to shift away from evaluating the authenticity of aDNA with a simple set of standard criteria as in the era of PCR, bacterial cloning, and Sanger sequencing (highly accurate single fragment sequencing) (Cooper and Poinar 2000; Willerslev and Cooper 2005), towards an approach today based in the bioinformatic identification of contaminants and background DNA, and their exclusion from subsequent analysis.

Research targeting ancient sedDNA is further challenged in this regard in establishing spatial-temporal and taxonomic authenticity without discrete materials (e.g. bone, teeth, soft tissues) for independent verification (Rawlence et al. 2014). For instance, when aDNA sequence data originates from a mammoth tusk, one can reasonably exclude sequenced reads that cannot be mapped to a *Mammuthus* reference genome (such as the associated microbiome, human, or plant DNA). However, when working with metagenomic samples, such as sediments or ice, the sample substrate consists of disseminated organics from all domains of life, and generally lacks any identifiable macro- or micro-tissues to ground-truth the metagenomic data. Therein, contamination can become difficult to identify and exclude with eDNA, except in cases of unambiguously nonsensical taxa. As such, the temporal authenticity of ancient eDNA becomes a critical concern as there are few means of identifying rogue molecules that are out of stratigraphic context due to processes such as leaching (the downward sedimentary movement of young DNA into older deposits) or reworking (the erosion and re-deposition of sediment and biological remains, resulting in old sedDNA being mixed with younger deposits) (Haile et al. 2007; Arnold et al. 2011). Establishing the contextual veracity of these data is critical—if those molecules are not reasonably demonstrated to be autochthonous (in the primary, originally deposited context) with the host deposit, there is a pervasive chance that the analysis will succumb to false-positive results and inaccurate palaeoecological reconstructions. This challenge necessitates a continued vigilance for establishing sedDNA authenticity, an awareness of pitfalls when working with sedDNA, and presents a growing need for additional bioinformatic tools, field strategies, reference genomes, and geomorphological support to avoid and identify deposits with decontextualized sedDNA.

Detecting Stratigraphically Decontextualized sedDNA

Andersen et al. (2012) were able to highlight a major challenge of sedDNA research, particularly on palaeo-soils: stratigraphically decontextualized DNA due to leaching, otherwise referred to as being allochthonous (in a secondary, or non-source context, this is opposed to being autochthonous or in primary, or original source context). Most notably, in a Copenhagen Zoo elephant enclosure, they were able to detect Elephantidae DNA some 70 cm below the surface. It is unclear how much sediment had accumulated at the sampling site since elephants were first introduced to the enclosure in 1914, or if any of these deposits had been reworked (by the elephants or landscapers), but these results suggest a substantial downward movement of elephant eDNA. The authors hypothesized that this was due to large volumes of urine deposited in a small area, with sediment structure (large particle size with little clay and low organic content) and eDNA conducive pH further contributing to the extent of leaching. They found similar leaching in the latrine portion of a tiger enclosure in Knuthenborg Park, although again it is unclear whether the 20 cm of detected tiger eDNA was caused entirely by leaching, or whether there had been sediment accumulation or reworking since tigers were introduced to the enclosure in 1969.

Such observations were echoed in work by Haile et al. (2007) who had similarly investigated the vertical movement of sedDNA at two cave sites on New Zealand's North Island. The sites, situated ~300 km apart, record the last $\gtrapprox$3000 years, including the European settlement of the study area around 1870 CE (80 BP) and are stratigraphically well defined with two large tephra deposits. Haile and colleagues used the discontinuous faunal record of the region to evaluate the extent of vertical eDNA movement, using the eDNA of extinct moa (Aves; order Dinornithiformes) for which the date of last appearance is well known (~550 years ago). The extinct moa eDNA was only expected to be found in deeper deposits prior to European contact, while non-indigenous sheep (*Ovis aries*) was expected in the upper strata. They observed a downward migration of sheep eDNA below the European layer at both sites, extending into the post-Taupo and Taupo ignimbrite (232 CE) deposits. Sheep DNA was found as much as 35 cm below the expected European layer. Haile et al. (2007) suggested the frequent deposition of large volumes of urine having likely saturated the sediments, thereby carrying sheep eDNA into deeper strata. Their DNA quantifications indicate that this decontextualized eDNA decreased in quantity with depth, supporting the urine percolation hypothesis. By contrast, moa eDNA was only found in expected deposits, implying there was no upward movement of avian DNA and that those avian sequences are ecologically informative of their respective time periods. This is also likely in part because birds produce far less DNA-containing urine capable of leaching into deeper layers, which was also attested for ostrich by Andersen et al. (2012). Together, Haile et al. (2007) and Andersen et al. (2012) highlight the potential for accurate ecological reconstructions with eDNA, but also the complicating pitfall of decontextualized sedDNA due to leaching. Their work cautions the necessity of careful stratigraphic analysis when interpreting a faunal sedDNA dataset.

In some stratigraphic settings, such as in permafrost conditions where the ground is frozen perennially, the potential for leaching is considered to be negligible (Arnold et al. 2011). Multiple studies have found synchronous palaeoecological shifts when comparing complementary palynological and macrofossil evidence with sedDNA reconstructions (Willerslev et al. 2003; Lydolph et al. 2005), indicating that in these perennially frozen conditions there has not been a detectable downward migration of sedDNA. Further, other permafrost studies report that the extent of DNA damage generally follows an age-dependent pattern (Willerslev et al. 2004; Hansen et al. 2006; Haile et al. 2007; Murchie et al. 2021a), as previously noted in some non-permafrost settings (Zavala et al. 2021). However, damage patterns such as cytosine deamination and DNA fragmentation have been observed to be more complexly influenced by a variety of environmental variables than simply their associated age (Kistler et al. 2017). As such, the unique sedimentary characteristics of each region and site should be considered when assessing age-related damage patterns. The lack of leaching potential in permafrost is due to ground ice functioning as an aquitard or aquiclude (Woo 2012), rendering these perennially frozen layers largely impervious to the transmission of groundwater, and hence limiting sedDNA leaching. A similar situation is thought to occur in lakes where the permanent saturation of sediments is thought to prevent the percolation of liquids, and as such limit leaching (Giguet-Covex et al. 2014, 2019). A tight continuity between palaeoecological proxies and a lack of evidence for temporally decontextualized sedDNA has been observed in multiple lake studies focused on plants and mammals (Giguet-Covex et al. 2014; Sjögren et al. 2017; Ficetola et al. 2018). However, are 'upward' eDNA movements also possible?

Arnold et al. (2011) investigated whether old eDNA can be mixed with younger deposits at seven permafrost sites, including four wetlands, around the Taimyr Peninsula in north-central Siberia. The authors used optically stimulated luminescence (OSL) and radiocarbon (^{14}C) dating to assess whether perennially frozen deposits had been exposed to reworking processes (i.e., if the ^{14}C ages of macrofossils were older than the associated sedimentary strata, which was dated with OSL, a technique that measures the time since quartz grains in those sediments were last exposed to sunlight). Three of their seven sites had discordant OSL and ^{14}C dates, suggesting that older macro-fossils—presumably along with well-preserved micro-fossils of eDNA—had been redeposited with younger sediments at those sites. This led the authors to emphasize that high-energy fluvial contexts and sedimentary "sinks" (as opposed to "sources") should be avoided when planning to recover permafrost samples for sedDNA (this is opposed to lake sedDNA research where high erosion rates favour the transfer of terrestrial extracellular DNA to lake basins for recovery [see Chap. 2 of this volume; Giguet-Covex et al. 2019]). As such, cold burial conditions can be both a benefit and a hindrance. While cold preserved sedDNA generally has some of the best metagenomic preservation (alongside caves), high densities of well-preserved sedDNA also can facilitate the survival of these biomolecules through erosion, transport, and redeposition alongside younger deposits with spatial-temporally discordant eDNA.

Reworking is, however, a potential problem with almost any sedDNA site type. As discussed by Giguet-Covex et al. (2019), the potential "time shift" of sedDNA is not only a result of erosion and redeposition, but can also be induced by delay in the release of eDNA. Pesticides, for example, as environmentally persistent molecules, can be released decades after their first introduction to an ecosystem (Sabatier et al. 2014). Likewise, it has been observed that eDNA from alpine crops could be found 50 years after cultivation (Yoccoz et al. 2012), and Andersen et al. (2012) were able to detect several animals' eDNA in the topsoil months to years after the animals had been removed from their enclosures. Giguet-Covex et al. (2019) argue that the influence of older sedDNA is likely minimal in the case of lakes given the number of studies with highly accurate records of ecological shifts compared with robust historic and proxy data. But the potential for leaching and reworking, along with the unknown accumulation time and delayed release of sedDNA (be it tens to perhaps even hundreds of years), highlights the need for cautious skepticism and corroborating evidence from complementary palaeoecological proxies (Parducci et al. 2017) when interpreting any sedDNA dataset. Ecological reconstructions using sedDNA are context dependent, like any other palaeontological or archaeological remains. Since individual DNA molecules cannot be dated, building a robust case to establish the chronological continuity between molecular proxy data and their host deposit is an important component of sedDNA research.

With these challenges in mind, we will shift here to considering the following questions: 1) Where should one look in order to best seek out ancient animal eDNA? 2) What characteristics can help inform that decision? and 3) What research has been done previously at these types of sampling localities, and how successful were they?

Where to Look for Ancient Animal eDNA?

Palaeobiogeography—i.e., the study of the past geographic distributions of differing species— is an area of research that can be particularly aided by sedDNA. Especially when dealing with spatial-temporal questions at the edges of our current understanding. For instance, attempts to detect the spatial distribution of small populations of pioneering or terminally dwindling fauna and/or humans. This research question typically lies beyond the detection limits of traditional methods given the unlikelihood of finding trace fossils. This challenge is encapsulated in the palaeontological principle known as the Signor-Lipps effect (Signor and Lipps 1982), along with its reverse, the Jaanusson effect (Jaanusson 1976). Therein, the first and last individuals of a population are essentially undetectable (probabilistically) due to their small population sizes and resulting small fossil output, which is compounded by the inherent incompleteness of even the best fossil records (Fig. 11.2).

Knowing the site-specific contexts to target in order to find genetic evidence of specific animals (along with a fair bit of luck) is important to maximize chances for detection. Factors such as mobility patterns, population densities, seasonality, and variable rates of biomass turnover all compound with the taphonomic characteristics inherent to specific sites (Hebsgaard et al. 2009; Yoccoz et al. 2012; Graham et al. 2016; Giguet-Covex et al. 2019; Lammers et al. 2019). For example, one would expect large herbivores to have far greater sedDNA representation in an open-air or lake setting compared to carnivores, rodents, or rare birds, whose eDNA may be better sought after beneath rocky outcrops or in caves. However, despite

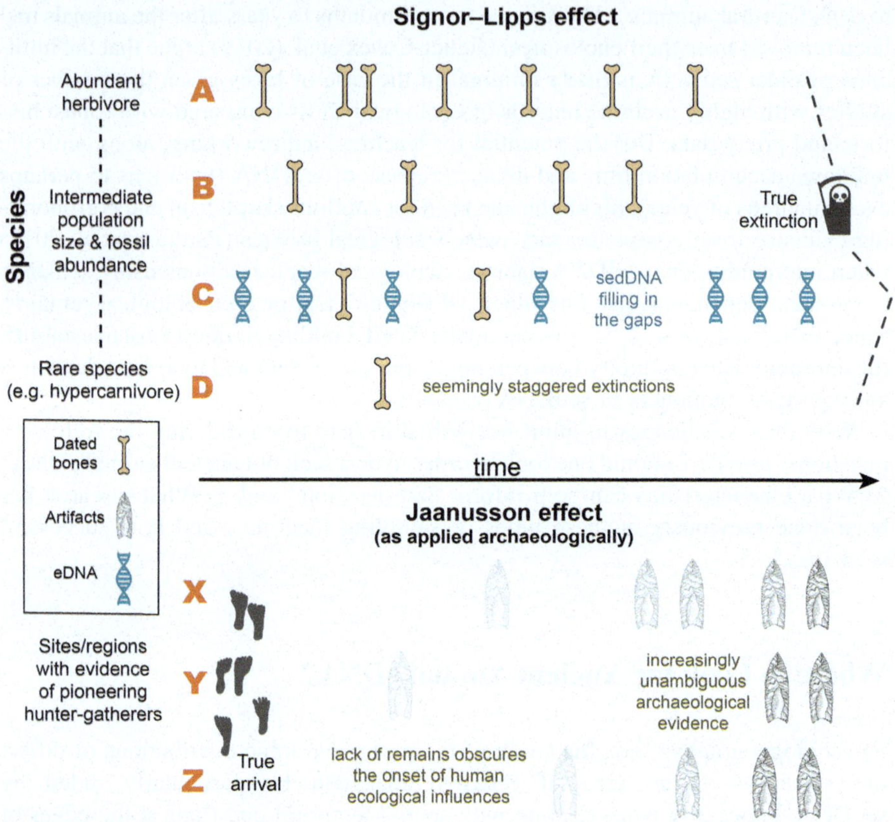

Fig. 11.2 The Signor-Lipps and Jaanusson effects obscuring biogeographic range shifts of humans and other animals. (Figure modified from Murchie 2021)

producing significantly more eDNA, large animals are also likely more transient on the landscape compared to small fauna or invertebrates that may be ever-present, like plants. These factors add unpredictability to site selection.

It is difficult to know whether herds of woolly mammoths, for instance, would have routinely travelled through one valley as opposed to the one adjacent, or passed near one rocky outcrop for millennia, while only seldomly, if ever, visiting another. Complementary isotopic research to reconstruct lifetime mobility routes can aid in this regard (Wooller et al. 2021), along with traditional excavations, surveys, and other complementary methods. Lakes may serve as ideal settings for collecting a catchment of sedDNA from a local area, but there is also a greater likelihood of eDNA degradation during erosion and transport to the lake basin as hydrolysis is the main driver in multiple forms of DNA damage (Dabney et al. 2013), along with the unpredictability of eDNA dispersal across the basin. Sites where the eDNA can essentially remain *in situ* after release, such as directly within a known archaeological/paleontological site, a cave, or an open-air setting, may have the highest densities of preserved sedDNA. But finding viable pockets of eDNA within these sites also necessitates knowing (or more likely, getting lucky with) the specific provenances of the sampling therein. As discussed previously, Massilani et al,'s (2022) work with sediment thin sections and millimeter-scale sampling demonstrated that sedDNA exists in micro-pockets and a dramatically different faunal spectrum can be found when shifting the subsampling locality by as little as one centimeter.

We may also expect other variables, such as site use intensity/activities, differing types of clothing, or other lifestyle factors, to impact human eDNA release patterns. For example, cold weather clothing may concentrate epidermal eDNA shedding to dwellings and latrines in colder areas (much like how clean room suits are intended to restrict eDNA shedding). To some degree, all sedDNA research is a "fishing expedition" as sedDNA provenance and preservation are difficult to detect or accurately predict without extensive molecular surveys or targeting previously studied sites. Even with abundant prior palaeobiogeographic knowledge, one only knows for certain once they have sequence data as to whether their target organisms were present or not.

With these factors in mind, we continue here with an overview the four main sedimentary site types that have been found to be the most productive for faunal and hominin sedDNA research to date: caves, lakes, permafrost/periglacial conditions, and known archaeological sites. We have left out other emerging site types for the sake of brevity, such as marine sedDNA, which is itself a burgeoning and exciting area of research with many of its own impressive highlights and associated challenges (Dell'Anno and Corinaldesi 2004; Sinniger et al. 2016; Armbrecht et al. 2019, 2022; Ogata et al. 2021; Selway et al. 2022). In subsequent sections, we discuss the general characteristics of these four main site types while highlighting published research efforts to date and discussing their findings. It is important to emphasize, though, that any given site needs to be individually evaluated within its own unique depositional, palaeo-ecological, and taphonomic contexts.

Caves and Rock Shelters

Caves, rock shelters, and other karst landforms (underground drainage systems formed through the dissolution of rock) have helped preserve many of the most biologically rich, diverse, and evolutionarily informative fossil and artifact deposits of any site types (Straus 1990; Schubert et al. 2003; Gunn 2004; Jass and George 2010; Schubert and Mead 2019). Caves are an exceptional resource for Quaternary science because they help protect biological remains from various forms of chemical, physical, and biological weathering (i.e. wind, water, sunlight, and decomposers). They often have relatively stable temperatures and humidity, allowing for the preservation of rarer remains, such as soft tissues. They are somewhat widespread even in areas that lack karst topography, and serve as catchments of many species, acting as natural traps where organisms repeatedly fall in and die, or as places of refuge where organisms are drawn for shelter (Jass and George 2010). While some very old karst fossil deposits exist, most cave records cover at most the past 2 million years due to geological weathering and erosion (Schubert and Mead 2019). Nonetheless, this timeframe is roughly equivalent to the Quaternary, which also happens to fall closely in line with the oldest sedDNA found to date (Kjær et al. 2022). Karst landforms sites often serve not only as exceptional Quaternary-aged archives of macro-fossils (referred to as Lagerstätten) (Seilacher 1970), but they also are ideal settings preserving and finding sedDNA.

The Quaternary science potential of caves is reflected in the history of sedDNA research where the earliest research papers reporting on the long-term survival of animal and plant DNA in sediments came from samples collected in caves. In the same year, one study (Willerslev et al. 2003) demonstrated the preservation of aDNA from three avian species in ca. 600 year old New Zealand cave sediments; and another (Hofreiter et al. 2003) described the retrieval of DNA of four cave-dwelling species from ca. 10,000 year old sediments in Arizona. Such cave focused sedDNA research has continued since then with varying success (e.g., see review by Pedersen et al. 2015 and references therein).

Ancient human genomics is one of the largest research areas within aDNA, with thousands of ancient individuals having been sequenced to date (Orlando et al. 2021). Nonetheless, investigations of the genetic makeup of Pleistocene-aged hominins are often limited by the scarcity of their skeletal remains. Retrieving ancient human DNA from sediments has the potential to circumvent that issue. In the early days of ancient sedDNA research, however, this objective was deemed "virtually impossible" (Hofreiter et al. 2003) due to difficulties in differentiating between authentic aDNA and contamination from modern human DNA that was shed by researchers themselves. The development of sample preparation strategies based on converting DNA fragments to libraries (Meyer and Kircher 2010; Gansauge and Meyer 2013), which directly allow for the distinction between ancient and modern DNA fragments based on patterns of DNA damage (Briggs et al. 2007; Krause et al. 2010; Ginolhac et al. 2011; Jónsson et al. 2013; Skoglund et al. 2014), drove forward the hunt for ancient hominin DNA in sediments deposited at Pleistocene cave sites.

This formerly 'impossible' goal was realized by Slon et al. (2017), who were able to recover hominin and other animal DNA (including mitochondrial genomes), from six out of seven of cave sites spread throughout Eurasia. They showed evidence of past human presence in depositional layers where no human remains had been found. This success was in part due to their development of a targeted enrichment approach intended to maximize the recovery of this sort of faunal and hominin DNA from sediment (Slon et al. 2016). Targeted enrichment is a method that will be discussed more fully later in the chapter, but briefly here this approach intentionally biases a DNA library that is otherwise dominated by prokaryotic (bacteria, archaea) and viral DNA towards a pre-selected set of target DNA sequences (be that individual genes or whole genomes from single or multiple species, simultaneously) (Carpenter et al. 2013). Perhaps unsurprisingly, in at least some of the cases, their hominin DNA recovered from a single sediment sample originated from a mixture of multiple individuals—what is now considered a common feature in sedDNA datasets (Slon et al. 2017; Murchie et al. 2022). Beyond the identification of taxa at the biological family level and their authentication based on patterns of ancient DNA-like damage (Dabney et al. 2013; Jónsson et al. 2013), analyses of certain taxa allowed for a more taxonomically specific distinction. For example, sufficient data were recovered to differentiate between brown bear (*Ursus arctos*) mitochondrial DNA (mtDNA) in sediments deposited at Denisova Cave (Altai region, Russia) from Eastern cave bear (*Ursus ingressus*) mtDNA in Vindija Cave (Croatia) (Slon et al. 2017). For nine of the sediment samples containing ancient hominin mtDNA, partial to nearly complete mtDNA genomes could be reconstructed, allowing for their placement within the phylogenetic diversity of Pleistocene hominins, and distinguishing between Neandertal-like and Denisovan-like mtDNA. This work demonstrated the potential of ancient sedDNA approaches to investigate the genetic signature of ancient hominins, even from deposits devoid of skeletal remains.

The potential of these approaches started to be fulfilled soon thereafter, when Zhang et al. (2020) applied a similar approach to sediments collected at Baishiya Karst Cave on the Tibetan Plateau, and detected Denisovan mtDNA outside of Denisova Cave (Krause et al. 2010) for the first time. A combination of phylogenetic analyses and dating of the sediments revealed that Denisovans (or Denisovan-like hominins) were present on the Tibetan Plateau at least 100,000, 60,000 and 45,000 years ago. More recently, in the largest-scale study of a single cave site to date, Zavala et al. (2021) analyzed 728 sediment samples from Pleistocene layers at Denisova Cave, in which they retrieved ancient faunal and hominin mtDNA from 685 and 175 samples, respectively. These data shed new light on the population history of hominins in the region, showing that the hominin mtDNA recovered in the earliest layers was exclusively of the Denisovan type, that individuals carrying Neandertal-like and Denisovan-like mtDNA later co-existed repeatedly, and that individuals carrying modern human-like mtDNA appeared at least 45,000 years ago. In addition, they revealed at least two turnovers in both faunal and human populations at the site: one occurring ca. 190,000 years ago, marked by a first shift in the relative proportions of large mammals, the transition from cave bears to brown bears, a change in hyaena mtDNA types, and the first appearance of Neandertal mtDNA at the site. The second, ca. 100,000 years ago, was denoted by a second shift in faunal taxonomic proportions, changes in mtDNA haplogroups of both ursids and hyenas, and the absence of Denisovan-like mtDNA. The coincidence of these shifts in the genetic signals with climatic events (i.e., transition from the interglacial Marine Isotope Stage [MIS] 7 to the glacial MIS 6 for the former, and from MIS 6 to the interglacial MIS 5 for the latter) suggested a possible connection between the turnovers in the human and faunal populations and changes in their surrounding environments.

Notably, inferences about the environment in which ancient humans lived can be made based on sedimentary aDNA analysis of archaeological cave sites, even when ancient human DNA is not recovered. In 2020, for example, Ardelean et al. (2020) combined an archaeological investigation of lithic artifacts discovered at Chiquihuite Cave (Mexico) with a tight chronological framework to argue for the presence of ancient humans in the region as early as 33,000–31,000 years ago. Accompanying sedDNA analyses showed the presence of a multitude of animals in the archaeological layers, including bats, bears, rodents, goats, sheep and birds; whereas plant DNA in the sediments pointed to shifts in plant communities over time. At Uluzzo Cave C (Italy), sedDNA analyses supported inferences made based on zoo-archaeological data regarding the environmental context of the cave during the transition from the Middle to the Upper Paleolithic between 50,000 and 40,000 years ago (Silvestrini et al. 2022). As methods to generate and analyze sedDNA data keep improving, the integration of sedimentary aDNA data into interdisciplinary studies of archaeological cave sites will likely become a more common practice in the future.

In 2021, three studies, all conducted at cave sites, took the field of sedimentary aDNA one step forward, as all reported on analyses of genome-wide nuclear DNA data recovered from sediments. Pedersen et al. (2021) reconstructed four low-coverage ursid genomes (three American black bear [*Ursus americanus*] and one giant short-faced bear [*Arctodus simus*]) by shotgun sequencing 14,000–16,000-year-old sediments collected at the above-mentioned Chiquihuite Cave (Mexico). Shotgun sequencing will be discussed further in the next major section of this chapter, but in short, it involves sequencing a DNA library to a chosen depth (i.e. 500,000 or 500 million sequence reads) without any attempt to bias the dataset towards any particular type of sequence data. By co-analyzing their shotgun sequencing data with newly-generated and existing genetic data from modern and ancient bears, Pedersen et al. (2021) showed that the ancient American black bear lineage found in Mexico is ancestral to the modern Eastern American black bear; and estimated the time of divergence of the now-extinct giant short-faced bear from extant bear species. Vernot et al. (2021) developed a targeted enrichment strategy adapted to the study of ancient hominin DNA from sediments. Applied to sediment samples collected at the site of Estatuas (Spain), these data revealed a replacement in the populations of Neandertals at the cave ca. 100,000 years ago. Compared to existing genome-wide data recovered from Neandertal skeletal remains, this turnover could be associated with two putative events of radiation of Neandertal populations, occurring ca. 35,000 years apart and possibly in relation to climate changes. Lastly, Gelabert et al. (2021) recovered genome-wide data from three species (human, wolf, and bison) from a single 25,000-year-old sediment sample collected at Satsurblia Cave (Georgia), demonstrating that these genomes belong to ancestral lineages compared to modern-day individuals, and shedding light on the population history of all three taxa in the region. Taken together, these three studies have heralded a new era in sedimentary aDNA analysis of "environmental genomes", which could become the key to filling in the gaps of the palaeogenetic record.

The extent of post-mortem degradation of DNA has been shown to be affected not only by age and temperature (Allentoft et al. 2012; Sawyer et al. 2012), but also by humidity and the extent of thermal fluctuations (Kistler et al. 2017). Thus, intuitively, caves may be particularly well-suited to preserve DNA over time, as they constitute a relatively sheltered micro-environment and where temperatures tend to be cooler and more stable than the open landscapes surrounding them. In geographical contexts less favourable for the long-term preservation of DNA (e.g., outside of permafrost), sampling cave sites can therefore be a good strategy to maximize the chances of ancient DNA recovery. More specifically, within each cave site, larger sampling efforts can help to determine areas more prone to preserve DNA over time. At Denisova Cave, for example, where ancient mammalian mtDNA was recovered in nearly 95% of samples tested, DNA preservation was the poorest in phosphatized and slightly acidic sediment samples (pH values between 6 and 6.5) (Zavala et al. 2021). In the Paleolithic layers of Sefunim Cave (Israel), where DNA preservation is expected to be very poor, only samples collected from underneath a large limestone boulder, which had been removed from the excavation area shortly prior to sampling, retained traces of ancient faunal mtDNA (Slon et al. 2022). While anecdotal, these reports could help guide future sampling efforts at other cave sites.

Overall, caves have proven to be an important source of sedimentary aDNA data. While freshly-excavated samples collected while wearing protective gear to minimize present-day contamination (Pedersen et al. 2021; Zavala et al. 2021; Slon et al. 2022) are probably best, experience has shown that samples collected at cave sites for other purposes (e.g., luminescence dating), even decades earlier, can preserve sedDNA as well (Slon et al. 2017). Beyond loose sediments, recent reports have also demonstrated the preservation of ancient eDNA in cave speleothems (Stahlschmidt et al. 2019), as well as in resin-impregnated sediment blocks generated for micro-morphology analyses where Massilani et al. (2022) demonstrated that sedDNA is not homogeneously distributed, but rather exists in micro-pockets that varies on the millimeter scale. Sarhan et al. (2021) likewise demonstrated that aDNA from human bones can diffuse into surrounding calcite stone in cave burials. Surprisingly, their calcite-bound ancient eDNA was found to have better preservation (in terms of DNA fragments lengths) than the aDNA preserved in the bones of the interred individual. This sort of research on diverse samples is helping to rapidly enlarge the potential scope of what materials are considered viable natural archives for eDNA preservation.

The primary downside to caves and karst landforms from an overall Quaternary science perspective, however, is that, while they often have some of the best preserved aDNA, we would only expect them to contain a relatively small fraction of the overall ecological diversity of an area given that they are generally isolated environments with sporadic distributions. Their depositional histories are also generally not entirely continuous with rock falls, other erosional events, and they are exposed to DNA leaching. This is where the next site type—lake sediments—comes in, as these have the potential to accumulate environmental DNA from a much more diverse swath of an ecosystem, and typically have much more continuous depositional records.

Lakes and Swamp Areas

From a Quaternary science perspective, lakes function as natural traps of minerogenic and organic matter from both terrestrial and aquatic ecosystems. This includes species from the overall catchment area (where streams and rivers carry eroded sediments and organic matter to the lake basin), along with the broader surrounding region (i.e. wind-transported pollen), as well as from the lake ecosystem itself and upstream rivers (i.e. aquatic organisms). Lakes are generally a focal point of human settlements and other animals, serving as a source of drinking water along with many other biotic and abiotic resources (Coolen and Gibson 2009; Pedersen et al. 2015; Parducci et al. 2017; Capo et al. 2021; Brown et al. 2022a). Lake sediments have therefore been used as a reliable sampling source for sedimentological, geochemical, and micro/macro-fossil based palaeoecology, palaeoenvironmental, and palaeoclimatological studies. Analysing specific molecular compounds (Dubois and Jacob 2016), especially sedDNA (Capo et al. 2021), has gained broad interest in recent years as an analytically powerful and complementary approach to help augment the datasets derived from other more traditional proxy methods in Quaternary science. Given the overall ubiquity of lakes in ecosystems with abundant animal life, lake sedimentary ancient DNA research has become the largest subfield within palaeo-environmental DNA (Rawlence et al. 2014; Von Eggers et al. 2022).

Compared to other site types, lake sediments have several benefits for sedDNA research. First, the chronology is often very well conserved (i.e., continuously increasing dates along the sediment depth without age reversals or gaps) due to the sedimentation processes, which are partly controlled by seasonal cycles (lake biogenic/bio-induced production and detrital clastic and/or organic inputs) and the anoxic condition rapidly developing in the deeper sediments. Furthermore, based on sedimentological/geochemical studies, potential reworking, sedimentation gaps or specific deposits related to instantaneous events of geological or climatic origins (such as volcanic eruptions, floods, earthquakes, avalanches) can be identified (Sabatier et al. 2022), and thus taken in consideration for sampling. Geomorphological processes, such as frost-heaving, cryoturbation, fluvial erosion, and redeposition, that disturbs other terrestrial sediment settings, are typically restricted or at least identifiable for lake sediments. Second, owing to the aforementioned factors and relatively high depositional rates, lake sediment profiles usually have long and continuous depositional histories, which can be accurately dated using a variety of chronological techniques. Third, lake sediments have the potential to integrate eDNA from relatively large areas, and potentially from a broader diversity of organisms relative to terrestrial samples, which can theoretically contain DNA from a smaller set of local species. Lake sedDNA records of terrestrial organisms by contrast may originate from the whole catchment, or only parts of it, depending on the hydrographic connectivity of the lake, along with the extent, magnitude, and erosional characteristics of the catchment area (Giguet-Covex et al. 2019). Fourth, lake sediments are minimally (if at all) affected by leaching processes. Although the remobilisation of old sedDNA and transfer to lake sediments through erosional processes are possible, lake sedDNA researchers make the reasonable assumption that this contribution should be relatively limited in comparison to the much greater biomass turnover produced by the organisms living in the catchment area at the time of sedimentary deposition. Finally, DNA molecules in lake sediments can be well preserved for long periods of time due to anoxic conditions that limit microbial processes involved in organic decomposition. For all of these reasons, lake sediments are often excellent archives for recording whole ecological eDNA from a regional catchment, allowing for the reconstruction of past changes in biotic composition with fine temporal resolution and over potentially long time-scales. Because mammals do not generally leave visible fossil remains in lake sedimentary archives, lake sedDNA analysis is a unique way to identify terrestrial animal DNA without any visible morphological remains.

The first study to successfully recover mammal DNA from lake sediments was conducted in the French Alps at a site above 2000 meters above sea leave (m asl) (Lake Anterne; Giguet-Covex et al. 2014), which followed three years after the first study on lake sedDNA studying plants (Anderson-Carpenter et al. 2011). Giguet-Covex et al. (2014) utilized a DNA metabarcoding approach (which will be detailed in the subsequent methods section of this chapter), to determine when farming activities intensified around the lake. Giguet-Covex et al. (2014), for the first time, were able to reconstruct the past composition of domestic animals from studied areas (which has poor bone preservation). The identification of abundant sheep and cattle DNA from the Late Iron Age and Roman Periods of the lake deposits, and later from the High Middle Age to the Early Modern Period (ca. 1000–1800 CE) were validated with complementary data sources. More specifically, zooarchaeological remains from a cabin dated to the Late Iron Age/Roman Period on the edge of the lake confirmed the mammal sedDNA data with sheep bones. Furthermore, the second phase of intense pastoral activities at the site was characterized by a change in animal composition (~12th–13th centuries CE) with the development of cattle herds and the decrease in sheep flocks—a shift also recorded in local historic archives. Whether their instances of low animal sedDNA detection were truly coeval with their suspected time periods (e.g. 1 sample with 2/8 positive sheep replicates in the Late Neolithic and 2 samples with 1/8 positive replicates dated to the mid-Bronze Age), or were the result of temporal mixing or contamination, remains an open question (Giguet-Covex et al. 2014). *Plantago* DNA, increased erosion, and other independent lines of evidence support the temporal authenticity of these results for the Bronze age, but such low detection rates are generally difficult to interpret with sedDNA datasets.

These results, combined with poor mammal DNA detection relative to plant DNA that has been reported in other lake metabarcoding studies (Giguet-Covex et al. 2019; Lammers et al. 2019), highlight the complexity to routinely recover faunal sedDNA from lake sediments compared with other terrestrial site types. This difficulty can be attributed to the depositional characteristics of lakes as being secondary contextual sources, relatively low biomass turnover of animals compared to other organismal groups such as plants and microbes (Fig. 11.3), in addition to other taphonomic issues specific to animals—especially their high mobility across a catchment region. This mobility is controlled by animal behaviour and human management strategies for domestic animals, hence increasing the chances of finding dense herds of domestic animals relative to wild fauna in low population densities. The presence of latrines, trails, and enclosures concentrating eDNA (Andersen et al. 2012), and the proximity of the source relative to the sink (i.e. how close and how many animals came to the lake) (Graham et al. 2016; Brown et al. 2021), are all factors that affect animal eDNA deposition in lakes (Giguet-Covex et al. 2019).

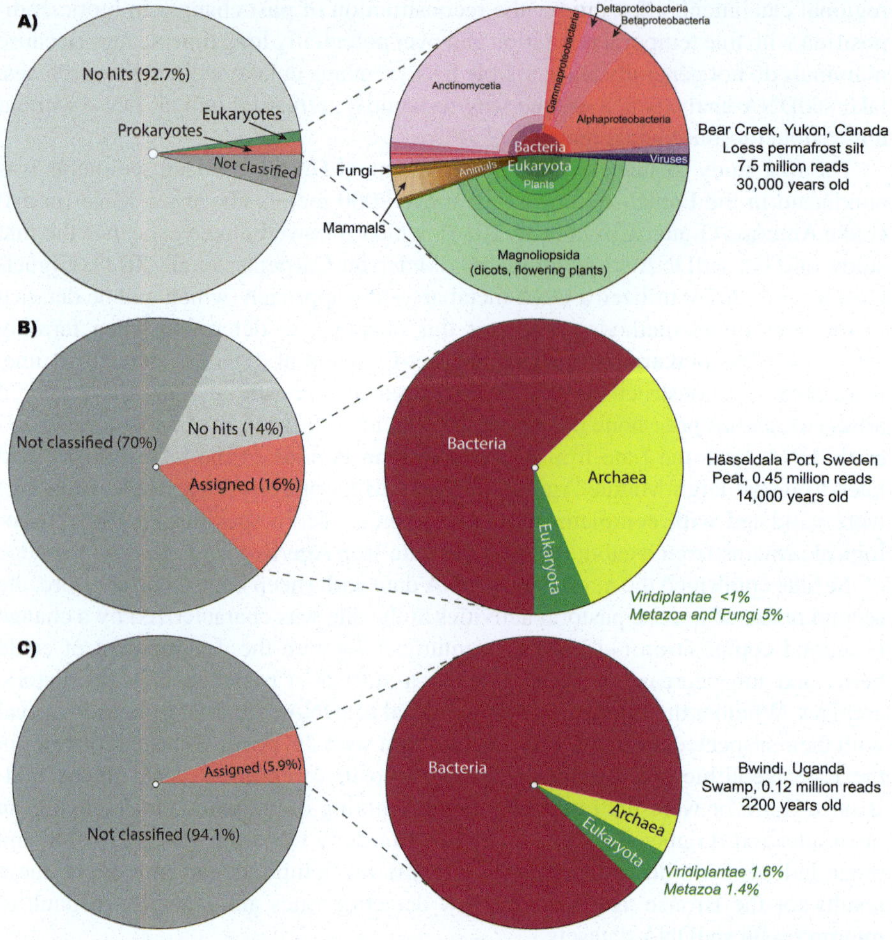

Fig. 11.3 Radial chart example of metagenomic classifications in shotgun sequenced sedDNA samples reprocessed from (**a**) Murchie et al. (2021b), (**b**) Ahmed et al. (2018), and (**c**) Dommain et al. (2020)

Methodological issues related to contamination may also contribute to the difficulty in detecting animal eDNA in lake sediments. Indeed, mammal eDNA is generally considered rare in sediment, not only due to their low biomass relative to plants (Fig. 11.3) and their differential mobility, but also in metabarcoding studies mammal eDNA is compounded with contamination from non-degraded human and domestic animal DNA that is thus preferentially amplified . This is the case even if human-blocking primers are used during the amplification process (Giguet-Covex et al. 2019). Increasing the number of PCR replicates in metabarcoding assays often aids in rare mammal eDNA detection and assessing modern contamination (Ficetola et al. 2015). As such, while lake sediments are often ideal for preserving the eDNA of plants continuously and with a secure stratigraphy, eDNA from terrestrial animals in lake basins (particularly wild animals with low population densities) remains a challenging research objective. Ongoing optimizations in site selection, paired with methodological innovations, are helping to improve animal lake sedDNA research.

In another small subalpine lake from the Swiss Alps (Lake Sulsseewli, 1921 m a.s.l.), wild animals such as ibex, chamois, and red deer were detected for the first time with lake sedDNA (Garcés-Pastor et al. 2022). This study also used a metabarcoding approach, for which the detection of wild animal signals is generally rare (Giguet-Covex et al. 2014, 2019; Bajard et al. 2017, 2020; Sabatier et al. 2017). Interestingly, however they used a newer DNA extraction procedure (i.e. the DNeasy PowerSoil kit [Qiagen, Germany] versus a NucleoSpin Soil kit [Macherey-Nagel, Germany] paired with a custom phosphate buffer) along with human-blocking primers. It is difficult currently to determine whether this improved detection of wild animals with sedDNA metabarcoding is due to site specific differences in land use, animal densities, erosion, or taphonomic processes between sites, or if this improved signal is specifically methodological, or some combination thereof. But it does highlight a growing ability to reconstruct the faunal spectrum of an area through lake sediments.

Messager et al. (2022) recently published a study on lake sedDNA from the Alps at a large (5.5 km^2) and deep (71 m depth) perialpine lake with a large catchment area (58.9 km^2). The authors found a significant signal from cattle and pig DNA, which would generally be identified as a common contaminant signal. Here, however, they argued that these are authentic signals because of their strong correlation with the associated record of *Sporormiella* sp. spores, a widely used fungal proxy for estimating herbivore abundance (Richardson 1972; Comandini and Rinaldi 2004; Davis and Shafer 2006; Raper and Bush 2009; Ejarque et al. 2011; Etienne et al. 2013; Baker et al. 2013; Conroy et al. 2020). The strong concordance between the number of positive PCR replicates and the accumulation rate of *Sporormiella* sp. spores is suggestive that their mammal DNA record may contain quantitative information related to animal biomass changes through time (Fig. 11.4). The debate regarding estimating past animal abundances with sedDNA is a topic we will come back to at the end of this chapter.

Two major lake sedDNA studies were published in 2016 with animal eDNA as a primary target, both highlighting the potential of animal eDNA preservation and recovery from lakes. Graham et al. (2016) specifically targeted woolly mammoth sedDNA to determine the timing and environmental drivers associated with the extirpation (local extinction) of mammoths on St. Paul Island, Alaska, in the Bering Sea. Using several ecological proxies from the island (lake sedDNA, diatoms, radiocarbon dated bones, dung fungal spore abundance, pollen grains, magnetic susceptibility, and estimates of island size/sea level), the authors were able to robustly estimate the timing when woolly mammoths disappeared from the island about 5600 ± 100 years ago. Their data clearly indicate that a synergistic combination of shrinking island size and a decline of freshwater resources due to a drier climate drove the local extinction of isolated woolly mammoths on this island. They were able to detect mammoth sedDNA in all of the lake core subsamples older than

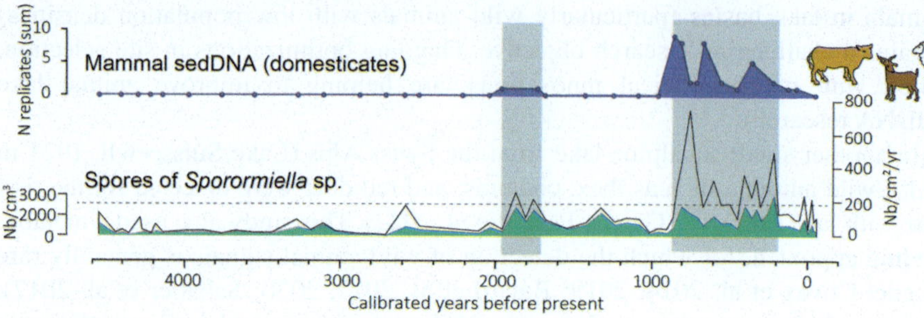

Fig. 11.4 Mammal sedDNA versus Sporormiella sp. abundance (a coprophilous fungi specific to herbivores) in Lake Gers in the Northern French Alps. Data from Messager et al. (2022). Mammal DNA is expressed as the sum of positive PCR replicates (with N reads >5) of the different animals detected (sheep, goats, and cattle). Sporormiella spore counts are expressed in volume concentration (count by cm^3) and in accumulation rate (count by cm^2 per year). The light blue areas highlight the phases of mammal DNA detection. The difference in the number of positive replicates between the two phases is reflected in the accumulation rate but not in the concentration of Sporormiella sp. spores. The dashed black line shows the potential threshold in this accumulation rate above which mammals are detected by DNA

the extinction, and a lack of mammoth DNA in samples that postdate the disappearance of mammoths from the island. This study underscores not only the analytic power of multiple overlapping proxies, but more directly of relevance to this chapter, that animal eDNA can be consistently found in lake sediments under the right conditions.

In the second major animal lake study from that year, Pedersen et al. (2016) were the first to successfully use a shotgun sequencing approach on lake sediments to recover a rich spectrum of ancient plant and animal DNA. Their goal was to assess the biotic viability for human migration along what would have been a ~1500 km long ice-free corridor route that opened ca. 15,000 years ago in western Canada along the Rocky Mountains. The ice-free corridor has long been considered the probable route ancient peoples took from northeast Asia to enter continental North America through the Bering land bridge at the end of the Pleistocene epoch, 11,700 years ago (Elias et al. 1996; Goebel et al. 2008; Kitchen et al. 2008; O'Rourke and Raff 2010; Raghavan et al. 2015; Heintzman et al. 2016; Potter et al. 2017, 2018; Scott et al. 2018). Based on the shifting detection of lake sedDNA from plants and animals (including woolly mammoth, bison, moose, elk, hare, eagle, bat, vole, and several species of fish), they reconstructed the ecological succession in part of the corridor from a closed passage (iced or flooded), towards an open corridor with steppe tundra, shrubs, and eventually the boreal forest. They combined this analysis with several palaeoecological proxies and challenged the ice-free corridor migration hypothesis—arguing that the corridor route was not viable early enough to explain Clovis period sites south of the ice sheets, and that a route along the Pacific coast is more likely. The debate is ongoing as to the route(s) and timings that the first Americans took to enter the "New World" (Potter et al. 2018), but this pioneering lake sedDNA work highlighted how environmental DNA can help reconceptualize and challenge long held models based on traditional buried records.

Most studies of animal sedDNA target sites in cold or temperate areas. For example, work on Arctic islands has found evidence for the introduction of invasive animal species (e.g. sheep and rabbits), with corresponding sedDNA data tracking the environmental impacts of these introductions in terms of changes to erosion and plant community composition (Ficetola et al. 2018; Curtin et al. 2021). In contrast, a notable study also successfully applied a shotgun sequencing approach to sediments from a tropical swamp in Uganda (Dommain et al. 2020). Dommain and colleagues found consistent sedDNA degradation patterns with time at the site (an exponential decline in the amount of DNA, along with an increase in DNA fragmentation and deamination rates with depth and age), as well as detecting both plants and animals—with even higher proportions and sedDNA diversity in older, more minerogenic layers of the deposit. They identified a range of sedDNA identified as nematodes, molluscs, insects, fish, amphibians, reptiles, birds, and mammals from the lake core that covers the past 2200 years. The results from this study are very promising and should encourage further work in tropical areas, which have been poorly studied to date by Quaternary scientists overall, and particularly less so from sedDNA researchers (Von Eggers et al. 2022) due to the supposed unfavourable preservation conditions (i.e. warm and wet).

Because lake sediments are able to record additional climatic, environmental, geological, and biological/ecological information, they are important archives for analyzing in parallel (i.e., on the same sample) changes in palaeoecological and environmental conditions. Importantly, lakes are typically not as significantly affected by dating uncertainties as other terrestrial site types. Rapid innovations in ancient DNA methodologies have been widely exploited by the studies cited above, especially to better understand human-environment interactions, biodiversity trajectories, and animal extirpations/extinctions. Several of these studies also highlight challenges for future research related to the sensitivity of animal sedDNA detection, potentially due to taphonomic issues and false negatives as a result of imperfectly optimized laboratory protocols, as well as absences in references in databases, especially in tropical areas where we find biodiversity hotspots.

Generally, the abundance of terrestrial animal eDNA in lake sediment is much smaller than has been found in caves, as caves are a primary depositional context for that eDNA, whereas lake sedDNA is in a secondary context for terrestrial organisms. The next broadly studied site type we discuss in this chapter splits this difference by targeting high abundance terrestrial eDNA in a primary context, but in an open-air, more ecologically diverse setting (relative to caves): perennially frozen ground or permafrost sediments.

Permafrost and Periglacial Sediments

Permafrost refers to ground that remains frozen (below 0 °C) for two years or more (see Dobinski 2011). Permafrost typically does not exist on the ground surface, but resides below the active layer: the uppermost ground layer that seasonally thaws in the summer and refreezes in the winter. Active layer thickness varies with location and season, and can be as little as ~10 cm thick or as much as several meters. Permafrost extends below this active layer to the depth where local geothermal heat and the mean annual temperature at the surface reach an equilibrium temperature of 0 °C. As such, permafrost can be as little as ~1 m thick, or thicker than 1000 m. The geographic continuity of permafrost varies from continuous in high-latitude/altitude areas, towards discontinuous and isolated patches in more southerly regions of the Holarctic (areas north of the Tropic of Cancer [23°27′ N]) (Fig. 11.5). 'Periglacial' refers to environments near glaciers where freezing and thawing significantly modify the landscape and characterised by features such as ice wedges, frost heaving, pingos, and polygon wedges (French 2007). Periglacial areas very often overlap with zones of permafrost, and as such, sedDNA research in these areas necessitates an understanding of the local geomorphological landforms that have been created by the periglacial conditions.

It has been repeatedly observed that cold and frozen conditions (especially perennially frozen ground) allows for optimal aDNA preservation over exceptionally long timescales (Smith et al. 2001; Mitchell et al. 2005). Permafrost conditions have helped preserve aDNA for more than 1 million years in bone (van der Valk et al. 2021) and eDNA has been found in ice cores as old as 450,000–800,000 years (Willerslev et al. 2007), in 1 million-year-old marine sediments (Armbrecht et al. 2022), and most recently, in sediments from Greenland dated to ca. 2 million years ago (Kjær et al. 2022).

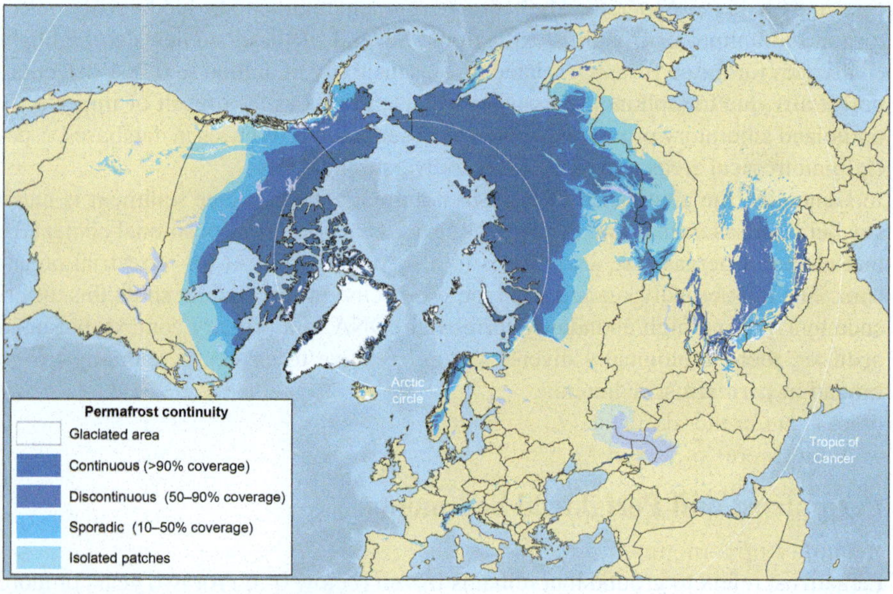

Fig. 11.5 The estimated extent of permafrost in the Holarctic using data from Brown et al. (1997, 1998, 2002; Heginbottom et al. 1993)

Permafrost was the first site type, alongside caves (Hofreiter et al. 2003), where animal sedDNA was successfully recovered and sequenced (Willerslev et al. 2003). Pioneering work from Willerslev et al. (2003) targeted mammalian mitochondrial genes (16S, 12S, cytochrome b, and the control region, alongside plant chloroplast genes) from permafrost cores dating between the Middle Pleistocene and Holocene, which were collected along the Arctic coast in Siberia. They used a combination of PCR and bacterial cloning with Sanger sequencing (reading single DNA molecules) and were able to successfully sequence aDNA from as many as 7 species of Holarctic fauna from 2 grams of permafrost sediments, including woolly mammoth (*Mammuthus primigenius*), horse (*Equus caballus*), lemming (*Lemus lemus*), hare (*Lepus europaeus*), bison (*Bison* spp.), musk ox (*Ovibos moschatus*), and caribou/reindeer (*Rangifer tarandus*). By also targeting chloroplast loci, Willerslev et al. (2003) were able to roughly track the transition from mammoth-steppe flora and fauna (herbs, grasses, and grazing megaherbivores) during part of the Middle and Late Pleistocene and their replacement with Holocene ecosystems.

Permafrost sediments are ideal for sedDNA research not only because frozen conditions significantly slow down the biological, chemical, and physical degradation of DNA molecules, and because the eDNA is in a primary context, enabling the recovery of genome-scale, whole ecosystem data, but also because permafrost functions as an aquitard/aquiclude (Woo 2012). Frozen sediments inhibit the downward migration of DNA in sedimentary strata (referred to as leaching) (Haile et al. 2007). This frozen barrier to liquid migration reduces mechanisms that can downwardly displace DNA from high eDNA outputs such as urine/faeces or the decomposition of a carcass. The often exceptional *in situ* preservation of permafrost sedDNA has many benefits, but such favorable preservation conditions in an open-air setting also presents the possibility of sedDNA remobilization if those sediments (and their molecularly bonded sedDNA) are reworked due to processes such as frost-heaving, cryoturbation, fluvial erosion, and other geomorphological processes that may move and mix old sedDNA into younger deposits (Arnold et al. 2011). However, this potential for redeposition mainly affects high energy fluvial settings, for which the deposition processes are more complex, and which can be observed through careful stratigraphic analyses to identify disturbances. Such remobilization events are usually evident by reversals and/or discontinuities in the age-depth chronologies, along with other geological measurements (i.e. magnetic susceptibility) of the sediment profile. Such datasets can therefore be utilized to aid in interpreting sedDNA reconstructions (Wang et al. 2021). Nevertheless, this potential for remobilization highlights the importance of an in-depth taphonomic and depositional assessment of the sedimentary profile during sedDNA sampling.

With careful site selection and keeping the above potential pitfalls in mind, faunal sedDNA has been able to help robustly assess palaeobiogeographic range constrictions of now extirpated fauna. Haile et al. (2009) was the first to demonstrate the utility of permafrost sedDNA for reassessing the late survival of Pleistocene megafauna. Haile and colleagues expanded on the study by Willerslev et al. (2003) with similar sedDNA methods, but while using those techniques to target a site in Alaska (Stevens Village) consisting almost entirely of loess (windborne sediments of sand and silt) during the Pleistocene-Holocene transition (ca. 11,700 years ago). Haile and colleagues used extensive on- and off-site control sampling to monitor for exogenous megafaunal eDNA that may be present in the environment. They found sedDNA sequences of woolly mammoth and North American caballine horse as young as 10,500 years old—species previously thought to have been extirpated (underwent a local extinction) from Yukon and Alaska between 15,000–13,000 years ago. Haile et al. (2009) exemplified the potential of sedDNA for expanding our understanding of palaeo-ecological turnover in the Holarctic, and the means by which confidence could be built in sedDNA's stratigraphic authenticity.

In 2021, Wang et al. (2021) published the largest permafrost sedDNA study to date, in which they shotgun sequenced 535 samples from 74 sites across the circumpolar. By utilizing such a large dataset, Wang and colleagues were able to reconstruct Arctic macro-ecological evolutionary histories over the last 50,000 years. Among their findings, they provided evidence for the late survival of many megafauna species, including woolly mammoths on the Arctic continental landmass until 3900 years ago, woolly rhinoceros until 9800 years ago, caballine horses in Alaska and Yukon to 7,900 years ago, and bison to as late as 6400 years ago in high-latitude localities of Siberia. Their data suggests that there were substantially later surviving populations of these species—sometimes by as much as several millennia—compared to previous estimates of their extinction and extirpation timings, which were modeled based on radiometrically dated bones, tusks, and megaherbivore dung fungal proxies (Burney et al. 2003; Comandini and Rinaldi 2004; Davis and Shafer 2006; Gill et al. 2009, 2012, 2013; Feranec et al. 2011; Etienne et al. 2013; Wood and Wilmshurst 2013; Stuart 2015; Monteath et al. 2021). These findings highlight the power of sedDNA for understanding the spatial chronology of fauna species, particularly for small remnant populations that may be palaeontologically invisible due to the Signor-Lipps effect (Signor and Lipps 1982) (Fig. 11.2). It is important to note, however, that even these very late populations may not mark their final extinction, as species could survive in small refugial areas well after their remains are last detected in macro/micro-fossil records. As previously discussed in this chapter, continuously disseminated eDNA with an extraordinarily high output presents the possibility of detecting lingering refugial populations where single skeletons may not ever preserve, or be found.

Beyond taxonomic identifications, Wang et al. (2021) also developed methods to place fauna species identified in shotgun sequencing sedDNA data onto phylogenetic trees based on mitochondrial haplogroups, and found that sedDNA and fossil-based phylogenies are highly consistent, both spatially and temporally. This further demonstrates the potential of metagenomic and sedDNA approaches to access novel information beyond presence/absence detection. With optimized methods and databases, such datasets can also be applied to investigate population dynamic histories. In addition to these, Wang and colleagues developed methods for modelling the broad distribution patterns of different species at varying scales based on their sedDNA data, along with their spatiotemporal integration with other complementary datasets such as palaeoclimate, archaeology, vegetation compositions, and megafaunal distributions. Through these modeling analyses, they argued that climate change and associated vegetation turnovers were likely the main drivers in the loss of Arctic megafaunal species in the late Quaternary, rather than as a result of human activities. This highlights another benefit of using sedDNA data as input for ecological modelling, since the different type of data that are required for those ecological models (for example palaeoclimate, floral and faunal compositions and functions, and ancient human distributions) can now all be reconstructed from the eDNA in a single sediment sample. This has the potential to significantly reduce the uncertainties (and potential errors) that are introduced by combining different palaeoecological, geophysical, palaeoclimatic, biogeographical, and archaeological datasets into the same ecological model, which are each generated or estimated through very different methods with their own sets of incongruent assumptions and differing spatiotemporal scales.

On a much smaller sedDNA scale, Murchie et al. (2021a, b, 2022) published a series of papers in parallel with Wang et al. (2021) working with permafrost sedDNA from four nearby sites in central Yukon, Canada, in an effort to reconstruct the ecological collapse of the mammoth-steppe. In their first paper of the series, Murchie et al. (2021b) tested multiple extraction protocols in order to increase the yield of plant and animal aDNA recovered from environmentalsamples. They found that it was important to find a balance between fully purifying a sedDNA extract through the removal of enzymatic inhibitor compounds common in open-air sediments (such as humic substances [partially degraded organic matter], which inhibit the attachment of sequencing adapters to aDNA during library preparation) and aDNA-specific methods designed to maximize the recovery of fragmented and damaged DNA. Total adherence to either extreme had drawbacks, highlighting the need for project-specific optimizations during demineralization/digestion and DNA purification to improve downstream recovery while minimizing the retention of inhibitory compounds. They also found that a targeted capture approach significantly improved the fraction of ecologically informative sequence data over PCR metabarcoding and shotgun sequencing. Finally, Murchie et al. (2021b) observed a similar late survival sedDNA signal as identified by Haile et al. (2009) for woolly mammoths and North American horses persisting until around 10,000 years ago on the mainland of eastern Beringia (Yukon and Alaska) despite local shrubification.

Murchie et al. (2021a) followed by analyzing the palaeoecology of central Yukon using additional samples from their four sites spanning 30,000–4000 years ago. They found: 1) a rich taxonomic breadth of plant and animal sedDNA that was temporally consistent across study sites within the region (taxa including grazing and browsing herbivores, small-to-medium mammals, rodents, birds, canids, felids, and bats); 2) a coeval turnover of grazing megafauna and mammoth-steppe flora (herbs and grasses) 'abruptly' transitioning towards woody shrubs and browsing/forest-dwelling fauna; 3) longer-term declines in megafaunal sedDNA preceding the environmental and vegetation transition, suggesting either taphonomic changes to eDNA preservation, or shifts in animal abundance in the region; and 4), a consistent signal of late megafaunal survival, postdating the disappearance of horses and woolly mammoths in the local fossil record by as much as 7000 years. Their abundance of faunal DNA allowed for assessing DNA fragmentation rates and deamination patterns to demonstrate that these sequences originated from ancient sources and not modern contaminants. Unlike many of the cores studied by Wang et al. (2021), the permafrost cores studied by Murchie et al. (2021a, b) were not originally collected for sedDNA analysis, as such, the authors warned about the possibility of decontextualized eDNA (Arnold et al. 2011). Murchie et al. (2021b) provided several justifications for stratigraphic authenticity of their late survival signal—e.g., thermal-age damage patterns through time, consistency across study sites, the lack of mixed plant signals, incomplete macro-fossil records, and similar findings as Haile et al. (2009), Wang et al. (2021), and late persistence on Arctic islands (Vartanyan et al. 2008; Graham et al. 2016). Nonetheless, they cautiously concluded that more research was needed to assess whether Haile et al.'s (2009) original late survival signal can robustly be spatiotemporally extended throughout Beringia and later into the Holocene.

In their third paper utilizing these samples, Murchie et al. (2022) explored the mitogenomic potential of faunal sedDNA from open-air permafrost contexts for reconstructing the evolutionary histories of individual species using maximum likelihood and Baysiean phylogenetics with mitochondrial genomes reconstructed solely from sediment. These were analyzed alongside other mitogenomes reassembled from ancient bones—a technique also utilized by Wang et al. (2021), Slon et al. (2017), Pedersen et al. (2021), and Gelabert et al. (2021). They observed two distinct haplogroups in their data originating from Eurasian and North American lineages of both woolly mammoths and steppe bison from the same core samples. In total, they were able to reassemble 18 ancient mitochondrial genomes from their permafrost sediments, including four from the small ground dwelling grouse, willow ptarmigan (*Lagopus lagopus*). Further, through deeper sequencing and uracil-DNA glycosylase (UDG) treatment to correct for cytosine deamination (Dabney et al. 2013), they were able to improve their faunal metagenomic resolution and use a Bayesian phylogenetic approach to accurately estimate the age of a 30,000 year old woolly mammoth mitogenome reconstructed from sediment. This ageing technique was less successful with their other samples that had not received deeper sequencing or UDG treatment, challenged by the need to disentangle inputs from diverse haplogroups of the same species (among many other mixed species) within individual DNA libraries. This third paper highlighted the genomic potential of sedDNA from open-air sediments for improving our understanding of past genetic diversity by augmenting databases of ancient genomes reconstructed from bone with those reassembled from sediment. Their work also demonstrated the potential of sedDNA for recovering genomes from understudied organisms whose bones are rarely preserved in fossil records.

Finally, two recent studies have massively extended the temporal range of sedDNA research. The first from Courtin et al. (2022), utilized sediments from western Beringian (Siberia) dating between ~23,000 and ~650,000 years ago to reconstruct the palaeoecology during two previous inter-glacial periods, MIS 17 and MIS 7 (Railsback et al. 2015). The authors used a combination of PCR metabarcoding and shotgun sequencing to evaluate their total metagenomic dataset, including plants, fungi, mammals, birds, insects, and prokaryotes (Bacteria, Archaea, and Viruses). The metagenomic data confirmed other lines of evidence that previous interglacial periods during the Middle Pleistocene (1.25–0.13 million years ago) (Clark et al. 2006) were forested with open grasslands but also were inhabited by megafauna like giant ground sloths, bison, and camels. The Late Pleistocene (130,000–11,700 years ago) samples recovered a sedDNA faunal spectrum typical of other parallel Holarctic research, such as reindeer, bison, horses, and mammoths. Unlike Wang et al. (2021) or Murchie et al. (2021a, b, 2022) who did not investigate their insect (or microbial) data in their original publications, Courtin et al. (2022) report a variety of insects including mosquitoes, crickets, moths, butterflies, beetles, mayflies, wasps, termites, and grasshoppers. The authors also found a range of bird sedDNA, including tits, barbets, warblers, sandpipers, rails, eagles, falcons, cranes, pigeons, and ground-dwelling birds. The reconstructed faunal spectrum changes in concert with expected ecological shifts between glacial and interglacial periods from steppe grasslands to woody forests and back again.

Courtin et al. (2022) also discussed the theoretical issue of highly curated reference databases to filter out supposedly ‘nonsensical’ (false positive) taxa. As sedDNA research reaches further back in time, our ability to discriminate between which organisms are sensical or not becomes increasingly problematic and can result in abundant false negatives and biased reconstructions. This problem was also discussed by Murchie et al. (2021b) in their attempts to disentangle which taxonomic hits were false positives (due to database biases and errors in reference sequences, inaccurate alignment/binning algorithms, DNA damage, etc.) or were simply previously unknown extended ranges of organisms that have yet to be identified with other palaeoecological proxies. Courtin et al.’s (2022) work underscores the massive palaeoecological potential of sedDNA for whole palaeoecosystem reconstruction—drawing attention to the need for more holistic efforts to track the coeval transition of organisms from all domains of life to better comprehend the interdependent network of variables driving change in the biosphere, past and present.

The last permafrost paper we will discuss here pushes the temporal needle of sedDNA set by Courtin et al. (2022) back far beyond what was thought possible (Allentoft et al. 2012). Kjær et al. (2022) were able to recover sedDNA from permafrost sediments in Greenland dating to approximately two million years ago. The authors assessed DNA preservation at the site and found that the mineralogy of the geologic formation they were targeting had high extracellular DNA absorption capacity. Their modeling of DNA depurination (fragmentation) based on the conditions at the site suggested that sedDNA preservation was plausible. This was confirmed using a molecular clock to age-estimate *Betula* (birch) chloroplast DNA that could be reassembled from the pool of sedDNA thanks to its high abundance. This molecular clock method tracks the number of accumulated nucleotide differences found in a reassembled genomic sequence relative to other samples of known ages given a specified rate of mutation (Drummond and Rambaut 2007; Lemey et al. 2012; Suchard et al. 2018). This analysis resulted in a molecular clock age estimate of 1.323 million years ago for their birch sedDNA, with a 95% highest probability density (HPD) ranging between 0.6786 and 2.0172 million years. This analysis falls in line with their complementary geologic age estimates, allowing Kjær et al. (2022) to independently age the sedDNA.

Kjær et al. (2022) were further able to taxonomically identify a rich sedDNA signal of plants and animals from the target deposits, reconstructing an open boreal forest ecosystem with a mix of shrubs and herbs, including many species that had not been previously detected with macrofossil or palynological evidence. With their plant sedDNA data, 39% of the taxonomically identified species no longer grow in Greenland, with most requiring far warmer temperatures and non-permafrost substrates that are found nowhere in Greenland today. This aspect underscores the difficulty of attempting to assess false positive/negative taxonomic calls from sedDNA data when the ecosystems being reconstructed are so far removed from our already incomplete understandings of palaeobiogeography.

Kjær et al. (2022) also identified mastodon (*Mammut* sp.) sedDNA and Bayesian age estimated their reassembled mitochondrial genome to be either between 191,000 and 3.27 million years old, or 1.64 and 10.1 million years. Mastodon remains have never been found on Greenland. This is in part due to ~80% of Greenland being overlain with ice, meaning that current palaeontological records of ancient Greenland are highly fragmented. The woody vegetation identified in their sedDNA supports the presence of mastodons in that sort of ecosystem as they were browsers—herbivores who mainly feed on the leaves and fruits of typically higher growing woody plants, shoots, and shrubs. This is contrasted by 'grazers' that feed primarily on grasses and other lower-lying herbaceous vegetation with a differing set of dentition and gastrointestinal mechanisms. Their mastodon signal is also supported by their co-identified faunal spectrum of reindeer/caribou, geese, and hare, all of which are found in similar boreal environments today. Beyond reconstructing the terrestrial species during this oldest component at the site, they were also able to identify a range of marine organisms. This landmark research serves as a critical example of how not only can sedDNA be used to study ecosystems as old as 2 million years, but that these data allow for the simultaneous recovery of whole ecosystem data that has no other analog with Quaternary proxy methods.

The permafrost studies we have outlined here each highlight how Holarctic conditions can facilitate exceptional sedDNA preservation over immense timescales. The *in situ* context of this eDNA (relative to eDNA in lake sediments) and open-air nature of these sites (relative to eDNA in caves) allows for both a wide spectrum of eDNA from all domains of life in an ecosystem, and can provide a large enough genomic signal to allow for detailed evolutionary analyses. However, as illustrated from Fig. 11.5, permafrost conditions only occur in limited regions of the planet. Furthermore, there are often no other indicators of what sorts of eDNA one may expect to find in a series of permafrost cores. Can we narrow the scope of expected organisms for a more targeted eDNA project, and in non-polar regions? The final site type we will discuss in this chapter are previously known archaeological sites.

Archaeological Sites

To date, eDNA studies associated with typical, open-air archaeological contexts have been surprisingly uncommon. This scarcity of archaeological sedDNA literature may be due to a higher potential for DNA leaching in some of these depositional contexts (Haile et al. 2007), poor preservation when sedDNA is not protected in caves, beneath lakes, or when consistently frozen, the omnipresent risk of human eDNA contamination, ethical concerns about working with ancient human DNA (Handsley-Davis et al. 2021), high lab costs, a lack of interdisciplinary collaborations, along with other potential factors.

To our knowledge, the first sedDNA study in an open-air archaeological context was led by Hebsgaard et al. (2009) in Greenland at a Norse farm dating between 1000–1400 CE in an area without permanently frozen sediments. Hebsgaard and colleagues' goal was to determine if ancient eDNA could be detected from such a context and to explore relative changes in animal sedDNA as compared with bones recovered from the site. The authors collected off-site sediment samples in fields connected with the farm and used general 16S mammal primers (Mam1 and Mam2) along with specific primers targeting pigs, horses, seals, and reindeer. They also applied real-time PCR analyses based on MGB (minor groove binder) probes to quantify sheep, cow, and goat DNA, and assessed the potential of sedDNA for providing a quantitative dataset in terms of past animal biomass. The authors mostly detected the presence of cattle, sheep, and goats in the archaeological layers corresponding to the Norse occupation, along with mice and reindeer in samples associated with the initial occupation. Human DNA was also detected, but mostly in the youngest samples, i.e. after the abandonment of the farm, suggesting that this DNA originated from modern humans—a common issue in mammalian aDNA research (Skoglund et al. 2014; Ushio et al. 2017; Harper et al. 2019; Orlando et al. 2021). Interestingly, the real-time PCR analyses showed a decrease in cattle DNA during the site occupation, which is consistent with their faunal bone record. Hebsgaard et al. (2009) argued that this supports their hypothesis that quantitative information regarding animal biomass can be recovered from sedDNA analyses (a point we will discuss later in the chapter). They also found that the distribution of cattle and sheep DNA at the site (increasing cattle sedDNA with depth paired with a consistent concentration of sheep sedDNA) suggested that no leaching occurred in this context.

Seersholm et al. (2016) likewise conducted archaeological sedDNA research at four midden sites (domestic waste deposits) in Greenland, tracking human ecological shifts in the area over the past ~4000 years from Saqqaq Paleo-Inuit (2500–800 BCE), to Dorset Paleo-Inuit (800 BCE to 1300 CE), to Norse (Viking) occupations (985–1452 CE), and finally Thule Neo-Inuit (1200 CE to present). They were able to observe shifting dietary fauna at the sites from predominantly wild animals (i.e. seals, whales, narwhal, walrus, reindeer/caribou, and arctic hare, but also including wolf/dog) during Paleo-Inuit periods, towards non-native domesticates in the Norse deposits (i.e. cattle, sheep, and goat). In total, they were able to recover sedDNA from 42 vertebrate taxa. They were also able to identify as many as 12 different helminth (parasite worm) species known to infect a variety of the fauna that they genetically identified, such as the tapeworms *Taenia multiceps* and *T. hydatigena* that are parasites of canids, ruminants, and other wild hosts (e.g., reindeer). They found a comparable biodiversity between their sedDNA identifications and zooarchaeologically excavated bones, however they observed that large animals (i.e. caribou, walrus, narwhal, and other whales) were underrepresented in bone records compared with their sedDNA abundances. They suggested that this may be indicative of hard tissues being discarded during initial field butchering, with only high value portions (blubber/fat and meat) being brought back to the living areas. This research highlights how sedDNA can be used to holistically reconstruct human ecologies through time when interpreted with complementary archaeological data. Seersholm et al. (2016) were also able to reassemble (and phylogenetically place) two bowhead whale mitochondrial genomes with relatively high coverage, along with a low coverage assembly of the helminth *Taenia hydatigena*, and provide evidence that sedDNA reads abundance can correlate with both expected animal biomass and excavated bone abundances of certain fauna (i.e., harp seal, ringed seal, birds, and fox). However, no correlations were found with the bone abundance of exceptionally large fauna (i.e., bowhead whale) and domestic animals, as their bones were presumably discarded off-site or have restricted movement and sedDNA dispersal potential.

Brown et al. (2021) utilized a suite of proxy methods to reconstruct a holistic molecular archive of human activity at a Medieval Celtic crannog settlement in Ireland: a site type built on small human-made islands in a lake or wetland system. The author's used a combination of ecological methods including sedDNA, palynology (pollen and coprophilous fungal spores), diatoms, chironomids, lipid biomarkers (bile acid), multiple geochemical indicators, mineralogy, and detailed stratigraphic analyses paired with archaeological data to carefully reconstruct human land-use patterns at the site. Brown et al. (2021) took sediment samples in direct connection with the Lough Yoan crannog (10–15 m into the lake) as a form of non-invasive archaeological sampling to provide a proof of concept for a lake-dwelling site that does not necessitate destructive archaeological excavation. Crannogs are found throughout Ireland (1200 known sites) and Scotland (600 known sites), but these archaeological sites are threatened by modern human activities. Although, in this case, the crannog itself is an open-air archaeological site, the samples they processed originate from adjacent lake sediments, and can thus be considered akin to "classical" lake sedimentary ancient DNA research, that were not likely affected by leaching issues (Giguet-Covex et al. 2014). This study provides the first independently-dated, continuous molecular archive of human activity from an open-air archeological site. Combined with coprophilous fungi spores and mineralogical analyses providing evidence of the presence of apatite (presumably from bones), the animal and plant DNA closely tracks the human occupation through crops stored and processed on site (barley, rye, and flax), animal farming (mostly cattle, sheep, pigs, goats, and horses) and on-site slaughtering (domesticated animals and red deer). The proximity of the core to an open-air archaeological site allowed for an abundance of biogeochemical trace analyses of human activities through time, along with significant quantities of mammal eDNA compared to "classical" lake sedDNA studies that have applied the same metabarcoding approach (same primers and human-blocking primers) (Giguet-Covex et al. 2019; Lammers et al. 2019) and sedDNA extraction method (Giguet-Covex et al. 2019).

Moving away from macro-animals, Oh et al. (2010) studied burial sediments recovered from an entombed coffin in Korea dated to 1755 ± 10 CE with an aim to determine if sedDNA from a parasitic roundworm (*Trichuris trichiura*), one of the most common eukaryotic parasites affecting humans, could be recovered. The authors used microscopy to identify *Trichuris* eggs from the sediments paired with PCRs targeting sedDNA in the small subunit ribosomal RNA (SSU rRNA) gene of *T. trichiura*. The authors were able to detect parasite DNA in 75% of samples containing eggs from *Trichuris* sp., and no parasite DNA was found in the soil surface sample used as a control sample for modern contamination. Eukaryotic parasites were also studied by Søe et al. (2018) who shotgun sequenced sediments from latrines, cesspits, or fill from eight sites across Northern Europe (Denmark, Lithuania, and the Netherlands) and the Middle East (Jordan and Bahrain). By using a combination of flotation to preconcentrate eggs, along with shotgun sequencing, the authors were able to identify sedDNA from a diverse set of helminths (parasite worms) known to infect humans, along with a number of other worms known to be host specific to domestic and wild fauna, including cattle, horses, dogs, cats, sheep, rats, fish, deer, rabbits, and birds. In addition to detecting the sedDNA of at least 11 unique helminth species, they also had sufficient data to reconstruct the mitochondrial genomes of whipworm (*Ascaris lumbricoides*) and two species of roundworms (*Trichuris* spp.), which allowed for the phylogenetic analyses of the worm sedDNA mitogenomes compared against other genomes reconstructed from modern helminths. Microscopy has been found to be an important tool when conducting sedDNA research on helminths, allowing for sediments to be pre-screened for eggs to maximize sedDNA success.

As a final interesting example here that admittedly stretches the scope of this chapter, Mahaney et al. (2017) used sedDNA to track the route that the Carthaginian Army (led by Hannibal) took to cross the Alps during the Second Punic war with Rome in 218 BCE. Hannibal's crossing is one of the most famous historical events of the classical world, although the actual route taken by Hannibal and his elephants through the European Alps remains controversial. One of these debated passes is the Traversette near Mount Viso in Italy, although it is very high in altitude, located at 2947 meters above sea level (masl). In a massive disturbed bed from a mire located just below the pass at 2580 m a.s.l. and dated to the period of the Punic war (2170 cal yr. BP), Mahaney et al. (2017) found the presence of fecal compounds and microbial DNA from Clostridia in higher concentrations than typically observed in other associated soils. Clostridia also grows in the guts of mammals such as horses. Consequently, combined with the detection of faecal compounds, the evidence of a disturbed sedimentary accumulation possibly due to trampling, and the dating of the deposit, the authors suggested that the detection of the microbial DNA could be related with the passage of Hannibal's Army. However, a pastoral origin for the presence of this deposit cannot be excluded, and the authors did not detect elephant, horse, or human sedDNA in these deposits. This ultimately leaves the question of Hannibal's route unresolved. While this study does not resolve its original research question, it does highlight the potential of sedDNA to investigate questions related to human mobility and activity in ambiguous archaeological contexts.

There are several ongoing projects using sedDNA at known, or suspected, archaeological sites that are well underway at this time. We hope that in 10 years' time, an equivalent section reviewing work on archaeological sites such as this will have many examples to highlight the successful application of sedDNA methods to holistically reconstruct human ecological interactions through time. Such an approach has immense promise to reconceptualize the information potential of sediments and soils at archaeological sites, and we look forward to seeing what may come from such a shift. For now, we will pivot our attention away from direct examples of site types and focus instead on various methodological approaches and their associated advantages and disadvantages.

Animal sedDNA Recovery, Targeting, and Sequencing

Along with taphonomic factors, method selection significantly impacts research success. This is probably even more critical for relatively rarer eDNA-releasing organisms such as animals (Fig. 11.3). It has been observed in microbial genetics that differing DNA extraction methods can bias the relative proportions of DNA from bacterial phyla, making the choice of extraction protocol an important aspect of microbiological research design that should vary based on objective (Holmsgaard et al. 2011; Wesolowska-Andersen et al. 2014; Young et al. 2014; Peng et al. 2020; Sui et al. 2020; Sahu et al. 2021). While it is known that factors such as GC bias can impact RNA hybridization during targeted enrichment (due to stronger hybridization of sequences rich in guanine or cytosine bases versus those that are AT-rich; discussed further below and in Chap. 3 of this volume), there has been limited testing as to how differing sedDNA extraction and library preparation procedures may differentially bias recovery of macro-ecological sedDNA (Rawlence et al. 2014). Ineffective extraction protocols are a larger problem across the field of ancient DNA, a problem known to result in substantial information loss (Barta et al. 2014). Each of the many stages of sedDNA processing—such as demineralization and digestion, purification, metabarcoding, library preparation, indexing amplification, capture enrichment, pooling, sequencing, bioinformatic read filtering, reference-based alignment, taxonomic binning, etc.—have a variety of potential protocols, each with their own sets of associated strengths and limitations (Orlando et al. 2021). These stages of processing are discussed in further depth in Chap. 3 of this volume and in other works (Buffalo 2015; Shapiro et al. 2019), but here our aim is to discuss considerations to be made when attempting to recover animal and human sedDNA in particular. In the end, the research question is the critical variable dictating whether to use a targeted (PCR metabarcoding or capture enrichment) or untargeted (shotgun) sequencing approach, along with other factors such as funding, wet-lab/computational resources, degree of authentication necessary, anticipated state of DNA preservation, and previous research.

Here, we first briefly discuss considerations to be made in early-stage sedDNA processing when looking for human and terrestrial animal eDNA. Afterward, we discuss examples from the literature of studies that have successfully applied these main approaches (shotgun sequencing, PCR metabarcoding, and capture enrichment) to sequencing libraries from environmental sources and how these methods impacted the results and conclusions.

Considerations to Be Made in the Lab When Processing Materials for Animal sedDNA

With most environmental sample types, human and other animal sedDNA is likely to constitute the tiniest fraction of metagenomic DNA. Most sedDNA will come from prokaryotes (Bacteria, Archaea, and Viruses), followed by fungi, nematodes, other micro-fauna, and plants. Most abundant of all will be unidentifiable DNA from organisms from which we have no current reference sequences (Fig. 11.3) (Slon et al. 2017; Murchie et al. 2021b). While the eDNA output from macro-animals is enormous, it is comparatively dwarfed by the astronomical quantities of DNA being constantly released into the surrounding environment by other organisms at all scales. As such, there are several considerations to be made during wet-lab processing when the end goal of the analysis is recovering sedDNA from the Kingdom of Animalia.

First and foremost, before even starting the sedDNA component of the project, it should be determined whether there are reference genomes available for the target organism(s), or at least closely related organisms. For example, as of September 2, 2022 there are no publicly available genome assemblies in FASTA format for *Canis dirus* (dire wolf) (although there is raw, unassembled sequence data in FASTQ format on the European Nucleotide Archive (ENA) in Project PRJEB31639 (Perri et al. 2021)). While there are abundant genomic reference sequences available for other canids in the genus, targeting and identifying *C. dirus* sedDNA is significantly limited by this absence when using public reference databases until either ancient genomes are reassembled through a *de novo* approach (Ekblom and Wolf 2014; Seitz and Nieselt 2017; Borry et al. 2021), or by using genomes from other species in the genus for a reference guided assembly, and then critically—the resulting reconstructed genome is made publicly available. Whether a reference genome is available in your database determines how much pre-emptive work is needed to taxonomically classify the target sedDNA, as well as identifying what false positive gaps one should expect in the results.

Second, an important consideration is whether sedDNA of the target organism(s) can be reasonably expected in the site type and region under consideration in the past. And if so, what sedDNA densities might you expect to find in those site types based on that animal's expected biomass turnover, mobility/transience on the landscape, whether their eDNA is deposited/preserved relatively *in situ* or is transported and deposited elsewhere, and so on. These factors determine site selection and targeting strategy, as well as perhaps the need for early-stage taxon-specific PCR assays to determine whether an extract or library is worth processing further or not (depending on the research question).

Third, eukaryotic sedDNA is in low quantities (relative to its initial release), is fragmented into small segments, contains miscoded nucleotides, and may have other forms of structural damage and blocking lesions (Dabney et al. 2013). While many kit-based extraction methods are available for different environmental sample types, almost all are designed for relatively modern eDNA, i.e., large DNA fragments in high abundance with minimal damage. Using kits on ancient samples has proven successful in many studies, but it is common when targeting faunal sedDNA for researchers to experimentally optimize in-house extraction protocols (demineralization/digestion and purification) tailored to the research project (Slon et al. 2017; Ardelean et al. 2020; Murchie et al. 2021b; Wang et al. 2021). These in-house methods are intended to help mitigate the challenge of co-eluted enzymatic inhibitors from sedimentary sources, while still maximizing the release and recovery of sedDNA from organo-mineral complexes (Morrissey et al. 2015). Soil bacteria have been found to have active repair mechanisms for up to half-a-million years (Johnson et al. 2007). When using purification methods that maximally recover larger, undamaged fragments, this can bias the metagenomic spectrum towards undamaged bacterial DNA sequences, much in the same way as modern contaminant eDNA may be preferentially recovered and/or adapter ligated and/or PCR amplified over ancient DNA where enzymatic processes are less efficient. Some kits can only reliably recover DNA fragments exceeding a certain base pairs (bp) size, such as Qiagen's Minelute PCR Purification kit, which has a minimum size recovery of 70 bp. If a purification kit with a high base pair cut-off is used prior to adapter ligation, this can lead to sedDNA loss and a bias towards larger, less damaged molecules (Glocke and Meyer 2017). Sediment type (mineral constituents, pH, grain size, etc.) can also significantly impact eDNA release during demineralization and digestion, which may need to be optimized on a site-by-site basis.

Fourth, false positives (modern and ancient contamination) are difficult to avoid entirely and can originate from desquamated eDNA of the people who had handled or were in proximity to the samples (i.e., in the lab, field, storage site, etc.) as well as people who manufactured the reagents and plasticware used during DNA processing, pets of these individuals (e.g. dogs and cats), animals of economic or scientific interest that are studied in adjacent labs or whose eDNA is common in the area (e.g. mice, fruit flies, chickens, cattle), among other potential sources. In the era of high-throughput sequencing, modern DNA contamination is somewhat of a lesser concern as these reads lack characteristic damage patterns and may originate from nonsensical taxa. With millions of sequenced reads, these contaminants can theoretically be identified and excluded from the dataset. But with small read counts from disseminated genetic sources and rare species, identifying deamination with software such as mapDamage (Jónsson et al. 2013) is not possible. It can also be difficult to confidently rule out 'nonsensical' taxa from ancient eDNA when our palaeobiogeographic knowledge is incomplete.

Fifth, after extracting sedDNA, one needs to decide whether or not to apply PCR metabarcoding the sample (contingent on whether the used barcodes provide a sufficient resolution for the research question); along with deciding which library preparation method is most suitable to their extracts. Double-stranded DNA methods (Meyer and Kircher 2010; Kircher et al. 2012; Carøe et al. 2018) are widely used, but can be somewhat time consuming and expensive, and are theoretically vulnerable to the loss of abundant single-stranded sedDNA molecules. Double-stranded DNA-targeting kits, such as the NEBNext® UltraTM II DNA Library Prep Kit (New England Biolabs, Ipswich, Massachusetts), should be used with caution as they can utilize uracil repair/excision as part of end repair or adapter ligation/clean up. Ancient DNA contains miscoded uracils that are useful for authentication. Kits that excise adapter uracils can inadvertently bias a library towards fragments without deamination. Uracil-DNA Glycosylase (UDG) treatment likewise removes deamination, which can be beneficial when calling consensus sequences in positions with single nucleotide polymorphisms (SNPs), but will also remove means of authenticating that the targeted molecules are ancient. Single-stranded methods such as ssDNA2.0 (Gansauge et al. 2017, 2020) and SCR (Kapp et al. 2021) are utilized much more frequently now in ancient DNA research as they have been reported to increase yields, although they come with their own associated challenges, such as increased costs (ssDNA2.0), and abundant adapter dimers (SCR).

Finally, sixth, deciding whether to apply shotgun sequencing to a library or utilize a targeted approach is an important question. The remainder of this section will discuss the three main sedDNA approaches—Shotgun metagenomics, PCR metabarcoding, and capture enrichment—in relation to their successful application to animal sedDNA research.

Shotgun Metagenomics

The most direct approach to sequencing eDNA from sediment samples is shotgun metagenomics, sometimes also referred to as shotgun sequencing, or simply metagenomics. Instead of PCR-amplifying short portions of marker genes as with metabarcoding, shotgun metagenomics randomly sequences all the DNA extracted from a sample as a pool, without any taxa-specific pre-clustering or amplification. The approach can be thought of analogously to sampling water from a pond, with the pond water being the sedDNA that was library adapted from a sediment sample in this analogy. A shallow shotgun sequencing approach is akin to using a cup or small bucket to remove water from the pond to analyze its molecular constituents. A deep shotgun approach involves sampling much larger portions, or the entirety, of the pond. How deeply one needs to sequence depends on the molecular complexity of the pond, the size of the pond, how common your sequences of interest are, and what you want to know in the end.

Shotgun metagenomics offers several advantages. First, without taxa-specific PCR amplification, the potential biases introduced by different PCR primer binding efficiencies are eliminated. Shotgun metagenomics as such allows for an improved quantification of the identified taxa. Second, whole genomic DNA fragments from a multitude of species are sequenced simultaneously, providing more opportunities for high resolution taxonomic calls, which is particularly useful in differentiating closely related organisms. Also, by using whole genome data, it provides higher sensitivities for detecting rare taxa when compared with targeting short marker genes. This is important when analysing animal sedDNA, since their eDNA is usually in very low concentrations compared to the eDNA of plants and microorganisms in sediments (Fig. 11.3). Shotgun metagenomics also allows for the analysis of functional genes and genetic structures through genotyping, making it possible to resolve phylogenies and evolutionary histories. Thirdly, shotgun metagenomics recovers DNA from all taxa, from virus, bacteria, fungi, protists, to plants, animals, and humans in a single process, making it the most holistic approach, extending the scope from taxonomic composition of certain taxa groups to comprehensive ecosystem reconstructions. Lastly, shotgun metagenomics also allows the separation and authentication ancient sedDNA signals from modern contaminants by modelling aDNA damage patterns of fragmentation and deamination (base misincorporations) (Dabney et al. 2013; Kistler et al. 2017).

Despite these many advantages, shotgun metagenomics also comes with several important limitations. First and most importantly, it requires whole genome (or whole-genome level) reference databases for aligning the eDNA reads, which are not available for most organisms. Incomplete reference databases result in a substantial number of false negatives, along with a higher chance of false positives since the read would otherwise be aligned onto the genome of an incorrect taxon that shares homologous genetic regions. This is, however, normally not a problem when targeting common fauna (e.g., horse or human), since high quality genomes already exist. This issue is most pronounced when studying whole ecosystems where most organisms have been understudied. The bioinformatic processing and analysis of shotgun metagenomic datasets are also more complicated, and usually require much more computing power compared with targeted approaches. A pool of DNA sequences can contain thousands to millions of different taxa, which ideally requires a comprehensive reference database that contains genomes of those thousands to millions of different organisms in order to minimize bias and alignment errors. Aligning an eDNA dataset against such a massive database can take months on a typical high performance computing platform (HPC) if not using faster alternative approaches for aligning metagenomes, such as k-mer based approaches like kraken2 (Wood et al. 2019) and CCMetagen (Marcelino et al. 2020). A notable drawback, however, is that these approaches normally further increase the error rate. These computational trade-offs, combined with high sequencing costs as a result of needing to sequence very deeply in order to find rare animal sedDNA, results in shotgun metagenomics being very costly when compared to other sedDNA approaches.

Despite shotgun sequencing being commonly utilized in modern environmental settings for studying microbiota, metagenomics is still in its infancy stage for sedDNA. Its applicability relative to that of metabarcoding for fragmented and degraded ancient eDNA remains largely unexplored. Pedersen et al. (2016) published the first sedDNA metagenomics study in which both floral and faunal communities were successfully reconstructed through aligning the shotgun sequenced reads against the NCBI nucleotide database (Clark et al. 2016; NCBI Resource Coordinators 2018). As an aside, the first to use this approach for just animal sedDNA was Graham et al. (2016), the first for plants was Smith et al. (2015), and the first for hominins was Slon et al. (2017). In the study from Pedersen et al. (2016), they initially attempted to apply a metabarcoding approach to investigate the faunal community. However, despite their use of three different PCR primer sets, no faunal taxa were genetically detected in any of the samples, which they presumed at the time was due to a high degree of DNA fragmentation. Nevertheless, by applying shotgun metagenomics, a diversified mammalian community was recovered, including mammoths, bison, moose, elk, hare, and vole, in addition to some fish and bird taxa, such as pike, stickleback, perch, brown bat, and eagle (Pedersen et al. 2016). Using the same approach, Seersholm et al. (2016) comprehensively reconstructed past subsistence economies through shotgun sequencing the sedDNA isolated from midden deposits in Greenland. The sedDNA identified animal data confirmed the presence of two seal species detected in the fossil record. They also found evidence of caribou, walrus, and whale species, which are known to have played an important role for the sustenance of Paleo-Inuit cultures.

Wang et al. (2021) assessed the effects of reference database availability and completeness on taxonomic identification based on their large sedDNA shotgun dataset. They found a remarkable improvement by adding whole genomes of target species to the reference panel. They also developed methods to construct the phylogenies of eDNA-identified fauna using the shotgun sequencing data, finding that eDNA and fossil-based mitochondrial phylogenies are highly consistent. Other examples of successful shotgun sedDNA research on animals include work from Jia et al. (2022) who isolated sedDNA from lake sediments on the Tibetan Plateau, and found a decline of large herbivores as a response to a decrease of alpine herbs. Courtin et al. (2022) detected several faunal taxa at the family level through a k-mer based method using sedDNA isolated from the Batagay megaslump in northern Yakutia, Siberia, including both extinct (mammoth, woolly rhinoceros, and giant ground sloths) and extant (such as reindeer and horse) species. Pedersen et al. (2021) reconstructed low-coverage (0.03x) whole genomes (environmental genomes) of American black bear (*Ursus americanus*) and extinct short-faced bear (*Arctodus simus*, at 0.04x) from cave sediments dated to 14,000–16,000 years ago from northern Mexico, and successfully performed a series of population genetics analysis on these ancient environmental genomes, such as whole-genome principal component analysis, genetic hamming distance, and admixture graphon. In the same year, Gelabert et al. (2021) recovered, through shotgun metagenomics, genome-wide data from human, wolf and bison from a single 25,000-year-old sediment sample collected at Satsurblia Cave (Georgia).

The above examples demonstrate the potential of using sedDNA shotgun metagenomics to study the phylogeny and evolution of individual species. This is particularly important for species without macro-fossil preservation. Notably, for most of the studies in this section, the development of DNA deamination damage models allowed the authors to provide an additional line of evidence for authenticating the identified animal taxa, when there were sufficient reads assigned.

PCR Metabarcoding

As animal DNA constitutes such a tiny fraction of the overall sedDNA pool (Fig. 11.3), targeted approaches that specifically amplify or capture animal sedDNA tend to be far more economical, efficient, and have a superior usable data yield. One such enrichment approach is metabarcoding, whereby a specific locus (or barcode) of a broad taxonomic group, such as mammals, is targeted using PCR (see Chap. 3 of this volume). Thus far, mammals have been the major focus of animal sedDNA metabarcoding experiments, with existing assays, such as Mamm1 and Mamm2 primers (Taberlet et al. 2018) (Table 11.1), providing excellent taxonomic resolution (86% and 90% identified to the species-level, respectively). Due to the ultralow abundance of animal sedDNA, studies typically use anywhere from 4–16 PCR replicates to maximize chance of detection (Ficetola et al. 2015).

The key advantage of metabarcoding is the sensitivity of PCR amplification, which allows for the recovery of templates represented by as few as 2–4 molecules (Forootan et al. 2017), and as such is currently unmatched capture enrichment approaches in terms of sensitivity. However, such high sensitivity combined with ultralow concentrations of animal sedDNA molecules makes metabarcoding vulnerable to laboratory and reagent contamination, including that derived from humans and livestock animals, such as cattle (*Bos taurus*) and pig (*Sus scrofa*), some of the most common laboratory contaminating organisms (Leonard et al. 2007). The ubiquity of human contamination can be particularly problematic, therefore locus-specific blocking primers have been designed to specifically inhibit the amplification of human template in animal sedDNA metabarcoding assays (Boessenkool et al. 2012; Giguet-Covex et al. 2014; Garcés-Pastor et al. 2022).

Authentication of sedDNA is a crucial step in the data quality control pipeline, and is usually achieved by assessing cytosine deamination damage patterns at the ends of DNA fragments (Renaud et al. 2019; Peyrégne and Prüfer 2020). However, these deamination damage patterns are not preserved during metabarcoding, as the original sequences at the ends of DNA fragments are replaced by artificial primer sequences (see Chap. 3 of this volume). This is a key weakness of metabarcoding, as inferring whether a detection represents ancient or contaminating DNA within a sample can become challenging or even impossible. Instead, authentication of sedDNA metabarcoding data relies on a combination of comparison to negative controls performed at each step of the analytical procedure (i.e. sampling, extraction, amplification), replicability of detections (either within samples or between

Table 11.1 Select primers for animal metabarcodes. List not exhaustive; see appendix 1 in Taberlet et al. (2018) for additional primers and taxonomic targets

Target	Primer names	Sequence (5′-3′)	Human-specific blocking probes (5′-3′)	Annealing temp. (°C)	Product length (bp)	References
Mammals	Mamm01F	CCGCCCGTCACCCTCCT	GGAGCTTTAATTTATTAATGCAAACAGTAGG	58	13–74	Taberlet et al. (2018); Garcés-Pastor et al. (2022)
	Mamm01R	GTAYRCTTACCWTGTTACGAC	CCCAACCGAAATTTTTAATGCAGGTTTGGTGA			
Mammals	MamP007F/Mamm02	CGAGAAGACCCTATGGAGCT	GGAGCTTTAATTTATTAATGCAAACAGTACCC	50	53–84	Giguet-Covex et al. (2014); (2019); Bajard et al. (2017); (2020); Sabatier et al. (2017); Ficetola et al. (2018); Lammers et al. (2019)
	MamP007R/Mamm02	CCGAGGTCRCCCCAACC				
Cattle	Bomito2-G	CATAGCAATTGCCATAGTCC		60	554	Garcés-Pastor et al. (2022)
	Bomito2-D	TTTTCGATTGTGCCGGCCGTT				
Cattle	BomitoN2-G	CCCTCTTACTAATTCTAGCTC		55	440	
	BomitoN2-D	TTAGCACTAGGATGAGGAGA				
Sheep	Ovmito2-G	GGAACCGTTACCACATGTtAC		59	689	
	Ovmito2-D	GGCTAGTAGGAAGATTAGGAGAAGA				
Sheep	OvmitoN2-G	TACACTGTTACAGGCATCAG		55	370	
	OvmitoN2-D	CGTGAAGTTAGTTAGGAGAGTA				
Mammals	16Smam1	CGGTTGGGGTGACCTCGGA	GAGCTACCTAAGAACAGCTA	60/55	91–142	Etienne et al. (2015)
	16Smam2	GCTGTTATCCCTAGGGTAACT	TTTGCTACATAGACGGGTGT			
Mammals	16Smam3	TGGGGTGACCTCGGAGAAY	GAGCTACCTAAGAACAGCTA	57	78	Willerslev et al. (2003); Haile et al. (2009); Hebsgaard et al. (2009)
	16Smam4	TCAACGGAMCAAGTTACCCTA	TTTGCTACATAGACGGGTGT			
Mammoth	Mammoth CRF1	CATGCTTATAAGCAAGTACTGT		56	165	
	Mammoth CRR1	TGAGAAATCTCTAGTCATCATG				

(continued)

Table 11.1 (continued)

Target	Primer names	Sequence (5′-3′)	Human-specific blocking probes (5′-3′)	Annealing temp. (°C)	Product length (bp)	References
Mammals	16SA&M Fv2	TCACTATTTTGCNACATAGA	GAGCTACCTAAGAACAGCTA	54	69–70	Haile et al. (2009)
	16SA&M Rv2	CCCCGAAACCAGACGAGCTA	TTTGCTACATAGACGGGTGT			
Mammoth	Mammoth Cb-EHL	TCAGTAGACAAAGCAACCTTAAAT		43	140	Rasmussen et al. (2009)
	Mammoth Cb-EHH	CTGAGTCTGAAGTGAGGCCC				
Mammoth	Mammoth Cb-EA	CCATCCAACATATCAACATGATGAA		55	270	
	Mammoth Cb-ED	GGTATTTCAAGTTTCCGAGTAT				
Bison	BisCR-16633f	GCCCCATGCATATAAGCAAG		56	150	
	BisCR-16810r	GCCT AGCGGGTTGCTGGTTTCACGC				
Moa	Moa262F	GCGAAGACTGACTAGAAGC		55	245	Willerslev et al. (2003)
	Moa441R	CGCATACCGGGTCTGTTTATGC				
Ratites	12SAFH1753	CTGGGATTAGATACCCCACTAT		60	280	
	12SHRb1985	CCTTGACCTGTCTTGTTAGC				
Sheep	Ovis CR 1F	CATAATGGTAAGCATGGGCAT		60	60	Haile et al. (2007)
	Ovis CR 1R	CACGATACAGCAGATATGTC				
Sheep	Sheep-87F	CCTGTCCATTAGATCACGAGCTTG		60	71	
	Sheep-157R	GAGGGATCCTTGCTAAGCGG				

Horses	EQ1	CGTGCATTAAATTGTTTGCC				Hebsgaard et al. (2009)
	EQ2	CATGGGAGGTGATATGCGTG				
Pig	Pig1	CTTTAAAACAAAAAAACCCATAAAAA				
	Pig2	TTAATGCACGACGTACATAGG				
Dog, Fox, Seal	MitMH-732 F	ATGGTTTCTCGAGGCATGGTG				
	MitMH-878 R	GCCCCATGCATATAAGCATGTAC				
Reindeer	L0624	ATCTCACCTAAAATCGCCCACT				
	H0682	CCAAATGTATGACACCACAGTTATG				
Seal	Phoca 975F	CTGCGTCGAGACCTTTACGG				
	Phoca 1071R	GGACTAATGACTAATCAGCC				
Diverse mammals	12So	GTCGATTATAGGACAGGTTCCTCTA				
	12SAFH1753	CTGGGATTAGATACCCCACTAT				
Sheep	Sheep probe	6FAM-CTCTGAGCGATTTTAGAG		60		
Cattle	Cattle probe	6FAM-TCCTCCGAGCGATTTT		60		
Goat	Goat probe	VIC-CAAGAGATCTTCCGAGCGA		60		
Eukaryotes	Euka01F	TGGTGCATGGCCGTTCTTAGT		59	48–777	Hardy et al. (2010)
	Euka01R	CATCTAAGGGCATCACAGACC				
Eukaryotes	Euka02F	TTTGTCTGGTTAATTCCG		45	36–892	Guardiola et al. (2015)
	Euka02R	CACAGACCTGTTATTGC				
Eukaryotes	Euka03F	CCCTTTGTACACACCGCC		55	49–264	Taberlet et al. (2018)
	Euka03R	CTTCYGCAGGTTCACCTAC				
Mollusca	Moll01F	CCAACATCGAGGTCRYAA		48	14–41	De Barba et al. (2014)
	Moll01R	ARYTACYNYAGGGATAACAG				
Gastropoda	Gast01	CCGGTCTGAACTCAGATCA		54	60–70	Taberlet et al. (2018)
	Gast01	TTTGTGACCTCGATGTTGGA				
Aves	Aves01F	GATTAGATACCCCACTATGC		54	39–63	Epp et al. (2012)
	Aves01R	GTTTTAAGCGTTTGTGCTCG				

(continued)

Table 11.1 (continued)

Target	Primer names	Sequence (5′-3′)	Human-specific blocking probes (5′-3′)	Annealing temp. (°C)	Product length (bp)	References
Aves	Aves02F	GAAAATGTAGCCCATTTCTTCC		56	80	Taberlet et al. (2018)
	Aves02R	CATACCGCCGTCGCCAG				
Aves	Aves03F	TATGATAAAGTGAACATRGAGG		51	22–46	Taberlet et al. (2018)
	Aves03R	GGTTCRAYTCCTRCTTTTCTA				
Amphibians	Batr01F	ACACCGCCCGTCACCCT		55	16–100	Valentini et al. (2016)
	Batr01R	GTAYACTTACCATGTTACGACTT				
Turtles	Test01F	AGACGAGAAGACCCTGTGGAA		54	69–83	Taberlet et al. (2018)
	Test01R	TCCGAGGTCRCCCCAACC				
Vertebrata	Vert01F	TTAGATACCCCACTATGC		49	56–132	Riaz et al. (2011)
	Vert01R	TAGAACAGGCTCCTCTAG				
Spiders	Aran01F	TTRYGACCTCGATGTTGAATT		51	56–160	Taberlet et al. (2018)
	Aran01R	CGGTYTGAACTCARATCATGT				
Beetles	Cole01F	TGCWAAGGTAGCATAATMATTAG		53	92–115	Taberlet et al. (2018)
	Cole01R	TCTATAGGGTCTTCTCGTC				

adjacent samples), *a priori* knowledge of a taxon's contamination potential, and, when existing, by comparison to other equivalent proxies (e.g. faecal biomarkers, coprophilous fungi spores) or data types (e.g. archaeological remains, historical archives). As discussed previously, human DNA has high contamination potential and is almost always found in negative controls. For these reasons, metabarcoding-derived human sedDNA detections should always be assumed to be contamination. Such techniques have thus not yet been used to reconstruct past human history from sedDNA samples. Conversely, sedDNA of wild animals has low contamination potential and is generally absent from negative controls. Therefore, and despite their generally low replicability and sporadic nature (Brown et al. 2022b; Garcés-Pastor et al. 2022), metabarcoding-derived wild animal detections are generally considered to be trustworthy. However, domesticated livestock detections should be treated with caution. In our experience, sporadic, unreplicated, and/or unexpected detections of livestock taxa should be considered unreliable, whereas detections that are replicable and occur in adjacent samples may be considered reliable (Garcés-Pastor et al. 2022).

Off-target amplification can occur in animal sedDNA metabarcoding experiments, whereby primer sets designed for a particular taxonomic group (e.g. 16S for mammals; Table 11.1) co-amplify other taxa. The ecological potential for exploiting such 'bycatch' detections was first suggested by a study that co-amplified clitellate worms during a mammalian metabarcoding assay (Lammers et al. 2019). More recently, and using a similar assay, bycatch taxa represented 29 out of 42 total taxa and included birds, amphibians, fish, and a diversity of invertebrates. Although detections were sporadic, the community compositions could be used to robustly determine ecological shifts that occurred as a post-glacial coastal lake developed over time from a marine beach to an aquatic-terrestrial influenced system (Brown et al. 2022b). We note that there are likely to be biases in the composition of taxa that appear as bycatch, and so we recommend that these taxa are not used for richness-based ecological inferences. In comparison with the record of aquatic plants detected by DNA analyses, a consistent ecological shift was also demonstrated for fish bycatch detections obtained with the same primer set on sediments from a peri-alpine lake (Giguet-Covex et al. 2023).

Animal sedDNA metabarcoding assays, particularly of mammals, are noteworthy for their high taxonomic resolution. For some organisms, however, there is also within-barcode variation that reflects known haplotypic diversity, such as with caribou/reindeer (Brown et al. 2022b; Seersholm et al. 2022). This opens the possibility for animal phylogeographic investigation using sedDNA metabarcoding data.

Lastly, metabarcoding requires intact templates of sedDNA molecules that cover the barcode region and both primer-binding sites (~60–150 bp; see Chap. 3 of this volume). The vast majority of sedDNA molecules are shorter (<50 bp; Dabney et al. 2013) than this requirement and are therefore unavailable to the PCR. Combined with the general ultra-low abundance of animal sedDNA molecules, this limits the utility of animal sedDNA metabarcoding to relatively well-preserved contexts, such as those of Holocene age (<11.7 ka) in Arctic, subarctic, and alpine environments. The sensitivity of metabarcoding makes it ideal for sediments with an ultra-complex mixture of sedDNA, such as those derived from lakes and open sites. Animal metabarcoding can therefore be a valuable tool in well-preserved and ultra-complex sedDNA contexts where there is a specific set of organisms under question.

Capture Enrichment

Targeted enrichment via hybridization capture relies on the binding of artificial probes (or "baits") of DNA or RNA that are complementary to the DNA fragments of interest. After probes hybridize to the target DNA, streptavidin-coated magnetic beads are added to the solution which bind to the biotinylated probes. Plate or rack magnets are then used to separate bead bound probe/target DNA while any unbound (off-target) DNA is washed away. As ancient DNA in general tends to be overwhelmed by prokaryotic sequences (e.g., of microorganisms colonizing remains post-mortem), capture enrichment techniques have been widely used in the field, particularly so when attempting to retrieve genetic data from specific taxa or when dealing with poor DNA preservation (Maricic et al. 2010; Ávila-Arcos et al. 2011; Horn 2012; Carpenter et al. 2013; Fu et al. 2014; Mathieson et al. 2015). The first application of capture enrichment to sediment samples was published in 2017, when Slon et al. (2017) applied probes spanning the full mitochondrial genomes of 242 extant and extinct mammalian species (Slon et al. 2016), as well as probes specifically targeting human mitochondrial DNA (Fu et al. 2014), to sediment samples from seven archaeological cave sites in Europe and Asia. Although mammalian DNA constituted only between 0.05% and 10% of the identifiable DNA fragments recovered via shotgun sequencing (with >80% of sequences being unidentifiable), targeted enrichment allowed the recovery of ancient sedDNA fragments pertaining to 12 mammalian families, including hominins, from a total of 61 samples dated to the Middle or Late Pleistocene.

A similar approach was used by Murchie et al. (2021a, b, 2022) to capture plant and animal sedDNA from permafrost sediments in subarctic Canada. Murchie et al. (2021b) built their RNA bait-set to capture the mitochondrial genomes of ~180 subarctic animals (predominantly megafauna) and three chloroplast regions of ~2100 arctic plant species (barcoding loci *rbc*L, *mat*K, and *trn*L). Compared with the proportion of animal and plant DNA recovered with shotgun sequencing using otherwise identical methods, they recovered on average 1224 times more identifiable animal sedDNA, and 1515 times more identifiable plant sedDNA (Murchie et al. 2021b). This proportional increase in on-target, eukaryotic, and identifiable sedDNA allowed for the high coverage reassembly of mitogenomes from multiple animal species (Murchie et al. 2022), along with the ecological reconstruction of flora and fauna from as little total sequencing depth as 1.6 million clusters (Murchie et al. 2021a). Assuming a similar factor scaling in on-target sedDNA, a shotgun sequencing approach would require a sequencing depth of approximately 2.4 billion clusters to achieve the same ecological breadth from these libraries.

Mitochondrial DNA (mtDNA) is often the first target of choice in ancient DNA studies. mtDNA is a small, non-recombining genetic locus that evolves faster than nuclear DNA and can be present in thousands of copies per cell. These factors make mtDNA an advantageous target in ancient DNA studies generally where DNA preservation over time is a major concern. This is also the case in the context of sedimentary DNA, where mtDNA can allow for relatively easier taxonomic distinctions between species compared with whole genome data where many organisms lack nuclear reference data. Yet, nuclear DNA, being inherited from both parents, carries far more information on the history of a population. Recently, Vernot et al. (2021) developed a capture-based approach to enrich hominin-specific SNPs across the genome. The probeset, targeting 1.6 million SNPs, was designed to include sites that are informative of the ancestry of hominin populations, while also being in regions of high diversity between hominins and non-primate mammals. This latter aspect was intended to reduce the risk of capturing DNA from other mammals and their mis-alignment to the human reference genome during data analysis. To evaluate this risk per library, the probeset also includes ca. 90,000 sites that are fixed-derived in hominins, chimpanzees, and bonobos, where the proportion of ancestral alleles likely to originate from non-hominin mammalian DNA can be quantified. Using this approach, the authors were able to demonstrate the preservation of archaic hominin DNA in sediment samples collected at three archaeological sites, and use these data to investigate Neandertal population history.

Targeted-enrichment approaches convey several advantages. Mainly, they allow for a time- and cost-efficient sequencing of target organisms, by reducing both sequencing costs in cases where the target organisms constitute only a small portion of the data that would be generated via shotgun sequencing, and by generating fewer data from 'off-target' organisms that would need to be filtered out through what can be lengthy and space-consuming bioinformatic processes. Additionally, capture-based approaches can theoretically generate fewer false-positive hits by proportionally constraining the set of sequenced molecules to those most likely to be contextually authentic with the sedimentary deposit. For example, if contamination is expected from a particular set of species, these taxa can be selectively excluded in the probe set, thereby reducing the proportion of false-positive reads in the final dataset. Likewise, single taxa of particular interest can be targeted in standalone essays. As such, capture enrichment strategies may be seen as a middle ground solution lying in between shotgun and PCR metabarcoding, as it allows for the generation of genomic-scale data that can be directly authenticated based on patterns of DNA damage patterns, while limiting the scope of off-target and unidentifiable DNA.

One downside of capture enrichment approaches is that it constitutes an additional wet lab procedure to be carried out prior to sequencing. As such, capture enrichment protocols, which vary in their length and complexity, necessitate reagents, personnel and lab space, and require practice to perform effectively. To some extent, the latter can be circumvented by performing protocols on automated liquid handling platforms (Slon et al. 2017; Zavala et al. 2022), but such facilities are not accessible everywhere. Furthermore, the design and testing of new probe sets may be time-consuming, and custom-made baits produced by commercial companies may be expensive—although approaches such as multiplex capture (where separately indexed samples are pooled into the same tube for simultaneous enrichment) may be helpful in alleviating costs for large datasets (Neiman et al. 2012; Zavala et al. 2022).

In addition, some aspects of the procedure, which may affect the downstream analysis of data, should be recognized when making use of capture enrichment techniques for data generation. One shortcoming of such an approach is the possibility of cross-capture (or incomplete off-target exclusion), which reduces the proportion of on-target reads—the end goal of capture enrichment. Paijmans et al. (2016) found that hybridization temperature (generally ranging between 45–65 °C) significantly impacts the proportion of on-target sequences. They found for degraded DNA that 65 °C resulted in the highest proportion of on-target reads, while lower temperatures (such as 50 °C) did not fully exclude the background metagenomic signal. Paijmans et al. (2016) were targeting discrete sample types (tissue/blood/bone) from modern and ancient remains however, and so the application of these findings to sedDNA is complex. The goals of capture enrichment on discrete tissues like bone, and disseminated materials like sediment, are not the same. On the one hand, the whole purpose of capture enrichment is to reduce background metagenomic noise and specifically recover DNA from a specific set of target organisms. Therefore, a hybridization temperature selected to minimize cross-capture is desirable. With sedDNA, however, being able to simultaneously capture a range of related species that share portions of their genome is beneficial for recovering a holistic spectrum of past fauna. Complete off-target exclusion would limit unexpected findings, and allowing some degree of cross-capture can also make up for reference database incompleteness during probe design. As such, a balance needs to be struck, which would likely benefit from further research. Slon et al. (2017) used a hybridization temperature of 65 °C for 24–72 h, whereas Vernot et al. (2021) used 62 °C for 48 h and Murchie et al. (2021a, b) used a combination of 55 °C and 60 °C for 24–72 h. To what degree these varying temperatures and durations matter is currently unclear.

Other downsides of capture enrichment include the reliance on incomplete reference databases for both bait-design and capture, along with downstream sequence classifications, which increases the proportion of false-negatives from overlooked and understudied organisms. Further, by intentionally targeting specific DNA fragments and not others (which comes with hybridization biases such as sequence similarity and GC content), this affects attempts to quantify relative abundances of sedDNA signals between species since the proportions of these molecules were altered during capture enrichment. By selectively targeting specific sedDNA and excluding others, it also further limits what can be said about taxonomic absences in the dataset driven by intentional methodological bias.

These three approaches—shotgun metagenomics, PCR metabarcoding, and capture enrichment—each come with their own sets of advantages and disadvantages. As Darwin stated there is no 'best' organism, only those better and worse adapted to their local environments. In the case of animal sedDNA there is no 'best' method.

The selection of the best approaches depends on a careful consideration of one's research question, budgetary and time resources, and the sedDNA taphonomy of the site type. We will move on from here towards considerations to be made during the computational stage of sedDNA analysis when studying animals.

Bioinformatic Considerations

Chapter 3 of this volume details the bioinformatic workflow of a sedDNA analysis; a topic that is also covered in-depth elsewhere (Buffalo 2015; Reuter et al. 2015; Renaud et al. 2019; Orlando et al. 2021). In this chapter, we briefly discuss a few faunal sedDNA considerations to be made during the computational data analysis stage. We encourage readers that are unfamiliar with high-throughput sequencing (HTS) data to seek out those other resources and to return to this section for further insights for how to best think about analyzing ancient animal sedDNA.

The two main considerations for faunal sedDNA are expected template fragment lengths and damage, along with the comprehensiveness (or lack thereof) of reference databases for the fauna under consideration. As mentioned earlier, ancient DNA (particularly animal aDNA) will tend to have an fragment length distribution (FLD) mode of $\lessapprox$50 base pairs (Dabney et al. 2013). As such, selecting a sequencing kit suitable for especially short read lengths is important when working with highly degraded DNA, as is utilizing a size selection procedure with sufficient resolution to remove short, artificial adapter sequences, but while retaining short animal sedDNA. Data processing after sequencing is similar for different taxa of interest, although gaps in animal reference databases remain a limitation, particularly for extinct or understudied animals.

Data Processing and Analysis

The DNA sequence generated from HTS technologies will include several components that need to be accounted for during data analysis. Each DNA sequence (or 'read') will contain a unique identifier (an index sequence) that allows reads to be assigned to its original sediment sample/library (Fig. 11.6). These indexes are separated into their associated samples in a step called demultiplexing. On Illumina platforms, demultiplexing is carried out using software packages that come with the sequencer. Afterwards, the artificial adapter sequences that were ligated onto the ancient fragments to allow for HTS need to be removed. Software has been designed with ancient DNA damage characteristics in mind for trimming away these adapter sequences, along with merging overlapping reads from paired-end sequencing. This is a type of sequencing that involves reading a DNA fragment from both directions to improve sequence quality as opposed to only reading from one direction where errors/ambiguities may be unresolved (Reuter et al. 2015). Two examples of aDNA

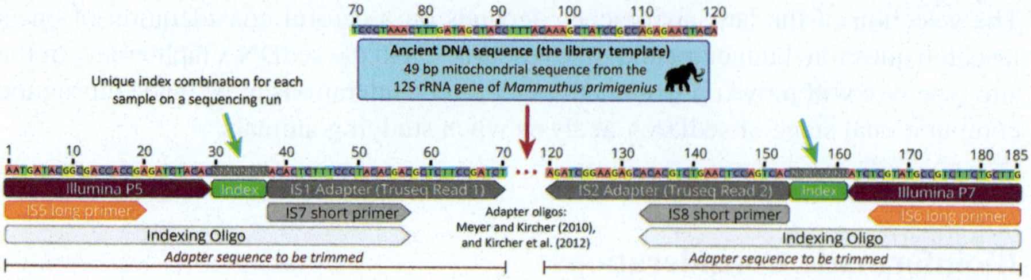

Fig. 11.6 A dual indexed library molecule with woolly mammoth 12S rRNA gene and TruSeq adapters. Sequence from Murchie et al. (2021b)

specific trimming/merging software are leeHom (Renaud et al. 2014) and AdapterRemoval v2 (Schubert et al. 2016). There are also publicly available pipelines designed specifically for the processing and filtering of ancient DNA sequence data such as EAGER (Peltzer et al. 2016) and metaBIT (Louvel et al. 2016), including others tailored to ancient eDNA such as Holi (Pedersen et al. 2016). PCR metabarcoding often follows a slightly different processing pathway using pipelines like OBITools (Boyer et al. 2016) that contain script packages for everything from trimming and filtering to taxonomic assignment.

After the early-stage steps with metagenomic data (shotgun or capture enriched), where reads are grouped by sample (demultiplexing), the artificial sequences attached to the ends of the fragments are removed (adapter trimming), and reads sequenced from both directions are combined (paired-end merging). The sequenced reads then need to be filtered for size (typically a minimum between 24–35 bp). Size filtering is intended to ensure correct alignment as shorter sequences have a higher chance of non-specific mapping. Thereafter, PCR duplicates (which are generated during amplification following library preparation, enrichment, and metabarcoding) are removed, and the reads are either run through a taxonomic classifier directly or are first mapped to a set of target reference sequences using tools such as BWA (Li and Durbin 2009) or Bowtie2 (Langmead and Salzberg 2012) to reduce the total dataset for taxonomic classification—a computationally intensive procedure that can take months with some software packages, even on high performance computers. Reads that pass this map-filtering step (Murchie et al. 2021b) can then used as the input for a classifier. There are several approaches to taxonomic sequence classification for metagenomic data, including BLASTn/MegaBLAST (Altschul et al. 1990), MALT (Herbig et al. 2016), DIAMOND (Buchfink et al. 2021), CENTRIFUGE (Kim et al. 2016), KRAKEN (Wood and Salzberg 2014), euka (Vogel et al. 2023), and ngsLCA (Wang et al. 2022). These classifiers can either use a curated reference database based on expected organisms, or a full public database such as NCBI Genbank (Clark et al. 2016). Classified reads are then binned using programs such as KRONA (Ondov et al. 2011), MEGAN (Huson et al. 2007), or PIA (Cribdon et al. 2020), where parameters can be set to filter for misclassified reads, such as using a minimum number of hits for a taxonomic node to be called, percent similarity to reference, bit score, and so on. Ancient DNA is expected to be damaged when generated with a shotgun or capture enrichment approach. It is therefore important to use a classifier that does not necessitate 100% identity as this will generate abundant false negative hits given that a 50 bp ancient DNA fragment may have 2–3 deaminated bases on its termini, along with the potential for sequencing errors. In software where these parameters can be specified, determining the appropriate values necessitates balancing between the possibility of false positive classifications from overly permissive parameters, and false negative classifications due to the parameters being overly restrictive.

Classifiers are also only as good as the reference database they are based on. If using a publicly available database, like GenBank, several complications arise that must be kept in mind, such as:

1) There can be adapter contamination in the reference sequence that causes false positives (the most well-known example of this is carp [NCBI accession: PRJEB7241], although adapter contamination exists in many other organisms). Additional issues may be caused by sequence errors from aDNA damage, during cluster generation, assembly, insufficient coverage, and undetected contamination that all can propagate into reference sequences that are then used to classify other DNA.

2) Some classifiers like BLAST allow the user to specify how many possible hits to previously sequenced organisms they allow to retrieve them from the dataset using flags like "-max_target_seqs", "-num_alignments", and "-max_hsps" (as files can become very large otherwise). However, these hit caps do not necessarily always report the best taxonomic hits. As reported by Shah et al. (2019), choosing to only report one taxonomic hit with "-max_target_seqs 1" will simply provide the first good hit (exceeding the specific e-value) and not the best hit after searching all possible options through several iterations. This limitation then necessitates recording many potential hits, which can rapidly balloon file sizes.

3) Aside from the non-redundant and curated NCBI RefSeq database (which is not currently released for download as a standalone nucleotide database, and is missing many organisms relative to GenBank), other databases are highly redundant for organisms of scientific or economic interest. If we take an example from the cattle genus *Bos*, there are currently 1958 sequences that match the search criteria "(Bos[Organism]) AND mitochondrion" on GenBank NCBI (as of September 15, 2022). If one has sedDNA from another organism in the Bovidae family from a relatively conserved region of the mitochondrial genome (such as the genes 12S or 16S), there is a high likelihood that regardless of the number of hit iterations, the sequence will be identified as *Bos taurus* (1446/1958 of those *Bos* sequences on GenBank). This causes a false positive species classification (but true positive genus assignment) if the environmentally recovered sequence originates from another *Bos* species that makes more biogeographic sense such as aurochs (*B. primigenius*) or yak (*B. grunniens*), or even from another species in the same family like American bison (*Bison bison*), steppe-bison (*Bison priscus*), or *Bison antiquus*. Murchie et al. (2021b, 2022) discuss this challenge with identifying hits within Bovidae, but this problem extends to other animal clades, and will likely become increasingly

problematic as more sequence data from economically important organisms is uploaded to public databases.

4) Genetic databases are, and will likely always be to some degree, incomplete. The 'true' organism from which eDNA was shed in the past may not have any available reference data. As such, there will either be a false positive gap in the record (it simply does not appear in the results despite having its sequenced DNA in the library that was binned as "unidentified"), or it will be taxonomically classified as a related organism (in the same genus or family), or called at a high taxonomic node (such as order, class, or phylum)—thereby obscuring its presence in the metagenomic faunal spectrum.

These four challenges combine to make species-specific hits with sedDNA difficult to authenticate. A more conservative approach is to collapse all hits at either the genus or family ranks, and/or use previous palaeobiogeographic knowledge to carefully interpret the classifications. Knowing the limitations of databases and classifiers help ensure that sedDNA datasets are interpreted cautiously. For example, knowing that a hit to a particular species may be inaccurate, and is more likely another closely related species that just so happens to have less genetic research dedicated to that lineage to date.

A workaround to limitations in large public databases is to use a manually curated reference database for the study region, or to pre-emptively map-filter reads to organisms that make palaeobiogeographic sense prior to classifying. Both of these approaches are likely to result in some false negatives by artificially stripping out overlooked organisms, or those with as yet unknown palaeobiogeographic ranges and genetic diversity, such as in instances discussed earlier with regards to the Middle Pleistocene work by Courtin et al. (2022). This *a priori* database filtering may be the only way to realistically remove false positive classifications from the dataset until we have better classifying software and more robust databases. Ideally, using a combination of curated and public reference databases, multiple classifiers, and multiple filtering approaches together in what is referred to as an ensemble approach within a single study is the most robust way to balance these challenges. An ensemble approach is, of course though, significantly more work that may not impact the results or interpretations in a meaningful way, particularly with regard to false negatives that are ultimately due to database incompleteness.

Another way to help authenticate taxonomic hits is investigating damage in those specific reads. Cytosine deamination results in nucleotide misincorporations on the termini of ancient DNA fragments, ultimately transitioning cytosine to thymine and guanine to adenine (Dabney et al. 2013). These can be quantified by mapping reads to a reference sequence and observing the frequency at which there is a mismatched thymine or adenine relative to the reference sequence on the termini of those reads. As this approach is frequency based, generally at least 200 or more reads mapping to the mitochondrial genome (~16,000 bp in length) are necessary to assess an aDNA typical deamination or "smile" pattern (Fig. 11.7). Authenticating taxonomic hits with abundant sedDNA representation from a particular organism can serve as a useful secondary confirmation, although rarer taxa with only a handful of hits cannot be authenticated through damage alone. Mapping to a closely related but

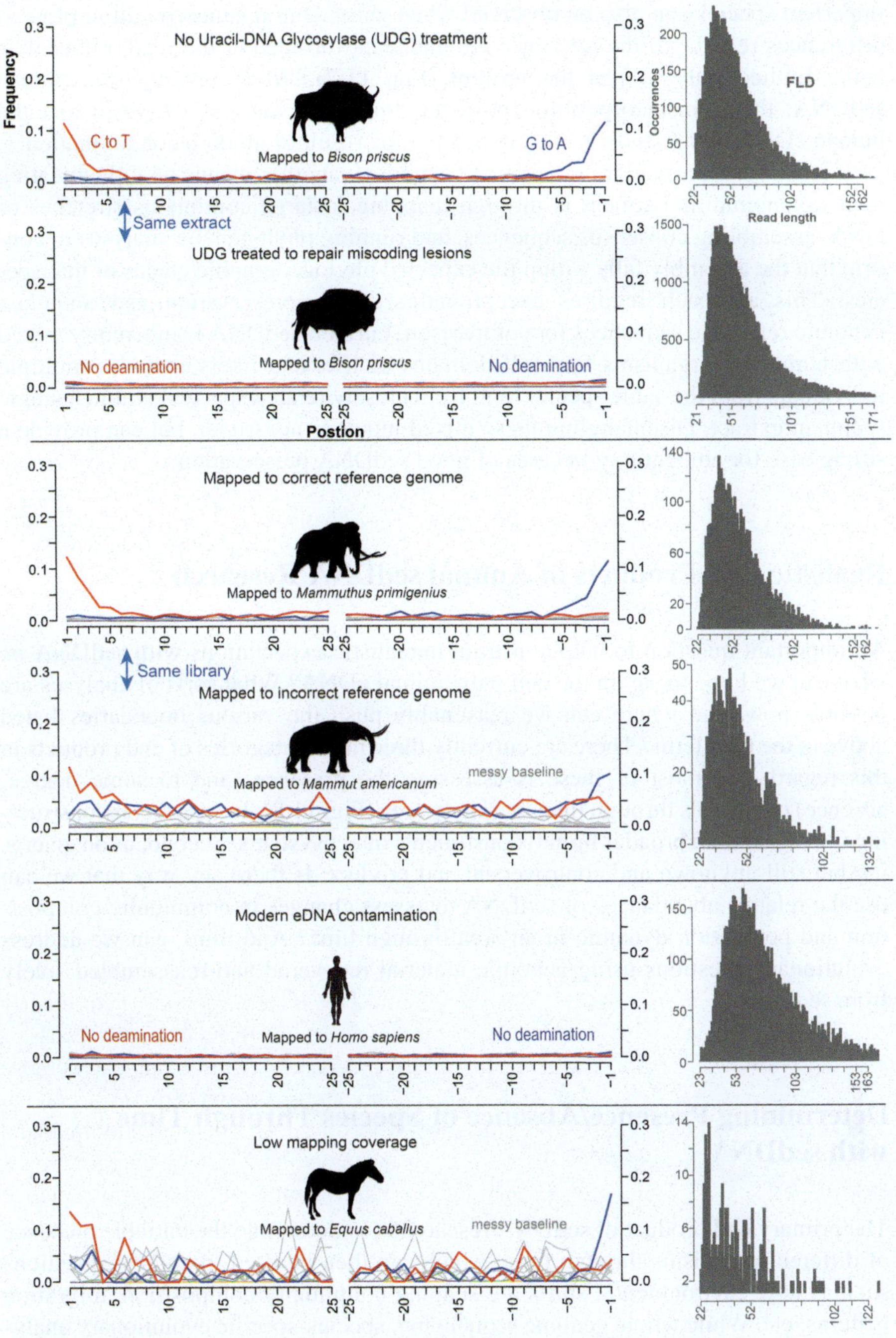

Fig. 11.7 Deamination and fragment length distribution (FLD) plots of sedDNA data library prepared with Meyer and Kircher (2010) and Kircher et al. (2012). Plots generated with mapDamage2.0 (Jónsson et al. 2013) with data reprocessed from Murchie et al. (2021a, 2022)

imperfect species can also be observed when generating misincorporation plots as differences relative to the reference are abundant throughout the read, rather than being limited only to just the termini (Fig. 11.7). When investigating human sedDNA, these misincorporation plots are important lines of evidence that the human eDNA under study is ancient and not the result of modern contamination.

If reads for a particular organism are abundant, another means of authenticating their taxonomic assignment is through mapping to large continuous stretches of DNA, assembling consensus sequences, and running phylogenetic analyses to confirm that the assembly falls within the expected phylogeographic clades of that species. This approach requires exceptional sedDNA preservation and abundant genomic reference sequences for comparison. Further, sedDNA is inherently mixed with numerous organisms from all domains of life, and likely includes multiple individuals from the same species (and from closely related species) within a single sediment extract. Disentangling these mixed inputs is not trivial, but can provide a strong case for authenticity in cases of good sedDNA preservation.

Realistic End Products of Animal sedDNA Research

An important question to help constrain unrealistic expectations with sedDNA is: what can we hope to say in the end with animal eDNA? What sorts of analyses are possible now, and where can we reasonably push the various boundaries listed above in the near term? There are currently three main categories of end products in this regard. Foremost of these is assessing the presence (and to some degree, absence) of animals through time in a study area (Fig. 11.8). In essence, reconstructing the faunal (and broader biotic) constituents of ecosystems. Second, as an emerging but still unknown and controversial end product: Is there any way that we can use the relative abundances of sedDNA to assess changes in community composition and population dynamic in an area through time? And third, can we address evolutionary questions using genomic material recovered and reassembled solely from sediments?

Determining Presence/Absence of Species Through Time with sedDNA

The primary end product of sedDNA research is to determine the variable presence of different organisms through time, which can then be used to address questions such as how environmental shifts are shaping community composition, migration patterns, etc. While whole genome sequencing, species-specific evolutionary analyses, and other up and coming end products are increasingly possible with cutting-edge sedDNA methods, identifying the cast of species present in a specific place

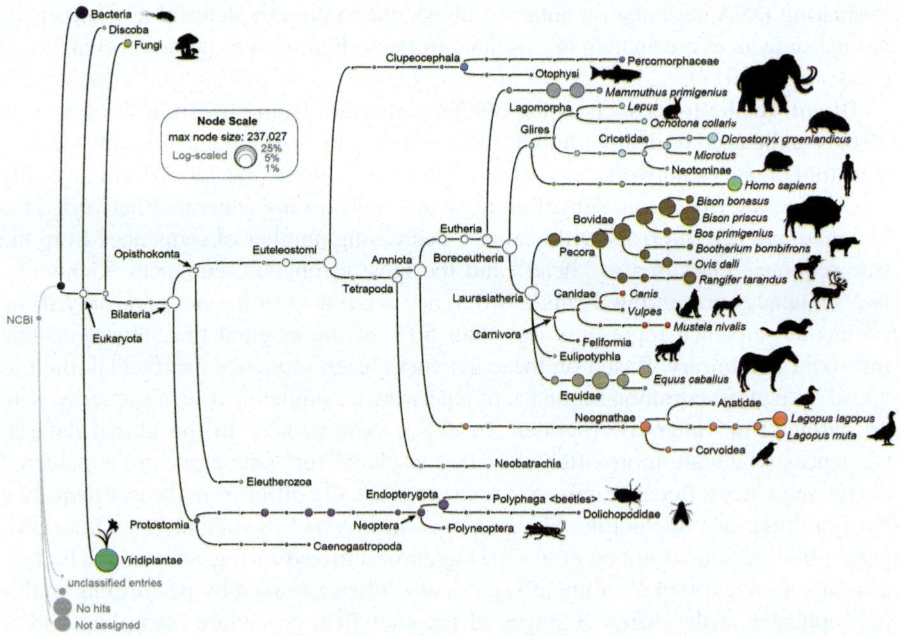

Fig. 11.8 Example reconstruction of animal sedDNA from permafrost sediments dated to 30,000 years ago in subarctic Canada; reprocessed from Murchie et al. (2022). Phylogram of taxonomically binned read counts (log-scaled), generated in MEGAN6 (Huson et al. 2007)

and how that changed through time—without needing any visibly diagnostic animal remains—is the most realistic and unique contribution of sedDNA research to Quaternary science. Determining the spatial-temporal range of humans and animals is limited in traditional palaeontological and archaeological approaches by taphonomic processes that leave an incomplete record of macro-biological remains (i.e., bones, teeth, soft tissues, artifacts). eDNA allows a perspective shift to the micro-scale of a study area, where environmentally released micro-fossils of DNA end up being far more abundant and continuous than what we could ever expect to unearth with traditional excavations. Despite this unique benefit, several challenges arise when considering the presence and absence of organisms through time that must be grappled with when interpreting any eDNA research.

First, does a detection reflect a true or a false positive? This question refers to contamination issues by modern DNA, taphonomic damage, PCR or sequencing errors, reference database biases, and otherwise inaccurate species identifications due to shortcomings and computational limitations in DNA classification algorithms. To address the issue of potential contamination, it is recommended to include negative controls at each stage of the laboratory procedure (i.e. in the field/during the sampling, during extraction, PCR, and other wet lab procedures), which may help identify sporadic contamination from reagents that tends to originate from domesticated species. In shotgun sequencing and capture enrichment approaches, evaluating DNA degradation patterns allows one to directly determine whether the animal sequences are ancient or originate from modern sources (Jónsson et al. 2013; Kistler et al. 2017).

Bioinformatic tools such as the obiclean program from OBITools (Boyer et al. 2016) can be used to deal with PCR or sequencing errors in metabarcoding studies. This tool creates a network for each PCR that connects sequences differing by only one nucleotide (insertion, substitution, or deletion). In the generated networks, the assumed cascade of errors will lead to a decreasing number of sequences from the true sequence classified as "head" and the most erroneous sequences. Generally, two sequences presenting a single difference are linked in the network only if the erroneous sequence represents less than 50% of the original one. However, this threshold is arbitrary. Based on these networks, each sequence can be classified as "head" (i.e. most common sequence of a network), "singleton" (i.e. a sequence with no variants) or "internal" (neither "head" or "singleton"). In the global dataset, sequences which are more often classified as "head" or "singleton" are considered as true sequences. Because chimeric sequences usually differ from the true sequence by more than one nucleotide, obiclean does not recognize such errors. Other programs like Perseus (Quince et al. 2011) can be utilized in this situation. The true presence of a taxon in a sediment layer is also often assessed by performing multiple replicates at the different stages of the analytical procedure (sampling and/or DNA extraction and/or amplification).

In lake sedDNA studies, our anecdotal experience suggests that amplification is often a key stage to repeat, especially for rare species, presumably due to the stochasticity of the PCR. A taxon is often considered as being present when more than one replicate is positive, i.e. with a number of reads higher than a chosen threshold (e.g. 5 or 10 reads). Also, if a taxon is detected in only one replicate but in consecutive samples (or samples with close ages), it can also be considered as being present. There are no strict rules to define the presence of a taxon—with the determination of presence often depending on personal experience, research contexts, scientific questions, and potentially independent knowledge building support in the calls that have been made. For a more statistical approach, occupancy models (Ficetola et al. 2015) or conditionally autoregressive occupancy models (i.e. integrating temporal autocorrelation; Chen and Ficetola 2019) can be used to estimate true prevalence and false-positive rates in eDNA studies. Ficetola et al. (2015) observed that model performance increases with the number of replicates. Finally, comparisons with independent proxies measured on the same samples or other sources of information such as archaeological remains can be used to validate a record. For instance with herbivores, one way is to compare their sedDNA presence with coprophilous fungi spores (Graham et al. 2016; Ficetola et al. 2018; Messager et al. 2022) and/or with faecal biomarkers such as bile acids (Brown et al. 2021). Comparisons with historical information (Messager et al. 2022) or bone records can also be very useful (Graham et al. 2016).

The second question we are faced with when analyzing an DNA dataset is: Does an absence of detection reflect a true negative; i.e., to what extent can the absence of an organism be interpreted as a true absence in the spatial-temporal setting under question? This question is not easy to address. One way to handle this issue, although imperfect, is to perform multiple replicates. Based on occupancy models, the optimal level of replication was shown to be strongly dependent on the detection probability. For rare species in ancient DNA studies based on metabarcoding approaches, at least eight replicates were recommended (Ficetola et al. 2015). Comparing sedDNA reconstructions with other proxies that have variable sensitivities to taphonomic/methodological biases is also a way to address the question of false negatives. Ultimately, assessing whether a negative is true or not may not be possible aside from building a robust case based on multiple lines of complementary evidence.

Finally, the third key question to address is related to taphonomy: Are the sequenced DNA fragments truly sourced from the place and time under question? How much eDNA accumulation time does a given deposit represent, and what spatial scale could these sequences reasonably originate from? Is there any possibility that sequenced reads are contaminants from another context? These questions directly refer to our earlier discussion on decontextualized sedDNA. DNA degradation patterns allows for the identification of ancient versus modern DNA sequences, but this approach does not give the age of a sequence and thus does not clearly establish the synchronicity between the DNA sequence and the sedimentary deposit considered in the study. Bayesian tip dating via genomic assembly and phylogenetics can help here, but this technique requires well-preserved, genomic-scale sedDNA, and will only reveal that specific organisms in the sample are age-authentic. One then must assume that this age applies to other taxa identified in the sample that do not have the necessary sequencing depth for similar Bayesian analyses (Drummond and Rambaut 2007; Lemey et al. 2012; Kjær et al. 2022). As discussed previously, independent reconstructions based on other lines of evidence testing for synchronous changes in the ecosystem are a useful way to assess the impact of decontextualized sedDNA. In lake sediments, because we assume that a large proportion of DNA is transferred by erosional processes, the question of sedDNA context changes with eDNA source area—a crucial question for comparing different time periods, especially in large catchments, where important changes in erosion sources and processes can be expected. Mineral and organic geochemical analyses as well as stable isotopic analyses can be very useful to identify such changes (Giguet-Covex et al. 2019; Rapuc et al. 2021).

Despite these three main challenges, many records with robust findings have been reported, as we have highlighted throughout this chapter. The reliability of a given interpretation relies heavily on multi-proxy and interdisciplinary approaches. A logical follow-up question from presence and absence is: How far can we use sedDNA data as quantitative or at least semi-quantitative proxy? What do differences in relative abundance—10 reads vs 1000 vs 100,000 within a metagenomic approach, or two positives vs eight positive replicates with metabarcoding—mean? Can these measures be used to help us understand how many animals were dispersing eDNA at different timepoints?

Can We Estimate Animal Biomass Abundance from sedDNA Data?

The ability to infer animal biomass abundance from sedDNA data would allow for the direct measurement of, for example, grazing intensity and relative population size changes through time. It is therefore of great interest to ecologists to infer these measures from molecular archives. SedDNA relative abundance data are generally derived from the relative counts of sequencing reads assigned to various taxa, although some studies alternatively use the relative proportion of detections in technical replicates. However, it is currently unknown whether sedDNA abundance data can be robustly utilized as a proxy for past animal biomass abundance, or whether there are too many variables impacting the release, preservation, recovery, and analysis of sedDNA for taxonomically assigned read counts to be at all informative of palaeo-biomass.

Seersholm et al. (2016) found that their sedDNA read counts did correlate with both archaeological bone abundance and expected animal biomass in the case of harp seal, ringed seal, birds, and foxes at their Greenlandic sites. By contrast, domestic species (cattle, sheep, goat, and dogs) were overrepresented in their genetic profiles compared with their expected biomass based on fossil abundances, whereas large wild animals (caribou, walrus, narwhal, and bowhead whale) were likely underrepresented in fossil bone records compared with sedDNA. They emphasised that DNA distribution likely does not always reflect the true biomass of the different species. For example, restricted defecation and urine is likely to inflate the sedDNA abundance of domesticated animals in certain areas while reducing it in others. Likewise in their study, they found high abundances of bowhead whale DNA, along with other large fauna, but few bones from those species were recovered. While this is likely the result of off-site disposal of hard tissues that biases the apparent subsistence record away from the studied animals, the continued use of some tissues such as blubber (which was used as lamp fuel) may also skew a sedDNA signal overtime. They encouraged the complementary combination of both sedDNA and zooarchaeological analyses to aid in interpreting shifting signals of abundance through time.

There is evidence from modern eDNA in controlled environments that it may be possible to derive animal biomass abundance (Hebsgaard et al. 2009; Andersen et al. 2012). However, how we translate genetic species hits to meaningful measures of palaeo-biomass abundance is currently unclear. This is due to the myriad of potential confounding factors that can impact the relative abundance of animal sedDNA in a sediment sample, many of which are currently uncharacterized and in need of further research. These can be classified as being both biological/geological and technical in nature. Biological confounders include the source, such as the differential release of waste products or discarded/deceased tissues, and differing rates of shedding of DNA into the environment between taxa (and through time), as well as other taphonomic and geologic factors such as differing transfer processes like erosion and deposition, variable organo-mineral complex formation, and differential degradation through time. There also may be thresholds of abundance/transfer that must be met for reliable detection. For instance, in a small high-altitude lake in the Alps, mammal DNA began to be detected when the accumulation rate of Sporormiella sp. spores started to exceed a threshold (around 200 Nb/cm^2/yr). Below this limit, the presence of mammals suggested by the presence of Sporormiella sp. spores was not detected by the DNA analyses (Messager et al. 2022) (Fig. 11.4). The value of this threshold is probably site- and method-dependent. Furthermore, in the same study, a higher number of positive PCR replicates, observed during the Middle Age compared to the Roman Period, was only reflected in the accumulation rate of *Sporormiella* sp. and not in the concentration of these fungal spores (Messager et al. 2022). These results argue towards the important role of erosion processes in the DNA transfer to the lake and not only of the DNA production by the animals. Consequently, the relationship between animal abundance in the catchment and the DNA quantity detected in the lake sediments is not straightforward, although one driver of the erosion may be the animal abundance, thus causing a positive feedback on the lake sedDNA record.

Technical confounders can occur during sample collection, or molecular biological and bioinformatics processing. Molecular biology techniques that involve a competitive enrichment step can lead to biases in relative abundance estimates, such as, for example, characteristics features of the template DNA (e.g. GC-content, length, extent of DNA damage, genetic similarity to the capture probes) can impact relative metabarcoding amplification efficiency or the taxonomic composition of a set of capture enrichment baits. PCR in general can theoretically bias library composition due to preferential amplification of certain molecules over others (such as those with less damage), even during indexing step. Other factors such as enzymatic inhibitors can also theoretically impact the relative abundances of sedDNA that could be library-adapted (Murchie et al. 2021b). During bioinformatics processing, the completeness of reference databases (relative to the true, but ultimately unknowable, biodiversity of an ancient ecosystem) that are required to identify sedDNA molecules can also impact relative abundance estimates.

As animal DNA constitutes an ultralow proportion of sedDNA mixtures (Fig. 11.3), especially in those derived from open depositional environments, we caution against using sedDNA relative abundance data as a proxy for palaeobiomass until the confounding factors listed above have been characterized and shown to be correctable.

Genome Assembly and Phylogenetics

A rapidly emerging benefit of metagenomic approaches (shotgun metagenomics and capture enrichment) is the ability to recover and reassemble genomes solely through eDNA. This allows environmentally disseminated genetic fragments to be processed and analyzed much like aDNA recovered from discrete tissues like bone or teeth where long stretches of DNA are reassembled and used to investigate evolutionary questions with phylogenetic inference (such as maximum likelihood and Bayesian analyses) (Lemey et al. 2012). The first application of this approach with sedDNA came from Slon et al. (2017) who assembled hominin mitochondrial genomes from Eurasian cave sediments. Similar organelle genomic reconstructions from sediment paired with phylogenetics (using both sedDNA and bone derived genomes) followed shortly thereafter with Zhang et al. (2020), Zavala et al. (2021), Wang et al. (2021), Murchie et al. (2022), and Kjær et al. (2022), who studied a variety of hominins and other animal mitogenomes from caves and permafrost. The high copy number per cell and small size of the mitochondrial genome (~16,000 bp) makes this approach readily feasible, where organellar genomes from multiple organisms can be reassembled simultaneously, thanks as well to an abundance of publicly available reference sequences.

While mitochondrial and chloroplast genomes have immense phylogenetic potential, the human mitogenome is four orders of magnitude smaller than the nuclear genome. With that scale and biparental inheritance comes far greater evolutionary resolution. It did not take long after whole organelle genomes were reconstructed from sedDNA for the nuclear genomic prospect to be actualized. In 2021 it was made clear by Pedersen et al. (2021), Vernot et al. (2021), and Gelabert et al. (2021)—who variously reassembled nuclear genomes of cave and short faced bears, Neandertals, modern humans, wolves, and bison—that sedDNA methods have the ability to recover whole palaeogenomes, irrespective of the preservation of identifiable remains.

The genomic potential of sedDNA means that not only can these methods be used to study the presence and absence of organisms through time, holistically reconstructing ecosystems across all domains of life, but so too can we study the evolutionary and even phenotypic characteristics of individual species within a single environmental sample. SedDNA will likely be used soon to reconstruct the coat colour, or musculature, or cold adaptations of ancient animals solely from sediments. Furthermore, Bayesian phylogenetic inference allows for reassembled genomes to be age estimated with a molecular clock (as it is thought that nucleotide mutations accrue at predictable rates) (Bromham and Penny 2003). Multiple studies have now used Bayesian approaches to molecularly date their reassembled sedDNA genomes (Zhang et al. 2020; Gelabert et al. 2021; Pedersen et al. 2021; Vernot et al. 2021; Zavala et al. 2021; Murchie et al. 2022). As discussed previously, this method provides another important means of stratigraphically authenticating sedDNA to determine whether these ages are in congruence with other methods like radiocarbon and optically stimulated luminescence (OSL) dating. Beyond these exciting applications is the fact that sedDNA is not relegated solely to megafauna, which to date constitutes the majority of animal aDNA research. Murchie et al. (2022) observed that even small ground-dwelling ptarmigans had genomic-scale preservation. The bones of small fauna are rarely recovered during excavations. As such, being able to investigate these understudied species via sedDNA not only allows their ecological presence to be properly accounted for, but sedDNA may also provide a unique window into their unique evolutionary histories that have otherwise been obscured by taphonomic and recovery biases.

A major caveat with metagenomic sedDNA, however, is that unlike with bone where there is known endogenous DNA from a single macro-organism, along with its associated pre/post-mortem microbiome, sedDNA contains mixed genomic contributions from many animals, both those closely and distantly related. Mapping tools (such as Bowtie2, BWA, and many others) (see Fonseca et al. 2012) can circumvent this challenge to some degree, as they are designed to align sequenced DNA reads to a reference genome by attempting to retain true genetic variation of the target organism whilst discarding background noise (sequencing errors and non-target DNA). The distinction between on/off-target with sedDNA becomes challenging however as inter/intra-related species in the same sediment sample will contain large homologous stretches of DNA that mapping tools are unlikely to accurately discriminate amongst. This means that even with a careful application of bioinformatic tools and approaches (such as competitive mapping; Feuerborn et al. 2020) to disentangle mixed inputs, along with experimentally tweaking mapping parameters, most regions with singleton SNPs (as opposed to loci with multiple nearby segregating polymorphisms) are unlikely to be completely separated amongst related organisms. To varying degrees, this creates "franken-genome" assemblies from the genetic inputs of related, but different, individuals (a sequence assembly that never existed in nature, but is an amalgamation of multiple similar genomes). This issue does not seem to significantly impact phylogenetic placement among major clades as segregating polymorphisms are both more conserved across individuals, and can be separated with competitive mapping between haplogroups of the same species (Murchie et al. 2022). It does however mask individually unique SNP variants, which impacts Bayesian age estimates due to the need for ambiguous bases (e.g. R [A/G], Y [T/C], K [G/T], or M [A/C]) in consensus sequences for these highly variable positions if they cannot be disentangled. Even with organellar DNA, allelic variation can exist in the same library due to such mixed inputs, creating a situation analogous to heteroplasmy (Kmiec et al. 2006). This challenge has been overcome in previous metagenomic sedDNA research, but it does highlight the need for caution and careful consideration when interpreting genomic assemblies and phylogenetics with sedDNA as multiple individuals may be represented in any given sedDNA genome. Despite this mixed input limitation that may yet be overcome with novel bioinformatic approaches in the near future, the genome-scale preservation potential of sedDNA is likely to facilitate high resolution explorations into the evolutionary histories of individual species at a scale that, until very recently, was only thought feasible with exceptionally well-preserved hard tissue remains.

Conclusion: the Future of Animal sedDNA Research

There are several avenues where sedDNA research related to animals is likely to see significant and much-needed improvements in the coming years. First among these is the expansion of reference databases. The study of environmental DNA relies on comprehensive reference sequences as there are generally no identifiable animal macro-tissues in sediments to aid taxonomic identification. Humans and megafauna are generally well represented in these databases (this is especially the case for animals in the Northern Hemisphere), but many other animals that happen to be of marginal scientific interest have little, if any, genetic references available to allow for their taxonomic identification with eDNA. Many larger sedDNA studies incorporate reference database sequencing as part of their analyses to improve taxonomic identification (Willerslev et al. 2014; Wang et al. 2021).

Second, faster, more accurate, and more customizable taxonomic classifiers are needed. False positive/negative taxonomic assignments may be inevitable to some degree. But this noise can theoretically be substantially reduced with further development of bioinformatic tools designed specifically for sedDNA to help alleviate the known limitations of available software packages.

Third, sedDNA extraction methodologies need improvement. Several labs develop in-house specific demineralization/digestion and purification protocols to improve upon the limitations of commercial kits designed with modern eDNA in mind (high quantity/quality DNA). Despite study-specific optimizations to maximize sedDNA recovery and purity, there is likely still substantial genetic loss during extraction, along with highly variable extraction success when using custom methods on different sediment types. Innovations in early-stage sedDNA processing (sedDNA release and isolation) may come from improved biogeochemical understandings of the mechanisms that promote sedDNA preservation in deep time.

Fourth, most sedDNA work to date has focused on organellar DNA (mitochondria for animals and plastids for plants). There are many reasons why aDNA research has been directed towards these organelle genomes, such as: high copy number, typically uniparental inheritance, elevated mutation rate aiding taxonomic discrimination, and a smaller overall size. But, given that there are 3–5 orders of magnitude more nuclear DNA in a given cell than organellar, it is likely that nuclear DNA constitutes the largest fraction of animal sedDNA in any given sedimentary sample. One limitation to utilizing this nuclear data is database incompleteness. Whole genome sequencing (WGS) continues to proliferate as sequencing technologies become more affordable, and as such the scope of reference databases available for eDNA classifications will likely continue to increase. The rapidly increasing size of reference databases (NCBI Genbank for instance doubles in size every 18 months [https://www.ncbi.nlm.nih.gov/genbank/statistics/]) will inevitably necessitate faster classifiers in order to realistically make use of these resources. Beyond taxonomic classification, however, other relatively untapped research paths include reconstructing phenotypic traits of animals solely from sediments, the simultaneous WGS of multiple species from single sediment extracts, and population genetic studies using genomes reassembled from sedimentary sources alongside others derived from macro-tissues.

Fifth, humans and iconic megafauna have been the focus of most sedDNA work. Other vertebrates (birds, rodents, reptiles, amphibians, fish, etc.), arthropods, molluscs, nematodes, and others within Animalia also shed eDNA and are amenable to sedDNA research. Much of this work is limited currently by database incompleteness, along with a comparative lack of scientific interest among sedDNA researchers. If this trend shifts, these understudied organisms could significantly expand reconstructed faunal spectra to more holistically reconstruct palaeo-ecosystems through time.

Sixth, most human sedDNA work to date has focused on caves. While caves can allow for optimal sedDNA preservation, there is no reason why human sedDNA research could not expand to other more typical open-air archaeological sites. This has the potential to revolutionize archaeological research, particularly from controversial sites like Monte Verde (Dillehay et al. 2015) and Bluefish Caves (Bourgeon et al. 2017), among many other sites where the anthropogenic component is unclear. Being able to recover ancient human DNA directly from sediments and palaeosols, quantify damage on the obtained reads, reassemble genomes, and derive haplogroups or run phylogenetic analyses all from material that would otherwise be screened and discarded would transform the perceived information value of the sedimentary matrix during an excavation. There are of course ethical concerns to be tackled when targeting disseminated human eDNA (Handsley-Davis et al. 2021), along with the omnipresent threat of contamination that must be considered carefully during analysis and interpretations. Regardless, human sedDNA research from typical archaeological sites would not only dramatically increase the palaeodemographic resolution of ancient peoples, it may help save human osteological remains from destructive analyses if that genetic information could instead be recovered from underlying sediments.

Summary

There is massive potential for animal sedimentary DNA research to expand far beyond its current scope. As we continue to better understand the abundance and distribution of preserved eDNA, and continue to improve wet lab and computation methods for recovering and understanding the information therein, it seems probable that sedDNA research could revolutionize our perceptions as to the resolution of information we can derive from buried records, even from otherwise unremarkable deposits. This new understanding is not only exciting for the future of Quaternary science, but it underscores a need for responsibility in safeguarding these vulnerable biomolecular archives. Not only do artifacts and bones contain dense, interwoven clues of a forgotten past, so does the entirety of the environmental matrix.

Acknowledgements TJM wishes to thank the CANA Foundation for their financial support. We are grateful for fruitful comments and revisions provided by Diyendo Massilani and Eric Capo.

References

Ahmed E, Parducci L, Unneberg P, Ågren R, Schenk F, Rattray JE, Han L, Muschitiello F, Pedersen MW, Smittenberg RH, Yamoah KA, Slotte T, Wohlfarth B (2018) Archaeal community changes in Lateglacial lake sediments: evidence from ancient DNA. Quat Sci Rev 181:19–29. https://doi.org/10.1016/j.quascirev.2017.11.037

Aitken MJ (1962) Physics applied to archaeology – Part I. Contemp Phys 3(3):161–176. https://doi.org/10.1080/00107516208221792

Allentoft ME, Collins M, Harker D, Haile J, Oskam CL, Hale ML, Campos PF, Samaniego JA, Gilbert MTP, Willerslev E, Zhang G, Scofield RP, Holdaway RN, Bunce M (2012) The half-life of DNA in bone: measuring decay kinetics in 158 dated fossils. Proc R Soc B Biol Sci 279(1748):4724–4733. https://doi.org/10.1098/rspb.2012.1745

Altschul SF, Gish W, Miller W, Myers EW, Lipman DJ (1990) Basic local alignment search tool. J Mol Biol 215(3):403–410. https://doi.org/10.1016/S0022-2836(05)80360-2

Andersen K, Bird KL, Rasmussen M, Haile J, Breuning-Madsen H, Kjaer KH, Orlando L, Gilbert MTP, Willerslev E (2012) Meta-barcoding of “dirt” DNA from soil reflects vertebrate biodiversity. Mol Ecol 21(8):1966–1979. https://doi.org/10.1111/j.1365-294X.2011.05261.x

Anderson-Carpenter LL, McLachlan JS, Jackson ST, Kuch M, Lumibao CY, Poinar HN (2011) Ancient DNA from lake sediments: Bridging the gap between paleoecology and genetics. BMC Evol Biol 11(1):30. https://doi.org/10.1186/1471-2148-11-30

Andruszkiewicz Allan E, Zhang WG, Lavery A, Govindarajan AF (2021) Environmental DNA shedding and decay rates from diverse animal forms and thermal regimes. Environmental DNA 3(2):492–514. https://doi.org/10.1002/edn3.141

Ardelean CF, Becerra-Valdivia L, Pedersen MW, Schwenninger JL, Oviatt CG, Macías-Quintero JI, Arroyo-Cabrales J, Sikora M, Ocampo-Díaz YZE, Rubio-Cisneros II, Watling JG, de Medeiros VB, De Oliveira PE, Barba-Pingarón L, Ortiz-Butrón A, Blancas-Vázquez J, Rivera-González I, Solís-Rosales C, Rodríguez-Ceja M, Gandy DA, Navarro-Gutierrez Z, De La Rosa-Díaz JJ, Huerta-Arellano V, Marroquín-Fernández MB, Martínez-Riojas LM, López-Jiménez A, Higham T, Willerslev E (2020) Evidence of human occupation in Mexico around the last glacial maximum. Nature 584(7819):87–92. https://doi.org/10.1038/s41586-020-2509-0

Armbrecht LH, Coolen MJL, Lejzerowicz F, George SC, Negandhi K, Suzuki Y, Young J, Foster NR, Armand LK, Cooper A, Ostrowski M, Focardi A, Stat M, Moreau JW, Weyrich LS (2019) Ancient DNA from marine sediments: precautions and considerations for seafloor coring, sample handling and data generation. Earth Sci Rev 196:102887. https://doi.org/10.1016/j.earscirev.2019.102887

Armbrecht L, Weber ME, Raymo ME, Peck VL, Williams T, Warnock J, Kato Y, Hernández-Almeida I, Hoem F, Reilly B, Hemming S, Bailey I, Martos YM, Gutjahr M, Percuoco V, Allen C, Brachfeld S, Cardillo FG, Du Z, Fauth G, Fogwill C, Garcia M, Glüder A, Guitard M, Hwang J-H, Iizuka M, Kenlee B, O'Connell S, Pérez LF, Ronge TA, Seki O, Tauxe L, Tripathi S, Zheng X (2022) Ancient marine sediment DNA reveals diatom transition in Antarctica. Nat Commun 13(1):5787. https://doi.org/10.1038/s41467-022-33494-4

Arnold LJ, Roberts RG, Macphee RDE, Haile JS, Brock F, Möller P, Froese DG, Tikhonov AN, Chivas AR, Gilbert MTP, Willerslev E (2011) Paper II – Dirt, dates and DNA: OSL and radiocarbon chronologies of perennially frozen sediments in Siberia, and their implications for sedimentary ancient DNA studies. Boreas 40(3):417–445. https://doi.org/10.1111/j.1502-3885.2010.00181.x

Ávila-Arcos MC, Cappellini E, Romero-Navarro JA, Wales N, Moreno-Mayar JV, Rasmussen M, Fordyce SL, Montiel R, Vielle-Calzada JP, Willerslev E, Gilbert MTP (2011) Application and comparison of large-scale solution-based DNA capture-enrichment methods on ancient DNA. Sci Rep 1(74). https://doi.org/10.1038/srep00074

Bajard M, Poulenard J, Sabatier P, Etienne D, Ficetola F, Chen W, Gielly L, Taberlet P, Develle A-L, Rey P-J, Moulin B, de Beaulieu J-L, Arnaud F (2017) Long-term changes in alpine pedogenetic processes: effect of millennial agro-pastoralism activities (French-Italian Alps). Geoderma 306:217–236. https://doi.org/10.1016/j.geoderma.2017.07.005

Bajard M, Poulenard J, Sabatier P, Bertrand Y, Crouzet C, Ficetola GF, Blanchet C, Messager E, Giguet-Covex C, Gielly L, Rioux D, Chen W, Malet E, Develle A-L, Arnaud F (2020) Pastoralism increased vulnerability of a subalpine catchment to flood hazard through changing soil properties. Palaeogeogr Palaeoclimatol Palaeoecol 538:109462. https://doi.org/10.1016/j.palaeo.2019.109462

Baker AG, Bhagwat SA, Willis KJ (2013) Do dung fungal spores make a good proxy for past distribution of large herbivores? Quat Sci Rev 62:21–31. https://doi.org/10.1016/j.quascirev.2012.11.018

Barta JL, Monroe C, Teisberg JE, Winters M, Flanigan K, Kemp BM (2014) One of the key characteristics of ancient DNA, low copy number, may be a product of its extraction. J Archaeol Sci 46:281–289. https://doi.org/10.1016/j.jas.2014.03.030

Bartlett JMS, Stirling D (2003) A Short History of the Polymerase Chain Reaction. In: Bartlett JMS, Stirling D (eds) PCR Protocols. Humana Press, Totowa, NJ, pp 3–6

Binford SR, Binford LR (1968) New perspectives in archeology. Aldine Publishing Company, Chicago

Blum SAE, Lorenz MG, Wackernagel W (1997) Mechanism of retarded DNA degradation and prokaryotic origin of DNases in nonsterile soils. Syst Appl Microbiol 20(4):513–521. https://doi.org/10.1016/S0723-2020(97)80021-5

Boessenkool S, Epp LS, Haile J, Bellemain E, Edwards M, Coissac E, Willerslev E, Brochmann C (2012) Blocking human contaminant DNA during PCR allows amplification of rare mammal species from sedimentary ancient DNA. Mol Ecol 21(8):1806–1815. https://doi.org/10.1111/j.1365-294X.2011.05306.x

Borry M, Hübner A, Rohrlach AB, Warinner C (2021) PyDamage: automated ancient damage identification and estimation for contigs in ancient DNA de novo assembly. PeerJ 9:e11845. https://doi.org/10.7717/peerj.11845

Bourgeon L, Burke A, Higham T (2017) Earliest human presence in North America dated to the last glacial maximum: new radiocarbon dates from Bluefish Caves, Canada. PLoS ONE 12(1):e0169486. https://doi.org/10.1371/journal.pone.0169486

Boyer F, Mercier C, Bonin A, Le Bras Y, Taberlet P, Coissac E (2016) obitools: a unix-inspired software package for DNA metabarcoding. Mol Ecol Resour 16(1):176–182. https://doi.org/10.1111/1755-0998.12428

Briggs AW, Stenzel U, Johnson PLF, Green RE, Kelso J, Prüfer K, Meyer M, Krause J, Ronan MT, Lachmann M, Pääbo S (2007) Patterns of damage in genomic DNA sequences from a Neandertal. Proc Natl Acad Sci U S A 104(37):14616–14621. https://doi.org/10.1073/pnas.0704665104

Bromham L, Penny D (2003) The modern molecular clock. Nat Rev Genet 4(3):216–224. https://doi.org/10.1038/nrg1020

Brown, TA, Barnes IM, (2015) The current and future applications of ancient DNA in Quaternary science. J Quat Sci 30:144–153. https://doi.org/10.1002/jqs.2770

Brown J, Ferrians OJ, Heginbottom JA, Melnikov ES (eds) (1997) Circum-Arctic map of permafrost and ground-ice conditions. U. S. Geological Survey in Cooperation with the Circum-Pacific Council for Energy and Mineral Resources. Washington, DC. Circum-Pacific Map Series CP-45, scale 1:10,000,000, 1 sheet

Brown J, Ferrians OJ, Heginbottom JA Jr, Melnikov ES (1998) Circum-arctic map of permafrost and ground ice conditions. National Snow and Ice Data Center/World Data Center for Glaciology, Boulder. Digital media. Revised February 2001

Brown J, Ferrians OJ, Heginbottom JA Jr, Melnikov ES (2002) Circum-Arctic map of permafrost and ground-ice conditions, version 2. NASA National Snow and Ice Data Center Distributed Active Archive Center, Boulder. https://doi.org/10.7265/skbgkf16

Brown AG, Van Hardenbroek M, Fonville T, Davies K, Mackay H, Murray E, Head K, Barratt P, McCormick F, Ficetola GF, Gielly L, Henderson ACG, Crone A, Cavers G, Langdon PG, Whitehouse NJ, Pirrie D, Alsos IG (2021) Ancient DNA, lipid biomarkers and palaeoecological evidence reveals construction and life on early medieval lake settlements. Sci Rep 11(1):11807. https://doi.org/10.1038/s41598-021-91057-x

Brown AG, Fonville T, van Hardenbroek M, Cavers G, Crone A, McCormick F, Murray E, Mackay H, Whitehouse NJ, Henderson ACG, Barratt P, Davies K, Head K, Langdon P, Alsos IG, Pirrie D (2022a) New integrated molecular approaches for investigating lake settlements in north-western Europe. Antiquity 96(389):1179–1199. https://doi.org/10.15184/aqy.2022.70

Brown T, Rijal DP, Heintzman PD, Clarke CL, Blankholm HP, Høeg HI, Lammers Y, Bråthen KA, Edwards M, Alsos IG (2022b) Paleoeconomy more than demography determined pre-historic human impact in Arctic Norway. PNAS Nexus 1(5):pgac209. https://doi.org/10.1093/pnasnexus/pgac209

Buchfink B, Reuter K, Drost H-G (2021) Sensitive protein alignments at tree-of-life scale using DIAMOND. Nat Methods 18(4):366–368. https://doi.org/10.1038/s41592-021-01101-x

Buffalo V (2015) Bioinformatics data skills: reproducible and robust research with open source tools. O'Reilly Media, Inc.

Bullen TF, Forrest S, Campbell F, Dodson AR, Hershman MJ, Pritchard DM, Turner JR, Montrose MH, Watson AJM (2006) Characterization of epithelial cell shedding from human small intestine. Lab Investig 86(10):1052–1063. https://doi.org/10.1038/labinvest.3700464

Burney DA, Robinson GS, Burney LP (2003) Sporormiella and the late Holocene extinctions in Madagascar. Proc Natl Acad Sci 100(19):10800–10805. https://doi.org/10.1073/pnas.1534700100

Cai P, Huang Q, Zhang X, Chen H (2006a) Adsorption of DNA on clay minerals and various colloidal particles from an Alfisol. Soil Biol Biochem 38(3):471–476. https://doi.org/10.1016/j.soilbio.2005.05.019

Cai P, Huang Q-Y, Zhang X-W (2006b) Interactions of DNA with clay minerals and soil colloidal particles and protection against degradation by DNase. Environ Sci Technol 40(9):2971–2976. https://doi.org/10.1021/es0522985

Capo E, Giguet-Covex C, Rouillard A, Nota K, Heintzman PD, Vuillemin A, Ariztegui D, Arnaud F, Belle S, Bertilsson S, Bigler C, Bindler R, Brown AG, Clarke CL, Crump SE, Debroas D, Englund G, Ficetola GF, Garner RE, Gauthier J, Gregory-Eaves I, Heinecke L, Herzschuh U, Ibrahim A, Kisand V, Kjær KH, Lammers Y, Littlefair J, Messager E, Monchamp M-E, Olajos F, Orsi W, Pedersen MW, Rijal DP, Rydberg J, Spanbauer T, Stoof-Leichsenring KR, Taberlet P, Talas L, Thomas C, Walsh DA, Wang Y, Willerslev E, van Woerkom A, Zimmermann HH, Coolen MJL, Epp LS, Domaizon I, G. Alsos I, Parducci L (2021) Lake sedimentary DNA research on past terrestrial and aquatic biodiversity: overview and recommendations. Quaternary 4(1):6. https://doi.org/10.3390/quat4010006

Carøe C, Gopalakrishnan S, Vinner L, Mak SST, Sinding MHS, Samaniego JA, Wales N, Sicheritz-Pontén T, Gilbert MTP (2018) Single-tube library preparation for degraded DNA. Methods Ecol Evol 9(2):410–419. https://doi.org/10.1111/2041-210X.12871

Carpenter ML, Buenrostro JD, Valdiosera C, Schroeder H, Allentoft ME, Sikora M, Rasmussen M, Gravel S, Guillén S, Nekhrizov G, Leshtakov K, Dimitrova D, Theodossiev N, Pettener D, Luiselli D, Sandoval K, Moreno-Estrada A, Li Y, Wang J, Gilbert MTP, Willerslev E, Greenleaf WJ, Bustamante CD (2013) Pulling out the 1%: whole-genome capture for the targeted enrichment of ancient DNA sequencing libraries. Am J Hum Genet 93(5):852–864. https://doi.org/10.1016/j.ajhg.2013.10.002

Chen W, Ficetola GF (2019) Conditionally autoregressive models improve occupancy analyses of autocorrelated data: an example with environmental DNA. Mol Ecol Resour 19(1):163–175. https://doi.org/10.1111/1755-0998.12949

Clare EL, Economou CK, Faulkes CG, Gilbert JD, Bennett F, Drinkwater R, Littlefair JE (2021) eDNAir: proof of concept that animal DNA can be collected from air sampling. PeerJ 9:e11030. https://doi.org/10.7717/peerj.11030

Clark PU, Archer D, Pollard D, Blum JD, Rial JA, Brovkin V, Mix AC, Pisias NG, Roy M (2006) The middle Pleistocene transition: characteristics, mechanisms, and implications for long-term changes in atmospheric pCO2. Quat Sci Rev 25(23):3150–3184. https://doi.org/10.1016/j.quascirev.2006.07.008

Clark K, Karsch-Mizrachi I, Lipman DJ, Ostell J, Sayers EW (2016) GenBank. Nucleic Acids Res 44(D1):D67–D72. https://doi.org/10.1093/nar/gkv1276

Clarke DL (1968) Analytical archaeology. Routledge, Abingdon

Comandini O, Rinaldi AC (2004) Tracing megafaunal extinctions with dung fungal spores. Mycologist 18(4):140–142. https://doi.org/10.1017/S0269-915X(04)00401-X

Conroy KJ, Baker AG, Jones VJ, van Hardenbroek M, Hopla EJ, Collier R, Lister AM, Edwards ME (2020) Tracking late-Quaternary extinctions in interior Alaska using megaherbivore bone remains and dung fungal spores. Quat Res 97:99–110. https://doi.org/10.1017/qua.2020.19

Coolen MJL, Gibson JAE (2009) Ancient DNA in lake sediment records. PAGES News 17:104–106. https://doi.org/10.22498/pages.17.3.104

Cooper A, Poinar HN (2000) Ancient DNA: do it right or not at all. Science 289(5482):1139. https://doi.org/10.1126/science.289.5482.1139b

Courtin J, Perfumo A, Andreev AA, Opel T, Stoof-Leichsenring KR, Edwards ME, Murton JB, Herzschuh U (2022) Pleistocene glacial and interglacial ecosystems inferred from ancient DNA analyses of permafrost sediments from Batagay megaslump, East Siberia. Environmental DNA 4(6):1265–1283. https://doi.org/10.1002/edn3.336

Cribdon B, Ware R, Smith O, Gaffney V, Allaby RG (2020) PIA: more accurate taxonomic assignment of metagenomic data demonstrated on sedaDNA from the North Sea. Front Ecol Evol:8

Curtin L, D'Andrea WJ, Balascio NL, Shirazi S, Shapiro B, de Wet GA, Bradley RS, Bakke J (2021) Sedimentary DNA and molecular evidence for early human occupation of the Faroe Islands. Commun Earth Environ 2(1):1–7. https://doi.org/10.1038/s43247-021-00318-0

Dabney J, Meyer M, Pääbo S (2013) Ancient DNA damage. Cold Spring Harb Perspect Biol 5(7):a012567. https://doi.org/10.1101/cshperspect.a012567

Davis OK, Shafer DS (2006) Sporormiella fungal spores, a palynological means of detecting herbivore density. Palaeogeogr Palaeoclimatol Palaeoecol 237(1):40–50. https://doi.org/10.1016/j.palaeo.2005.11.028

De Barba M, Miquel C, Boyer F, Mercier C, Rioux D, Coissac E, Taberlet P (2014) DNA metabarcoding multiplexing and validation of data accuracy for diet assessment: application to omnivorous diet. Mol Ecol Resour 14(2):306–323. https://doi.org/10.1111/1755-0998.12188

Deetz J (1977) In Small Things Forgotten: An Archaeology of Early American Life, Expanded,Subsequent edition. Anchor, New York

Dell'Anno A, Corinaldesi C (2004) Degradation and turnover of extracellular DNA in marine sediments: ecological and methodological considerations. Appl Environ Microbiol 70(7):4384. https://doi.org/10.1128/AEM.70.7.4384-4386.2004

Dillehay TD, Ocampo C, Saavedra J, Sawakuchi AO, Vega RM, Pino M, Collins MB, Cummings LS, Arregui I, Villagran XS, Hartmann GA, Mella M, González A, Dix G (2015) New archaeological evidence for an early human presence at Monte Verde, Chile. PLoS ONE 10(11):e0141923. https://doi.org/10.1371/journal.pone.0141923

Dobinski W (2011) Permafrost. Earth Sci Rev 108(3):158–169. https://doi.org/10.1016/j.earscirev.2011.06.007

Dommain R, Andama M, McDonough MM, Prado NA, Goldhammer T, Potts R, Maldonado JE, Nkurunungi JB, Campana MG (2020) The challenges of reconstructing tropical biodiversity with sedimentary ancient DNA: a 2200-year-long metagenomic record from Bwindi impenetrable forest, Uganda. Front Ecol Evol 8

Drummond AJ, Rambaut A (2007) BEAST: Bayesian evolutionary analysis by sampling trees. BMC Evol Biol 7(1):214. https://doi.org/10.1186/1471-2148-7-214

Dubois N, Jacob J (2016) Molecular biomarkers of anthropic impacts in natural archives: a review. Front Ecol Evol 4

Edwards ME, Alsos IG, Yoccoz N, Coissac E, Goslar T, Gielly L, Haile J, Langdon CT, Tribsch A, Binney HA, von Stedingk H, Taberlet P (2018) Metabarcoding of modern soil DNA gives a highly local vegetation signal in Svalbard tundra. The Holocene 28(12):2006–2016. https://doi.org/10.1177/0959683618798095

Ejarque A, Miras Y, Riera S (2011) Pollen and non-pollen palynomorph indicators of vegetation and highland grazing activities obtained from modern surface and dung datasets in the eastern Pyrenees. Rev Palaeobot Palynol 167(1–2):123

Ekblom R, Wolf JBW (2014) A field guide to whole-genome sequencing, assembly and annotation. Evol Appl 7(9):1026–1042. https://doi.org/10.1111/eva.12178

Elias SA, Short SK, Nelson CH, Birks HH (1996) Life and times of the Bering land bridge. Nature 382(6586):60–63. https://doi.org/10.1038/382060a0

Epp LS, Boessenkool S, Bellemain EP, Haile J, Esposito A, Riaz T, Erséus C, Gusarov VI, Edwards ME, Johnsen A, Stenøien HK, Hassel K, Kauserud H, Yoccoz NG, Bråthen KA, Willerslev E, Taberlet P, Coissac E, Brochmann C (2012) New environmental metabarcodes for analysing soil DNA: potential for studying past and present ecosystems. Mol Ecol 21(8):1821–1833. https://doi.org/10.1111/j.1365-294X.2012.05537.x

Etienne D, Wilhelm B, Sabatier P, Reyss J-L, Arnaud F (2013) Influence of sample location and livestock numbers on Sporormiella concentrations and accumulation rates in surface sediments of Lake Allos, French Alps. J Paleolimnol 49(2):117–127. https://doi.org/10.1007/s10933-012-9646-x

Etienne D, Destas M, Lyautey E, Marti R, Ruffaldi P, Georges-Leroy M, Dambrine E, Topp E (2015) Two thousand-year reconstruction of livestock production intensity in France using sediment-archived fecal Bacteroidales and source-specific mitochondrial markers. The Holocene 25(9):1384–1393. https://doi.org/10.1177/0959683615585836

Feranec RS, Miller NG, Lothrop JC, Graham RW (2011) The Sporormiella proxy and end-Pleistocene megafaunal extinction: A perspective. Quat Int 245(2):333–338. https://doi.org/10.1016/j.quaint.2011.06.004

Feuerborn TR, Palkopoulou E, van der Valk T, von Seth J, Munters AR, Pečnerová P, Dehasque M, Ureña I, Ersmark E, Lagerholm VK, Krzewińska M, Rodríguez-Varela R, Götherström A, Dalén L, Díez-del-Molino D (2020) Competitive mapping allows for the identification and exclusion of human DNA contamination in ancient faunal genomic datasets. BMC Genomics 21(1):844. https://doi.org/10.1186/s12864-020-07229-y

Ficetola GF, Pansu J, Bonin A, Coissac E, Giguet-Covex C, De Barba M, Gielly L, Lopes CM, Boyer F, Pompanon F, Rayé G, Taberlet P (2015) Replication levels, false presences and the estimation of the presence/absence from eDNA metabarcoding data. Mol Ecol Resour 15(3):543–556. https://doi.org/10.1111/1755-0998.12338

Ficetola GF, Poulenard J, Sabatier P, Messager E, Gielly L, Leloup A, Etienne D, Bakke J, Malet E, Fanget B, Støren E, Reyss J-L, Taberlet P, Arnaud F (2018) DNA from lake sediments reveals long-term ecosystem changes after a biological invasion. Science Advances 4(5):eaar4292. https://doi.org/10.1126/sciadv.aar4292

Fonseca NA, Rung J, Brazma A, Marioni JC (2012) Tools for mapping high-throughput sequencing data. Bioinformatics 28(24):3169–3177. https://doi.org/10.1093/bioinformatics/bts605

Forootan A, Sjöback R, Björkman J, Sjögreen B, Linz L, Kubista M (2017) Methods to determine limit of detection and limit of quantification in quantitative real-time PCR (qPCR). Biomol Detect Quantif 12:1–6. https://doi.org/10.1016/j.bdq.2017.04.001

French HM (2007) Periglacial Environment 3e, 3rd edition. John Wiley &Sons, Chichester, England ; Hoboken, NJ

Frostegård Å, Courtois S, Ramisse V, Clerc S, Bernillon D, Le Gall F, Jeannin P, Nesme X, Simonet P (1999) Quantification of bias related to the extraction of DNA directly from soils. Appl Environ Microbiol 65(12):5409–5420

Fu GK, Xu W, Wilhelmy J, Mindrinos MN, Davis RW, Xiao W, Fodor SPA (2014) Molecular indexing enables quantitative targeted RNA sequencing and reveals poor efficiencies in standard library preparations. Proc Natl Acad Sci U S A 111(5):1891–1896. https://doi.org/10.1073/pnas.1323732111

Gallo RL (2017) Human Skin Is the Largest Epithelial Surface for Interaction with Microbes. J Invest Dermatol 137(6):1213–1214. https://doi.org/10.1016/j.jid.2016.11.045

Gansauge M-T, Meyer M (2013) Single-stranded DNA library preparation for the sequencing of ancient or damaged DNA. Nat Protoc 8(4):737–748. https://doi.org/10.1038/nprot.2013.038

Gansauge M-T, Gerber T, Glocke I, Korlevic P, Lippik L, Nagel S, Riehl LM, Schmidt A, Meyer M (2017) Single-stranded DNA library preparation from highly degraded DNA using T4 DNA ligase. Nucleic Acids Res 45(10):e79. https://doi.org/10.1093/nar/gkx033

Gansauge M-T, Aximu-Petri A, Nagel S, Meyer M (2020) Manual and automated preparation of single-stranded DNA libraries for the sequencing of DNA from ancient biological remains and other sources of highly degraded DNA. Nat Protoc 15(8):2279–2300. https://doi.org/10.1038/s41596-020-0338-0

Garcés-Pastor S, Coissac E, Lavergne S, Schwörer C, Theurillat J-P, Heintzman PD, Wangensteen OS, Tinner W, Rey F, Heer M, Rutzer A, Walsh K, Lammers Y, Brown AG, Goslar T, Rijal DP, Karger DN, Pellissier L, Heiri O, Alsos IG (2022) High resolution ancient sedimentary DNA shows that alpine plant diversity is associated with human land use and climate change. Nat Commun 13(1):6559. https://doi.org/10.1038/s41467-022-34010-4

Gelabert P, Sawyer S, Bergström A, Margaryan A, Collin TC, Meshveliani T, Belfer-Cohen A, Lordkipanidze D, Jakeli N, Matskevich Z, Bar-Oz G, Fernandes DM, Cheronet O, Özdoğan KT, Oberreiter V, Feeney RNM, Stahlschmidt MC, Skoglund P, Pinhasi R (2021) Genome-scale sequencing and analysis of human, wolf, and bison DNA from 25,000-year-old sediment. Curr Biol 31(16):3564–3574.e9. https://doi.org/10.1016/j.cub.2021.06.023

Giguet-Covex C, Pansu J, Arnaud F, Rey P-J, Griggo C, Gielly L, Domaizon I, Coissac E, David F, Choler P, Poulenard J, Taberlet P (2014) Long livestock farming history and human landscape shaping revealed by lake sediment DNA. Nat Commun 5(1):3211. https://doi.org/10.1038/ncomms4211

Giguet-Covex C, Ficetola GF, Walsh K, Poulenard J, Bajard M, Fouinat L, Sabatier P, Gielly L, Messager E, Develle AL, David F, Taberlet P, Brisset E, Guiter F, Sinet R, Arnaud F (2019) New insights on lake sediment DNA from the catchment: importance of taphonomic and analytical issues on the record quality. Sci Rep 9:14676. https://doi.org/10.1038/s41598-019-50339-1

Giguet-Covex C, Bajard M, Chen W, Walsh KJ, Rey PJ, Messager E, Etienne D, Sabatier P, Ficetola FG, Gielly L, Blanchet C (2023) Long-term trajectories of mountain agro-ecosystems in the North-Western Alps. Reg Environ Chang 23(58). https://doi.org/10.1007/s10113-023-02030-5

Gill JL, Williams JW, Jackson ST, Lininger KB, Robinson GS (2009) Pleistocene megafaunal collapse, novel plant communities, and enhanced fire regimes in North America. Science 326(5956):1100–1103. https://doi.org/10.1126/science.1179504

Gill JL, Williams JW, Jackson ST, Donnelly JP, Schellinger GC (2012) Climatic and megaherbivory controls on late-glacial vegetation dynamics: a new, high-resolution, multi-proxy record from Silver Lake, Ohio. Quat Sci Rev 34:66–80. https://doi.org/10.1016/j.quascirev.2011.12.008

Gill JL, McLauchlan KK, Skibbe AM, Goring S, Zirbel CR, Williams JW (2013) Linking abundances of the dung fungus Sporormiella to the density of bison: implications for assessing grazing by megaherbivores in palaeorecords. J Ecol 101(5):1125–1136. https://doi.org/10.1111/1365-2745.12130

Ginolhac A, Rasmussen M, Gilbert MTP, Willerslev E, Orlando L (2011) mapDamage: testing for damage patterns in ancient DNA sequences. Bioinformatics 27(15):2153–2155. https://doi.org/10.1093/bioinformatics/btr347

Glocke I, Meyer M (2017) Extending the spectrum of DNA sequences retrieved from ancient bones and teeth. Genome Res 27(7):1230–1237. https://doi.org/10.1101/gr.219675.116

Goebel T, Waters MR, O'Rourke DH (2008) The late pleistocene dispersal of modern humans in the Americas. Science 319(5869):1497–1502. https://doi.org/10.1126/science.1153569

Goodwin S, McPherson JD, McCombie WR (2016) Coming of age: ten years of next-generation sequencing technologies. Nat Rev Genet 17(6):333–351. https://doi.org/10.1038/nrg.2016.49

Graham RW, Belmecheri S, Choy K, Culleton BJ, Davies LJ, Froese D, Heintzman PD, Hritz C, Kapp JD, Newsom LA, Rawcliffe R, Saulnier-Talbot É, Shapiro B, Wang Y, Williams JW, Wooller MJ (2016) Timing and causes of mid-Holocene mammoth extinction on St. Paul Island, Alaska. Proc Natl Acad Sci 113(33):9310–9314. https://doi.org/10.1073/pnas.1604903113

Green RE, Krause J, Ptak SE, Briggs AW, Ronan MT, Simons JF, Du L, Egholm M, Rothberg JM, Paunovic M, Pääbo S (2006) Analysis of one million base pairs of Neanderthal DNA. Nature 444(7117):330–336. https://doi.org/10.1038/nature05336

Guardiola M, Uriz MJ, Taberlet P, Coissac E, Wangensteen OS, Turon X (2015) Deep-sea, deep-sequencing: metabarcoding extracellular DNA from sediments of marine canyons. PLoS ONE 10(10):e0139633. https://doi.org/10.1371/journal.pone.0139633

Gunn J (ed) (2004) Encyclopedia of Caves and Karst Science. Routledge, New York

Haile J (2012) Ancient DNA extraction from soils and sediments. Methods Mol Biol 840:57–63. https://doi.org/10.1007/978-1-61779-516-9_8

Haile J, Holdaway R, Oliver K, Bunce M, Gilbert MTP, Nielsen R, Munch K, Ho SYW, Shapiro B, Willerslev E (2007) Ancient DNA chronology within sediment deposits: are paleobiological reconstructions possible and is DNA leaching a factor? Mol Biol Evol 24(4):982–989. https://doi.org/10.1093/molbev/msm016

Haile J, Froese DG, Macphee RDE, Roberts RG, Arnold LJ, Reyes AV, Rasmussen M, Nielsen R, Brook BW, Robinson S, Demuro M, Gilbert MTP, Munch K, Austin JJ, Cooper A, Barnes I, Möller P, Willerslev E (2009) Ancient DNA reveals late survival of mammoth and horse in interior Alaska. Proc Natl Acad Sci U S A 106(52):22352–22357. https://doi.org/10.1073/pnas.0912510106

Handsley-Davis M, Kowal E, Russell L, Weyrich LS (2021) Researchers using environmental DNA must engage ethically with Indigenous communities. https://doi.org/10.1038/s41559-020-01351-6

Hansen AJ, Mitchell DL, Wiuf C, Paniker L, Brand TB, Binladen J, Gilichinsky DA, Rønn R, Willerslev E (2006) Crosslinks rather than strand breaks determine access to ancient DNA sequences from frozen sediments. Genetics 173(2):1175–1179. https://doi.org/10.1534/genetics.106.057349

Hardy CM, Krull ES, Hartley DM, Oliver RL (2010) Carbon source accounting for fish using combined DNA and stable isotope analyses in a regulated lowland river weir pool. Mol Ecol 19(1):197–212. https://doi.org/10.1111/j.1365-294X.2009.04411.x

Harper LR, Lawson Handley L, Carpenter AI, Ghazali M, Di Muri C, Macgregor CJ, Logan TW, Law A, Breithaupt T, Read DS, McDevitt AD, Hanfling B (2019) Environmental DNA (eDNA) metabarcoding of pond water as a tool to survey conservation and management priority mammals. Biol Conserv 238(108225):UNSP 108225. https://doi.org/10.1016/j.biocon.2019.108225

Harrison JB, Sunday JM, Rogers SM (2019) Predicting the fate of eDNA in the environment and implications for studying biodiversity. Proc R Soc B Biol Sci 286(1915):20191409. https://doi.org/10.1098/rspb.2019.1409

Hebsgaard MB, Gilbert MTP, Arneborg J, Heyn P, Allentoft ME, Bunce M, Munch K, Schweger C, Willerslev E (2009) 'The Farm Beneath the Sand' – an archaeological case study on ancient 'dirt' DNA. Antiquity 83(320):430–444. https://doi.org/10.1017/S0003598X00098537

Hedges SB, Schweitzer MH (1995) Detecting dinosaur DNA. Science 268(5214):1191–1192.; author reply 1194. https://doi.org/10.1126/science.7761839

Heginbottom JA, Brown J, Melnikov ES, Ferrians Jr, OJ (1993) Circumarctic map of permafrost and ground ice conditions. In Proceedings of the Sixth International Conference on Permafrost (Vol. 2, pp. 1132–1136). Wushan Guangzhou, China: South China University of Technology Press.

Heintzman PD, Froese D, Ives JW, Soares AER, Zazula GD, Letts B, Andrews TD, Driver JC, Hall E, Hare PG, Jass CN, MacKay G, Southon JR, Stiller M, Woywitka R, Suchard MA, Shapiro B (2016) Bison phylogeography constrains dispersal and viability of the Ice Free Corridor in western Canada. Proc Natl Acad Sci 113(29):8057–8063. https://doi.org/10.1073/pnas.1601077113

Herbig A, Maixner F, Bos KI, Zink A, Krause J, Huson DH (2016) MALT: fast alignment and analysis of metagenomic DNA sequence data applied to the Tyrolean Iceman:050559

Higuchi R, Bowman B, Freiberger M, Ryder OA, Wilson AC (1984) DNA sequences from the quagga, an extinct member of the horse family. Nature 312(5991):282–284. https://doi.org/10.1038/312282a0

Hofreiter M, Mead JI, Martin P, Poinar HN (2003) Molecular caving. Curr Biol 13(18):R693–R695. https://doi.org/10.1016/j.cub.2003.08.039

Holmsgaard PN, Norman A, Hede SC, PHB P, Al-Soud WA, Hansen LH, Sørensen SJ (2011) Bias in bacterial diversity as a result of Nycodenz extraction from bulk soil. Soil Biol Biochem 43(10):2152–2159. https://doi.org/10.1016/j.soilbio.2011.06.019

Horn S (2012) Target enrichment via DNA hybridization capture. Methods Mol Biol 840:177–188. https://doi.org/10.1007/978-1-61779-516-9_21

Huson DH, Auch AF, Qi J, Schuster SC (2007) MEGAN analysis of metagenomic data. Genome Res 17(3):377–386. https://doi.org/10.1101/gr.5969107

Jaanusson V (1976). Faunal dynamics in the middle Ordovician (Viruan) of Balto-Scandia. In: The Ordovician System: Proceedings of a Palaeontological Association Symposium. pp. 301œ326. University of Wales Press and National Museum of Wales, Cardiff, pp. 301–326.

Jass CN, George CO (2010) An assessment of the contribution of fossil cave deposits to the Quaternary paleontological record. Quat Int 217(1):105–116. https://doi.org/10.1016/j.quaint.2009.11.008

Jelavic S, Lanson B, Findling N, Lanson M (2023) A route for long-term DNA preservation through nanoconfinement in smectites

Jia W, Anslan S, Chen F, Cao X, Dong H, Dulias K, Gu Z, Heinecke L, Jiang H, Kruse S, Kang W, Li K, Liu S, Liu X, Liu Y, Ni J, Schwalb A, Stoof-Leichsenring KR, Shen W, Tian F, Wang J, Wang Y, Wang Y, Xu H, Yang X, Zhang D, Herzschuh U (2022) Sedimentary ancient DNA reveals past ecosystem and biodiversity changes on the Tibetan Plateau: Overview and prospects. Quat Sci Rev 293:107703. https://doi.org/10.1016/j.quascirev.2022.107703

Johnson SS, Hebsgaard MB, Christensen TR, Mastepanov M, Nielsen R, Munch K, Brand T, Gilbert MTP, Zuber MT, Bunce M, Rønn R, Gilichinsky D, Froese D, Willerslev E (2007) Ancient bacteria show evidence of DNA repair. Proc Natl Acad Sci U S A 104(36):14401–14405. https://doi.org/10.1073/pnas.0706787104

Jónsson H, Ginolhac A, Schubert M, Johnson PLF, Orlando L (2013) mapDamage2.0: fast approximate Bayesian estimates of ancient DNA damage parameters. Bioinformatics 29(13):1682–1684. https://doi.org/10.1093/bioinformatics/btt193

Kanbar HJ, Olajos F, Englund G, Holmboe M (2020) Geochemical identification of potential DNA-hotspots and DNA-infrared fingerprints in lake sediments. Appl Geochem 122:104728. https://doi.org/10.1016/j.apgeochem.2020.104728

Kapp JD, Green RE, Shapiro B (2021) A fast and efficient single-stranded genomic library preparation method optimized for ancient DNA. J Hered 112(3):241–249. https://doi.org/10.1093/jhered/esab012

Kim D, Song L, Breitwieser FP, Salzberg SL (2016) Centrifuge: rapid and sensitive classification of metagenomic sequences. Genome Res 26(12):1721–1729. https://doi.org/10.1101/gr.210641.116

Kircher M, Sawyer S, Meyer M (2012) Double indexing overcomes inaccuracies in multiplex sequencing on the Illumina platform. Nucleic Acids Res 40(1):e3. https://doi.org/10.1093/nar/gkr771

Kirkpatrick J, Walsh E, D'Hondt S (2016) Fossil DNA persistence and decay in marine sediment over hundred-thousand-year to million-year time scales. Graduate School of Oceanography Faculty Publications. https://doi.org/10.1130/G37933.1

Kistler L, Ware R, Smith O, Collins M, Allaby RG (2017) A new model for ancient DNA decay based on paleogenomic meta-analysis. Nucleic Acids Res 45(11):6310–6320. https://doi.org/10.1093/nar/gkx361

Kitchen A, Miyamoto MM, Mulligan CJ (2008) A three-stage colonization model for the peopling of the Americas. PLoS ONE 3(2):e1596. https://doi.org/10.1371/journal.pone.0001596

Kjær KH, Winther Pedersen M, De Sanctis B, De Cahsan B, Korneliussen TS, Michelsen CS, Sand KK, Jelavić S, Ruter AH, Schmidt AMA, Kjeldsen KK, Tesakov AS, Snowball I, Gosse JC, Alsos IG, Wang Y, Dockter C, Rasmussen M, Jørgensen ME, Skadhauge B, Prohaska A, Kristensen JÅ, Bjerager M, Allentoft ME, Coissac E, Rouillard A, Simakova A, Fernandez-Guerra A, Bowler C, Macias-Fauria M, Vinner L, Welch JJ, Hidy AJ, Sikora M, Collins MJ, Durbin R, Larsen NK, Willerslev E (2022) A 2-million-year-old ecosystem in Greenland uncovered by environmental DNA. Nature 612(7939):283–291. https://doi.org/10.1038/s41586-022-05453-y

Kmiec B, Woloszynska M, Janska H (2006) Heteroplasmy as a common state of mitochondrial genetic information in plants and animals. Curr Genet 50(3):149–159. https://doi.org/10.1007/s00294-006-0082-1

Krause J, Fu Q, Good JM, Viola B, Shunkov MV, Derevianko AP, Pääbo S (2010) The complete mitochondrial DNA genome of an unknown hominin from southern Siberia. Nature 464(7290):894–897. https://doi.org/10.1038/nature08976

Lammers Y, Clarke CL, Erséus C, Brown AG, Edwards ME, Gielly L, Haflidason H, Mangerud J, Rota E, Svendsen JI, Alsos IG (2019) Clitellate worms (Annelida) in lateglacial and Holocene sedimentary DNA records from the Polar Urals and northern Norway. Boreas 48(2):317–329. https://doi.org/10.1111/bor.12363

Lammers Y, Heintzman PD, Alsos IG (2021) Environmental palaeogenomic reconstruction of an Ice Age algal population. Commun Biol 4(1):1–11. https://doi.org/10.1038/s42003-021-01710-4

Landon DB (2005) Zooarchaeology and historical archaeology: progress and prospects. J Archaeol Method Theory 12(1):1–36

Langmead B, Salzberg SL (2012) Fast gapped-read alignment with Bowtie 2. Nat Methods 9(4):357–359. https://doi.org/10.1038/nmeth.1923

Lemey P, Salemi M, Vandamme A-M (eds) (2012) The phylogenetic handbook: a practical approach to phylogenetic analysis and hypothesis testing, 2nd edn. Cambridge, Cambridge University Press

Leonard JA, Shanks O, Hofreiter M, Kreuz E, Hodges L, Ream W, Wayne RK, Fleischer RC (2007) Animal DNA in PCR reagents plagues ancient DNA research. J Archaeol Sci 34(9):1361–1366. https://doi.org/10.1016/j.jas.2006.10.023

Levy-Booth DJ, Campbell RG, Gulden RH, Hart MM, Powell JR, Klironomos JN, Pauls KP, Swanton CJ, Trevors JT, Dunfield KE (2007) Cycling of extracellular DNA in the soil environment. Soil Biol Biochem

Li H, Durbin R (2009) Fast and accurate short read alignment with Burrows-Wheeler transform. Bioinformatics 25(14):1754–1760. https://doi.org/10.1093/bioinformatics/btp324

Lloyd-Jones G, Hunter DWF (2001) Comparison of rapid DNA extraction methods applied to contrasting New Zealand soils. Soil Biol Biochem 33(15):2053–2059. https://doi.org/10.1016/S0038-0717(01)00133-X

Lorenz MG, Wackernagel W (1994) Bacterial gene transfer by natural genetic transformation in the environment. Microbiol Rev 58(3):563–602. https://doi.org/10.1128/mr.58.3.563-602.1994

Louvel G, Der Sarkissian C, Hanghøj K, Orlando L (2016) metaBIT, an integrative and automated metagenomic pipeline for analysing microbial profiles from high-throughput sequencing shotgun data. Mol Ecol Resour 16(6):1415–1427. https://doi.org/10.1111/1755-0998.12546

Lydolph MC, Jacobsen J, Arctander P, Gilbert MTP, Gilichinsky DA, Hansen AJ, Willerslev E, Lange L (2005) Beringian paleoecology inferred from permafrost-preserved fungal DNA. Appl Environ Microbiol 71(2):1012–1017. https://doi.org/10.1128/AEM.71.2.1012-1017.2005

Lynggaard C, Bertelsen MF, Jensen CV, Johnson MS, Frøslev TG, Olsen MT, Bohmann K (2022) Airborne environmental DNA for terrestrial vertebrate community monitoring. Curr Biol 32(3):701–707.e5. https://doi.org/10.1016/j.cub.2021.12.014

Mahaney WC, Allen CCR, Pentlavalli P, Kulakova A, Young JM, Dirszowsky RW, West A, Kelleher B, Jordan S, Pulleyblank C, O'Reilly S, Murphy BT, Lasberg K, Somelar P, Garneau M, Finkelstein SA, Sobol MK, Kalm V, Costa PJM, Hancock RGV, Hart KM, Tricart P, Barendregt RW, Bunch TE, Milner MW (2017) Biostratigraphic evidence relating to the age-old question of Hannibal's invasion of Italy, II: chemical biomarkers and microbial signatures. Archaeometry 59(1):179–190. https://doi.org/10.1111/arcm.12228

Marcelino VR, Clausen PTLC, Buchmann JP, Wille M, Iredell JR, Meyer W, Lund O, Sorrell TC, Holmes EC (2020) CCMetagen: comprehensive and accurate identification of eukaryotes and prokaryotes in metagenomic data. Genome Biol 21(1):103. https://doi.org/10.1186/s13059-020-02014-2

Marciniak S, Klunk J, Devault A, Enk J, Poinar HN (2015) Ancient human genomics: the methodology behind reconstructing evolutionary pathways. J Hum Evol 79:21–34. https://doi.org/10.1016/j.jhevol.2014.11.003

Maricic T, Whitten M, Pääbo S (2010) Multiplexed DNA Sequence Capture of Mitochondrial Genomes Using PCR Products. PLoS ONE 5(11):e14004. https://doi.org/10.1371/journal.pone.0014004

Massilani D, Morley MW, Mentzer SM, Aldeias V, Vernot B, Miller C, Stahlschmidt M, Kozlikin MB, Shunkov MV, Derevianko AP, Conard NJ, Wurz S, Henshilwood CS, Vasquez J, Essel E, Nagel S, Richter J, Nickel B, Roberts RG, Pääbo S, Slon V, Goldberg P, Meyer M (2022) Microstratigraphic preservation of ancient faunal and hominin DNA in Pleistocene cave sediments. Proc Natl Acad Sci 119(1):e2113666118. https://doi.org/10.1073/pnas.2113666118

Mathieson I, Lazaridis I, Rohland N, Mallick S, Patterson N, Roodenberg SA, Harney E, Stewardson K, Fernandes D, Novak M, Sirak K, Gamba C, Jones ER, Llamas B, Dryomov S, Pickrell J, Arsuaga JL, de Castro JMB, Carbonell E, Gerritsen F, Khokhlov A, Kuznetsov P, Lozano M, Meller H, Mochalov O, Moiseyev V, Guerra MAR, Roodenberg J, Vergès JM, Krause J, Cooper A, Alt KW, Brown D, Anthony D, Lalueza-Fox C, Haak W, Pinhasi R, Reich D (2015) Genome-wide patterns of selection in 230 ancient Eurasians. Nature 528(7583):499–503. https://doi.org/10.1038/nature16152

Mauvisseau Q, Harper LR, Sander M, Hanner RH, Kleyer H, Deiner K (2022) The multiple states of environmental DNA and what is known about their persistence in aquatic environments. Environ Sci Technol 56(9):5322–5333. https://doi.org/10.1021/acs.est.1c07638

Messager E, Giguet-Covex C, Doyen E, Etienne D, Gielly L, Sabatier P, Banjan M, Develle A-L, Didier J, Poulenard J, Julien A, Arnaud F (2022) Two millennia of complexity and variability in a perialpine socioecological system (Savoie, France): the contribution of palynology and sedaDNA analysis. Front Ecol Evol 10

Meyer M, Kircher M (2010) Illumina sequencing library preparation for highly multiplexed target capture and sequencing. Cold Spring Harb Protoc 2010(6):pdb.prot5448. https://doi.org/10.1101/pdb.prot5448

Mitchell D, Willerslev E, Hansen A (2005) Damage and repair of ancient DNA. Mutat Res 571:265–276. https://doi.org/10.1016/j.mrfmmm.2004.06.060

Monteath AJ, Gaglioti BV, Edwards ME, Froese D (2021) Late Pleistocene shrub expansion preceded megafauna turnover and extinctions in eastern Beringia. Proc Natl Acad Sci 118(52):e2107977118. https://doi.org/10.1073/pnas.2107977118

Morrissey EM, McHugh TA, Preteska L, Hayer M, Dijkstra P, Hungate BA, Schwartz E (2015) Dynamics of extracellular DNA decomposition and bacterial community composition in soil. Soil Biol Biochem 86:42–49. https://doi.org/10.1016/j.soilbio.2015.03.020

Mullis K, Faloona F, Scharf S, Saiki R, Horn G, Erlich H (1986) Specific enzymatic amplification of DNA in vitro: the polymerase chain reaction. Cold Spring Harb Symp Quant Biol 51(Pt 1):263–273. https://doi.org/10.1101/sqb.1986.051.01.032

Murchie TJ (2021) Ancient environmental DNA as a means of understanding ecological restructuring during the Pleistocene-Holocene transition in Yukon, Canada. McMaster University

Murchie TJ, Monteath AJ, Mahony ME, Long GS, Cocker S, Sadoway T, Karpinski E, Zazula G, MacPhee RDE, Froese D, Poinar HN (2021a) Collapse of the mammoth-steppe in central Yukon as revealed by ancient environmental DNA. Nat Commun 12(1):7120. https://doi.org/10.1038/s41467-021-27439-6

Murchie TJ, Kuch M, Duggan AT, Ledger ML, Roche K, Klunk J, Karpinski E, Hackenberger D, Sadoway T, MacPhee R, Froese D, Poinar H (2021b) Optimizing extraction and targeted capture of ancient environmental DNA for reconstructing past environments using the PalaeoChip Arctic-1.0 bait-set. Quat Res 99:305–328. https://doi.org/10.1017/qua.2020.59

Murchie TJ, Karpinski E, Eaton K, Duggan AT, Baleka S, Zazula G, MacPhee RDE, Froese D, Poinar HN (2022) Pleistocene mitogenomes reconstructed from the environmental DNA of permafrost sediments. Curr Biol 32(4):851–860.e7. https://doi.org/10.1016/j.cub.2021.12.023

NCBI Resource Coordinators (2018) Database resources of the National Center for Biotechnology Information. Nucleic Acids Res 46:8–13. https://doi.org/10.1093/nar/gkx985

Neiman M, Sundling S, Grönberg H, Hall P, Czene K, Lindberg J, Klevebring D (2012) Library preparation and multiplex capture for massive parallel sequencing applications made efficient and easy. PLoS ONE 7(11):e48616

Nguyen TH, Elimelech M (2007) Adsorption of plasmid DNA to a natural organic matter-coated silica surface: kinetics, conformation, and reversibility. Langmuir 23(6):3273–3279. https://doi.org/10.1021/la0622525

O'Rourke DH, Raff JA (2010) The human genetic history of the Americas: the final frontier. Curr Biol 20(4):R202–R207. https://doi.org/10.1016/j.cub.2009.11.051

Ogata M, Masuda R, Harino H, Sakata MK, Hatakeyama M, Yokoyama K, Yamashita Y, Minamoto T (2021) Environmental DNA preserved in marine sediment for detecting jellyfish blooms after a tsunami. Sci Rep 11(1):16830. https://doi.org/10.1038/s41598-021-94286-2

Oh CS, Seo M, Chai JY, Lee SJ, Kim MJ, Park JB, Shin DH (2010) Amplification and sequencing of Trichuris trichiura ancient DNA extracted from archaeological sediments. J Archaeol Sci 37(6):1269–1273. https://doi.org/10.1016/j.jas.2009.12.029

Ondov BD, Bergman NH, Phillippy AM (2011) Interactive metagenomic visualization in a Web browser. BMC Bioinformat 12(1):385. https://doi.org/10.1186/1471-2105-12-385

Orlando L, Allaby R, Skoglund P, Der Sarkissian C, Stockhammer PW, Ávila-Arcos MC, Fu Q, Krause J, Willerslev E, Stone AC, Warinner C (2021) Ancient DNA analysis. Nat Rev Methods Primers 1(1):1–26. https://doi.org/10.1038/s43586-020-00011-0

Pääbo S (1984) Über den Nachweis von DNA in altägyptischen Mumien. Das Altertum 30:213–218

Paijmans JLA, Fickel J, Courtiol A, Hofreiter M, Förster DW (2016) Impact of enrichment conditions on cross-species capture of fresh and degraded DNA. Mol Ecol Resour 16(1):42–55. https://doi.org/10.1111/1755-0998.12420

Parducci L, Bennett KD, Ficetola GF, Alsos IG, Suyama Y, Wood JR, Pedersen MW (2017) Ancient plant DNA in lake sediments. New Phytol 214(3):924–942. https://doi.org/10.1111/nph.14470

Pedersen MW, Overballe-Petersen S, Ermini L, Sarkissian CD, Haile J, Hellstrom M, Spens J, Thomsen PF, Bohmann K, Cappellini E, Schnell IB, Wales NA, Carøe C, Campos PF, Schmidt AMZ, Gilbert MTP, Hansen AJ, Orlando L, Willerslev E (2015) Ancient and modern environmental DNA. Philos Trans Royal Soc B Biol Sci 370(1660):20130383. https://doi.org/10.1098/rstb.2013.0383

Pedersen MW, Ruter A, Schweger C, Friebe H, Staff RA, Kjeldsen KK, Mendoza MLZ, Beaudoin AB, Zutter C, Larsen NK, Potter BA, Nielsen R, Rainville RA, Orlando L, Meltzer DJ, Kjær KH, Willerslev E (2016) Postglacial viability and colonization in North America's ice-free corridor. Nature 537(7618):45–49. https://doi.org/10.1038/nature19085

Pedersen MW, De Sanctis B, Saremi NF, Sikora M, Puckett EE, Gu Z, Moon KL, Kapp JD, Vinner L, Vardanyan Z, Ardelean CF, Arroyo-Cabrales J, Cahill JA, Heintzman PD, Zazula G, MacPhee RDE, Shapiro B, Durbin R, Willerslev E (2021) Environmental genomics of Late Pleistocene black bears and giant short-faced bears. Current Biology 31(12):2728–2736.e8. https://doi.org/10.1016/j.cub.2021.04.027

Peltzer A, Jäger G, Herbig A, Seitz A, Kniep C, Krause J, Nieselt K (2016) EAGER: efficient ancient genome reconstruction. Genome Biol 17(1):60. https://doi.org/10.1186/s13059-016-0918-z

Peng Z, Zhu X, Wang Z, Yan X, Wang G, Tang M, Jiang A, Kristiansen K (2020) Comparative analysis of sample extraction and library construction for shotgun metagenomics. Bioinform Biol Insights 14:1177932220915459. https://doi.org/10.1177/1177932220915459

Perri AR, Mitchell KJ, Mouton A, Álvarez-Carretero S, Hulme-Beaman A, Haile J, Jamieson A, Meachen J, Lin AT, Schubert BW, Ameen C, Antipina EE, Bover P, Brace S, Carmagnini A, Carøe C, Samaniego Castruita JA, Chatters JC, Dobney K, dos Reis M, Evin A, Gaubert P, Gopalakrishnan S, Gower G, Heiniger H, Helgen KM, Kapp J, Kosintsev PA, Linderholm A, Ozga AT, Presslee S, Salis AT, Saremi NF, Shew C, Skerry K, Taranenko DE, Thompson M, Sablin MV, Kuzmin YV, Collins MJ, Sinding M-HS, Gilbert MTP, Stone AC, Shapiro B, Van Valkenburgh B, Wayne RK, Larson G, Cooper A, Frantz LAF (2021) Dire wolves were the last of an ancient New World canid lineage. Nature 591(7848):87–91. https://doi.org/10.1038/s41586-020-03082-x

Peyrégne S, Prüfer K (2020) Present-Day DNA Contamination in Ancient DNA Datasets. BioEssays 42(9):2000081. https://doi.org/10.1002/bies.202000081

Poinar HN, Schwarz C, Qi J, Shapiro B, Macphee RDE, Buigues B, Tikhonov A, Huson DH, Tomsho LP, Auch A, Rampp M, Miller W, Schuster SC (2006) Metagenomics to paleogenomics: large-scale sequencing of mammoth DNA. Science 311(5759):392–394. https://doi.org/10.1126/science.1123360

Potter BA, Reuther JD, Holliday VT, Holmes CE, Miller DS, Schmuck N (2017) Early colonization of Beringia and Northern North America: Chronology, routes, and adaptive strategies. Quat Int 444:36–55. https://doi.org/10.1016/j.quaint.2017.02.034

Potter BA, Baichtal JF, Beaudoin AB, Fehren-Schmitz L, Haynes CV, Holliday VT, Holmes CE, Ives JW, Kelly RL, Llamas B, Malhi RS, Miller DS, Reich D, Reuther JD, Schiffels S, Surovell TA (2018) Current evidence allows multiple models for the peopling of the Americas. Sci Adv 4(8):eaat5473. https://doi.org/10.1126/sciadv.aat5473

Quince C, Lanzen A, Davenport RJ, Turnbaugh PJ (2011) Removing noise from pyrosequenced amplicons. BMC Bioinformat 12(1):38. https://doi.org/10.1186/1471-2105-12-38

Raghavan M, Steinrücken M, Harris K, Schiffels S, Rasmussen S, DeGiorgio M, Albrechtsen A, Valdiosera C, Ávila-Arcos MC, Malaspinas A-S, Eriksson A, Moltke I, Metspalu M, Homburger JR, Wall J, Cornejo OE, Moreno-Mayar JV, Korneliussen TS, Pierre T, Rasmussen M, Campos PF, de Damgaard P B, Allentoft ME, Lindo J, Metspalu E, Rodríguez-Varela R, Mansilla J, Henrickson C, Seguin-Orlando A, Malmström H, Stafford T, Shringarpure SS, Moreno-Estrada A, Karmin M, Tambets K, Bergström A, Xue Y, Warmuth V, Friend AD, Singarayer J, Valdes P, Balloux F, Leboreiro I, Vera JL, Rangel-Villalobos H, Pettener D, Luiselli D, Davis LG, Heyer E, Zollikofer CPE, Ponce de León MS, Smith CI, Grimes V, Pike K-A, Deal M, Fuller BT, Arriaza B, Standen V, Luz MF, Ricaut F, Guidon N, Osipova L, Voevoda MI, Posukh OL, Balanovsky O, Lavryashina M, Bogunov Y, Khusnutdinova E, Gubina M, Balanovska E, Fedorova S, Litvinov S, Malyarchuk B, Derenko M, Mosher MJ, Archer D, Cybulski J, Petzelt B, Mitchell J, Worl R, Norman PJ, Parham P, Kemp BM, Kivisild T, Tyler-Smith C, Sandhu MS, Crawford M, Villems R, Smith DG, Waters MR, Goebel T, Johnson JR, Malhi RS, Jakobsson M, Meltzer DJ, Manica A, Durbin R, Bustamante CD, Song YS, Nielsen R, Willerslev E (2015) Genomic evidence for the Pleistocene and recent population history of Native Americans. Science 349(6250):aab3884. https://doi.org/10.1126/science.aab3884

Railsback LB, Gibbard PL, Head MJ, Voarintsoa NRG, Toucanne S (2015) An optimized scheme of lettered marine isotope substages for the last 1.0 million years, and the climatostratigraphic nature of isotope stages and substages. Quat Sci Rev 111:94–106. https://doi.org/10.1016/j.quascirev.2015.01.012

Raper D, Bush M (2009) A test of Sporormiella representation as a predictor of megaherbivore presence and abundance. Quat Res 71(3):490–496. https://doi.org/10.1016/j.yqres.2009.01.010

Rapuc W, Bouchez J, Sabatier P, Genuite K, Poulenard J, Gaillardet J, Arnaud F (2021) Quantitative evaluation of human and climate forcing on erosion in the alpine Critical Zone over the last 2000 years. Quat Sci Rev 268:107127. https://doi.org/10.1016/j.quascirev.2021.107127

Rasmussen M, Cummings LS, Gilbert MTP, Bryant V, Smith C, Jenkins DL, Willerslev E (2009) Response to Comment by Goldberg et al. on "DNA from Pre-Clovis Human Coprolites in Oregon, North America." Science 325(5937):148–148. https://doi.org/10.1126/science.1167672

Rawlence NJ, Lowe DJ, Wood JR, Young JM, Churchman GJ, Huang Y-T, Cooper A (2014) Using palaeoenvironmental DNA to reconstruct past environments: progress and prospects. J Quat Sci 29(7):610–626. https://doi.org/10.1002/jqs.2740

Renaud G, Stenzel U, Kelso J (2014) leeHom: adaptor trimming and merging for Illumina sequencing reads. Nucleic Acids Res 42(18):e141. https://doi.org/10.1093/nar/gku699

Renaud G, Schubert M, Sawyer S, Orlando L (2019) Authentication and Assessment of Contamination in Ancient DNA. Methods Mol Biol 1963:163–194. https://doi.org/10.1007/978-1-4939-9176-1_17

Reuter JA, Spacek DV, Snyder MP (2015) High-throughput sequencing technologies. Mol Cell 58(4):586–597. https://doi.org/10.1016/j.molcel.2015.05.004

Riaz T, Shehzad W, Viari A, Pompanon F, Taberlet P, Coissac E (2011) ecoPrimers: inference of new DNA barcode markers from whole genome sequence analysis. Nucleic Acids Res 39(21):e145. https://doi.org/10.1093/nar/gkr732

Richardson MJ (1972) Coprophilous ascomycetes on different dung types. Trans Br Mycol Soc 58(1):37–48. https://doi.org/10.1016/S0007-1536(72)80069-X

Roberts D, Marks R (1980) The determination of regional and age variations in the rate of desquamation: a comparison of four techniques. J Invest Dermatol 74(1):13–16. https://doi.org/10.1111/1523-1747.ep12514568

Roger F, Ghanavi HR, Danielsson N, Wahlberg N, Löndahl J, Pettersson LB, Andersson GKS, Boke Olén N, Clough Y (2022) Airborne environmental DNA metabarcoding for the monitoring of terrestrial insects—a proof of concept from the field. Environmental DNA 4(4):790–807. https://doi.org/10.1002/edn3.290

Rogers Y-H, Venter JC (2005) Massively parallel sequencing. Nature 437(7057):326–327. https://doi.org/10.1038/437326a

Romanowski G, Lorenz MG, Wackernagel W (1991) Adsorption of plasmid DNA to mineral surfaces and protection against DNase I. Appl Environ Microbiol 57(4):1057–1061. https://doi.org/10.1128/aem.57.4.1057-1061.1991

Sabatier P, Poulenard J, Fanget B, Reyss J-L, Develle A-L, Wilhelm B, Ployon E, Pignol C, Naffrechoux E, Dorioz J-M, Montuelle B, Arnaud F (2014) Long-term relationships among pesticide applications, mobility, and soil erosion in a vineyard watershed. Proc Natl Acad Sci 111(44):15647–15652. https://doi.org/10.1073/pnas.1411512111

Sabatier P, Wilhelm B, Ficetola GF, Moiroux F, Poulenard J, Develle A-L, Bichet A, Chen W, Pignol C, Reyss J-L, Gielly L, Bajard M, Perrette Y, Malet E, Taberlet P, Arnaud F (2017) 6-kyr record of flood frequency and intensity in the western Mediterranean Alps – interplay of solar and temperature forcing. Quat Sci Rev 170:121–135

Sabatier P, Moernaut J, Bertrand S, Van Daele M, Kremer K, Chaumillon E, Arnaud F (2022) A review of event deposits in lake sediments. Quaternary 5(3):34. https://doi.org/10.3390/quat5030034

Sahu S, Kaushik SR, Goswami B, Dasgupta A, Guha H, Das R, Saha S, Das A, Nanda RK (2021) Fecal genomic DNA extraction method impacts outcome of MinION based metagenome profile of tuberculosis patients. 2021.11.15.21266154

Sarhan MS, Lehmkuhl A, Straub R, Tett A, Wieland G, Francken M, Zink A, Maixner F (2021) Ancient DNA diffuses from human bones to cave stones. iScience 24(12):103397. https://doi.org/10.1016/j.isci.2021.103397

Sawyer S, Krause J, Guschanski K, Savolainen V, Pääbo S (2012) Temporal patterns of nucleotide misincorporations and DNA fragmentation in ancient DNA. PLoS ONE 7(3):e34131. https://doi.org/10.1371/journal.pone.0034131

Schubert BW, Mead JI (2019) Chapter 96 – Paleontology of caves. In: Culver DC, Pipan T (eds) White WB, 3rd edn. Academic, Encyclopedia of caves, pp 794–805
Schubert BW, Mead JI, Graham RW (2003) Ice age cave faunas of North America. Indiana University Press, Bloomington
Schubert M, Lindgreen S, Orlando L (2016) AdapterRemoval v2: rapid adapter trimming, identification, and read merging. BMC Res Notes 9(1):88. https://doi.org/10.1186/s13104-016-1900-2
Scott GR, Schmitz K, Heim KN, Paul KS, Schomberg R, Pilloud MA (2018) Sinodonty, Sundadonty, and the Beringian Standstill model: Issues of timing and migrations into the New World. Quat Int 466:233–246. https://doi.org/10.1016/j.quaint.2016.04.027
Seersholm FV, Pedersen MW, Søe MJ, Shokry H, Mak SST, Ruter A, Raghavan M, Fitzhugh W, Kjær KH, Willerslev E, Meldgaard M, Kapel CMO, Hansen AJ (2016) DNA evidence of bowhead whale exploitation by Greenlandic Paleo-Inuit 4,000 years ago. Nat Commun 7(1):13389. https://doi.org/10.1038/ncomms13389
Seersholm FV, Harmsen H, Gotfredsen AB, Madsen CK, Jensen JF, Hollesen J, Meldgaard M, Bunce M, Hansen AJ (2022) Ancient DNA provides insights into 4,000 years of resource economy across Greenland. Nat Hum Behav 6(12):1723–1730. https://doi.org/10.1038/s41562-022-01454-z
Seilacher A (1970) Begriff und Bedeutung der Fossil-Lagerstätten: Neues Jahrbuch fur Geologie und Paläontologie. Monatshefte:34–39
Seitz A, Nieselt K (2017) Improving ancient DNA genome assembly. PeerJ 5:e3126. https://doi.org/10.7717/peerj.3126
Selway CA, Armbrecht L, Thornalley D (2022) An outlook for the acquisition of marine sedimentary ancient DNA (sedaDNA) From North Atlantic Ocean Archive Material. Paleoceanogr Paleoclimatol 37(5):e2021PA004372. https://doi.org/10.1029/2021PA004372
Shah N, Nute MG, Warnow T, Pop M (2019) Misunderstood parameter of NCBI BLAST impacts the correc of bioinformatics workflows. Bioinformatics 35(9):1613–1614. https://doi.org/10.1093/bioinformatics/bty833
Shapiro B, Barlow A, Heintzman PD, Hofreiter M, Paijmans JLA, Soares AER (eds) (2019) Ancient DNA: methods and protocols. Springer, New York
Shepard AO (1956) Ceramics for Archaeologists. Carnegie Institution of Washington, Washington, D.C.
Shokralla S, Spall JL, Gibson JF, Hajibabaei M (2012) Next-generation sequencing technologies for environmental DNA research. Mol Ecol 21(8):1794–1805. https://doi.org/10.1111/j.1365-294X.2012.05538.x
Signor PW III, Lipps JH (1982) Sampling bias, gradual extinction patterns and catastrophes in the fossil record. In: Silver LT, Schultz PH (eds) Geological Implications of Impacts of Large Asteroids and Comets on the Earth. Geological Society of America, p 0
Silvestrini S, Romandini M, Marciani G, Arrighi S, Carrera L, Fiorini A, López-García JM, Lugli F, Ranaldo F, Slon V, Tassoni L, Higgins OA, Bortolini E, Curci A, Meyer M, Meyer MC, Oxilia G, Zerboni A, Benazzi S, Spinapolice EE (2022) Integrated multidisciplinary ecological analysis from the Uluzzian settlement at the Uluzzo C Rock Shelter, south-eastern Italy. JQUATSCI 37(2):235–256. https://doi.org/10.1002/jqs.3341
Sinniger F, Pawlowski J, Harii S, Gooday AJ, Yamamoto H, Chevaldonné P, Cedhagen T, Carvalho G, Creer S (2016) Worldwide analysis of sedimentary DNA reveals major gaps in taxonomic knowledge of deep-sea benthos. Front Mar Sci 3
Sjögren P, Edwards ME, Gielly L, Langdon CT, Croudace IW, Merkel MKF, Fonville T, Alsos IG (2017) Lake sedimentary DNA accurately records 20th Century introductions of exotic conifers in Scotland. New Phytol 213(2):929–941. https://doi.org/10.1111/nph.14199
Skoglund P, Northoff BH, Shunkov MV, Derevianko AP, Pääbo S, Krause J, Jakobsson M (2014) Separating endogenous ancient DNA from modern day contamination in a Siberian Neandertal. Proc Natl Acad Sci 111(6):2229–2234. https://doi.org/10.1073/pnas.1318934111
Slon V, Glocke I, Barkai R, Gopher A, Hershkovitz I, Meyer M (2016) Mammalian mitochondrial capture, a tool for rapid screening of DNA preservation in faunal and undiagnostic remains,

and its application to Middle Pleistocene specimens from Qesem Cave (Israel). Quat Int 398:210–218. https://doi.org/10.1016/j.quaint.2015.03.039

Slon V, Hopfe C, Weiß CL, Mafessoni F, de la Rasilla M, Lalueza-Fox C, Rosas A, Soressi M, Knul MV, Miller R, Stewart JR, Derevianko AP, Jacobs Z, Li B, Roberts RG, Shunkov MV, de Lumley H, Perrenoud C, Gušić I, Kućan Ž, Rudan P, Aximu-Petri A, Essel E, Nagel S, Nickel B, Schmidt A, Prüfer K, Kelso J, Burbano HA, Pääbo S, Meyer M (2017) Neandertal and Denisovan DNA from Pleistocene sediments. Science 356(6338):605–608. https://doi.org/10.1126/science.aam9695

Slon V, Clark JL, Friesem DE, Orbach M, Porat N, Meyer M, Kandel AW, Shimelmitz R (2022) Extended longevity of DNA preservation in Levantine Paleolithic sediments, Sefunim Cave, Israel. Sci Rep 12(1):14528. https://doi.org/10.1038/s41598-022-17399-2

Smith CI, Chamberlain AT, Riley MS, Cooper A, Stringer CB, Collins MJ (2001) Not just old but old and cold? Nature 410(6830):771–772. https://doi.org/10.1038/35071177

Smith O, Momber G, Bates R, Garwood P, Fitch S, Pallen M, Gaffney V, Allaby RG (2015) Archaeology. Sedimentary DNA from a submerged site reveals wheat in the British Isles 8000 years ago. Science 347(6225):998–1001. https://doi.org/10.1126/science.1261278

Søe MJ, Nejsum P, Seersholm FV, Fredensborg BL, Habraken R, Haase K, Hald MM, Simonsen R, Højlund F, Blanke L, Merkyte I, Willerslev E, Kapel CMO (2018) Ancient DNA from latrines in Northern Europe and the Middle East (500 BC–1700 AD) reveals past parasites and diet. PLoS ONE 13(4):e0195481. https://doi.org/10.1371/journal.pone.0195481

Stahlschmidt MC, Collin TC, Fernandes DM, Bar-Oz G, Belfer-Cohen A, Gao Z, Jakeli N, Matskevich Z, Meshveliani T, Pritchard JK, McDermott F, Pinhasi R (2019) Ancient Mammalian and Plant DNA from Late Quaternary Stalagmite Layers at Solkota Cave, Georgia. Sci Rep 9(1):6628. https://doi.org/10.1038/s41598-019-43147-0

Straus LG (1990) Underground archaeology: perspectives on caves and rockshelters. Archaeol Method Theory 2:255–304

Stuart AJ (2015) Late quaternary megafaunal extinctions on the continents: a short review. Geol J 50(3):338–363. https://doi.org/10.1002/gj.2633

Suchard MA, Lemey P, Baele G, Ayres DL, Drummond AJ, Rambaut A (2018) Bayesian phylogenetic and phylodynamic data integration using BEAST 1.10 Virus Evolut 4(1):vey016. https://doi.org/10.1093/ve/vey016

Sui H, Weil AA, Nuwagira E, Qadri F, Ryan ET, Mezzari MP, Phipatanakul W, Lai PS (2020) Impact of DNA extraction method on variation in human and built environment microbial community and functional profiles assessed by shotgun metagenomics sequencing. Front Microbiol:11

Swift JA, Bunce M, Dortch J, Douglass K, Faith JT, Fellows Yates JA, Field J, Haberle SG, Jacob E, Johnson CN, Lindsey E, Lorenzen ED, Louys J, Miller G, Mychajliw AM, Slon V, Villavicencio NA, Waters MR, Welker F, Wood R, Petraglia M, Boivin N, Roberts P (2019) Micro methods for megafauna: novel approaches to late quaternary extinctions and their contributions to faunal conservation in the anthropocene. Bioscience 69(11):877–887. https://doi.org/10.1093/biosci/biz105

Taberlet P, Bonin A, Zinger L, Coissac E, Taberlet P, Bonin A, Zinger L, Coissac E (2018) Environmental DNA: for biodiversity research and monitoring. Oxford University Press, Oxford/New York

Thomsen PF, Willerslev E (2015) Environmental DNA – An emerging tool in conservation for monitoring past and present biodiversity. Biol Conserv 183:4–18. https://doi.org/10.1016/j.biocon.2014.11.019

Ushio M, Fukuda H, Inoue T, Makoto K, Kishida O, Sato K, Murata K, Nikaido M, Sado T, Sato Y, Takeshita M, Iwasaki W, Yamanaka H, Kondoh M, Miya M (2017) Environmental DNA enables detection of terrestrial mammals from forest pond water. Mol Ecol Resour 17(6):e63–e75. https://doi.org/10.1111/1755-0998.12690

Valentini A, Taberlet P, Miaud C, Civade R, Herder J, Thomsen PF, Bellemain E, Besnard A, Coissac E, Boyer F, Gaboriaud C, Jean P, Poulet N, Roset N, Copp GH, Geniez P, Pont D, Argillier C, Baudoin J-M, Peroux T, Crivelli AJ, Olivier A, Acqueberge M, Le Brun M,

Møller PR, Willerslev E, Dejean T (2016) Next-generation monitoring of aquatic biodiversity using environmental DNA metabarcoding. Mol Ecol 25(4):929–942. https://doi.org/10.1111/mec.13428

van der Valk T, Pečnerová P, Díez-Del-Molino D, Bergström A, Oppenheimer J, Hartmann S, Xenikoudakis G, Thomas JA, Dehasque M, Sağlıcan E, Fidan FR, Barnes I, Liu S, Somel M, Heintzman PD, Nikolskiy P, Shapiro B, Skoglund P, Hofreiter M, Lister AM, Götherström A, Dalén L (2021) Million-year-old DNA sheds light on the genomic history of mammoths. Nature 591(7849):265–269. https://doi.org/10.1038/s41586-021-03224-9

van Dijk EL, Auger H, Jaszczyszyn Y, Thermes C (2014) Ten years of next-generation sequencing technology. Trends Genet 30(9):418–426. https://doi.org/10.1016/j.tig.2014.07.001

Vartanyan SL, Arslanov KA, Karhu JA, Possnert G, Sulerzhitsky LD (2008) Collection of radiocarbon dates on the mammoths (Mammuthus Primigenius) and other genera of Wrangel Island, northeast Siberia, Russia. Quat Res 70(1):51–59. https://doi.org/10.1016/j.yqres.2008.03.005

Vernot B, Zavala EI, Gómez-Olivencia A, Jacobs Z, Slon V, Mafessoni F, Romagné F, Pearson A, Petr M, Sala N, Pablos A, Aranbur A, De Castro JMB, Carbonell E, Li B, Krajcarz MT, Krivoshapkin AI, Kolobova KA, Kozlikin MB, Shunkov MV, Derevianko AP, Viola B, Grote S, Essel E, Herraéz DL, Nagel S, Nickel B, Richter J, Schmidt A, Peter B, Kelso J, Roberts RG, Arsuaga JL, Meyer M (2021) Unearthing Neanderthal population history using nuclear and mitochondrial DNA from cave sediments. Science 372(6542). https://doi.org/10.1126/science.abf1667

Vogel NA, Rubin JD, Swartz M, Vlieghe J, Sackett PW, Pedersen AG, Pedersen MW, Renaud G (2023) euka: Robust detection of eukaryotic taxa from modern and ancient environmental DNA using pangenomic reference graphs. 2023.04.04.535531

Von Eggers J, Monchamp M-E, Capo E, Giguet-Covex C, Spanbauer T, Heintzman PD (2022) Inventory of ancient environmental DNA from sedimentary archives: locations, methods, and target taxa. Zenodo. https://doi.org/10.5281/ZENODO.6847522

Wales N, Carøe C, Sandoval-Velasco M, Gamba C, Barnett R, Samaniego JA, Madrigal JR, Orlando L, Gilbert MTP (2015) New insights on single-stranded versus double-stranded DNA library preparation for ancient DNA. Biotechniques 59(6):368–371. https://doi.org/10.2144/000114364

Wang Y, Pedersen MW, Alsos IG, De Sanctis B, Racimo F, Prohaska A, Coissac E, Owens HL, Merkel MKF, Fernandez-Guerra A, Rouillard A, Lammers Y, Alberti A, Denoeud F, Money D, Ruter AH, McColl H, Larsen NK, Cherezova AA, Edwards ME, Fedorov GB, Haile J, Orlando L, Vinner L, Korneliussen TS, Beilman DW, Bjørk AA, Cao J, Dockter C, Esdale J, Gusarova G, Kjeldsen KK, Mangerud J, Rasic JT, Skadhauge B, Svendsen JI, Tikhonov A, Wincker P, Xing Y, Zhang Y, Froese DG, Rahbek C, Bravo DN, Holden PB, Edwards NR, Durbin R, Meltzer DJ, Kjær KH, Möller P, Willerslev E (2021) Late Quaternary dynamics of Arctic biota from ancient environmental genomics. Nature 600(7887):86–92. https://doi.org/10.1038/s41586-021-04016-x

Wang Y, Korneliussen TS, Holman LE, Manica A, Pedersen MW (2022) ngsLCA—A toolkit for fast and flexible lowest common ancestor inference and taxonomic profiling of metagenomic data. Methods Ecol Evol 13(12):2699–2708. https://doi.org/10.1111/2041-210X.14006

Wei N, Nakajima F, Tobino T (2018) A microcosm study of surface sediment environmental DNA: decay observation, abundance estimation, and fragment length comparison. Environ Sci Technol 52(21):12428–12435. https://doi.org/10.1021/acs.est.8b04956

Wesolowska-Andersen A, Bahl MI, Carvalho V, Kristiansen K, Sicheritz-Pontén T, Gupta R, Licht TR (2014) Choice of bacterial DNA extraction method from fecal material influences community structure as evaluated by metagenomic analysis. Microbiome 2(1):19. https://doi.org/10.1186/2049-2618-2-19

Willerslev E, Cooper A (2005) Review paper. Ancient DNA. Proc Royal Soc Biol Sci 272(1558):3–16. https://doi.org/10.1098/rspb.2004.2813

Willerslev E, Hansen AJ, Binladen J, Brand TB, Gilbert MTP, Shapiro B, Bunce M, Wiuf C, Gilichinsky DA, Cooper A (2003) Diverse plant and animal genetic records from Holocene and Pleistocene sediments. Science 300(5620):791–795. https://doi.org/10.1126/science.1084114

Willerslev E, Hansen AJ, Poinar HN (2004) Isolation of nucleic acids and cultures from fossil ice and permafrost. Trends Ecol Evol 19(3):141–147. https://doi.org/10.1016/j.tree.2003.11.010

Willerslev E, Cappellini E, Boomsma W, Nielsen R, Hebsgaard MB, Brand TB, Hofreiter M, Bunce M, Poinar HN, Dahl-Jensen D, Johnsen S, Steffensen JP, Bennike O, Schwenninger J-L, Nathan R, Armitage S, de Hoog C-J, Alfimov V, Christl M, Beer J, Muscheler R, Barker J, Sharp M, Penkman KEH, Haile J, Taberlet P, Gilbert MTP, Casoli A, Campani E, Collins MJ (2007) Ancient biomolecules from deep ice cores reveal a forested Southern Greenland. Science 317(5834):111–114. https://doi.org/10.1126/science.1141758

Willerslev E, Davison J, Moora M, Zobel M, Coissac E, Edwards ME, Lorenzen ED, Vestergård M, Gussarova G, Haile J, Craine J, Gielly L, Boessenkool S, Epp LS, Pearman PB, Cheddadi R, Murray D, Bråthen KA, Yoccoz N, Binney H, Cruaud C, Wincker P, Goslar T, Alsos IG, Bellemain E, Brysting AK, Elven R, Sønstebø JH, Murton J, Sher A, Rasmussen M, Rønn R, Mourier T, Cooper A, Austin J, Möller P, Froese D, Zazula G, Pompanon F, Rioux D, Niderkorn V, Tikhonov A, Savvinov G, Roberts RG, MacPhee RDE, Gilbert MTP, Kjær KH, Orlando L, Brochmann C, Taberlet P (2014) Fifty thousand years of Arctic vegetation and megafaunal diet. Nature 506(7486):47–51. https://doi.org/10.1038/nature12921

Woo M (2012) Permafrost hydrology. Springer, Berlin/Heidelberg

Wood DE, Salzberg SL (2014) Kraken: ultrafast metagenomic sequence classification using exact alignments. Genome Biol 15(3):R46. https://doi.org/10.1186/gb-2014-15-3-r46

Wood JR, Wilmshurst JM (2013) Accumulation rates or percentages? How to quantify Sporormiella and other coprophilous fungal spores to detect late Quaternary megafaunal extinction events. Quat Sci Rev 77:1–3. https://doi.org/10.1016/j.quascirev.2013.06.025

Wood DE, Lu J, Langmead B (2019) Improved metagenomic analysis with Kraken 2. Genome Biol 20(1):257. https://doi.org/10.1186/s13059-019-1891-0

Wooller MJ, Bataille C, Druckenmiller P, Erickson GM, Groves P, Haubenstock N, Howe T, Irrgeher J, Mann D, Moon K, Potter BA, Prohaska T, Rasic J, Reuther J, Shapiro B, Spaleta KJ, Willis AD (2021) Lifetime mobility of an Arctic woolly mammoth. Science 373(6556):806–808. https://doi.org/10.1126/science.abg1134

Yoccoz NG, Bråthen KA, Gielly L, Haile J, Edwards ME, Goslar T, Von STEDINGKH, Brysting AK, Coissac E, Pompanon F, Sønstebø JH, Miquel C, Valentini A, De BELLOF, Chave J, Thuiller W, Wincker P, Cruaud C, Gavory F, Rasmussen M, Gilbert MTP, Orlando L, Brochmann C, Willerslev E, Taberlet P (2012) DNA from soil mirrors plant taxonomic and growth form diversity. Mol Ecol 21(15):3647–3655. https://doi.org/10.1111/j.1365-294X.2012.05545.x

Young JM, Rawlence NJ, Weyrich LS, Cooper A (2014) Limitations and recommendations for successful DNA extraction from forensic soil samples: a review. Sci Justice 54(3):238–244. https://doi.org/10.1016/j.scijus.2014.02.006

Zavala EI, Jacobs Z, Vernot B, Shunkov MV, Kozlikin MB, Derevianko AP, Essel E, de Fillipo C, Nagel S, Richter J, Romagné F, Schmidt A, Li B, O'Gorman K, Slon V, Kelso J, Pääbo S, Roberts RG, Meyer M (2021) Pleistocene sediment DNA reveals hominin and faunal turnovers at Denisova Cave. Nature 595(7867):399–403. https://doi.org/10.1038/s41586-021-03675-0

Zavala EI, Aximu-Petri A, Richter J, Nickel B, Vernot B, Meyer M (2022) Quantifying and reducing cross-contamination in single- and multiplex hybridization capture of ancient DNA. Mol Ecol Resour 22(6):2196–2207. https://doi.org/10.1111/1755-0998.13607

Zhang D, Xia H, Chen F, Li B, Slon V, Cheng T, Yang R, Jacobs Z, Dai Q, Massilani D, Shen X, Wang J, Feng X, Cao P, Yang MA, Yao J, Yang J, Madsen DB, Han Y, Ping W, Liu F, Perreault C, Chen X, Meyer M, Kelso J, Pääbo S, Fu Q (2020) Denisovan DNA in Late Pleistocene sediments from Baishiya Karst Cave on the Tibetan Plateau. Science 370(6516):584–587. https://doi.org/10.1126/SCIENCE.ABB6320

Zischler H, Höss M, Handt O, von Haeseler A, van der Kuyl AC, Goudsmit J (1995) Detecting dinosaur DNA. Science 268(5214):1192–1193.; author reply 1194. https://doi.org/10.1126/science.7605504

Chapter 12
An Overview of Biodiversity and Network Modeling Approaches: Applications to Sedimentary DNA Records

Zofia E. Taranu, Irene Gregory-Eaves, and Marie-Eve Monchamp

Keywords Sedimentary DNA · Modelling approaches · Statistical tools · Biodiversity · Networks · Interactive food web · Regime shifts

Introduction

The response of aquatic organisms to anthropogenic drivers varies over space and time and, as a result, the cumulative effects of these drivers is often heterogeneous and context-dependent (De Senerpont Domis et al. 2013; Nogués-Bravo et al. 2018; Barouillet et al. 2023). By quantifying trends in abiotic and biotic factors before, during, and after major anthropogenic events, we can better identify the timing and severity of these cumulative effects on species responses (e.g., whether driver-response relationships are indicative of an abrupt ecosystem transition; Vermaire et al. 2017; Zhao et al. 2022), and possibly curtail delayed and unforeseeable outcomes (Eastwood et al. 2022).

In this chapter, we present statistical approaches that can improve our understanding of how organismal responses to anthropogenic drivers percolate across their interactive web. Building on recent work that reviewed ways in which

---

Z. E. Taranu (✉)
Environment and Climate Change Canada, Aquatic Contaminants Research Division, McGill University, Montreal, QC, Canada

Groupe de Recherche Interuniversitaire en Limnologie, Montreal, QC, Canada

Department of Biology, McGill University, Montreal, QC, Canada
e-mail: zofia.taranu@ec.gc.ca

I. Gregory-Eaves · M.-E. Monchamp
Groupe de Recherche Interuniversitaire en Limnologie, Montreal, QC, Canada

Department of Biology, McGill University, Montreal, QC, Canada

E. Capo et al. (eds.), *Tracking Environmental Change Using Lake Sediments*, Developments in Paleoenvironmental Research 21,
https://doi.org/10.1007/978-3-031-43799-1_12

numerical ecology (Legendre and Legendre 2012) and sedimentary DNA (sedDNA) could be used to track long-term changes in biodiversity (Chen and Ficetola 2020), we comment on new insights that can be gained from the application of emerging statistical models to sedDNA records. In particular, we reflect on how network analyses can help track changes across different food web levels and compartments, and determine whether certain taxa respond more rapidly to a given stressor, or combination of stressors. We also show how these modeling approaches can shed some light on whether functional traits have evolved over time and if/how they have modulated environmental niche dimensions. Finally, we comment on the importance of tracking both changes in dominant taxa within a network perspective and measuring changes across the entire community via biodiversity metrics, as food-web interactions and biodiversity patterns are not directly scalable.

To provide a bit of background on the sections that follow, we note that network theory offers a variety of metrics used to quantify potential interactions within a trophic web (Fortin et al. 2021). Schematically, an individual taxon or functional group of taxa are referred to and represented as nodes in a network, while the potential interactions or links between taxa are referred to as edges (Fig. 12.1a). From there, one can quantify the structure of the web by measuring network properties such as connectance (i.e., number of edges per node squared), the average shortest path length between nodes (or characteristic path length) or the clustering

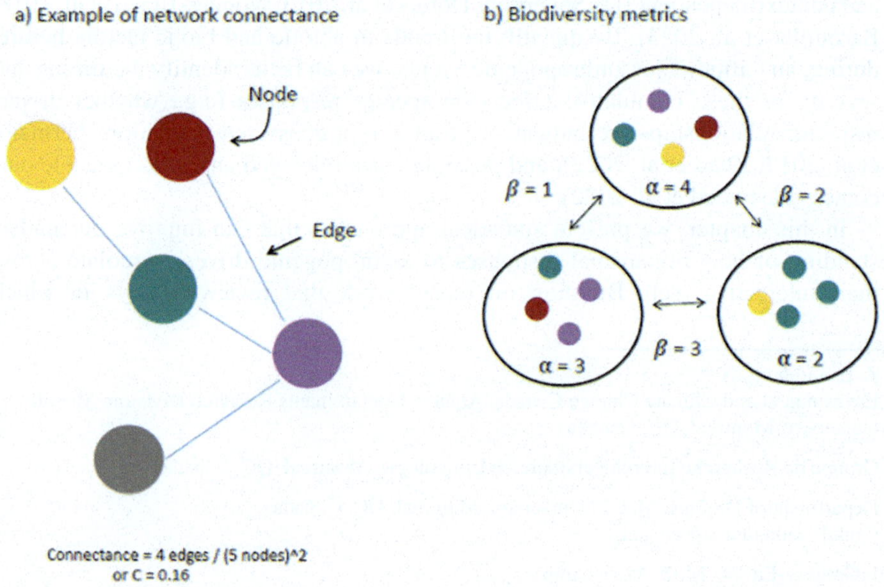

Fig. 12.1 Schematic representation of: (**a**) a network graph and example property, where nodes of different colours represent different species and blue edges (links) represent interactions between species; and (**b**) alpha and beta diversity metrics, where large open circles represent a community within which each circle of different colour represents a different species

coefficient (i.e., the ratio of existing edges connecting neighboring nodes to the maximum possible number of edges). These network metrics provide an indication of how closely species within a community interact, though it is recommended to track more than one metric as each one provides a different lens through which the system and hypothesis are evaluated (Fortin et al. 2021). Network properties complement biodiversity metrics, such as alpha diversity (within-community variation) and beta diversity (among-communities variation) (Magurran and McGill 2011). In particular, network metrics provide an indication of how the composition of species, and thus potential interactions, change over space and time (Fig. 12.1b). For example, a drastic reorganization of a community in response to an environmental pressure in one location may in turn alter species interactions and network complexity.

Food-web Structures

Network Analyses

Traditional paleolimnological techniques focusing on obtaining data from biological remains (e.g., subfossil analyses of diatoms, cladocerans and chironomids), while insightful, can be time consuming, and as a result few studies integrate multiple levels in a trophic network (also referred to as food webs). In contrast, high-throughput sequencing of sedDNA is uniquely positioned to measure responses of taxa to anthropogenic drivers across trophic groups. At the same time, open access of ecological datasets are increasingly facilitating the application of network analyses in ecology to explore co-occurrence patterns among and within trophic levels and compartments, to test how biotic and abiotic interactions change through space and time (e.g., Barberán et al. 2012). This multi-trophic approach may be particularly valuable to evaluate whether different trophic levels respond in the same way to the set of drivers. Using a decade-long time series, Kara et al. (2013) found that natural seasonal dynamics best explained bacterioplankton co-occurrence networks, whereas the suite of 32 physico-chemical and biological water quality parameters evaluated in their study did not drive the network complexity of this group. It would be worthwhile to explore whether this pronounced effect of seasonality dissipates across the trophic web, or if higher trophic levels are more strongly restricted by biotic interactions and resource limitation.

Network analysis has recently been extended to paleolimnological records (Eastwood et al. 2022). For example, Capo et al. (2017) were the first to apply weighted correlation network analysis to sedDNA data (amplicon sequencing of 18S rRNA genes) revealing that certain clusters (also known as network modules; i.e., highly correlated taxa) of the microbial eukaryotic community responded significantly to climate warming, while other clusters were more sensitive to changes in nutrient concentrations. Their study also made use of Generalized Additive Model (GAM) contribution plots, which illustrate how the magnitude and significance of the smooth slope coefficients vary over time (Simpson and Anderson 2009), to help pinpoint when nutrients and temperature were significant drivers of network clusters. Overall, Capo et al. (2017) found that the effect of rising nutrient concentrations on the community was greatest ca. 1960, while temperature increase became the prevalent stressor post 1980. Likewise, the correlation-based network analysis from Tse et al. (2018) showed that the dominance of harmful algal bloom taxa was significantly correlated to an increase in cyanotoxin-producing genes (microcystin synthetase; *mcy*A) as well as other paleolimnological proxies (i.e., concentrations of nutrients, pigments, stanols, sterols, and abundance of certain diatom species). More recently, Li et al. (2019) used network analyses of sedDNA to demonstrate that bacterial (16S rRNA genes) and microbial eukaryotic (18S rRNA genes) community network structures were more sensitive than diversity metrics to anthropogenic change. Specifically, the authors were able to show a loss in ecosystem stability in recent decades by demonstrating an increase in the variability of closeness (i.e., proximity of a node to other nodes) and betweenness (i.e., importance of a node as a connector) centrality indices, and how changes dissipated among trophic groups and levels. Thus, assessing new network parameters helped the authors build on prior work that solely examined species and diversity loss.

Traits, Phylogeny and Unmeasured Predictors

Building on the above correlative relationships between biotic and abiotic factors, Joint Species Distribution Models (JSDMs), in particular Latent Variable Models (LVMs; Hui 2021; Warton et al. 2015; van der Veen et al. 2023) and model-based ordination with copulas (Popovic et al. 2022), offer a promising approach to examine the co-response of species to spatial and/or temporal environmental gradients. Furthermore, such approaches examine co-responses to unmeasured predictors, which are also known as latent variables (Fig. 12.2). However, the LVMs and copula ordination differ in that the latter is a marginal model, whereas the LVM is a conditional model. That is, in the copula ordination model, we first regress the response (Y) on to measured predictors (X), and then calculate the latent variables (Z) using the model residuals (by running an unconstrained ordination of the residuals). In doing so, the marginal effect of the predictors is examined first, then we evaluate the additional effect of the latent variables. In contrast, with LVMs, the ordination of the response matrix comes first, then we regress the response on to the measured predictors, partialling out the effect of the latent variables. Thus, the effect of the measured predictors is conditional on the latent variables. Typically, only the first few latent variables are examined, and depending on their spatial or temporal extent, they can be used to assess the response of species to unmeasured environmental gradients (typically broad scale) or species interactions (typically fine scale). Latent variables are treated as random effects, as they are unobserved, and their factor loadings (λ_j)

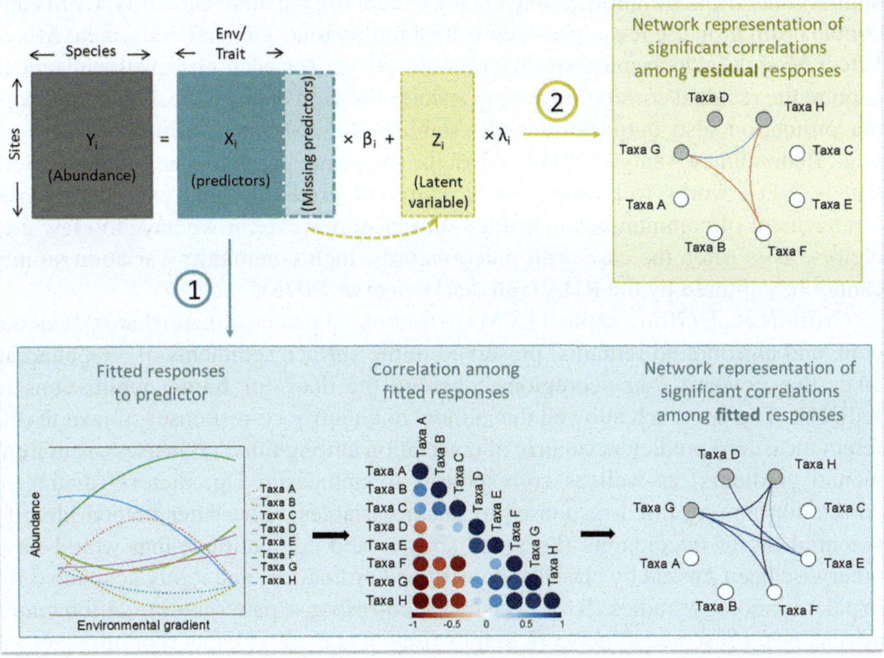

Fig. 12.2 Schematic representation of Latent Variable Models, modified from Warton et al. (2015). The approach incorporates both responses to measured environmental predictors (β_j) as well as responses to unmeasured environmental gradients or interactions among species (λ_j). For the former, one examines (1) the co-response of taxa to each measured environmental predictor (only one shown here) by first regressing each taxa versus each environmental variable, retaining the fitted values (as shown for Taxa A to H), then the correlation between fitted taxa responses is calculated – e.g., correlating the fitted response of Taxa A (green line) to the fitted response of Taxa H (blue line). This provides, for each environmental variable, a co-response correlation matrix (i.e., correlation among fitted responses). These correlation matrices can then be visualized as correlation plots, where large red circles indicate strong negative co-responses (also known as niche separation), whereas large blue circles indicate strong positive co-responses. Alternatively, correlation matrices can be visualized as network graphs (*sensu* Letten et al. 2015), where line colour and thickness indicate the strength of the correlation among fitted responses. The network graph shown here focuses on the positive co-responses (blue circles from the corrplot). A network graph illustrating negative co-responses (red circles from the corrplot) can also be made to focus instead on niche separation. Finally, after fitting a full model with all environmental variables, one can (2) extract a residual co-response network, representing the co-responses to the unmeasured predictors, or unmeasured species interactions (i.e., correlation among fitted responses to latent variables). The example shown here indicates that there remain both positive (blue line) and negative (red lines) residual co-responses, suggesting that at least one important variable was not included in the model. Note that had the residual co-responses been solely positive, then it would have suggested that the environmental variables included in the model captured most of the fine scale niche separation (i.e., no remaining negative co-responses in the residuals, as observed in the Letten et al. (2015) study)

approximate the correlation among taxa. By reducing the dimensionality, LVMs and copula ordination use fewer parameters than multivariate Generalized Linear Mixed Effect Models, which must specify random effects for each observational unit to capture the residual correlation among species (Warton et al. 2015). LVMs and copula ordination also outperform well-established constrained ordination methods (e.g., Redundancy Analysis; RDA) when the important predictors are not measured; that is, RDA works well when we have lots of predictors that explain well the arrangement of communities in ordination space; however, if we have too few predictors, as is often the case with paleo-records, then community variation simply cannot be captured by the RDA (van der Veen et al. 2023).

Griffiths et al. (2021) applied LVMs to traditional paleo-proxies (diatom, cladoceran, and chironomid remains) preserved in the surface sediments of 33 Canadian lakes that covered four ecoregions, opening the door for future applications to sedDNA. The approach allowed the authors to quantify co-responses of taxa to different measured predictors (matrix of correlation among fitted responses to environmental gradients) as well as co-responses to unmeasured predictors (matrix of correlation among fitted responses to latent variables). The latter helped identify potential biotic interactions (keystone species and competition) that would have otherwise been missed by classical constrained ordination, and rarely accounted for in paleolimnology studies. To help visualize co-response patterns, correlation matrices can be represented as network graphs (Letten et al. 2015), incorporating metrics from network theory such as the number of species needed to remove negative or positive pairwise co-responses (gray nodes in Fig. 12.2). Exploring how many species could be removed over time before the co-response network collapses could help identify keystone species and assess ecosystem robustness to environmental change. In lieu of this random removal of species (nodes), the topological role of nodes, based on the ratio of within- to among-module connectivity, can be used to classify species as peripherals (specialists with few links, nearly exclusively from their module), connectors (generalists with many links across several modules), module hubs (generalists with many links within their module), and network hubs (super generalists with many links within and across modules) (Deng et al. 2012). Finally, if applied to sedDNA data records, we suggest using methods such as Latent Dirichlet Allocation (LDA) or other topic models that reduce the dimensionality by grouping taxa into community-types (i.e., guilds or groups of species that tend to rise and fall together; Christensen et al. 2018) prior to running an LVM or copula ordination. In doing so, a handful of community-types are used as input as opposed to a very large matrix of thousands of amplicon sequence variants (ASVs) or operational taxonomic units (OTUs). The LDA dimension reduction could help reduce computing time and aid in the interpretation of results.

Certain LVM R packages (e.g., Hmsc; Table 12.1) can also incorporate trait and phylogeny data to examine how both modulate the environmental niche of species. That is, whether taxa with a given trait have higher (steeper regression slope) or lower (flatter regression slope) than average responses to a given environmental gradient. Unique to paleolimnology, resurrection ecology (i.e., hatching of resting

Table 12.1 Information on Latent Variable Models (LVMs) highlighted in this chapter

Name of method	R package	Output	Reference
Bayesian Ordination and Regression Analysis	boral	Correlation among fitted responses to measured and unmeasured predictors (as shown in Fig. 12.2). Biplot (latent variables 1 and 2) of unconstrained and of residual ordination to provide an indication of the reduction in variation after accounting for predictors.	Hui et al. (2015)
Hierarchical Modeling of Species Community	Hmsc	Heat plot of regression coefficients (β_j) for each taxa-environment combination. Variation partitioning barplot of the proportion of taxa variation explained by each environmental variable and random effects (spatial or temporal). Heatplot of γ_j coefficients quantifying how environmental niche dimensions are modulated by traits and phylogeny.	Ovaskainen et al. (2017)

The reference cited corresponds to the original paper in which the R package was described

eggs from lake sediments to conduct genetic analyses; Orsini et al. 2013; Ellegaard et al. 2020) can potentially be used to assess how traits have evolved over time. For example, changes in resistance to contaminants between animals hatched from older eggs collected downcore and descendant eggs collected in surface sediments, would allow one to conclude that the trait matrix has varied between historical (bottom) and contemporary (top) core intervals. Thus, using these data, one could examine how certain trait-niche interactions have evolved over the last ca. 200 years by comparing the output of a LVM fit with a contemporary trait matrix and the community from the upper sediment core intervals to the output of a LVM fit with a historical trait matrix (determined via resurrection ecology) and the community reconstructed from bottom (older) sediment core intervals. Given that only a subset of the community produces resting eggs (e.g., ephippia of *Daphnia*), this research avenue will likely be restricted to a few model organisms. However, we believe that considering co-responses of even just a handful of species or functional groups can be highly informative (e.g., Griffiths et al. 2021).

Finally, space and time can be incorporated into LVMs to test for spatial or temporal autocorrelation. For example, if the abundance of species from prior years (downcore) dictate more modern assemblages, or if neighboring lakes have more similar species occurrences than more distal sites. By comparing the degree of spatial and temporal autocorrelation among trophic levels (e.g., from microbes to predators), one could test interesting questions such as whether the degree of spatial autocorrelation relates to different dispersal rates (that is, do we observe greater spatial autocorrelation in species or trophic levels with strong dispersal capabilities), or whether some trophic levels are more resilient to environmental change over time (temporal autocorrelation). As some LVM approaches provide a means to partition the variation among drivers (see Ovaskainen et al. 2017 and the Hmsc package), one can then examine the proportion of variation explained by space, time and the environment, and how this varies among trophic levels.

Biodiversity

Diverging Responses Among Trophic Levels

When working on data from multiple trophic levels, questions regarding biodiversity and community scalability can also be addressed to test what factors affect each trophic level with varying dispersal capabilities, life cycles and responses to anthropogenic stressors. For example, using sedDNA records, Li et al. (2019) found that the bacterial community responded most to eutrophication while micro-eukaryotes responded most to heavy metal pollution. In a different context, Eastwood et al. (2022) found that prokaryotic community turnover (beta diversity) was shaped by the redox state of the sediment, whereas the eukaryotic community changed with different eutrophication phases. These studies highlight that, when examining biodiversity, the effect of a given environmental gradient may not be scalable across trophic levels.

Threshold Responses and Abrupt Changes

Another interesting line of investigation is to examine how the pool of taxa and biodiversity metrics change among key time periods. Using a Venn diagram, Kisand et al. (2018) showed that the number of pairwise shared OTUs between the Anthropocene (defined in their study as 2000 calendar years before present; cal. yr. BP) and the Mid-Holocene (8000–4000 cal. yr. BP; which corresponds to the Holocene Thermal Maximum) was greater than between any other period (i.e., Late-Pleistocene, Early- and Late-Holocene). Thus, the current change in biodiversity under a warmer climate resembles changes in biodiversity that occurred during a past warming period. Interestingly, the authors also used a type of Discriminant Function Analysis (i.e., a between-groups Principal Component Analysis (bgPCA)) to compare the variation within and between their predefined climatic periods, and identified taxa that contributed most to each time period. The bgPCA approach ordinates and arranges the predefined groups, rather than individual samples, so as to maximize among-group variance; samples and variables are then projected onto the bgPC axes to obtain their scores. As a follow up, it could be worthwhile to compare these results to network analyses in order to examine whether the contributor taxa identified in discriminant function analyses also represent key network nodes. Importantly, for high-dimensional data such as datasets produced via sedDNA sequencing, a cross-validation method is recommended when applying bgPCA, as the approach inflates spurious group differences when the number of taxa is large. In particular, Cardini and Polly (2020) detected pronounced distortion of patterns and R^2 when more than 60 variables were used in simulations (testing a range of variables between 4 and 720).

In addition to comparing known periods of change, several approaches have been applied to explore and identify when rapid changes in biodiversity occur over time.

For example, a Random Forest modeling approach helped Tse et al. (2018) uncover threshold responses in their sedDNA metric (i.e., α-diversity of cyanobacteria community) to traditional paleo-proxies (i.e., organic matter and lake trophic status). Furthermore, time-constrained multivariate regression trees allowed Li et al. (2019) to show that micro-eukaryotes tended to experience earlier breaks in community composition (by 1–5 years) relative to bacteria, and that community-level thresholds occurred earlier than the threshold based on diversity indices.

Biodiversity metrics can also be sensitive at detecting critical transitions, defined as a catastrophic transition from one equilibrium state to another. Doncaster et al. (2016) demonstrated how compositional disorder, which measures decreased nestedness in species composition over time, provided an earlier warning of a critical transition using traditional paleolimnological biodiversity datasets. Furthermore, the authors showed that the sign of the correlation between compositional disorder and alpha diversity (richness and Hill's index) during a given time slice helped identify if the system was composed of pervasive, keystone species (negative correlation, or low compositional disorder combined with high biodiversity) or a transition to an ephemeral, weedy species phase (positive correlation, or high disorder and high biodiversity). These studies echo the utility of high-dimensional data and its sensitivity to detect change. In this regard, sedDNA can be a key method to detect critical transitions (Taranu et al. 2018; Monchamp et al. 2021), as it could highlight whether a given community, trophic level, genera or species is the "canary in the mine", detecting critical transitions earlier than any one species or proxy on its own. Given that certain modeling approaches used to detect critical transitions (i.e., multivariate time-varying autoregressive models (MARSS) and online dynamic linear models (DLMs)) require long computing time when the response matrix is high dimensional, one can apply the models to either the dominant OTUs or ASVs (Monchamp et al. 2021) or to community-types identified using LDA. Bayesian change point (bcp) analysis and multivariate regression trees (MRT) could also be used to identify change points, and estimates of their posterior probabilities, in multivariate sedDNA datasets (Capo et al. 2016) or their derived community-types (Heathcote et al. 2023). Finally, Generalized Additive Mixed effect Models (GAMMs) can help detect a rapid rise in variance or the mean of dominant OTUs/ASVs or community-types over time, and have the added value of incorporating an autocorrelation structure for irregular time domains as well as the error of the ^{210}Pb-based age model; both of which are common features of paleolimnological records (Simpson 2018).

Integrating Analyses of Food-web Structures and Biodiversity

Ultimately, food-web interactions and biodiversity patterns are not directly related to each other, even when sampled at the same spatial and temporal scales, yet both endpoints must be considered to fully predict the stability, functioning and resilience of ecosystems (Blackman et al. 2022). Indeed, in their evaluation of a broad-scale riverine network (based on modern, water-column data), Blackman et al. (2022) found contrasting biodiversity patterns between major taxonomic groups, with opposing trajectories among different trophic levels. In particular, they found an increase in local richness in fish along a downstream gradient, yet a decreasing trend in richness for bacteria, and no change for invertebrates. Network structures likewise changed spatially and temporally (seasonal sampling) but did not mirror those of biodiversity indices. Within a classical paleolimnological context, Mayfield et al. (2022) measured richness, beta diversity, and network skewness of simulated chironomid networks to show that beta diversity is most sensitive to species loss (e.g., loss of cold-water species due to climate warming), while network metrics are more sensitive to species gains (e.g., arrival of strongly connected species in response to habitat homogenization), and both beta diversity and network skewness tracked species replacement. Future sedDNA studies can address these questions, across many indicator groups, by examining how biodiversity patterns compare to network characteristics: Do we first see a turnover of species towards a redundant community, which is only later picked up by network indices? Likewise, one can evaluate whether trajectories and tipping points differ from microbes to predators, as well as from southern to northern regions of the world (e.g., higher stress in extreme environments). Finally, critical transition modeling can be combined with network theory to examine the restructuring of communities pre-, during, and post- abrupt transitions and the impacts this may have on network stability. For instance, McKay et al. (2020) combined early warning signal metrics with a neural network-based proxy of ecological memory (i.e., how the correlation in the activity between two network nodes leads to an increase in the connection strength between them, potentially increasing ecosystem resilience) to traditional paleo-records, to test whether past memorized states helped the system recover more rapidly from perturbations.

Summary

By applying the numerical methods highlighted above, one might uncover new patterns that could uniquely be explored with sedDNA records. For instance, applying regime shift modeling to sedDNA records on a global scale may answer where, when and why hotspots of change have occurred. Then, by applying JSDMs and network analyses to these hotspots, one could determine which links have been lost, which species niche dimensions have changed, as well as how these niche dimensions are constrained by traits or phylogeny (i.e., determine if certain traits accentuate or downplay the effect of a given environmental gradient). If using community-types as the response (obtained from a LDA analysis), one could likewise examine if certain communities have been lost in different places of the world, thereby uncovering spatial patterns of change. From there, emergent properties like the ability of a community to withstand habitat degradation, rising temperatures, invasive species or other stressors could be examined, properties that cannot be understood simply by looking at individual species.

Acknowledgements We acknowledge the following funding sources for salary support: Environment and Climate Change Canada (supporting ZET), Canada Research Chairs program (awarded to IGE), and an NSERC Post-Doctoral Fellowship (awarded to MEM). We are also grateful for fruitful comments and revisions provided by Drs. Susanne McGowan, Luke Holman and Eric Capo, John Smol and Cécilia Barouillet.

References

Barberán A, Bates ST, Casamayor EO, Fierer N (2012) Using network analysis to explore co-occurrence patterns in soil microbial communities. ISME J 6(2):343–351. https://doi.org/10.1038/ismej.2011.119

Barouillet C, Monchamp M-E, Bertilsson S, Brasell K, Domaizon I, Epp LS, Ibrahim A, Mejbel H, Nwosu EC, Pearman JK, Picard M, Thomson-Laing G, Tsugeki N, Von Eggers J, Gregory-Eaves I, Pick FR, Wood SA, Capo E (2023) Investigating the effects of anthropogenic stressors on lake biota using sedimentary DNA. Freshw Biol n/a(n/a). https://doi.org/10.1111/fwb.14027

Blackman RC, Ho H-C, Walser J-C, Altermatt F (2022) Spatio-temporal patterns of multi-trophic biodiversity and food-web characteristics uncovered across a river catchment using environmental DNA. Commun Biol 5(1):1–11. https://doi.org/10.1038/s42003-022-03216-z

Capo E, Debroas D, Arnaud F, Guillemot T, Bichet V, Millet L, Gauthier E, Massa C, Develle A-L, Pignol C, Lejzerowicz F, Domaizon I (2016) Long-term dynamics in microbial eukaryotes communities: a palaeolimnological view based on sedimentary DNA. Mol Ecol 25(23):5925–5943. https://doi.org/10.1111/mec.13893

Capo E, Debroas D, Arnaud F, Perga M-E, Chardon C, Domaizon I (2017) Tracking a century of changes in microbial eukaryotic diversity in lakes driven by nutrient enrichment and climate warming. Environ Microbiol 19(7):2873–2892. https://doi.org/10.1111/1462-2920.13815

Cardini A, Polly PD (2020) Cross-validated between group PCA scatterplots: a solution to spurious group separation? Evol Biol 47(1):85–95. https://doi.org/10.1007/s11692-020-09494-x

Chen W, Ficetola GF (2020) Numerical methods for sedimentary-ancient-DNA-based study on past biodiversity and ecosystem functioning. Environ DNA 2(2):115–129. https://doi.org/10.1002/edn3.79

Christensen EM, Harris DJ, Ernest SKM (2018) Long-term community change through multiple rapid transitions in a desert rodent community. Ecology 99(7):1523–1529. https://doi.org/10.1002/ecy.2373

de Senerpont Domis LN, Elser JJ, Gsell AS, Huszar VLM, Ibelings BW, Jeppesen E, Kosten S, Mooij WM, Roland F, Sommer U, van Donk E, Winder M, Lürling M (2013) Plankton dynamics under different climate conditions in tropical freshwater systems (a reply to the comment by Sarmento, Amado & Descy). Freshw Biol 58(10):2211–2213. https://doi.org/10.1111/fwb.12203

Deng Y, Jiang Y-H, Yang Y, He Z, Luo F, Zhou J (2012) Molecular ecological network analyses. BMC Bioinf 13(1):113. https://doi.org/10.1186/1471-2105-13-113

Doncaster CP, Alonso Chávez V, Viguier C, Wang R, Zhang E, Dong X, Dearing JA, Langdon PG, Dyke JG (2016) Early warning of critical transitions in biodiversity from compositional disorder. Ecology 97(11):3079–3090. https://doi.org/10.1002/ecy.1558

Eastwood N, Stubbings WA, Abou-Elwafa Abdallah MA, Durance I, Paavola J, Dallimer M, Pantel JH, Johnson S, Zhou J, Hosking JS, Brown JB, Ullah S, Krause S, Hannah DM, Crawford SE, Widmann M, Orsini L (2022) The Time Machine framework: monitoring and prediction of biodiversity loss. Trends Ecol Evol 37(2):138–146. https://doi.org/10.1016/j.tree.2021.09.008

Ellegaard M, Clokie MRJ, Czypionka T, Frisch D, Godhe A, Kremp A, Letarov A, McGenity TJ, Ribeiro S, John Anderson N (2020) Dead or alive: sediment DNA archives as tools for tracking aquatic evolution and adaptation. Commun Biol 3(1):1–11. https://doi.org/10.1038/s42003-020-0899-z

Fortin M-J, Dale MRT, Brimacombe C (2021) Network ecology in dynamic landscapes. Proc R Soc B Biol Sci 288(1949):20201889. https://doi.org/10.1098/rspb.2020.1889

Griffiths K, Jeziorski A, Paquette C, Taranu ZE, Baud A, Antoniades D, Beisner B, Hamilton PB, Smol JP, Gregory-Eaves I (2021) Multi-trophic level responses to environmental stressors over the past ~150 years: insights from a lake-rich region of the world. Ecol Indic 127:107700. https://doi.org/10.1016/j.ecolind.2021.107700

Heathcote AJ, Taranu ZE, Tromas N, MacIntyre-Newell M, Leavitt PR, Pick FR (2023) Sedimentary DNA and pigments show increasing abundance and toxicity of cyanoHABs during the Anthropocene. Freshw Biol n/a(n/a). https://doi.org/10.1111/fwb.14069

Hui F (2021) boral: Bayesian Ordination and Regression AnaLysis. R package version 2.0

Hui FKC, Taskinen S, Pledger S, Foster SD, Warton DI (2015) Model-based approaches to unconstrained ordination. Methods Ecol Evol 6(4):399–411. https://doi.org/10.1111/2041-210X.12236

Kara EL, Hanson PC, Hu YH, Winslow L, McMahon KD (2013) A decade of seasonal dynamics and co-occurrences within freshwater bacterioplankton communities from eutrophic Lake Mendota, WI, USA. ISME J 7(3):680–684. https://doi.org/10.1038/ismej.2012.118

Kisand V, Talas L, Kisand A, Stivrins N, Reitalu T, Alliksaar T, Vassiljev J, Liiv M, Heinsalu A, Seppä H, Veski S (2018) From microbial eukaryotes to metazoan vertebrates: wide spectrum paleo-diversity in sedimentary ancient DNA over the last ~14,500 years. Geobiology 16(6):628–639. https://doi.org/10.1111/gbi.12307

Legendre P, Legendre LF (2012) Numerical Ecology. (Vol. 20). Elsevier, The Netherlands, pp. 209–213, 406–411

Letten AD, Keith DA, Tozer MG, Hui FKC (2015) Fine-scale hydrological niche differentiation through the lens of multi-species co-occurrence models. J Ecol 103(5):1264–1275. https://doi.org/10.1111/1365-2745.12428

Li F, Zhang X, Xie Y, Wang J (2019) Sedimentary DNA reveals over 150 years of ecosystem change by human activities in Lake Chao, China. Environ Int 133:105214. https://doi.org/10.1016/j.envint.2019.105214

Magurran AE, McGill BJ (2011) Biological diversity: frontiers in measurement and assessment, 1st edn. Oxford University Press, Oxford/New York

Mayfield RJ, Dearing JA, Doncaster CP, Langdon PG (2022) Stability of chironomid community structure during historic climatic and environmental change in subarctic Alaska. Limnol Oceanogr 67(S1):S444–S460. https://doi.org/10.1002/lno.12007

McKay DIA, Dyke JG, Doncaster CP, Dearing JA, Wang R (2020) Network-based metrics of resilience and ecological memory in lake ecosystems. https://doi.org/10.1101/810762

Monchamp M-È, Bruel R, Frossard V, McGowan S, Lavrieux M, Muschick M, Perga M-É, Dubois N (2021) Paleoecological evidence for a multi-trophic regime shift in a perialpine lake (Lake Joux, Switzerland). Anthropocene 35(100301). https://doi.org/10.1016/j.ancene.2021.100301

Nogués-Bravo D, Rodríguez-Sánchez F, Orsini L, de Boer E, Jansson R, Morlon H, Fordham DA, Jackson ST (2018) Cracking the code of biodiversity responses to past climate change. Trends Ecol Evol 33(10):765–776. https://doi.org/10.1016/j.tree.2018.07.005

Orsini L, Schwenk K, De Meester L, Colbourne JK, Pfrender ME, Weider LJ (2013) The evolutionary time machine: using dormant propagules to forecast how populations can adapt to changing environments. Trends Ecol Evol 28(5):274–282. https://doi.org/10.1016/j.tree.2013.01.009

Ovaskainen O, Tikhonov G, Norberg A, Guillaume Blanchet F, Duan L, Dunson D, Roslin T, Abrego N (2017) How to make more out of community data? A conceptual framework and its implementation as models and software. Ecol Lett 20(5):561–576. https://doi.org/10.1111/ele.12757

Popovic GC, Hui FKC, Warton DI (2022) Fast model-based ordination with copulas. Methods Ecol Evol 13(1):194–202. https://doi.org/10.1111/2041-210X.13733

Simpson GL (2018) Modelling palaeoecological time series using generalised additive models. Front Ecol Evol 6. https://doi.org/10.3389/fevo.2018.00149

Simpson GL, Anderson NJ (2009) Deciphering the effect of climate change and separating the influence of confounding factors in sediment core records using additive models. Limnol Oceanogr 54(6 part 2):2529–2541. https://doi.org/10.4319/lo.2009.54.6_part_2.2529

Taranu ZE, Carpenter SR, Frossard V, Jenny J-P, Thomas Z, Vermaire JC, Perga M-E (2018) Can we detect ecosystem critical transitions and signals of changing resilience from paleo-ecological records? Ecosphere 9(10):e02438. https://doi.org/10.1002/ecs2.2438

Tse TJ, Doig LE, Tang S, Zhang X, Sun W, Wiseman SB, Feng CX, Liu H, Giesy JP, Hecker M, Jones PD (2018) Combining high-throughput sequencing of sedaDNA and traditional paleo-limnological techniques to infer historical trends in cyanobacterial communities. Environ Sci Technol 52(12):6842–6853. https://doi.org/10.1021/acs.est.7b06386

van der Veen B, Hui FKC, Hovstad KA, O'Hara RB (2023) Concurrent ordination: simultaneous unconstrained and constrained latent variable modelling. Methods Ecol Evol 14(2):683–695. https://doi.org/10.1111/2041-210X.14035

Vermaire JC, Taranu ZE, MacDonald GK, Velghe K, Bennett EM, Gregory-Eaves I (2017) Extrinsic vs. intrinsic regimes shifts in shallow lakes: long-term response of cyanobacterial blooms to historical catchment phosphorus loading and climate warming. Front Ecol Evol 5. https://doi.org/10.3389/fevo.2017.00146

Warton DI, Blanchet FG, O'Hara RB, Ovaskainen O, Taskinen S, Walker SC, Hui FKC (2015) So many variables: joint modeling in community ecology. Trends Ecol Evol 30(12):766–779. https://doi.org/10.1016/j.tree.2015.09.007

Zhao Y, Wang R, Yang X, Dearing JA, Doncaster CP, Langdon P, Dong X (2022) Reconstruction of ecological transitions in a temperate shallow lake of the middle Yangtze River Basin in the last century. Water 14(7):1136. https://doi.org/10.3390/w14071136

Chapter 13
Perspectives and Future Developments Within Sedimentary DNA Research

Luke E. Holman, Yi Wang, Rikai Sawafuji, Laura S. Epp, Kristine Bohmann, and Mikkel Winther Pedersen

Keywords Future perspectives · Ancient environmental DNA · Metabarcoding · Shotgun sequencing · Ancient environmental genomics

Introduction

How does nature change through time and across space? How might it change in the future? While time travel has long been the domain of science fiction, scientists have approached these questions using techniques that examine the traces of the past remaining in today's environment. The use of ancient environmental DNA (ancient eDNA) is the latest method for paleoenvironmental reconstruction in a line of ancient biomolecular research, with the first research into DNA barcoding appearing near the turn of the last century (Willerslev et al. 2003). This field has principally advanced through a combination of milestone discoveries and a technological revolution that ultimately transformed genetic and genomic analyses into the Big Data Science we recognise today (Fig. 13.1). But what has been the trajectory

L. E. Holman (✉) · K. Bohmann
Section for Molecular Ecology and Evolution, Globe Institute, University of Copenhagen, Copenhagen, Denmark
e-mail: luke.holman@sund.ku.dk

Y. Wang · L. S. Epp
Limnological Institute, Department of Biology, University of Konstanz, Konstanz, Germany

R. Sawafuji
School of Advanced Sciences, The Graduate University for Advanced Studies (SOKENDAI), Kanagawa, Japan

M. W. Pedersen
Centre for Ancient Environmental Genomics, Globe Institute, University of Copenhagen, Copenhagen, Denmark

E. Capo et al. (eds.), *Tracking Environmental Change Using Lake Sediments*, Developments in Paleoenvironmental Research 21,
https://doi.org/10.1007/978-3-031-43799-1_13

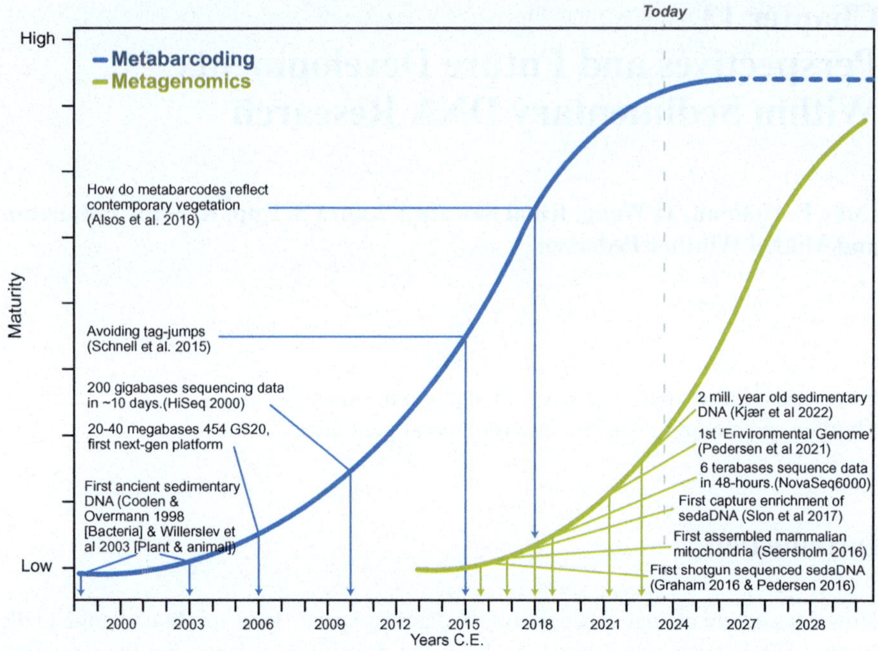

Fig. 13.1 Milestones and method maturity that have driven the development in ancient environmental metabarcoding and metagenomics

of this development, and perhaps more importantly, what future potential lies ahead for this research area?

In this chapter, we detail the expected progress within ancient eDNA research in the coming years but also outline where we might hope to arrive in the longer term. We do so from the perspective of researchers working principally on eukaryotic species but many of our reflections are also relevant for those working on prokaryotes. Furthermore, we do not discriminate between ancient eDNA obtained from samples in aquatic sediments and cave sediments, nor from other geological and archaeological sedimentary deposits. We argue that all ancient eDNA samples can be defined as a sample containing a great complexity of fragmented and damaged DNA molecules derived from multiple taxonomic groups and individuals. We therefore deduce that all future potential and limitations are confined by the information across the individual genome and the aggregate nature of ancient eDNA samples.

Today the field can be split into two broad methodological categories. The first relies on PCR amplification of genetic markers (DNA barcodes) for either single species identification, or amplification of a gene fragment flanked by conserved regions to target a defined phylogenetic group of organisms (metabarcoding) (Taberlet et al. 2012). These approaches were the defining methodology for ancient eDNA between 2003–2016 until it was demonstrated that randomly sequencing a pool of DNA (shotgun sequencing) extracted from ancient lake sediments can produce high resolution species diversity data, and that these molecules had similar DNA damage characteristics (Graham et al. 2016; Pedersen et al. 2016) to observations of ancient genomes obtained from fossils (Dabney et al. 2013). This second category, 'shotgun sequencing' or 'metagenomics' (in which we also place enrichment technologies as discussed below), randomly sequences DNA molecules from a sample of unknown organismal composition, and subsequent bioinformatic analyses identify the origin of the DNA sequences. These are in fact two different, but linked, approaches within ancient eDNA, hence their history, current state, future development and more important future perspectives have different trajectories.

Trajectories

Why are metabarcoding and metagenomic sequencing of sedaDNA at different stages of maturity for use in research (Fig. 13.1)? Firstly, technological advancement is tightly bound to economic costs. In the earliest phases of ancient eDNA research, it was relatively expensive to sequence DNA molecules, and the technology required numerous time-consuming cloning steps to isolate individual DNA sequences. By the mid-2000s, next-generation sequencing platforms emerged that parallelized sequencing of DNA molecules, in contrast to previous instruments that sequenced a single DNA molecule. This dramatically lowered the cost and time required for the production of sequence data, and was further boosted with the invention of nucleotide tagged PCR primers that enabled high-throughput sequencing of multiple samples simultaneously (Binladen et al. 2007). In response, the number of studies using ancient DNA metabarcoding increased in subsequent years with publications using this approach to illuminate past community compositions in many different environments such as permafrost soil (Haile et al. 2009; Jørgensen et al. 2012) and lakes (Parducci et al. 2012; Pedersen et al. 2013). However, this increase in popularity came with a number of obstacles for the field. A critical issue was the inability to disentangle contamination and false-positives from true ancient taxa (Ficetola et al. 2015; Alsos et al. 2018). These problems arose in part due to the damaged and fragmented target DNA isolated from samples, which are present at such low concentrations that it was necessary to use PCR to generate a large number of copies of the target region and high-throughput sequencing to generate many millions of sequences for analysis. While powerful, these techniques in combination are prone to stochastic amplification of sequences with non-ancient origin, either through contamination of non-modern DNA across the workflow or through sequencing of random errors incorporated during PCR and on sequencing platforms. At the time, the only possible resolution to this problem was to increase the stringency for a positive detection by incorporating many control samples and replicates across the workflow, resulting in a need for more laboratory work and more sequence data.

A second key hurdle was the realisation that common metabarcoding laboratory workflows to attach nucleotide barcode tags that match sequences to samples could result in incorrect assignments, so-called 'tag jumping' (Schnell et al. 2015). An initial solution was to exclusively use unique nucleotide tags in library designs, such as twin-tagging (Kircher et al. 2012). In doing so, tag-jumps would result in unused tag-combinations and thereby not falsely assign sequences to samples. Later, the cause of the issue was determined, and a library preparation protocol developed, which eliminated the detrimental steps and thereby the occurrence of tag-jumps (Carøe and Bohmann 2020). From this point it seemed that the path was paved for a rapid increase in publications, both in the number of environments analysed and in the variety of organisms studied. While there is still huge scope for improvement, the potential for continued rapid technological development within metabarcoding is relatively low. This does not mean it has become obsolete; on the contrary, it is now a well-established method for retrieving information on past organismal communities. In summary, metabarcoding is a powerful tool to uncover ancient and modern biodiversity. During the first few decades of its inception, research has established appropriate thresholds for PCR replication, required negative controls and determined how to efficiently plan the experimental as well as the laboratory setup. The field has further advanced with new high-throughput sequencing platforms, that allow for even larger numbers of labelled amplicons to be sequenced in parallel.

In contrast, metagenomics for ancient eDNA still has a high potential to develop previously unanticipated applications, break records for the oldest eDNA sequenced, isolate eDNA from novel sample types, and glean new kinds of information from samples (Fig. 13.1). This is important considering that metabarcoding with ancient DNA is of limited use with ultra-short sequences, i.e. <60 base pairs (bp), common in highly degraded samples. In contrast, metagenomics can provide useful data with sequences as short as 30 bp long, which allows the analysis of highly degraded samples from older environments with poorer preservation. This approach was initially driven by the need to overcome limited amplification of highly fragmented plant and animal DNA and was made possible by the high-throughput parallelised sequencing platforms such as the Illumina HiSeq 2500 to deliver the vast amount of sequences required for metagenomic analyses. To date, we have only scratched the surface of the potential of metagenomics.

Two suggestions for the future are common in metagenomic and metabarcoding papers: (1) the much-bemoaned incompleteness of DNA reference databases; (2) the potential for greater insights driven by cheaper sequencing or a greater number of DNA reads per sample. Here, wherever possible, we avoid repeating these well discussed themes that are responsible for much recent research progress. Instead, we focus here on novel future insights or step-changes in our understanding across technologies. We begin by outlining advances in parts of the workflow common to both metabarcoding and metagenomics before detailing what the future may hold for each method in turn and finish by exploring the potential unknown unknowns of the far-flung future.

Experimental Design, Taphonomy, Sampling and DNA Extraction

Planning

We envision that sampling of sedaDNA in the future will be guided by much improved study designs as we gain a better understanding of provenance and the taphonomic process involved in the deposition and preservation of sedaDNA. Transport and distribution models of sedaDNA will develop to incorporate both sedimentological properties of the sampled environment and biodiversity records to predict DNA presence and preservation. As we better understand where to look for which types of DNA, research hypotheses and organisms of interest will play an increasingly important role at the planning stages of a research project.

Taphonomy

DNA degrades in the environment through both spontaneous non-enzymatic degradation (physico-chemical processes) and biological enzymatic digestion (Dabney et al. 2013; Pedersen et al. 2015; Harrison et al. 2019; Mauvisseau et al. 2022). Therefore, sedaDNA isolated from ancient environmental samples represents a fraction of the total present at the moment of deposition, which itself represents a fraction of the total eDNA released by organisms in the environment. A better understanding of DNA degradation (which is an important part of its taphonomy) will be crucial to help guide where and what to sample. The composition of sediments themselves has been suggested to play an important role for long-term DNA preservation and are likely linked to mineralogy and the adsorption potential of different mineral surfaces (Huang et al. 2012; Freeman et al. 2023; Sand et al. 2023). However, equally important are the abiotic factors such as oxidation conditions, pH, salinity, water content and temperature that control physico-chemical damage as well as the abundance of microorganisms and their enzymatic degradation of DNA molecules (Cai et al. 2006; Gilbert et al. 2003; Levy-Booth et al. 2007). While *in situ* observations of eDNA degradation are lacking, successes and failures in the retrieval of DNA from different sedimentary deposits provide some evidence of what favours DNA preservation. Good quality data have been produced from lake sediments (Parducci et al. 2017; Capo et al. 2021), marine sediments (Armbrecht et al. 2022; Zimmermann et al. 2023), cave deposits (Vernot et al. 2021; Pedersen et al. 2021), and permafrost soil (Murchie et al. 2022; Wang et al. 2021), and are all characterised by limited exposure to air and relatively low constant temperatures. These stable environments can contain minerals of high binding capacity such as carbonates, clays and abundant organic material that are ideal for ancient DNA preservation. Acidic soils, peat environments, and generally warm oxic environments do not favour DNA preservation (Dommain et al. 2020). A flurry of recent studies into the ecology of 'modern' eDNA (Barnes and Turner 2016) has helped us better understand the origin, state, fate and transport of contemporary eDNA. However, we have comparatively less information regarding the ecology of ancient DNA molecules. While manipulative experiments to show the real-world effect of degradation experiments are unlikely to produce ecologically relevant results in a human lifetime, future correlative work will dramatically increase our knowledge of eDNA taphonomy allowing us to better target analyses of certain regions or sedimentary contexts.

Sampling

Sources of sedaDNA are often relatively inaccessible. Marine sediment beds can be far out in inclement seas, sampling deep lentic systems may require researchers to access materials hundreds or thousands of metres below the water surface, and archaeological material may be deep within complex cave systems or below hundreds of metres of sediment. To meet these challenges, sampling of sedimentary DNA will be better automated, and some studies have already demonstrated the efficiency of using autonomous drones for sampling modern water eDNA (McQuillan and Robidart 2017; Yamahara et al. 2019). While there are many engineering challenges associated with clean autonomous sediment core collection (discussed in Armbrecht et al. 2019), these methods will open up novel harsh and hard-to-reach environments, while also making current fieldwork safer and potentially less costly. Integration of geophysical sensing technologies in terrestrial, marine and freshwater sampling equipment (Steinbacher et al. 2015) will improve future accuracy and speed in which to target sedimentary or archaeological records. There will always be a need for scientists to visit and understand first-hand the environments they study and we anticipate developments in manned submersible research vessels and coring equipment to develop in tandem to autonomous solutions. These solutions will enable those who dive deeper into sea, lake and sediment to experience and better understand the material collected and environment sampled.

Community Driven Science

Sampling campaigns may cover hundreds, or even thousands, of sites but, except for government monitoring, few academic projects can incorporate such a large number of samples taken contemporaneously. Community science is scientific research conducted, in part, by amateur scientists, and has been used in a growing number of eDNA studies (Kopf et al. 2015; Agersnap et al. 2022). We envision that community science will enable the collection of ancient eDNA samples at an unprecedented scale, but, like modern eDNA community science programs, attention is required to ensure protocols are simple and provide good sample integrity. This is of particular importance in sedaDNA studies, as there is an elevated risk of sample contamination with modern material during sampling.

Age Estimation

The construction of an accurate age-depth model for material from which sedaDNA is sampled is essential to understand the temporal changes in biodiversity when investigating a question of interest. This is important as sedimentation rates are often not homogeneous, as the processes driving deposition can change with time or climate (Huggett 2016). Additionally, sedimentary archives can have low resolution; i.e., a large period of time covered in a relatively short sequence of sediments, which complicates accurate time-series analysis. Future methods for dating material, or improvements to existing methods, will enable expedited analyses at a resolution and scale currently unimaginable. Researchers could analyse more points along a time series, enabled by automation, allowing them to precisely target their period of interest and more closely tie their biological observations to data collected from other sources. As the resolution of sampling increases, using smaller samples from a greater number of locations, sub-centimetre sampling for proxies, dating and potential sedaDNA novel methods of dating may become widely available. For example, varve analysis uses the yearly deposition patterns to establish an exact annual chronology. A potential next step might be ancient eDNA varves, distinguishing seasonally dynamic ecosystems using the distinct eDNA compositional footprint or proxies recorded in sedimentary contexts by biological processes (Capo et al. 2019), to resolve a continuous annual pattern used for dating that can be anchored to calendar dates using other methods.

DNA Source

The rapid growth in eDNA methods for biodiversity monitoring has been accompanied by an increase in the number of different contexts from which eDNA can be isolated. Beyond sediment, soil and water, the most common substrates (Jarman et al. 2018), are honey (Hawkins et al. 2015), flowers (Thomsen and Sigsgaard 2019), marine sponges (Mariani et al. 2019) and most recently air (Lynggaard et al. 2022; Clare et al. 2022), which have proved to be useful for biodiversity analyses. Similarly, the bulk of ancient eDNA work has focussed on sedimentary archives such as lake sediments and marine cores (Pedersen et al. 2015), but time-series data has been shown to be recoverable from unexpected materials such as speleothems (Stahlschmidt et al. 2019), coral cores (Del Carmen Gomez Cabrera et al. 2019) and marine mollusc shells (Der Sarkissian et al. 2017). We can, therefore, expect that novel types of material will provide useful biodiversity time series data in the future.

DNA Extraction

Once samples are in the laboratory, researchers have a plethora of DNA extraction methods that range from expensive commercial kits to cheaper, academically-developed open source protocols. It has unfortunately not been possible to find a single protocol that generally works across all ancient eDNA sample types, however we expect that a smaller number of protocols that produce consistent DNA extraction results at a reasonable cost will become commonplace in the near future. Studies comparing the effectiveness and biases of different DNA extraction kits for isolating both modern and ancient eDNA have shown minor but significant differences in the detected biodiversity (Capo et al. 2021). Fewer studies of this type exist for sedaDNA, but these studies show similar results; with greater recovery for some methods. While the varied nature of sediments makes it unlikely that a single protocol will perform well across all contexts, we expect some standardisation according to material input type. The standardisation of these methods will enable better cross-comparison and enable meta-analyses between studies. This work has already been completed by the palynological community, which went through a pioneer phase (1916–1950) and a build-up phase (1951–1973) before entering its current mature phase (1974-today) (Birks and Berglund 2018) and work has commenced to standardise methods in contemporary eDNA sampling (Minamoto et al. 2021; Thalinger et al. 2021). Beyond common methods, laboratory automation will dramatically decrease human error, and at the same time increase throughput (Buchner et al. 2021), enabling researchers to tackle larger sample sizes and questions in ecology and evolution with greater scope.

Moving beyond DNA extraction, we envision that methods to acquire ancient DNA will increasingly incorporate a physical enrichment step, for example by simply performing a size selection that filters for DNA molecules shorter than 100 bp. Alternatively, the use of capture probes (i.e., biotinylated oligonucleotide baits of target organism sequence) on the sedaDNA extract could be used to enrich target taxa or attach to deaminated sites - one of the damage characteristics of ancient DNA. While capture probes for targeted enrichment have begun to be implemented, no work to date has attempted to increase the ratio of ancient to modern sedaDNA using biotinylated probes (Wales et al. 2014; Rohland et al. 2018; Heintzman et al. 2019; Murchie et al. 2021).

Finally, automated wet-lab workflows have taken the first steps into extracting ancient DNA from sediments. However, this technology is still developing and only a small number of well-supported laboratories have been able to support such facilities (Marx 2017; Slon et al. 2017). We predict that this will become a standard operational procedure in nearly all laboratories, as the cost of automation decreases, the cost for trained researchers in the wet lab increases, and as the technology matures and becomes more readily available. Not all laboratories handling ancient eDNA will have automated workflows and, as protocols become increasingly standardised, organisations will routinely offer these services routinely. Automated workflows will limit the biases introduced by the handling personnel, batch variations will decrease, and datasets will become more standardised and comparable among studies and laboratories. This will have a pronounced effect on the competencies required by the students and trainees, while simultaneously easing the burden of large, long and uncertain laboratory experiments.

Metabarcoding

Metabarcoding has been widely implemented in both contemporary and ancient eDNA studies, thanks to its cost-effectiveness in data generation, the increasing availability of robust metabarcoding primers (Alsos et al. 2020; Wang et al. 2021; Alsos et al. 2022; Garcés-Pastor et al. 2022; Taberlet et al. 2007; Epp et al. 2012; Stoof-Leichsenring et al. 2012; Giguet-Covex et al. 2014), and comparatively well curated and complete reference databases. For sedaDNA, reference databases for plants were quite well established early on (e.g., Willerslev et al. 2014; Boessenkool et al. 2014), and have been improved with the more recent genomic reference libraries (Alsos et al. 2020; Garcés-Pastor et al. 2022). Research with DNA metabarcoding has become particularly important for biodiversity monitoring of contemporary ecosystems (Deiner et al. 2017), but is also widely used in paleoecological and archaeological reconstruction (Capo et al. 2021; Brown et al. 2022), including addressing questions dealing with species histories, traits, and evolution (Epp et al. 2018; Alsos et al. 2022). We consider the field to have matured to a degree where novelty and innovation will predominantly arise from new scientific questions, the analysis of new materials or the application of DNA metabarcoding in novel environments or disciplines.

A major area for improvement in metabarcoding studies is the development of metabarcoding primers. So far, robust metabarcoding primers have been developed for certain groups of organisms, such as for the P6 loop of the *trn*L intron of plants (Taberlet et al. 2007), which shows high within-group sequence variability while being flanked by highly conserved primers. For this locus, comprehensive reference databases have been developed, making accurate biodiversity reconstructions possible for the target group. However, for taxa that have low within-group variability in the genes commonly used for metabarcoding, or are lacking reference databases (e.g., Cladocerans or water fleas; Meredith et al. (2021)), identification can be challenging. Moreover, limited genomic references can further impede the discovery of new genetic regions for metabarcoding. The design of better metabarcoding primers will enable taxonomic identification at higher resolution with greater specificity. Critically, the targeted DNA fragments should be short enough so that they may be applied to older, more fragmented DNA molecules.

Beyond better primers, future studies will incorporate a greater number of primer sets, targeting many different genetic regions across taxa. Some current work has already shown the benefit of sequencing regions across the tree of life to understand biotic responses across taxonomic kingdoms (Holman et al. 2021). Additionally, advances in multiplexing may allow these markers to be amplified in a single PCR, reducing the amount of DNA required for analysis and decreasing the amount of reagents required for sequencing experiments (Ficetola and Taberlet 2023).

In parallel to the development of more efficient reactions, we expect improvements in wet-lab techniques and computational tools that will enable better delineation of true sequence variants from errors introduced in the laboratory. Novel (or repurposed existing) polymerases can help reduce false positive rates with lower PCR error and chimaera formation (Nichols et al. 2018). Unique Molecular Identifiers (UMIs), which are short molecular DNA barcodes that are incorporated during PCR to uniquely tag each DNA molecule individually, can be integrated into the lab workflow and will allow for better error correction and increased sequencing accuracy (Yoshitake et al. 2021).

Following data generation, future improvements in metabarcoding are also expected with innovations in data analysis and interpretation. In a typical metabarcoding dataset, two values can be measured for each amplicon sequence within a sample: the number of detections across replicate PCRs and the number of reads assigned to each replicate. These two values, directly associated with the quantity of DNA molecules in the extracted DNA sample, are also potentially linked to the abundance and prevalence of organisms in the sampled environment. However, the interpretation of amplicon abundance and replication level has not always been satisfactory. A positive but noisy relationship between amplicon count and replication level has been observed (Skelton et al. 2022), and a positive relationship between amplicon count and taxa abundance has also been acknowledged but not systematically quantified due to practical difficulty (Alsos et al. 2018). While there remain some issues with quantitatively interpreting read counts per sequence (Lamb et al. 2019; Skelton et al. 2022), some recent innovations in both the laboratory and in bioinformatic analyses may alleviate these issues (Ji et al. 2020; Shelton et al. 2023; McLaren et al. 2022) and true quantitative metabarcoding may become more routine. In the future, both read proportion and replicate positive-rate might be incorporated in GIS-informed occupancy modelling to better predict the presence and abundance of taxa of interest, even for low-abundance and scarcely distributed organisms.

Better sampling, dating and wet-lab techniques, in addition to reduced costs, means that future metabarcoding datasets will be able to cover higher spatial and temporal resolutions. As many studies have shown, eDNA is naturally heterogeneously distributed in the environment from millimetre to kilometre scales (Wang et al. 2023; Jeunen et al. 2019; Kuwae et al. 2020; Massilani et al. 2022). Therefore, generating datasets with greater spatial resolution is not only necessary for making valid spatiotemporal inferences, but also to address global-scale questions in biodiversity across the earth's surface (Williams et al. 2023; Bush et al. 2017; Dornelas et al. 2019)). In addition, many quantitative time series methods used to study the earth system, climate and population dynamics require highly resolved-temporal resolution. In the future, eDNA will be sampled at a finer temporal resolution across the environment to describe the dynamics of ecosystems and biological communities in response to intricate environmental variations, such as natural ocean current circulation and solar activity changes or shifts in response to anthropogenic stressors such as ocean acidification or climate change-induced heat waves.

Metagenomics

What can shotgun-sequenced metagenomes deliver that metabarcoding cannot? If we, for a moment, neglect the limitations and current lack of analytical capabilities, we envision that entire genomes from species and individuals might be untangled from ancient metagenomes extracted from environmental samples. Population genetics at the individual level might then be possible, allowing us to describe how the frequency of an allele controls trait changes over time and even relate this to how the factors lead to changes in gene (allele) frequencies. Ultimately this would enable an understanding of how changes in gene (allele) frequencies affect evolution and speciation.

In the past few years, there have been successful attempts to assemble genome-scale data from sedimentary DNA (Seersholm et al. 2016; Slon et al. 2017; Lammers et al. 2021; Vernot et al. 2021; Wang et al. 2021; Zavala et al. 2022; Murchie et al. 2022; Schulte et al. 2022), with the initial studies targeting organellar genomes. In this work, phylogenetic placements have been inferred under the assumption that the retrieved organelle genomes are derived from a representative population, this composite has since been defined as an "environmental genome" (Pedersen et al. 2021; Gelabert et al. 2021). Moreover, Lammers et al. (2021) and Murchie et al. (2022) specifically separated and independently analysed different haplogroups.

Most recently, pioneering studies have shown that nuclear genome-wide phylogenetic placements can be performed on humans and animals by using the nuclear regions of the environmental genomes. Interestingly, much of the data has been obtained from cave sediments, which likely is explained by the fact that the sedimentary ancient DNA in caves is less biologically diverse than other sources (Pedersen et al. 2021; Vernot et al. 2021; Gelabert et al. 2021) and that caves are a particularly good source for sedaDNA of mammals in comparison to other substrates (Seeber and Epp 2022). Intriguingly, Slon and co-authors (2017) note that most of their samples showed signs of single individuals as sources of the mitochondrial DNA, with only a single sample of mixed origin - a sample from the Èl Sidron cave (Spain) where two individuals seemed to have contributed to that environmental genome.

A similar situation has been shown for plant DNA obtained from lake sediments in Siberia, which was enriched for full chloroplasts of larch trees (*Larix*) (Schulte et al. 2021), and where single samples yielded data that were attributed to two different larch species. With a larger dataset, including nuclear DNA of *Larix,* Schulte and co-authors (2022) could trace shifting species distributions since the Late Pleistocene in relation to climatic changes. Interestingly, the initial dataset, based on specific enrichment of *Larix* chloroplast DNA, also yielded sufficient information on alders, birches and willows to study the past 7000 years of population dynamics for these shrubs (Meucci et al. 2021). This work highlights the importance of the initial DNA concentration in the sediments, which was very high for the shrubs. A further important aspect, enabling such analyses, was the assembly of genomic reference data for the larch species (Zimmermann et al. 2019), without which the analyses would not have been possible.

To overcome part of the limitations inherent in complex environmental samples, it is clear that capture enrichment will become widely applicable and has a huge potential when studying ancient metagenomes. Capture enrichment involves the physical enrichment of DNA in the laboratory relevant to the question of interest from a total pool of DNA, often a metagenomic library. However, improvements in efficiency, yields and costs are needed before the capture probes (small nucleotide regions used to enrich target DNA) can compare favourably with the primers used for metabarcoding in a typical research project. Additionally, efforts are also needed to develop robust statistical tools that will allow us to discern individual identities within environmental samples even when in low abundance. We could imagine that this technology would provide the ability to determine the composition of groups, in terms of relatedness and sex, from DNA extractions of larger sediment samples. Such work could provide unprecedented insight into ancient hominin group dynamics or changes in the composition of animal populations. It may also allow us to look for genetic disorders and even epigenome responses or variations between different populations and among individuals.

An outstanding challenge today in capture enrichment is that the majority of the captured DNA molecules that are enriched and sequenced are from non-target organisms, even in successful reactions. However, with the speed and advancement seen by sequencing technology, it might even be a future possibility to perform adaptive sequencing of only ancient DNA molecules (Edwards et al. 2019). This technology has currently been demonstrated with modern samples on the Oxford Nanopore platform (Loose et al. 2016), wherein non-target DNA molecules are selectively rejected from reactions during sequencing. This idea might even be extended to include only the target taxa of interest, saving time and resources for sequencing and analysis (Su et al. 2023).

From a paleo-ecologist's point of view, abundance estimates or biomass reconstruction of plants and animals in the given landscapes could revise and refine the interpretation of biodiversity composition. With the experience gained from research into pollen counts (Broström et al. 1998) and by incorporating knowledge between modern eDNA and species (McLaren et al. 2022; Yates et al. 2019), numerical models could be developed to explore the DNA/landscape relationships and to identify correlations between modern sedimentary DNA spectra and the quantitative landscape variables. This is particularly relevant for shotgun-metagenomics as the technique provides more accurate quantitative estimates of the relative concentration of DNA in a given sample compared to other techniques (Bista et al. 2018). This accuracy aside, there are outstanding questions about how eDNA in modern samples reflects biomass in the environment, and how taphonomic processes subsequently preserve any quantitative measure of biomass after deposition (Alsos et al. 2018; Giguet-Covex et al. 2019; Wang et al. 2023).

In the future, geologically older and older samples will be collected and sequenced, pushing our knowledge past the current limits and into the fog of the past. This will enable us to directly observe evolutionary processes, not just on a microbial level but on the macro-organismal level (e.g., plants and animals).

Standards and Ethics

In tandem with the increase in the technical power for sedaDNA research, we must also improve our data and metadata standards, while ensuring the field has rigorous shared research ethics. As we advance scientific questions, an important future development will be the integration of FAIR data principles (see in glossary) for data produced from sedimentary ancient DNA (Wilkinson et al. 2016). FAIR data are findable, accessible, interoperable and reusable by design and we expect sedaDNA to increasingly comply with these principles. Some good progress has already been made in developing reporting guidelines for eDNA (Goldberg et al. 2016) and metadata reporting requirements for ancient metagenomes (Fellows Yates et al. 2021). We anticipate broad uptake of these community-driven efforts.

Besides data reporting, scientific institutions and researchers in the field of sedimentary DNA should also be guided by ethical considerations when designing research, conducting science, and sharing research outcomes. As the scope of research expands and sampling scale enlarges, sedimentary DNA research will increasingly involve collaboration with local communities, and on indigenous or pristine lands (Handsley-Davis et al. 2021). With methodological advances, research outcomes will likely include more data that are ethically sensitive, such as human DNA (Kaestle and Horsburgh 2002); data that are economically valuable, such as DNA of agricultural crops (Hansson 2019); or data critical for human and ecosystem health, such as DNA of disease pathogens or extinct/endangered animals (Campbell and Whittle 2017; (Yarzábal et al. 2021). The Nagoya protocol, for example, sets out obligations for contracting parties to promote the fair benefit-sharing arising from the use of genetic resources. Furthermore, the FAIR guidelines have been expanded into the CARE principles (Carroll et al. 2020), which include properties that reflect Indigenous values for data governance: (1) collective benefit; (2) authority to control; (3) responsibility and (4) ethics. In addition, a recent paper published by Alpaslan-Roodenberg et al. (2021) provides guidelines for DNA research on human remains. In the future, we expect the guidance for sedimentary DNA research to also develop and mature alongside the science, with ethical considerations that address the benefits not only for human society but also for ecosystem health and sustainability. We advocate the scientific community to contribute to the establishment of such guidance and conduct research with the highest ethical standards.

The Unknown Future

While most of the future methods and analytical possibilities above seem within grasp in the coming decades, the ideas presented in this section could easily belong in the science fiction worlds of Arthur C. Clarke and Stanley Kubrick's *"2001: A Space Odyssey"* or Isaac Asimov's 'Foundation' novels. Science fiction ideas that seemed unthinkable at the time, but more than 50 years later in the present day, are increasingly found in common technology. Here, we attempt not to be limited by the boundaries set by the technology or analytical methods possible today. In this section, the imagination is truly free.

Automated drones in the future might help us reach and automate sampling of hard-to-reach and even completely inaccessible places, across air, water, ice, and even rock/earth crust matrices. Perhaps, sampling autonomous robots could collect a myriad of information such as elemental composition, isotopic concentrations, organic content and residues, temperatures, salinities and DNA sequences. With such technologies, we could understand unreachable places in the ocean, caves, lake sediments, ice caps, and the driest deserts. Swarms of miniaturised drones with real time sequencing could be sent through deposits to retrieve high-resolution, spatially distributed ancient genomic landscapes, analysed immediately to inform fieldwork dynamically while guided by strict ethical protocols. Imagine such swarms of self-repairing drones being deployed extraterrestrially to find past evidence of life on other planets and in foreign solar systems.

Or what if we could extract and single-cell sequence DNA from ancient dead but intact cells preserved in sediments and other deposits? Although it would likely be fragmented, we could be certain that the DNA would have originated from a single individual and hence allow for higher detail and precision in population analysis, by maintaining the link between ancient eDNA strands physically close to one another. Such technology would allow us to understand evolution from the perspective of single individuals, observing the exact change of alleles, and even traits through time.

While today the isolation and analysis of biomolecules are lengthy and complicated, the complete non-destructive characterisation of biomolecular composition and organisation may be commonplace in the future. A combined analysis of ancient RNA, DNA, lipids, proteins and carbohydrates from a sample would provide unparalleled insight into questions in ecology, evolution and archaeology, while also leaving the sample *in situ* or in storage for subsequent analyses or as a reference.

Imagine the possibilities offered by advanced artificial intelligence. While the phenotype of an organism is, to some extent, affected by the environment, it may be possible to reconstruct the phenotype of organisms from genomic data. Metagenomes contain a huge amount of genomic information, and if complete genomes can be reconstructed, a visual representation of the species found in the sample might be possible. Today artists create reconstructions of ancient humans from limited genomic information provided by their ancient DNA such as eye colour or skin pigmentation. Future technologies may go further allowing us to fully reconstruct the physical appearance of ancient hominids and look into the face of the past.

Summary

In real-time, foreseeable future and through our wild imagination, technological advances will allow us to understand our environment in finer and richer detail. These technologies will decipher the secrets of past life on Earth and better inform the future trajectory of the biosphere. The perspectives outlined in this chapter are reflections of today's achievements in sedimentary DNA research, and our hope is that they serve as a way marker, mapping out one possible route into the unknown future.

Acknowledgements We thank Eric Capo, Pete Heintzmann, Inger Alsos, Kathleen Stoof-Leichsenring and Charlene Giguet-Covex for comments on this manuscript. M.W.P. was supported by the ERC project LASTJOURNEY (ERC_Adv_834514) and the Danish National Research Foundation (DNRF174). L.E.H. and K.B. were supported by the European Research Council (ERC) under the European Union's Horizon 2020 Research and Innovation Programme (grant agreement no. 856488). L.S.E. was supported by the Elite Program for Postdocs of the Baden-Württemberg Foundation, which also supported the work of Y.W. Y.W. is furthermore funded by the graduate school scholarship programme (57450037) of the German Academic Exchange Service, the Research Training Group R3 (GRK 2272, DFG award number 298726046) as well as the Doctoral Fund and the Konstanzia MEiN programme at the University of Konstanz, Germany. R.S. was supported by JSPS KAKENHI Grant Number JP20J00078.

References

Agersnap S, Sigsgaard EE, Jensen MR, Avila MDP, Carl H, Møller PR, Krøs SL, Knudsen SW, Wisz MS, Thomsen PF (2022) A national scale "BioBlitz" using citizen science and eDNA metabarcoding for monitoring coastal marine fish. Front Mar Sci 9. https://doi.org/10.3389/fmars.2022.824100

Alpaslan-Roodenberg S, Anthony D, Babiker H, Bánffy E, Booth T, Capone P, Deshpande-Mukherjee A, Eisenmann S, Fehren-Schmitz L, Frachetti M, Fujita R, Frieman CJ, Fu Q, Gibbon V, Haak W, Hajdinjak M, Hofmann KP, Holguin B, Inomata T, Kanzawa-Kiriyama H, Keegan W, Kelso J, Krause J, Kumaresan G, Kusimba C, Kusimba S, Lalueza-Fox C, Llamas B, MacEachern S, Mallick S, Matsumura H, Morales-Arce AY, Matuzeviciute GM, Mushrif-Tripathy V, Nakatsuka N, Nores R, Ogola C, Okumura M, Patterson N, Pinhasi R, Prasad SPR, Prendergast ME, Punzo JL, Reich D, Sawafuji R, Sawchuk E, Schiffels S, Sedig J, Shnaider S, Sirak K, Skoglund P, Slon V, Snow M, Soressi M, Spriggs M, Stockhammer PW, Szécsényi-Nagy A, Thangaraj K, Tiesler V, Tobler R, Wang C-C, Warinner C, Yasawardene S, Zahir M (2021) Ethics of DNA research on human remains: five globally applicable guidelines. Nature 599(7883):41–46. https://doi.org/10.1038/s41586-021-04008-x

Alsos IG, Lammers Y, Yoccoz NG, Jørgensen T, Sjögren P, Gielly L, Edwards ME (2018) Plant DNA metabarcoding of lake sediments: how does it represent the contemporary vegetation. PLoS One 13(4):e0195403. https://doi.org/10.1371/journal.pone.0195403

Alsos IG, Lavergne S, Merkel MKF, Boleda M, Lammers Y, Alberti A, Pouchon C, Denoeud F, Pitelkova I, Puşcaş M, Roquet C, Hurdu B-I, Thuiller W, Zimmermann NE, Hollingsworth PM, Coissac E (2020) The treasure vault can be opened: large-scale genome skimming works well using herbarium and silica gel dried material. Plan Theory 9(4):432. https://doi.org/10.3390/plants9040432

Alsos IG, Rijal DP, Ehrich D, Karger DN, Yoccoz NG, Heintzman PD, Brown AG, Lammers Y, Pellissier L, Alm T, Bråthen KA, Coissac E, Merkel MKF, Alberti A, Denoeud F, Bakke J, Phylonorway Consortium (2022) Postglacial species arrival and diversity buildup of northern ecosystems took millennia. Sci Adv 8(39):eabo7434. https://doi.org/10.1126/sciadv.abo7434

Armbrecht LH, Coolen MJL, Lejzerowicz F, George SC, Negandhi K, Suzuki Y, Young J, Foster NR, Armand LK, Cooper A, Ostrowski M, Focardi A, Stat M, Moreau JW, Weyrich LS (2019) Ancient DNA from marine sediments: precautions and considerations for seafloor coring, sample handling and data generation. Earth Sci Rev 196:102887. https://doi.org/10.1016/j.earscirev.2019.102887

Armbrecht L, Weber ME, Raymo ME, Peck VL, Williams T, Warnock J, Kato Y, Hernández-Almeida I, Hoem F, Reilly B, Hemming S, Bailey I, Martos YM, Gutjahr M, Percuoco V, Allen C, Brachfeld S, Cardillo FG, Du Z, Fauth G, Fogwill C, Garcia M, Glüder A, Guitard M, Hwang J-H, Iizuka M, Kenlee B, O'Connell S, Pérez LF, Ronge TA, Seki O, Tauxe L, Tripathi S, Zheng X (2022) Ancient marine sediment DNA reveals diatom transition in Antarctica. Nat Commun 13(1):5787. https://doi.org/10.1038/s41467-022-33494-4

Barnes MA, Turner CR (2016) The ecology of environmental DNA and implications for conservation genetics. Conserv Genet 17(1):1–17. https://doi.org/10.1007/s10592-015-0775-4

Birks HJB, Berglund BE (2018) One hundred years of Quaternary pollen analysis 1916–2016. Veg Hist Arch 27:271–309. https://doi.org/10.1007/s00334-017-0630-2

Binladen J, Gilbert MTP, Bollback JP, Panitz F, Bendixen C, Nielsen R, Willerslev E (2007) The use of coded PCR primers enables high-throughput sequencing of multiple homolog amplification products by 454 parallel sequencing. PLoS One 2(2):e197. https://doi.org/10.1371/journal.pone.0000197

Bista I, Carvalho GR, Tang M, Walsh K, Zhou X, Hajibabaei M, Shokralla S, Seymour M, Bradley D, Liu S, Christmas M, Creer S (2018) Performance of amplicon and shotgun sequencing for accurate biomass estimation in invertebrate community samples. Mol Ecol Resour 18(5):1020–1034. https://doi.org/10.1111/1755-0998.12888

Boessenkool S, Mcglynn G, Epp LS, Taylor D, Pimentel M, Gizaw A, Nemomissa S, Brochmann C, Popp M (2014) Use of ancient sedimentary DNA as a novel conservation tool for high-altitude tropical biodiversity. Conserv Biol 28(2):446–455. https://doi.org/10.1111/cobi.12195

Broström A, Gaillard M-J, Ihse M, Odgaard B (1998) Pollen-landscape relationships in modern analogues of ancient cultural landscapes in southern Sweden — a first step towards quantification of vegetation openness in the past. Veget Hist Archaebot 7(4):189–201. https://doi.org/10.1007/BF01146193

Brown T, Rijal DP, Heintzman PD, Clarke CL, Blankholm H-P, Høeg HI, Lammers Y, Bråthen KA, Edwards M, Alsos IG (2022) Paleoeconomy more than demography determined prehistoric human impact in Arctic Norway. PNAS Nexus 1(5):pgac209. https://doi.org/10.1093/pnasnexus/pgac209

Buchner D, Macher T-H, Beermann AJ, Werner M-T, Leese F (2021) Standardized high-throughput biomonitoring using DNA metabarcoding: strategies for the adoption of automated liquid handlers. Environ Sci Ecotechnol 8:100122. https://doi.org/10.1016/j.ese.2021.100122

Bush A, Sollmann R, Wilting A, Bohmann K, Cole B, Balzter H, Martius C, Zlinszky A, Calvignac-Spencer S, Cobbold CA, Dawson TP, Emerson BC, Ferrier S, Gilbert MTP, Herold M, Jones L, Leendertz FH, Matthews L, Millington JDA, Olson JR, Ovaskainen O, Raffaelli D, Reeve R, Rödel M-O, Rodgers TW, Snape S, Visseren-Hamakers I, Vogler AP, White PCL, Wooster MJ, Yu DW (2017) Connecting Earth observation to high-throughput biodiversity data. Nat Ecol Evol 1(7):1–9. https://doi.org/10.1038/s41559-017-0176

Cai P, Huang Q-Y, Zhang X-W (2006) Interactions of DNA with clay minerals and soil colloidal particles and protection against degradation by DNase. Environ Sci Technol 40(9):2971–2976. https://doi.org/10.1021/es0522985

Campbell DI, Whittle PM (2017) Resurrecting extinct species: ethics and authenticity. Palgrave Macmillan, London

Capo E, Rydberg J, Tolu J, Domaizon I, Debroas D, Bindler R, Bigler C (2019) How does environmental inter-annual variability shape aquatic microbial communities? A 40-year annual record of sedimentary DNA from a boreal lake (Nylandssjön, Sweden). Front Ecol Evol 7. https://doi.org/10.3389/fevo.2019.00245

Capo E, Giguet-Covex C, Rouillard A, Nota K, Heintzman PD, Vuillemin A, Ariztegui D, Arnaud F, Belle S, Bertilsson S, Bigler C, Bindler R, Brown AG, Clarke CL, Crump SE, Debroas D, Englund G, Ficetola GF, Garner RE, Gauthier J, Gregory-Eaves I, Heinecke L, Herzschuh U, Ibrahim A, Kisand V, Kjær KH, Lammers Y, Littlefair J, Messager E, Monchamp M-E, Olajos

F, Orsi W, Pedersen MW, Rijal DP, Rydberg J, Spanbauer T, Stoof-Leichsenring KR, Taberlet P, Talas L, Thomas C, Walsh DA, Wang Y, Willerslev E, van Woerkom A, Zimmermann HH, Coolen MJL, Epp LS, Domaizon IG, Alsos I, Parducci L (2021) Lake sedimentary DNA research on past terrestrial and aquatic biodiversity: overview and recommendations. Quaternary 4(1):6. https://doi.org/10.3390/quat4010006

Carøe C, Bohmann K (2020) Tagsteady: a metabarcoding library preparation protocol to avoid false assignment of sequences to samples. Mol Ecol Resour 20(6):1620–1631. https://doi.org/10.1111/1755-0998.13227

Carroll SR, Garba I, Figueroa-Rodríguez OL, Holbrook J, Lovett R, Materechera S, Parsons M, Raseroka K, Rodriguez-Lonebear D, Rowe R, Sara R, Walker JD, Anderson J, Hudson M (2020) The CARE principles for indigenous data. Governance 19(1):43. https://doi.org/10.5334/dsj-2020-043

Clare EL, Economou CK, Bennett FJ, Dyer CE, Adams K, McRobie B, Drinkwater R, Littlefair JE (2022) Measuring biodiversity from DNA in the air. Curr Biol 32(3):693–700.e5. https://doi.org/10.1016/j.cub.2021.11.064

Dabney J, Meyer M, Pääbo S (2013) Ancient DNA damage. Cold Spring Harb Perspect Biol 5(7):a012567. https://doi.org/10.1101/cshperspect.a012567

Deiner K, Bik HM, Mächler E, Seymour M, Lacoursière-Roussel A, Altermatt F, Creer S, Bista I, Lodge DM, de Vere N, Pfrender ME, Bernatchez L (2017) Environmental DNA metabarcoding: transforming how we survey animal and plant communities. Mol Ecol 26(21):5872–5895. https://doi.org/10.1111/mec.14350

del Carmen Gomez Cabrera M, Young JM, Roff G, Staples T, Ortiz JC, Pandolfi JM, Cooper A (2019) Broadening the taxonomic scope of coral reef palaeoecological studies using ancient DNA. Mol Ecol 28(10):2636–2652. https://doi.org/10.1111/mec.15038

Der Sarkissian C, Pichereau V, Dupont C, Ilsøe PC, Perrigault M, Butler P, Chauvaud L, Eiríksson J, Scourse J, Paillard C, Orlando L (2017) Ancient DNA analysis identifies marine mollusc shells as new metagenomic archives of the past. Mol Ecol Resour 17(5):835–853. https://doi.org/10.1111/1755-0998.12679

Dommain R, Andama M, McDonough MM, Prado NA, Goldhammer T, Potts R, Maldonado JE, Nkurunungi JB, Campana MG (2020) The challenges of reconstructing tropical biodiversity with sedimentary ancient DNA: a 2200-year-long metagenomic record from Bwindi impenetrable forest, Uganda. Front Ecol Evolut 8:218

Dornelas M, Madin EMP, Bunce M, DiBattista JD, Johnson M, Madin JS, Magurran AE, McGill BJ, Pettorelli N, Pizarro O, Williams SB, Winter M, Bates AE (2019) Towards a macroscope: leveraging technology to transform the breadth, scale and resolution of macroecological data. Glob Ecol Biogeogr 28(12):1937–1948. https://doi.org/10.1111/geb.13025

Edwards HS, Krishnakumar R, Sinha A, Bird SW, Patel KD, Bartsch MS (2019) Real-time selective sequencing with RUBRIC: read until with basecall and reference-informed criteria. Sci Rep 9(1):11475. https://doi.org/10.1038/s41598-019-47857-3

Epp LS, Boessenkool S, Bellemain EP, Haile J, Esposito A, Riaz T, Erséus C, Gusarov VI, Edwards ME, Johnsen A, Stenøien HK, Hassel K, Kauserud H, Yoccoz NG, Bråthen KA, Willerslev E, Taberlet P, Coissac E, Brochmann C (2012) New environmental metabarcodes for analysing soil DNA: potential for studying past and present ecosystems. Mol Ecol 21(8):1821–1833. https://doi.org/10.1111/j.1365-294X.2012.05537.x

Epp LS, Kruse S, Kath NJ, Stoof-Leichsenring KR, Tiedemann R, Pestryakova LA, Herzschuh U (2018) Temporal and spatial patterns of mitochondrial haplotype and species distributions in Siberian larches inferred from ancient environmental DNA and modeling. Sci Rep 8(1):17436. https://doi.org/10.1038/s41598-018-35550-w

Fellows Yates JA, Andrades Valtueña A, Vågene ÅJ, Cribdon B, Velsko IM, Borry M, Bravo-Lopez MJ, Fernandez-Guerra A, Green EJ, Ramachandran SL, Heintzman PD, Spyrou MA, Hübner A, Gancz AS, Hider J, Allshouse AF, Zaro V, Warinner C (2021) Community-curated and standardised metadata of published ancient metagenomic samples with AncientMetagenomeDir. Sci Data 8(1):31. https://doi.org/10.1038/s41597-021-00816-y

Ficetola GF, Taberlet P (2023) Towards exhaustive community ecology via DNA metabarcoding. Mol Ecol. https://doi.org/10.1111/mec.16881

Ficetola GF, Pansu J, Bonin A, Coissac E, Giguet-Covex C, De Barba M, Gielly L, Lopes CM, Boyer F, Pompanon F, Rayé G, Taberlet P (2015) Replication levels, false presences and the estimation of the presence/absence from eDNA metabarcoding data. Mol Ecol Resour 15(3):543–556. https://doi.org/10.1111/1755-0998.12338

Freeman CL, Dieudonné L, Agbaje OBA, Zure M, Sanz JQ, Collins M, Sand KK (2023) Survival of environmental DNA in sediments: mineralogic control on DNA taphonomy. https://doi.org/10.1101/2020.01.28.922997

Garcés-Pastor S, Coissac E, Lavergne S, Schwörer C, Theurillat J-P, Heintzman PD, Wangensteen OS, Tinner W, Rey F, Heer M, Rutzer A, Walsh K, Lammers Y, Brown AG, Goslar T, Rijal DP, Karger DN, Pellissier L, Heiri O, Alsos IG (2022) High resolution ancient sedimentary DNA shows that alpine plant diversity is associated with human land use and climate change. Nat Commun 13(1):6559. https://doi.org/10.1038/s41467-022-34010-4

Gelabert P, Sawyer S, Bergström A, Margaryan A, Collin TC, Meshveliani T, Belfer-Cohen A, Lordkipanidze D, Jakeli N, Matskevich Z, Bar-Oz G, Fernandes DM, Cheronet O, Özdoğan KT, Oberreiter V, Feeney RNM, Stahlschmidt MC, Skoglund P, Pinhasi R (2021) Genome-scale sequencing and analysis of human, wolf, and bison DNA from 25,000-year-old sediment. Curr Biol 31(16):3564–3574.e9. https://doi.org/10.1016/j.cub.2021.06.023

Giguet-Covex C, Pansu J, Arnaud F, Rey P-J, Griggo C, Gielly L, Domaizon I, Coissac E, David F, Choler P, Poulenard J, Taberlet P (2014) Long livestock farming history and human landscape shaping revealed by lake sediment DNA. Nat Commun 5(1):3211. https://doi.org/10.1038/ncomms4211

Giguet-Covex C, Ficetola GF, Walsh K, Poulenard J, Bajard M, Fouinat L, Sabatier P, Gielly L, Messager E, Develle AL, David F, Taberlet P, Brisset E, Guiter F, Sinet R, Arnaud F (2019) New insights on lake sediment DNA from the catchment: importance of taphonomic and analytical issues on the record quality. Sci Rep 9:14676. https://doi.org/10.1038/s41598-019-50339-1

Gilbert MTP, Hansen AJ, Willerslev E, Rudbeck L, Barnes I, Lynnerup N, Cooper A (2003) Characterization of genetic miscoding lesions caused by postmortem damage. Am J Hum Genet 72(1):48–61. https://doi.org/10.1086/345379

Goldberg CS, Turner CR, Deiner K, Klymus KE, Thomsen PF, Murphy MA, Spear SF, McKee A, Oyler-McCance SJ, Cornman RS, Laramie MB, Mahon AR, Lance RF, Pilliod DS, Strickler KM, Waits LP, Fremier AK, Takahara T, Herder JE, Taberlet P (2016) Critical considerations for the application of environmental DNA methods to detect aquatic species. Methods Ecol Evol 7(11):1299–1307. https://doi.org/10.1111/2041-210X.12595

Graham RW, Belmecheri S, Choy K, Culleton BJ, Davies LJ, Froese D, Heintzman PD, Hritz C, Kapp JD, Newsom LA, Rawcliffe R, Saulnier-Talbot É, Shapiro B, Wang Y, Williams JW, Wooller MJ (2016) Timing and causes of mid-Holocene mammoth extinction on St. Paul Island, Alaska. Proc Natl Acad Sci 113(33):9310–9314. https://doi.org/10.1073/pnas.1604903113

Haile J, Froese DG, Macphee RDE, Roberts RG, Arnold LJ, Reyes AV, Rasmussen M, Nielsen R, Brook BW, Robinson S, Demuro M, Gilbert MTP, Munch K, Austin JJ, Cooper A, Barnes I, Möller P, Willerslev E (2009) Ancient DNA reveals late survival of mammoth and horse in interior Alaska. Proc Natl Acad Sci U S A 106(52):22352–22357. https://doi.org/10.1073/pnas.0912510106

Handsley-Davis M, Kowal E, Russell L, Weyrich LS (2021) Researchers using environmental DNA must engage ethically with indigenous communities. https://doi.org/10.1038/s41559-020-01351-6

Hansson SO (2019) A science-informed ethics for agricultural biotechnology. Crop Breed Genet Genom 1(1). https://doi.org/10.20900/cbgg20190006

Harrison JB, Sunday JM, Rogers SM (2019) Predicting the fate of eDNA in the environment and implications for studying biodiversity. Proc R Soc B 286:20191409. http://doi.org/10.1098/rspb.2019.1409

Hawkins J, de Vere N, Griffith A, Ford CR, Allainguillaume J, Hegarty MJ, Baillie L, Adams-Groom B (2015) Using DNA metabarcoding to identify the floral composition of honey: a new tool for investigating honey bee foraging preferences. PLoS One 10(8):e0134735. https://doi.org/10.1371/journal.pone.0134735

Heintzman P, Rijal DP, Brown AG, Pitelkova I, Ancin Murguzur FJ, Clarke CL, Edwards ME, Alsos IG (2019) Optimization of efficient ancient DNA extraction from lake sediment. Curr Res Inf Syst Norway 2019:B53O-2611

Holman LE, de Bruyn M, Creer S, Carvalho G, Robidart J, Rius M (2021) Animals, protists and bacteria share marine biogeographic patterns. Nat Ecol Evol 5(6):738–746. https://doi.org/10.1038/s41559-021-01439-7

Huang DY-T, Churchman J, Lowe D, Rawlence N, Schipper L, Cooper A (2012) Evaluating the character and preservation of DNA within allophane clusters in buried soils on Holocene tephras, northern New Zealand

Huggett R (2016) Fundamentals of geomorphology, 4th edn. Routledge, London/New York

Jarman SN, Berry O, Bunce M (2018) The value of environmental DNA biobanking for long-term biomonitoring. Nat Ecol Evol 2(8):1192–1193. https://doi.org/10.1038/s41559-018-0614-3

Jeunen G-J, Knapp M, Spencer HG, Lamare MD, Taylor HR, Stat M, Bunce M, Gemmell NJ (2019) Environmental DNA (eDNA) metabarcoding reveals strong discrimination among diverse marine habitats connected by water movement. Mol Ecol Resour 19(2):426–438. https://doi.org/10.1111/1755-0998.12982

Ji Y, Huotari T, Roslin T, Schmidt NM, Wang J, Yu DW, Ovaskainen O (2020) SPIKEPIPE: a metagenomic pipeline for the accurate quantification of eukaryotic species occurrences and intraspecific abundance change using DNA barcodes or mitogenomes. Mol Ecol Resour 20(1):256–267. https://doi.org/10.1111/1755-0998.13057

Jørgensen T, Kjaer KH, Haile J, Rasmussen M, Boessenkool S, Andersen K, Coissac E, Taberlet P, Brochmann C, Orlando L, Gilbert MTP, Willerslev E (2012) Islands in the ice: detecting past vegetation on Greenlandic nunataks using historical records and sedimentary ancient DNA metabarcoding. Mol Ecol 21(8):1980–1988. https://doi.org/10.1111/j.1365-294X.2011.05278.x

Kaestle FA, Horsburgh KA (2002) Ancient DNA in anthropology: methods, applications, and ethics. Am J Phys Anthropol 119(S35):92–130. https://doi.org/10.1002/ajpa.10179

Kircher M, Sawyer S, Meyer M (2012) Double indexing overcomes inaccuracies in multiplex sequencing on the Illumina platform. Nucleic Acids Res 40(1):e3. https://doi.org/10.1093/nar/gkr771

Kopf A, Bicak M, Kottmann R, Schnetzer J, Kostadinov I, Lehmann K, Fernandez-Guerra A, Jeanthon C, Rahav E, Ullrich M, Wichels A, Gerdts G, Polymenakou P, Kotoulas G, Siam R, Abdallah RZ, Sonnenschein EC, Cariou T, O'Gara F, Jackson S, Orlic S, Steinke M, Busch J, Duarte B, Caçador I, Canning-Clode J, Bobrova O, Marteinsson V, Reynisson E, Loureiro CM, Luna GM, Quero GM, Löscher CR, Kremp A, Delorenzo ME, Ovreas L, Tolman J, Laroche J, Penna A, Frischer M, Davis T, Barker K, Meyer CP, Ramos S, Magalhaes C, Jude-Lemeilleur F, Aguirre-Macedo ML, Wang S, Poulton N, Jones S, Collin R, Fuhrman JA, Conan P, Alonso C, Stambler N, Goodwin K, Yakimov MM, Baltar F, Bodrossy L, van de Kamp J, Frampton DM, Otrowski M, van Ruth P, Malthouse P, Claus S, Deneudt K, Mortelmans J, Pitois S, Wallom D, Salter I, Costa R, Schroeder DC, Kandil MM, Amaral V, Biancalana F, Santana R, Pedrotti M-L, Yoshida T, Ogata H, Ingleton T, Munnik K, Rodriguez-Ezpeleta N, Berteaux-Lecellier V, Wecker P, Cancio I, Vaulot D, Bienhold C, Ghazal H, Chaouni B, Essayeh S, Ettamimi S, Zaid EH, Boukhatem N, Bouali A, Chahboune R, Barrijal S, Titinouni M, Otmani FE, Bennani M, Mea M, Todorova N, Karamfilov V, ten Hoopen P, Cochrane G, L'Haridon S, Bizsel KC, Vezzi A, Lauro FM, Martin P, Jensen RM, Hinks J, Gebbels S, Rosselli R, de Pascale F, Schiavon R, dos Santos A, Villar E, Pesant S, Cataletto B, Malfatti F, Edirisinghe R, Silveira JAH, Barbier M, Turk V, Tinta T, Fuller WJ, Salihoglu I, Serakinci N, Ergoren MC, Bresnan E, Iriberri J, Nyhus PAF, Bente E, Karlsen HE, Golyshin PN, Gasol JM, Moncheva S, Dzhembekova N, Johnson Z, Sinigalliano CD, Gidley ML, Zingone A, Danovaro R, Tsiamis G, Clark MS, Costa AC, Bour ME, Martins AM, Collins RE, Ducluzeau A-L, Martinez J, Costello

MJ, Amaral-Zettler LA, Gilbert JA, Davies N, Field D, Glöckner FO (2015) The ocean sampling day consortium. GigaScience 4(1):27. https://doi.org/10.1186/s13742-015-0066-5
Kuwae M, Tamai H, Doi H, Sakata MK, Minamoto T, Suzuki Y (2020) Sedimentary DNA tracks decadal-centennial changes in fish abundance. Commun Biol 3(1):1–12. https://doi.org/10.1038/s42003-020-01282-9
Lamb PD, Hunter E, Pinnegar JK, Creer S, Davies RG, Taylor MI (2019) How quantitative is metabarcoding: a meta-analytical approach. Mol Ecol 28(2):420–430. https://doi.org/10.1111/mec.14920
Lammers Y, Heintzman PD, Alsos IG (2021) Environmental palaeogenomic reconstruction of an Ice Age algal population. Commun Biol 4(1):1–11. https://doi.org/10.1038/s42003-021-01710-4
Levy-Booth DJ, Campbell RG, Gulden RH, Hart MM, Powell JR, Klironomos JN, Pauls KP, Swanton CJ, Trevors JT, Dunfield KE (2007) Cycling of extracellular DNA in the soil environment. Soil Biol Biochem 39(12):2977–2991
Loose M, Malla S, Stout M (2016) Real-time selective sequencing using nanopore technology. Nat Methods 13(9):751–754. https://doi.org/10.1038/nmeth.3930
Lynggaard C, Bertelsen MF, Jensen CV, Johnson MS, Frøslev TG, Olsen MT, Bohmann K (2022) Airborne environmental DNA for terrestrial vertebrate community monitoring. Curr Biol 32(3):701–707.e5. https://doi.org/10.1016/j.cub.2021.12.014
Mariani S, Baillie C, Colosimo G, Riesgo A (2019) Sponges as natural environmental DNA samplers. Curr Biol 29(11):R401–R402. https://doi.org/10.1016/j.cub.2019.04.031
Marx V (2017) Genetics: new tales from ancient DNA. Nat Methods 14(8):771–774. https://doi.org/10.1038/nmeth.4367
Massilani D, Morley MW, Mentzer SM, Aldeias V, Vernot B, Miller C, Stahlschmidt M, Kozlikin MB, Shunkov MV, Derevianko AP, Conard NJ, Wurz S, Henshilwood CS, Vasquez J, Essel E, Nagel S, Richter J, Nickel B, Roberts RG, Pääbo S, Slon V, Goldberg P, Meyer M (2022) Microstratigraphic preservation of ancient faunal and hominin DNA in Pleistocene cave sediments. Proc Natl Acad Sci 119(1):e2113666118. https://doi.org/10.1073/pnas.2113666118
Mauvisseau Q, Harper LR, Sander M, Hanner RH, Kleyer H, Deiner K (2022) The multiple states of environmental DNA and what is known about their persistence in aquatic environments. Environ Sci Technol 56(9):5322–5333. https://doi.org/10.1021/acs.est.1c07638
McLaren MR, Nearing JT, Willis AD, Lloyd KG, Callahan BJ (2022) Implications of taxonomic bias for microbial differential-abundance analysis. https://doi.org/10.1101/2022.08.19.504330
McQuillan JS, Robidart JC (2017) Molecular-biological sensing in aquatic environments: recent developments and emerging capabilities. Curr Opin Biotechnol 45:43–50. https://doi.org/10.1016/j.copbio.2016.11.022
Meredith C, Hoffman J, Trebitz A, Pilgrim E, Okum S, Martinson J, Cameron ES (2021) Evaluating the performance of DNA metabarcoding for assessment of zooplankton communities in Western Lake Superior using multiple markers. Metabarcod Metagen 50:83–97. https://doi.org/10.3897/mbmg.5.64735
Meucci S, Schulte L, Zimmermann HH, Stoof-Leichsenring KR, Epp L, Bronken Eidesen P, Herzschuh U (2021) Holocene chloroplast genetic variation of shrubs (Alnus alnobetula, Betula nana, Salix sp.) at the siberian tundra-taiga ecotone inferred from modern chloroplast genome assembly and sedimentary ancient DNA analyses. Ecol Evol 11(5):2173–2193. https://doi.org/10.1002/ece3.7183
Minamoto T, Miya M, Sado T, Seino S, Doi H, Kondoh M, Nakamura K, Takahara T, Yamamoto S, Yamanaka H, Araki H, Iwasaki W, Kasai A, Masuda R, Uchii K (2021) An illustrated manual for environmental DNA research: water sampling guidelines and experimental protocols. Environ DNA 3(1):8–13. https://doi.org/10.1002/edn3.121
Murchie TJ, Kuch M, Duggan AT, Ledger ML, Roche K, Klunk J, Karpinski E, Hackenberger D, Sadoway T, MacPhee R, Froese D, Poinar H (2021) Optimizing extraction and targeted capture of ancient environmental DNA for reconstructing past environments using the PalaeoChip Arctic-1.0 bait-set. Quat Res 99:305–328. https://doi.org/10.1017/qua.2020.59

Murchie TJ, Karpinski E, Eaton K, Duggan AT, Baleka S, Zazula G, MacPhee RDE, Froese D, Poinar HN (2022) Pleistocene mitogenomes reconstructed from the environmental DNA of permafrost sediments. Curr Biol 32(4):851–860.e7. https://doi.org/10.1016/j.cub.2021.12.023

Nichols RV, Vollmers C, Newsom LA, Wang Y, Heintzman PD, Leighton M, Green RE, Shapiro B (2018) Minimizing polymerase biases in metabarcoding. Mol Ecol Resour 18(5):927–939. https://doi.org/10.1111/1755-0998.12895

Parducci L, Jørgensen T, Tollefsrud MM, Elverland E, Alm T, Fontana SL, Bennett KD, Haile J, Matetovici I, Suyama Y, Edwards ME, Andersen K, Rasmussen M, Boessenkool S, Coissac E, Brochmann C, Taberlet P, Houmark-Nielsen M, Larsen NK, Orlando L, Gilbert MTP, Kjær KH, Alsos IG, Willerslev E (2012) Glacial survival of boreal trees in northern Scandinavia. Science 335(6072):1083–1086. https://doi.org/10.1126/science.1216043

Parducci L, Bennett KD, Ficetola GF, Alsos IG, Suyama Y, Wood JR, Pedersen MW (2017) Ancient plant DNA in lake sediments. New Phytol 214(3):924–942. https://doi.org/10.1111/nph.14470

Pedersen MW, Ginolhac A, Orlando L, Olsen J, Andersen K, Holm J, Funder S, Willerslev E, Kjær KH (2013) A comparative study of ancient environmental DNA to pollen and macrofossils from lake sediments reveals taxonomic overlap and additional plant taxa. Quat Sci Rev 75:161–168. https://doi.org/10.1016/j.quascirev.2013.06.006

Pedersen MW, Overballe-Petersen S, Ermini L, Sarkissian CD, Haile J, Hellstrom M, Spens J, Thomsen PF, Bohmann K, Cappellini E, Schnell IB, Wales NA, Carøe C, Campos PF, Schmidt AMZ, Gilbert MTP, Hansen AJ, Orlando L, Willerslev E (2015) Ancient and modern environmental DNA. Philos Trans R Soc B 370(1660):20130383. https://doi.org/10.1098/rstb.2013.0383

Pedersen MW, Ruter A, Schweger C, Friebe H, Staff RA, Kjeldsen KK, Mendoza MLZ, Beaudoin AB, Zutter C, Larsen NK, Potter BA, Nielsen R, Rainville RA, Orlando L, Meltzer DJ, Kjær KH, Willerslev E (2016) Postglacial viability and colonization in North America's ice-free corridor. Nature 537(7618):45–49. https://doi.org/10.1038/nature19085

Pedersen MW, De Sanctis B, Saremi NF, Sikora M, Puckett EE, Gu Z, Moon KL, Kapp JD, Vinner L, Vardanyan Z, Ardelean CF, Arroyo-Cabrales J, Cahill JA, Heintzman PD, Zazula G, MacPhee RDE, Shapiro B, Durbin R, Willerslev E (2021) Environmental genomics of Late Pleistocene black bears and giant short-faced bears. Curr Biol 31(12):2728–2736.e8. https://doi.org/10.1016/j.cub.2021.04.027

Rohland N, Glocke I, Aximu-Petri A, Meyer M (2018) Extraction of highly degraded DNA from ancient bones, teeth and sediments for high-throughput sequencing. Nat Protoc 13(11):2447–2461. https://doi.org/10.1038/s41596-018-0050-5

Sand KK, Jelavić S, Kjær KH, Prohaska A (2023) Importance of eDNA taphonomy and provenance for robust ecological inference: insights from interfacial geochemistry. https://doi.org/10.1101/2023.01.24.525431

Schnell IB, Bohmann K, Gilbert MTP (2015) Tag jumps illuminated--reducing sequence-to-sample misidentifications in metabarcoding studies. Mol Ecol Resour 15(6):1289–1303. https://doi.org/10.1111/1755-0998.12402

Schulte L, Bernhardt N, Stoof-Leichsenring K, Zimmermann HH, Pestryakova LA, Epp LS, Herzschuh U (2021) Hybridization capture of larch (Larix Mill.) chloroplast genomes from sedimentary ancient DNA reveals past changes of Siberian forest. Mol Ecol Resour 21(3):801–815. https://doi.org/10.1111/1755-0998.13311

Schulte L, Meucci S, Stoof-Leichsenring KR, Heitkam T, Schmidt N, von Hippel B, Andreev AA, Diekmann B, Biskaborn BK, Wagner B, Melles M, Pestryakova LA, Alsos IG, Clarke C, Krutovsky KV, Herzschuh U (2022) Larix species range dynamics in Siberia since the Last Glacial captured from sedimentary ancient DNA. Commun Biol 5(1):1–11. https://doi.org/10.1038/s42003-022-03455-0

Seeber PA, Epp LS (2022) Environmental DNA and metagenomics of terrestrial mammals as keystone taxa of recent and past ecosystems. Mammal Rev 52(4):538–553. https://doi.org/10.1111/mam.12302

Seersholm FV, Pedersen MW, Søe MJ, Shokry H, Mak SST, Ruter A, Raghavan M, Fitzhugh W, Kjær KH, Willerslev E, Meldgaard M, Kapel CMO, Hansen AJ (2016) DNA evidence of bowhead whale exploitation by Greenlandic Paleo-Inuit 4,000 years ago. Nat Commun 7(1):13389. https://doi.org/10.1038/ncomms13389

Shelton AO, Gold ZJ, Jensen AJ, D'Agnese E, Allan EA, Cise AV, Gallego R, Ramón-Laca A, Garber-Yonts M, Parsons K, Kelly RP (2023) Toward quantitative metabarcoding. https://doi.org/10.1101/2022.04.26.489602

Skelton J, Cauvin A, Hunter ME (2022) Environmental DNA metabarcoding read numbers and their variability predict species abundance, but weakly in non-dominant species. Environ DNA n/a(n/a). https://doi.org/10.1002/edn3.355

Slon V, Hopfe C, Weiß CL, Mafessoni F, de la Rasilla M, Lalueza-Fox C, Rosas A, Soressi M, Knul MV, Miller R, Stewart JR, Derevianko AP, Jacobs Z, Li B, Roberts RG, Shunkov MV, de Lumley H, Perrenoud C, Gušić I, Kućan Ž, Rudan P, Aximu-Petri A, Essel E, Nagel S, Nickel B, Schmidt A, Prüfer K, Kelso J, Burbano HA, Pääbo S, Meyer M (2017) Neandertal and Denisovan DNA from Pleistocene sediments. Science 356(6338):605–608. https://doi.org/10.1126/science.aam9695

Stahlschmidt MC, Collin TC, Fernandes DM, Bar-Oz G, Belfer-Cohen A, Gao Z, Jakeli N, Matskevich Z, Meshveliani T, Pritchard JK, McDermott F, Pinhasi R (2019) Ancient mammalian and plant DNA from late quaternary stalagmite layers at Solkota cave, Georgia. Sci Rep 9(1):6628. https://doi.org/10.1038/s41598-019-43147-0

Steinbacher F, Bodmer T, Baran R (2015) Airborne hydromapping – bathymetry of lake constance and new possibilities for the ecological assessment of bodies of water. Österreichische Wasser- und Abfallwirtschaft

Stoof-Leichsenring KR, Epp LS, Trauth MH, Tiedemann R (2012) Hidden diversity in diatoms of Kenyan Lake Naivasha: a genetic approach detects temporal variation. Mol Ecol 21(8):1918–1930. https://doi.org/10.1111/j.1365-294X.2011.05412.x

Su J, Lui WW, Lee Y, Zheng Z, Siu GK-H, Ng TT-L, Zhang T, Lam TT-Y, Lao H-Y, Yam W-C, Tam KK-G, Leung KS-S, Lam T-W, Leung AW-S, Luo R (2023) Evaluation of Mycobacterium Tuberculosis enrichment in metagenomic samples using ONT adaptive sequencing and amplicon sequencing for identification and variant calling. https://doi.org/10.1101/2022.12.17.520855

Taberlet P, Coissac E, Pompanon F, Gielly L, Miquel C, Valentini A, Vermat T, Corthier G, Brochmann C, Willerslev E (2007) Power and limitations of the chloroplast trnL (UAA) intron for plant DNA barcoding. Nucleic Acids Res 35(3):e14. https://doi.org/10.1093/nar/gkl938

Taberlet P, Coissac E, Hajibabaei M, Rieseberg LH (2012) Environmental DNA. Mol Ecol 21(8):1789–1793. https://doi.org/10.1111/j.1365-294X.2012.05542.x

Thalinger B, Deiner K, Harper LR, Rees HC, Blackman RC, Sint D, Traugott M, Goldberg CS, Bruce K (2021) A validation scale to determine the readiness of environmental DNA assays for routine species monitoring. Environ DNA 3(4):823–836. https://doi.org/10.1002/edn3.189

Thomsen PF, Sigsgaard EE (2019) Environmental DNA metabarcoding of wild flowers reveals diverse communities of terrestrial arthropods. Ecol Evol 9(4):1665–1679. https://doi.org/10.1002/ece3.4809

Vernot B, Zavala EI, Gómez-Olivencia A, Jacobs Z, Slon V, Mafessoni F, Romagné F, Pearson A, Petr M, Sala N, Pablos A, Aranbur A, De Castro JMB, Carbonell E, Li B, Krajcarz MT, Krivoshapkin AI, Kolobova KA, Kozlikin MB, Shunkov MV, Derevianko AP, Viola B, Grote S, Essel E, Herraéz DL, Nagel S, Nickel B, Richter J, Schmidt A, Peter B, Kelso J, Roberts RG, Arsuaga JL, Meyer M (2021) Unearthing Neanderthal population history using nuclear and mitochondrial DNA from cave sediments. Science 372(6542). https://doi.org/10.1126/science.abf1667

Wales N, Andersen K, Cappellini E, Ávila-Arcos MC, Gilbert MTP (2014) Optimization of DNA recovery and amplification from non-carbonized archaeobotanical remains. PLoS One 9(1):e86827. https://doi.org/10.1371/journal.pone.0086827

Wang Y, Pedersen MW, Alsos IG, De Sanctis B, Racimo F, Prohaska A, Coissac E, Owens HL, Merkel MKF, Fernandez-Guerra A, Rouillard A, Lammers Y, Alberti A, Denoeud F, Money

D, Ruter AH, McColl H, Larsen NK, Cherezova AA, Edwards ME, Fedorov GB, Haile J, Orlando L, Vinner L, Korneliussen TS, Beilman DW, Bjørk AA, Cao J, Dockter C, Esdale J, Gusarova G, Kjeldsen KK, Mangerud J, Rasic JT, Skadhauge B, Svendsen JI, Tikhonov A, Wincker P, Xing Y, Zhang Y, Froese DG, Rahbek C, Bravo DN, Holden PB, Edwards NR, Durbin R, Meltzer DJ, Kjær KH, Möller P, Willerslev E (2021) Late quaternary dynamics of Arctic biota from ancient environmental genomics. Nature 600(7887):86–92. https://doi.org/10.1038/s41586-021-04016-x

Wang Y, Wessels M, Pedersen MW, Epp LS (2023) Spatial distribution of sedimentary DNA is taxon-specific and linked to local occurrence at intra-lake scale. Commun Earth Environ 4(1):1–13. https://doi.org/10.1038/s43247-023-00829-y

Wilkinson MD, Dumontier M, Aalbersberg IJ, Appleton G, Axton M, Baak A, Blomberg N, Boiten J-W, da Silva Santos LB, Bourne PE, Bouwman J, Brookes AJ, Clark T, Crosas M, Dillo I, Dumon O, Edmunds S, Evelo CT, Finkers R, Gonzalez-Beltran A, Gray AJG, Groth P, Goble C, Grethe JS, Heringa J, Hoen PAC't, Hooft R, Kuhn T, Kok R, Kok J, Lusher SJ, Martone ME, Mons A, Packer AL, Persson B, Rocca-Serra P, Roos M, van Schaik R, Sansone S-A, Schultes E, Sengstag T, Slater T, Strawn G, Swertz MA, Thompson M, van der Lei J, van Mulligen E, Velterop J, Waagmeester A, Wittenburg P, Wolstencroft K, Zhao J, Mons B (2016) The FAIR guiding principles for scientific data management and stewardship. Sci Data 3(1):160018. https://doi.org/10.1038/sdata.2016.18

Willerslev E, Hansen AJ, Binladen J, Brand TB, Gilbert MTP, Shapiro B, Bunce M, Wiuf C, Gilichinsky DA, Cooper A (2003) Diverse plant and animal genetic records from Holocene and Pleistocene sediments. Science 300(5620):791–795. https://doi.org/10.1126/science.1084114

Willerslev E, Cappellini E, Boomsma W, Nielsen R, Hebsgaard MB, Brand TB, Hofreiter M, Bunce M, Poinar HN, Dahl-Jensen D, Johnsen S, Steffensen JP, Bennike O, Schwenninger J-L, Nathan R, Armitage S, de Hoog C-J, Alfimov V, Christl M, Beer J, Muscheler R, Barker J, Sharp M, Penkman KEH, Haile J, Taberlet P, Gilbert MTP, Casoli A, Campani E, Collins MJ (2007) Ancient biomolecules from deep ice cores reveal a forested southern Greenland. Science 317(5834):111–114. https://doi.org/10.1126/science.1141758

Willerslev E, Davison J, Moora M, Zobel M, Coissac E, Edwards ME, Lorenzen ED, Vestergård M, Gussarova G, Haile J, Craine J, Gielly L, Boessenkool S, Epp LS, Pearman PB, Cheddadi R, Murray D, Bråthen KA, Yoccoz N, Binney H, Cruaud C, Wincker P, Goslar T, Alsos IG, Bellemain E, Brysting AK, Elven R, Sønstebø JH, Murton J, Sher A, Rasmussen M, Rønn R, Mourier T, Cooper A, Austin J, Möller P, Froese D, Zazula G, Pompanon F, Rioux D, Niderkorn V, Tikhonov A, Savvinov G, Roberts RG, MacPhee RDE, Gilbert MTP, Kjær KH, Orlando L, Brochmann C, Taberlet P (2014) Fifty thousand years of Arctic vegetation and megafaunal diet. Nature 506(7486):47–51. https://doi.org/10.1038/nature12921

Williams JW, Spanbauer TL, Heintzman PD, Blois J, Capo E, Goring SJ, Monchamp M-E, Parducci L, Eggers JMV, Alsos IG, Bowler C, Coolen MJL, Cullen N, Crump S, Epp LS, Fernandez-Guerra A, Grimm E, Herzschuh U, Mereghetti A, Meyer RS, Nota K, Pedersen MW, Pérez V, Shapiro B, Stoof-Leichsenring KR, Wood J (2023) Strengthening global-change science by integrating aeDNA with paleoecoinformatics. Trends Ecol Evol 0(0). https://doi.org/10.1016/j.tree.2023.04.016

Yamahara KM, Preston CM, Birch J, Walz K, Marin R, Jensen S, Pargett D, Roman B, Ussler W, Zhang Y, Ryan J, Hobson B, Kieft B, Raanan B, Goodwin KD, Chavez FP, Scholin C (2019) In situ autonomous acquisition and preservation of marine environmental DNA using an autonomous underwater vehicle. Front Mar Sci 6. https://doi.org/10.3389/fmars.2019.00373

Yarzábal LA, Salazar LMB, Batista-García RA (2021) Climate change, melting cryosphere and frozen pathogens: should we worry…? Environ Sustain 4(3):489–501. https://doi.org/10.1007/s42398-021-00184-8

Yates MC, Fraser DJ, Derry AM (2019) Meta-analysis supports further refinement of eDNA for monitoring aquatic species-specific abundance in nature. Environ DNA 1(1):5–13. https://doi.org/10.1002/edn3.7

Yoshitake K, Fujiwara A, Matsuura A, Sekino M, Yasuike M, Nakamura Y, Nakamichi R, Kodama M, Takahama Y, Takasuka A, Asakawa S, Nishikiori K, Kobayashi T, Watabe S (2021) Estimation of tuna population by the improved analytical pipeline of unique molecular identifier-assisted HaCeD-Seq (haplotype count from eDNA). Sci Rep 11(1):7031. https://doi.org/10.1038/s41598-021-86190-6

Zavala EI, Aximu-Petri A, Richter J, Nickel B, Vernot B, Meyer M (2022) Quantifying and reducing cross-contamination in single- and multiplex hybridization capture of ancient DNA. Mol Ecol Resour 22(6):2196–2207. https://doi.org/10.1111/1755-0998.13607

Zimmermann H, Raschke E, Epp L, Stoof-Leichsenring K, Schwamborn G, Schirrmeister L, Herzschuh U (2019) Past treeline dynamics at Bol'shoy Lyakhovsky Island (new Siberian archipelago) since the last interglacial inferred from sedimentary ancient DNA. In: EPIC320th congress of the International Union for Quaternary Research (INQUA), Dublin, 2019-07-25-2019-07-31. Dublin

Zimmermann HH, Stoof-Leichsenring KR, Dinkel V, Harms L, Schulte L, Hütt M-T, Nürnberg D, Tiedemann R, Herzschuh U (2023) Marine ecosystem shifts with deglacial sea-ice loss inferred from ancient DNA shotgun sequencing. Nat Commun 14(1):1650. https://doi.org/10.1038/s41467-023-36845-x

Correction to: Terrestrial Plant DNA from Lake Sediments

Sandra Garcés-Pastor, Kevin Nota, Dilli P. Rijal, Sisi Liu, Weihan Jia, Maria Leunda, Christoph Schwörer, Sarah E. Crump, Laura Parducci, and Inger G. Alsos

Correction to:
Chapter 10 in: E. Capo et al. (eds.), *Tracking Environmental Change Using Lake Sediments*, Developments in Paleoenvironmental Research 21, https://doi.org/10.1007/978-3-031-43799-1_10

The original version of this book was inadvertently published with affiliation correction in Chapter 10. This has been updated in the mentioned chapter in this version.

The updated version of this chapter can be found at
https://doi.org/10.1007/978-3-031-43799-1_10

E. Capo et al. (eds.), *Tracking Environmental Change Using Lake Sediments*,
Developments in Paleoenvironmental Research 21,
https://doi.org/10.1007/978-3-031-43799-1_14

Glossary, Acronyms and Abbreviations

12S rRNA gene Gene encoding for the 12S ribosomal RNA (rRNA). It is found in the mitochondrial genome of eukaryotic organisms, including humans.

^{13}C Stable isotope of carbon with an atomic mass of 13. It is less abundant compared to the more common ^{12}C isotope. The measurement and analysis of the ratio of ^{13}C to ^{12}C in a sample are often used to gain insights into carbon cycling, tracing organic matter sources, and studying ecological processes.

^{14}C Radioactive isotope of carbon with an atomic mass of 14. It is a variant of carbon that undergoes radioactive decay over time. The presence of ^{14}C in organic materials, such as fossils or artifacts, can be used in radiocarbon dating to determine their age by measuring the remaining amount of ^{14}C compared to the stable isotope ^{12}C.

16S rRNA gene Gene encoding for the 16S ribosomal RNA (rRNA). It is found in the genomes of bacteria and archaea.

18S rRNA gene Gene encoding for the 18S ribosomal RNA. It is found in the genomes of eukaryotic organisms, including humans.

^{15}N Stable isotope of nitrogen with an atomic mass of 15. It is a less abundant variant of nitrogen compared to the more common ^{14}N isotope. The measurement and analysis of the ratio of ^{15}N to ^{14}N in a sample are often used in ecological and biogeochemical studies to trace nitrogen cycling, study food webs, and understand nutrient dynamics in ecosystems.

^{210}Pb Radioactive isotope of lead with an atomic mass of 210. It is a product of the radioactive decay of uranium and thorium isotopes present in rocks and soil. In paleolimnology, ^{210}Pb dating is used to determine the age of sediment layers in lakes and other aquatic environments over approximately the last 150 years.

Age-depth model Numerical model of the relationship between the age of sediment (typically determined using radiometric dating such as ^{210}Pb or ^{14}C) and its stratigraphic depth. The model is usually defined by a small number of dated points and then used to provide estimates of the age of intermediate levels by interpolation.

E. Capo et al. (eds.), *Tracking Environmental Change Using Lake Sediments*, Developments in Paleoenvironmental Research 21,
https://doi.org/10.1007/978-3-031-43799-1

Airborne eDNA Environmental DNA isolated from air samples.

AIV Avian Influenza Virus.

Akinetes Specialized dormant cells produced by certain types of cyanobacteria. These cells are resistant to adverse environmental conditions and serve as a survival strategy, allowing the organism to withstand unfavorable periods until conditions become more favorable for growth and reproduction.

Alpha-diversity Measure of species diversity within a particular habitat or ecosystem. It quantifies the number of different species (species richness) and their relative abundance (species evenness) in a specific area or sample.

Amplicon Sequence Variant (ASV) Unique genetic sequence derived from high-throughput sequencing of DNA typically obtained from environmental or microbial samples.

Anammox bacteria Annamox bacteria (anaerobic ammonium-oxidizing bacteria) possess the capacity to convert ammonium (NH_4^+) and nitrite (NO_2^-) into nitrogen gas (N_2) under anaerobic conditions.

Ancient DNA Genetic material extracted from ancient remains such as bones, teeth or preserved tissues and environmental matrices, which can be hundreds to million years old.

Ancient DNA damage Deterioration and degradation of DNA molecules over time due to chemical and physical processes.

Anthropocene Proposed geological epoch characterized by the significant impact of human activities on Earth's ecosystems and geological processes. The year 1950 was proposed by some as the start of the Anthropocene as it corresponds to the dramatic increase in human activity affecting the planet, an event known as the Great Acceleration.

Anthropogenic Processes, activities or phenomena that are derived from or influenced by human actions.

Archaeology Study of human history and prehistory through the excavation and analysis of artifacts, structures and other physical remains.

Autochthonous Organisms or elements that originate or naturally occur within a particular habitat or ecosystem, without external introduction or migration.

Allochthonous Organisms or elements that are introduced or originate from outside of a particular habitat or ecosystem.

Barcoding Molecular biology technique based on the sequencing of short DNA sequences, typically from a specific gene or genomic region, to identify and classify organisms.

Base See Nucleotide.

Bayesian analysis Statistical approach that combines prior knowledge or assumptions with observed data to make probabilistic inferences and estimate parameters or hypotheses.

Bayesian change point (bcp) Statistical approach used to detect and analyze abrupt changes or shifts in data over time and space.

BCE Before Common Era i.e., before the year Jesus Christ was born.

Benthos Community of organisms that inhabit the bottom or substrates of aquatic ecosystems. Benthic organisms include a wide variety of organisms such as worms, mollusks, crustaceans and plants.

Beta-diversity Measure of the variation in species composition or diversity between different habitats or locations within a larger ecosystem. It quantifies the differences in species identities and abundance among distinct sites or samples.

Between-groups Principal Component Analysis (bgPCA) Statistical approach used to analyze and compare multivariate data across different groups or categories. It is a variation of the standard PCA that takes into account the between-group differences while identifying the principal components.

Biodiversity The wide range and abundance of living organisms and their genetic variations within a specific region or ecosystem.

Biofilm Colonies of microorganisms (bacteria, microeukaryotes) embedded into mucilage (an extracellular matrix that is composed of extracellular polymeric substance) and living attached to organic or inorganic substrate.

Biogeochemical cycles Natural processes through which elements, such as carbon, nitrogen and phosphorus, are cycled between the lithosphere, the hydrosphere, the atmosphere and the Earth's biomes.

Biogeography Study of the geographic distribution of organisms and hence investigate the dispersal, adaptability and endemicity of organisms.

Bioturbation Disturbance and mixing (vertical and/or horizontal) of sedimentary deposits by living organisms.

Blooms Algal blooms refer to excessive growth of algae in aquatic environments, resulting in dense populations of algae of one or a few dominant species. Harmful algal blooms (HABs) can produce toxins that can be harmful to aquatic life and pose a risk to human health.

BOLD Barcode of Life Data system.

bp base pairs.

BP Before Present. Denote the timing of the events in relation to the events that occurred relative to the origin of practical radiocarbon dating in the 1950.

Brackish waters Mixture of freshwater and seawater exhibiting salinity levels intermediate between freshwater and marine environments.

Bronze Age Period in human history characterized by the widespread use of bronze, a metal alloy composed primarily of copper and tin. It arose between approximately 3000 BCE and 1200 BCE.

Bryophytes Group of non-vascular plants that includes mosses, liverworts and hornworts.

CE Common Era i.e., after the year Jesus Christ was born.

Chloroplast genome circular double stranded DNA located in the chloroplasts of photosynthetic organisms. The chloroplast genome is characterized by its ability for independent replication and transcription.

Chironomids Group of small flying insects belonging to the family Chironomidae. The chitinized head capsules of their larvae are common indicators in paleolimnological investigations.

Cladoceran Group of small crustaceans belonging to the class Cladocera.

COI gene Gene encoding for the cytochrome *c* oxidase subunit I.

Copepods Group of tiny crustaceans belonging to the subclass Copepoda.

CPUE Catch per unit effort.

Cryptic species Distinct biological species that are morphologically similar or indistinguishable from each other but have distinct genetic differences.

Cyanobacteria Group of photosynthetic organisms capable of producing oxygen through photosynthesis and that are often referred to as blue-green algae. They belong to the Bacteria domain.

Cyanotoxin Toxic compounds produced by certain species of cyanobacteria.

Cylindrospermopsin Hepatotoxic cyanotoxin produced by certain species of cyanobacteria, known to cause liver damage and potential health risks to humans and other animals upon exposure to contaminated water sources.

Cysts Resting or dormant cells of an organism characterized by low metabolic activity and a thick cell wall that provides resistance to adverse environmental conditions (e.g., nutrient deficiency, desiccation, adverse pH, and low levels of oxygen).

Daphnia Genus of small, planktonic cladoceran crustaceans (water fleas) commonly found in freshwater ecosystems where they serve as primary consumers and play a role in nutrient cycling.

ddPCR (droplet digital PCR) Molecular biology technique used to quantify and analyze target DNA or RNA sequences. It partitions the DNA or RNA sample into thousands of individual droplets allowing for absolute quantification of low-abundance targets with high accuracy and precision.

Deamination Chemical process that involves the removal of an amino group from a molecule, typically an amino acid or a nitrogenous base in nucleic acids.

Denaturing Gel Gradient Electrophoresis (DGGE) Molecular biology technique used to analyze and differentiate DNA fragments based on their sequence variations.

Denitrifiers Group of microorganisms able to convert nitrate (NO_3^-) to nitrogen gas (N_2) through a process called denitrification.

Deoxyribonucleic Acid (DNA) Molecule formed by a long- double-stranded helical structure composed of nucleotides that is found in the cells of all living organisms, It carries the genetic material necessary for the development, functioning and reproduction of living organisms.

Depurination Chemical process that involves the hydrolysis or removal of a purine base (adenine or guanine) from the sugar-phosphate backbone of DNA or RNA molecules.

Diagenesis Physical, chemical and biological changes that occur in sediment or rock after its initial deposition and burial.

Diatoms Group of microalgae that belong to the phylum Bacillariophyta. They are single-celled organisms encased in silica cell walls called frustules or valves.

Diatom frustules See diatoms.

Diatom valves See diatoms.

Diazotrophs Microorganisms that have the ability to fix atmospheric nitrogen gas (N_2) into a biologically useful form, such as ammonia (NH_3) or nitrate (NO_3^-).

Dissolved organic matter (DOC) Organic compounds in the water column that are in a dissolved form, including complex mixtures of carbon-based substances.

DNA conformation A- and B-DNA (form) represent a tertiary structure of DNA, i.e. the three-dimensional arrangement of atoms in DNA polymer that defines how DNA is twisted. A-DNA has a larger diameter but it is more compact than B-DNA because the twist frequency is greater. The transition between A- and B-forms is a function of solvent composition and hydration. Quaternary DNA structure describes how DNA polymer coils in three dimensions either around itself, other DNA molecules or proteins. For example, linear DNA contains free polymer ends and can be tightly coiled and packed inside of a nucleus but also fully spread-out once adsorbed on a surface while circular DNA represents a close-end conformation that only minimally modifies its 3D arrangement upon adsorption to a surface.

Dynamic linear model (DLM) Statistical framework used to model and analyze time series data with underlying dynamic structure.

ENA European Nucleotide Archive.

Endemicity Related to endemic species, which are species whose distribution is restricted to a small geographic area.

Environmental DNA (eDNA) DNA present in environmental matrices such as soil, water or sediment.

Ephippia Specialized dormant eggs produced by certain species of crustaceans such as cladocerans.

Eutrophication Process in which a body of water becomes overly enriched with nutrients, primarily nitrogen and phosphorus.

Excavation Process of digging, removing, and relocating soil, rocks, or other materials from a site, typically to uncover archaeological artifacts.

exDNA Extracellular DNA, DNA external to cells, typically referring to a component of environmental DNA.

FAIR principles A set of guiding principles for data management, FAIR is an acronym for Findability, Accessibility, Interoperability and Reuse (of digital assets).

FASTQ format File format commonly used in bioinformatics to store nucleotide sequences and corresponding quality scores of DNA or RNA sequences obtained from high-throughput sequencing technologies.

Food webs Complex, interconnected systems that represent the feeding relationships and energy transfer among organisms within an ecosystem.

Fragment length distribution (FLD) Pattern or distribution of DNA or RNA fragments lengths generated during a sequencing. It represents the range and abundance of fragments of different sizes in a sample.

GC content Proportion of guanine (G) and cytosine (C) nucleotide bases in a DNA or RNA sequence, often used as an indicator of various biological characteristics or for primer design in molecular biology.

Generalized Additive Mixed effect Models (GAMM) Statistical model used to analyze complex relationships between variables, allowing for non-linear and non-parametric modeling.

Generalized Linear Model (GLM) Statistical model that extends the linear regression framework to accommodate various types of dependent variables, such as binary, count or categorical data.

Genome Complete set of genetic information present in the cell of an organism.

Genome skimming High-throughput sequencing approach that captures and sequences a broad range of genomic regions, providing a snapshot of the genetic diversity within a sample.

Genotype Genetic makeup of an individual organism, typically referring to the specific combination of alleles or genetic variants present at a particular gene or set of genes.

Haplogroup Group of individuals who share a common ancestor and specific sets of genetic variation.

Haplotypes A particular variation of a gene or of a genomic region. Haplotypes represent a set of genetic markers that are closely linked on a chromosome and are inherited from a single parent.

HAB Harmful Algal Bloom. See blooms.

Heterocytes Specialized cells found in some cyanobacteria that fix atmospheric nitrogen into a form usable by the organisms. Also named heterocysts.

Holarctic Biogeographic region that encompasses the northern parts of the globe including the Arctic, subarctic and temperate zones.

Holocene Current geological epoch, starting around 11,700 years ago after the last major ice age, characterized by a relatively stable climate and the development of human civilization.

Homopolymeric sequences Repeated sequences of a single nucleotide within a DNA or RNA molecule.

High Resolution Melting curve analysis (HRM) Molecular biology technique that utilizes the melting behavior of double-stranded DNA during a temperature ramp to detect sequence variations, mutations, or polymorphisms in a target DNA sample.

High-Throughput Sequencing (HTS) Set of techniques used to sequence large amounts of DNA or DNA. They allow for the simultaneous sequencing of thousands to millions of DNA or RNA fragments. It is also referred to as next-generation sequencing (NGS).

Hybridization capture A molecular biology technique that uses complementary probes to selectively capture and enrich specific DNA or RNA sequences from a complex mixture. Also referred to as capture enrichment.

Individual Sequence Unit (ISU) Short DNA sequences that can be used to identify and distinguish individual organisms within a population or community.

inDNA Intracellular DNA, DNA inside of cells, typically referring to a component of environmental DNA.

Iron oxidizers Microorganisms capable of oxidizing ferrous iron (Fe^{2+}) into its ferric form (Fe^{3+}).

Jaanuson effect Phenomenon where certain tree species exhibit delayed bud burst after winter due to cold temperature.

Joint Species Distribution Models (JSDM) Statistical model used to analyze and predict the distribution patterns of multiple species based on environmental and ecological variables.

ka kiloanni, equivalent to thousands of years.

Last Common Ancestor (LCA) Most recent individual or species from which two or more different lineages or species have evolved.

Last Interglacial period ca. 129,000–116,000 before present (BP).

Latent Dirichlet Allocation (LDA) Machine learning algorithm used in natural language processing to identify the latent topics or themes within a collection of text documents.

Latent Variable Model (LVM) Statistical model that incorporates unobservable or hidden variables to explain the relationships between observed variables.

Leaching Process by which soluble elements are removed from sediment particles and transported by water through the sediment column.

Locus Specific location or position on a chromosome where a particular gene or DNA sequence is located.

Lysis (cell) Refer to the process by which a cell membrane is broken down or degraded by viral, enzymic, or osmotic mechanisms often leading to the death of the cell.

Ma million years

Macroalgae Large multicellular photosynthetic marine algae, commonly known as seaweeds.

Macrofossils Relatively large and identifiable remnants or remains of ancient organisms such as shells, bones, or plant parts.

Macrophytes Large aquatic plants, typically rooted in the substrate, that are visible to the naked eye.

Marine Isotope Stage (MIS) Specific period of time within the Earth's geological history, particularly during the Quaternary period, which spans from approximately 2.6 million years ago to the present. MIS is defined by variations in the isotopic composition of oxygen preserved in the calcium carbonate shells of marine microorganisms such as foraminifera.

Maximum Likelihood analysis Statistical method that involves finding the values of model parameters that maximize the likelihood of observing the given data, assuming a specific probability distribution for the data.

Melting curve A curve representing the change in fluorescence observed when double-stranded DNA (dsDNA) with incorporated dye molecules dissociates, or 'melts', into single-stranded DNA (ssDNA) as the reaction temperature is increased.

Metabarcoding Method based on the high-throughput sequencing of DNA markers, typically specific gene or genomic regions, to simultaneously identify and characterize multiple species within a complex environmental sample.

Metagenomics Method based on the sequencing of all DNA fragments present in a given DNA extract without the use of prior targeted approaches (e.g. PCR, hybridization capture). It is also referred to as shotgun sequencing or shotgun metagenomics.

Metatranscriptomics Method based on the reverse transcription of all RNA fragments and sequencing of complementary DNA (cDNA) from a given DNA extract without the use of prior targeted approaches (e.g. PCR, hybridization capture).

Methanogens Microorganisms that produce methane as a metabolic byproduct in anaerobic environments.

Methylotrophs Microorganisms that can utilize organic compounds without carbon-carbon bonds as their sole carbon and energy source.

Micro-computed tomography (μCT) scanning Non-destructive imaging technique that utilizes X-rays to generate high-resolution three-dimensional (3D) images of the internal structure of an object or specimen at a microscale level.

Microstratigraphy Study of microscopic layers within sediment or rock formations to understand depositional processes and environmental changes.

Mitochondrial genome (mtDNA) Circular double-stranded DNA located in the mitochondria of eukaryotic cells.

Molecular clock Method used to estimate the timing of evolutionary events based on the accumulation of genetic changes over time.

Multivariate Autoregressive State-Space (MARSS) Statistical framework used to estimate hidden states and parameters in multivariate time series data.

Multivariate Regression Trees analysis (MRT) Statistical method that combines regression analysis with decision trees to analyze relationships among multiple variables.

NCBI National Center for Biotechnology Information.

NGS Next-Generation Sequencing.

Nitrifiers Microorganisms that convert ammonia (NH_3) to nitrite (NO_2-) and nitrate (NO_3-) in the process of nitrification.

non-Metric Multidimensional Scaling (nMDS) Statistical method used for visualizing and analyzing similarities or dissimilarities among objects or samples.

Nucleotides It is the basic structural unit of nucleic acids, such as DNA and RNA. It consists of three components: a nitrogenous base (adenine, thymine, cytosine, guanine in DNA; adenine, uracil, cytosine, guanine in RNA); a sugar molecule (deoxyribose in DNA; ribose in RNA); and a phosphate group.

Optically Stimulated Luminescence (OSL) Dating technique that measures the light energy emitted from minerals when they are exposed to light, providing information about the time since they were last exposed to sunlight.

Ostracods Small, bivalved crustaceans that are commonly found as fossils and are used as indicators of past environmental conditions.

OTU (Operational Taxonomic Unit) Method used to define groups of organisms based on sequence similarity, often used in microbial ecology to classify and analyze genetic data. Also sometimes referred to as MOTUs for Molecular Operational Taxonomic Units.

Paleobiogeography Study of the geographical distribution of organisms in the past and how it has changed over time.

Paleolakes Ancient lakes that existed in the past but have since dried up or significantly reduced in size.

Paleontology Study of prehistoric life, including the study of fossils and their interpretation.

Pelagic organisms Community of organisms living in the open water, distant from the shore or bottom.

Periphyton Community of organisms, including algae, bacteria, and other microorganisms, attached to surfaces in aquatic ecosystems.

Permafrost Layer of soil, sediment or rock that remains frozen for more than two consecutive years, typically found in high-latitude or altitude regions.

pH Measure of the acidity or alkalinity of a solution, representing the concentration of hydrogen ions.

Phenotype Observable physical or biochemical characteristics of an organism, resulting from the interaction between its genotype and the environment.

Phytoplankton Microscopic, photosynthetic organisms that float in aquatic environments and form the base of many food chains.

pK_a Measure of the acidity or basicity of a compound and represents the pH at which it is half dissociated.

Plankton Diverse group of organisms, including phytoplankton and zooplankton, that drift in the water column.

Pleistocene Geological epoch that occurred from about 2.6 million to 11,700 years ago, characterized by repeated glaciations and interglacial periods.

Pollen The male gametes of flowering plants that are dispersed by wind or animals and can be used to study past vegetation.

Polymerase Chain Reaction (PCR) Laboratory technique used to amplify a specific DNA sequence, allowing for the production of multiple copies for further analysis.

Primary producers Organisms, such as plants and algae, that produce organic matter through photosynthesis, forming the basis of the food chain.

Primers Short DNA sequences that serve as starting points for DNA synthesis in PCR reactions.

Principal Component Analysis (PCA) Statistical technique used to reduce the dimensionality of multivariate data and identify underlying patterns or relationships.

Prokaryotes Single-celled organisms that lack a nucleus and other membrane-bound organelles. They are found in the domains Bacteria and Archaea.

Propagules Structures or cells capable of giving rise to new individuals or organisms.

Protists Diverse groups of eukaryotes that are all unicellular. They include several groups of microalgae, amoebas, and ciliates.

Proxy Measurable or observable entity that is used as a substitute or indicator for another variable or phenomenon.

Point of Zero Charge (PZC) pH at which a solid surface has no net electrical charge.

Quantitative PCR (qPCR) a technique that allows for the amplification and quantification of specific DNA sequences.

Quaternary Current geological period, which includes the Pleistocene and the Holocene, characterized by the presence of extensive ice ages and interglacial periods.

***rbc*L gene** Gene encoding for the large subunit of the ribulose-1,5-biphosphate carboxylase/oxygenase /Rubisco) enzyme.

Redundancy Analysis (RDA) A multivariate statistical approach used to explore the relationship between a set of response variables and a set of explanatory variables.

Reworking Process by which sediments or fossils are transported and redeposited, potentially altering their original context or age.

Rotifers A group of small multicellular animals commonly found in freshwater environments.

RNA Stands for Ribonucleic Acid, a biological molecule that can play various roles in cellular processes, particularly for protein synthesis.

rRNA Stands for ribosomal RNA, plays a role in ribosomes, cellular structures that are responsible for protein synthesis.

Secondary producers Organisms that obtain energy by consuming primary producers or other consumers in a food chain.

Sedimentary ancient DNA (sedaDNA) Fraction of the sedimentary DNA that can be identified as ancient DNA.

Sedimentary DNA (sedDNA) Environmental DNA preserved in the sediments e.g., of aquatic systems, caves, permafrost and soils.

Shotgun sequencing See metagenomics.

Signor-Lipps effect Phenomenon where the abundance of a fossil taxon shows fluctuations or oscillations over time instead of a smooth evolutionary pattern.

Single nucleotide polymorphism (SNP) Variation in a single nucleotide at a specific position in the genome, often used as genetic markers for population studies.

Site occupancy detection modeling (SODM) Statistical approach used to estimate the presence or absence of a species at a particular site based on detection/non-detection data.

Spores Reproductive structures produced by certain organisms, such as fungi and some protists, which can develop into new individuals.

Sporormiella Genus of fungal spores commonly found in fossil records and used as a proxy for past herbivore activity.

Steric hindrances Refers to intramolecular interactions that are a consequence of a size of a molecule. Steric hindrance prevents reactions between larger molecules that are otherwise observed in smaller ones.

Sub-fossils Organic remains that are not completely fossilized, commonly found in more recent sediment or archaeological contexts.

Sulfate reducers Microorganisms that use sulfate as an electron acceptor in anaerobic respiration, producing hydrogen sulfide as a metabolic byproduct.

Sulfur oxidizers Microorganisms that obtain energy by oxidizing sulfur compounds, such as sulfide or elemental sulfur.

Syntrophs Microorganisms that rely on a metabolic partnership with other organisms to carry out complex metabolic reactions.

Transfer function Mathematical model used to estimate past environmental conditions or variables based on the relationship between biological or ecological data and environmental data.

***trn*L gene** Region of the chloroplast DNA commonly used as a molecular marker in biological research.

totDNA All the DNA extracted from a given sample.

Trophic Diatom Index (TDI) Measure used in some ecological assessments to evaluate the nutrient status or trophic condition of freshwater environments based on the composition and abundance of diatom species present.

Trophic network Complex interactions and flow of energy and nutrients between different organisms in an ecosystem, depicting the feeding relationships and ecological dynamics.

TSAR Telonemids, Stramenopiles, Alveolates, and Rhizaria.

Unique Molecular Identifier (UMI) Short, barcode-like sequences that are added to individual molecules during library preparation, allowing for accurate quantification and removal of PCR duplicates, thereby enhancing the accuracy and sensitivity of sequencing-based analyses.

Whole Genome Sequencing (WGS) Comprehensive genomic analysis technique that involves sequencing the complete DNA sequence of an organism's genome, providing a wealth of information about its genetic makeup, variations, and potential functional elements.

Zooplankton Small animals that drift or swim in aquatic environments, including tiny crustaceans, rotifers and larvae.

Index

E. Capo et al. (eds.), *Tracking Environmental Change Using Lake Sediments*, Developments in Paleoenvironmental Research 21,
https://doi.org/10.1007/978-3-031-43799-1

N

O

P

Q

GPSR Compliance

The European Union's (EU) General Product Safety Regulation (GPSR) is a set of rules that requires consumer products to be safe and our obligations to ensure this.

If you have any concerns about our products, you can contact us on ProductSafety@springernature.com

In case Publisher is established outside the EU, the EU authorized representative is:

Springer Nature Customer Service Center GmbH
Europaplatz 3
69115 Heidelberg, Germany

The manufacturer's authorised representative in the EU is Springer Nature Customer Service Centre GmbH, Europaplatz 3, 69115 Heidelberg, Germany. If you have any concerns regarding our products, please contact ProductSafety@springernature.com

Printed and bound by CPI Group (UK) Ltd, Croydon, CR0 4YY
01/10/2026
02225902-0017